K. Goerttler G. E. Feichter S. Witte (Eds.)

New Frontiers in Cytology

Modern Aspects of Research and Practice

With Contributions by Numerous Authors

With 151 Figures

Springer-Verlag Berlin Heidelberg New York
London Paris Tokyo

Professor Dr. Klaus Goerttler
Deutsches Krebsforschungszentrum
Im Neuenheimer Feld 280
D-6900 Heidelberg
Federal Republic of Germany

Priv.-Doz. Dr. Georg E. Feichter
Institut für Vergleichende und
Experimentelle Pathologie der Universität
Im Neuenheimer Feld 220-221
D-6900 Heidelberg
Federal Republic of Germany

Professor Dr. Siegfried Witte
Diakonissenhospital
Abteilung Innere Medizin
D-7500 Karlsruhe
Federal Republic of Germany

Library of Congress Cataloging-in-Publication Data
New Frontiers in cytology : modern aspects of research and practice / K. Goerttler,
G. Feichter, S. Witte (eds.). p. cm. Includes index.
ISBN-13: 978-3-642-73598-1 e-ISBN-13: 978-3-642-73596-7
DOI: 10.1007/ 978-3-642-73596-7
1. Cytodiagnosis. 2. Immunocytochemistry. 3. Human cytogenetics. I. Goerttler, Klaus.
II. Feichter, G. (Georg), 1946- . III. Witte, Siegfried. [DNLM: 1. Cytodiagnosis-trends.
2. Cytological Technics-trends. OY 95 N523] RB43.N48 1988 616.07'582-dc19
DNLM/DLC for Library of Congress 88-24770 CIP

Softcover reprint of the hardcover 1st edition 1988

Typesetting: Appl, Wemding
2123/3145-543210 - Printed on acid-free paper

Preface

The 15th European Congress of Cytology took place in Baden-Baden on 21-24 September 1987. The primary aim of the organizers was to select topics covering recently developed methods and their applications in cytology. Some do not even belong to the traditional domain of cytology. Stimulated by the success of this Congress and by many suggestions, we decided to compile a review of these new developments.

Following the spirit of the Congress, this book goes beyond simply reporting recent developments, however. It undertakes the task of describing the evolution of the cytology of tomorrow. The title *New Frontiers in Cytology* reflects this consideration.

The main concern of the section on *upcoming methods* is to describe the relationships to molecular biology. The relations between cytology and cell tissue culture were less intensive in the past, and this important contact to scientific cytology should be restored. This is also true for cytogenetics, which is a young and steadily developing branch of science.

Viral infections have gained increasing importance in the last years, especially in connection with our knowledge of oncogenes. New insights have resulted from the application of new methods, such as in situ hybridization, to virology and to cancer research as a whole.

The fundamentals of the *quantitative methods* were elaborated decades ago. Modern engineering and the applications of computer technology, with the possibilities it provides to process data, have enormously widened the scope to apply cytological techniques in diagnosis, prognosis, and therapy in clinical work.

Immunocytochemistry has produced a real revolution in diagnostic cell typing. After a period of rapid accumulation of knowledge on new monoclonal antibodies and on their applications to cells from various tissues, the time has come to clarify the practicability of immunocytochemistry for routine cytologic diagnosis. The importance of this topic can been seen in the number of chapters on immunocytology in this volume. For didactic reasons it was necessary to subdivide this subject into three separate parts.

Years ago transmission and scanning *electron microscopy* provided completely new insights into the structure and the function of the cell. Even now, after a period of really new developments, these methods still furnish important information and insights, especially in combination with other methods, such as immunoelectronmicroscopy.

The differential diagnosis of *reactive changes* and neoplastic transformations is a permanent source of error and of concern for cytologists. This problem, an "evergreen" of cytopathology, makes the limits of cytology evident; the latter continue to exist even in the era of monoclonal antibodies.

Both *cytology and histology* belong to the pool of methods used by the morphologist. The cytologist traditionally has more intensive connections to clinical work than the histopathologist, but cytology must resist being controlled by histology. Favorable circumstances for cytologic diagnosis are encountered especially in those pathology laboratories in which both methods are practiced in close connection.

Last but not least, *epidemiology* has made the great successes of cytology visible. Cytologic examinations as a part of cancer prevention require noninvasive methods, such as exfoliative cytology. Epidemiology indicates new trends in morbidity, helps to identify high-risk populations, and reveals the epidemiclike character of diseases such as cervical cancer. The results of epidemiology lead us to reflect on modifications of strategies for early cancer detection, such as changes in screening intervals and the recognition of new population groups at high risk. Epidemiologic data represent a permanent challenge to the cytologist in his work to lower morbidity and search for new methods and ideas.

Cytology is a science in motion. Some of the new frontiers of today will become the routine methods of tomorrow. The aim of this book is to sharpen our eyes for what is to come.

We wish to express our gratitude to all of our collegues who have furnished ideas and supported this enterprise, as well as to all the authors. We adress our special thanks to Springer Verlag for its interest in this subject, the outstanding quality of the print and the speedy production of the book, and to its staff for the constructive cooperation and excellent help in the stylistic revision of the texts.

Heidelberg, June 1988

K. Goerttler
G. E. Feichter
S. Witte

Table of Contents

List of Contributors

(the number in parentheses gives the page on which the article begins; page numbers in italics indicate the senior author of the article)

Åkerman, M. (p. 302)
Cytodiagnostik Department, University Hospital, 221 85 Lund, Sweden

Arentz, P. W. (p. 340)
Department of Pathology, University Hospital, Leiden, The Netherlands

Aste, H. (p. 104)
Endoscopic Service Gastrointestinal Unit of the National Cancer Institute (IST), Viale Benedetto XV n. 10, 16132 Genova, Italy

Auer, G. (p. 90)
Patologiska Institutionen, Avd. för tumörpatologi, Karolinska Sjukhuset, 104 01 Stockholm 60, Sweden

Beekhuis-Brussee, J. A. M. (p. 340)
Department of Urology, University Hospital, Leiden, The Netherlands

Berghof, B. (p. 386)
Zentralinstitut für die kassenärztliche Versorgung in der Bundesrepublik Deutschland, 5000 Köln, Federal Republic of Germany

Biesterfeld, St. (p. 130)
Abteilung Pathologie der RWTH, Pauwelsstraße 1, 5100 Aachen, Federal Republic of Germany

Bittel, D. (p. 192)
Institut für Pathologie der Universität, Schönbeinstraße 40, 4003 Basel, Switzerland

Blonk, D. I. (p. 206)
Dr. Daniel den Hoed Kliniek, Rotterdamsch Radio-Therapeutisch Instituut, Postbus 5201, 3008 AE Rotterdam, The Netherlands

Bloss, W.H. (p.140)
Institut für Physikalische Elektronik der Universität, 7000 Stuttgart, Federal Republic of Germany

Blümcke, S. (p.284)
Institut für Pathologie, Univ.-Klinikum Rudolf Virchow, Spandauer Damm 130, 1000 Berlin 19, Federal Republic of Germany

Boccato, P. (p.365)
Servizio di Citodiagnostica II, Ospedale Civile, ULSS N.21, 35100 Padova, Italy

Böcking, A. (pp.108, *130*)
Abteilung Pathologie der RWTH, Pauwelsstraße 1, 5100 Aachen, Federal Republic of Germany

Boghaert, E. (p.27)
Laboratory of Experimental Cancerology, Department of Radiotherapy and Nuclear Medicine, University Hospital, 9000 Ghent, Belgium

Boon, M.E. (pp.108, 425)
Leids Cytologisch en Patologisch Laboratorium, Postbus 16084, 2301 GB Leiden, The Netherlands

Bostofte, E. (p.421)
Department of Gynecology, Herlev Hospital, Herlev Ringvej, 2730 Herlev, Denmark

Bross, K.J. (pp.241, 250)
Medizinische Univ.-Klinik, Hugstetter Straße 55, 7800 Freiburg, Federal Republic of Germany

Bruins, J.L. (p.340)
Department of Urology, University Hospital, Leiden, The Netherlands

Bruno, S. (p.104)
Laboratory of Biophysics of the National Cancer Institute (IST), Viale Benedetto XV n.10, 16132 Genova, Italy

Burger, G. (p.135)
GSF München, Ingolstädter Landstraße 1, 8042 Neuherberg, Federal Republic of Germany

Busch, R. (p. 404)
Institut für Medizinische Statistik und Epidemiologie, Technische Universität München, 8000 München, Federal Republic of Germany

Campo, S. (pp. 45, 50)
Cattedra di Genetica Medica, Università di Roma, "La Sapienza", 00161 Roma, Italy

Cecchini, S. (p. 392)
Centro per lo Studio e la Prevenzione Oncologica (CSPO), Viale A. Volta 171, 50131 Firenze, Italy

Chiappino, G. (p. 440)
Istituto di Medicina del Lavoro, Clinica del Lavoro "L. Devoto", Via S. Barnaba 8, 20122 Milano, Italy

Chiara, G. (p. 355)
1 Divisione Chirurgica, Unità Sanitaria Locale, n° 10, Treviso, Italy

Cianciulli, A. M. (pp. 50, *233*)
Istituto Nazionale Tumori "Regina Elena", Viale Regina Elena, 291, 00161 Roma, Italy

Ciarmiello, G. (p. 355)
Istituto di Anatomia Patologica, Unità Sanitaria Locale, n° 10, Treviso, Italy

Costabel, U. (pp. 241, 250)
Ruhrlandklinik, Tüschener Weg 40, 4300 Essen 16, Federal Republic of Germany

Dal Pozzo, A. (p. 355)
4 Divisione Chirurgica, Unità Sanitaria Locale, n° 10, Treviso, Italy

Dalquen, P. (p. 192)
Institut für Pathologie der Universität, Schönbeinstraße 40, 4003 Basel, Switzerland

De Graaff Guilloud, J. C. (p. 425)
Leids Cytologisch en Patologisch Laboratorium, Postbus 16084, 2301 GB Leiden, The Netherlands

De Jong, A. (p. 265)
Department of Pathology, Erasmus University, Burg. Oudln 50, Rotterdam, The Netherlands

Delahaye, M. (p. 265)
Department of Pathology, Erasmus University, Burg. Oudln 50, Rotterdam, The Netherlands

Del Porto, G. (pp. 41, 45, 50)
Cattedra di Genetica Medica, Università di Roma, "La Sapienza", 00161 Roma, Italy

Deuscher, P. (p. 241)
Medizinische Univ.-Klinik, Hugstetter Straße 55, 7800 Freiburg, Federal Republic of Germany

De Villiers, E.-M. (p. 71)
Deutsches Krebsforschungszentrum, Im Neuenheimer Feld 280, 6900 Heidelberg, Federal Republic of Germany

Devonec, M. (p. 350)
Urology Department, Hôpital Antiquaille, 69321 Lyon Cedex 05, France

Di Falco, G. (p. 355)
3 Divisione Chirurgica, Unità Sanitaria Locale, n° 10, Treviso, Italy

Di Fusco, C. (p. 41)
Chair of Medical Genetics, Università di Roma, "La Sapienza", 00161 Roma, Italy

Di Natale, I. (p. 355)
4 Divisione Chirurgica, Unità Sanitaria Locale, n° 10, Treviso, Italy

Dinges, H. P. (p. 151)
Pathologisches Institut der Universität, Auenbruggerplatz 25, 8036 Graz, Austria

Dirk, T. (p. 199)
Abteilung Zytopathologie, Zentrum Pathologie der Universität, Postfach 884, 3400 Göttingen, Federal Republic of Germany

Di Vinci, A. (p. 104)
Laboratory of Biophysics of the National Cancer Institute (IST), Viale Benedetto XV n. 10, 16132 Genova, Italy

Domagala, W. (p. 178)
Department of Tumor Pathology, Medical Academy, ul. Unii Lubelskiej, 71-344 Szczecin, Poland

Dominici, C. (p. 45)
II Clinic of Pediatry, University of Rome, Italy

D'Orazi, G. (pp. 45, 289)
Dipartimento di Medicina Sperimentale, Università di Roma, "La Sapienza", Viale Regina Elena, 324, 00161 Roma, Italy

Droese, M. (p. 199)
Abteilung Zytopathologie, Zentrum Pathologie der Universität, Postfach 884, 3400 Göttingen, Federal Republic of Germany

Eiermann, W. (p. 135)
Universitäts-Frauenklinik im Klinikum Großhadern, Marchioninistraße 15, 8000 München 70, Federal Republic of Germany

Eulderink, F. (p. 340)
Department of Pathology, University Hospital, Leiden, The Netherlands

Feichter, G. E. (p. 95)
Institut für Vergleichende und Experimentelle Pathologie der Universität, Im Neuenheimer Feld 220-221, 6900 Heidelberg, Federal Republic of Germany

Fischer, G. F. (p. 32)
Immunologisches Institut der Universität Wien, Borschkegasse 8a, 1090 Wien, Austria

Fleuren, G. J. (p. 245)
Department of Pathology, State University of Leiden, P. O. Box 9603, 2300 RC Leiden, The Netherlands

Forni, A. (p. 440)
Istituto di Medicina del Lavoro, Clinica del Lavoro "L. Devoto", Via S. Barnaba 8, 20122 Milano, Italy

Gais, P. (p. 135)
GSF München, Ingolstädter Landstraße 1, 8042 Neuherberg, Federal Republic of Germany

Gallo-Curcio, C. (pp. 233, 237, 359)
Istituto Nazionale Tumori "Regina Elena", Divisione di Oncologia Medica II, Servizio di Citopatologia, Viale Regina Elena, 291, 00161 Roma, Italy

Geido, E. (p. 104)
Laboratory of Biophysics of the National Cancer Institute (IST), Viale Benedetto XV n. 10, 16132 Genova, Italy

Gerdes, J. (p. 212)
Pathologisches Institut, Klinikum Steglitz der Freien Universität, Hindenburgdamm 30, 1000 Berlin 45, Federal Republic of Germany

Giaretti, W. (p. 104)
Laboratory of Biophysics of the National Cancer Institute (IST), Viale Benedetto XV n. 10, 16132 Genova, Italy

Gissmann, L. (p. 55)
Deutsches Krebsforschungszentrum, Im Neuenheimer Feld 280, 6900 Heidelberg, Federal Republic of Germany

Glenthøj, A. (p. 421)
Institute of Pathology, Herlev Hospital, Herlev Ringvej, 2730 Herlev, Denmark

Goecke, C. (p. 130)
Luisen-Hospital, Abt. Gynäkologie und Geburtshilfe, Boxgraben 99, 5100 Aachen, Federal Republic of Germany

Goerttler, K. (p. 87)
Deutsches Krebsforschungszentrum, Im Neuenheimer Feld 280, 6900 Heidelberg, Federal Republic of Germany

Grammatico, P. (pp. 45, 50)
Cattedra di Genetica Medica, Università di Roma, "La Sapienza", 00161 Roma, Italy

Gross, G. (p. 63)
Universitäts Hautklinik und -Poliklinik, Martinistraße 52, 2000 Hamburg 20, Federal Republic of Germany

Grundmann, E. (p. 385)
Pathologisches Institut der Universität, Domagkstraße 17, 4400 Münster, Federal Republic of Germany

Grunert, F. (p. 241)
Institut für Immunbiologie der Universität, Stefan-Meier-Straße 8, 7800 Freiburg, Federal Republic of Germany

Guzman, J. (pp. *241*, 250)
Ludwig-Aschoff-Haus der Universität, Albertstraße 19, 7800 Freiburg, Federal Republic of Germany

Halbach, M. (p. 78)
Psychiatrische Klinik der Universität, Bergische Landstraße 2, 4000 Düsseldorf, Federal Republic of Germany

Haye, C. (p.371)
Division of Ophthalmology, Institut Curie, 26, Rue d'Ulm, 75231 Paris Cedex 05, France

Heidenreich, W. (p.212)
Frauenklinik der MHH, Podbielskistraße 380, 3000 Hannover 51, Federal Republic of Germany

Hilgarth, M. (p.250)
Universitäts-Frauenklinik, Hugstetter Straße 55, 7800 Freiburg, Federal Republic of Germany

Hjerpe, A. (p.434)
Department of Pathology II, Karolinska Institute, Huddinge University Hospital, 141 86 Huddinge, Sweden

Hoder, D. (p.284)
Institut für Pathologie, Univ.-Klinikum Rudolf Virchow, Spandauer Damm 130, 1000 Berlin 19, Federal Republic of Germany

Höfler, H. (p.151)
Pathologisches Institut der Universität, Auenbruggerplatz 25, 8036 Graz, Austria

Hoeltermann, W. (p.306)
Zentrum für Pathologie und Biophysik der Universität, Abt. für Zytopathologie, Moorenstraße 5, 4000 Düsseldorf, Federal Republic of Germany

Holm, R. (p.257)
Department of Pathology, The Norwegian Radium Hospital and Institute for Cancer Research, Montebello, 0310 Oslo 3, Norway

Hoogsteden, C.H. (p.265)
Department of Pulmonology, Academic Hospital Dijkzigt, Rotterdam, The Netherlands

Huijsmans, A. (p.159)
Department of Pathology, University Hospital Nijmegen, Geert Grooteplein Zuid 24, 6525 GA Nijmegen, The Netherlands

Ioannidou, M. (p.320)
Kerasountos 17, Athens 11528, Greece

Johannessen, J. V. (p. 257)
The Norwegian Radium Hospital and Institute for Cancer Research, Montebello, 0310 Oslo 3, Norway

Jütting, U. (p. 135)
GSF München, Ingolstädter Landstraße 1, 8042 Neuherberg, Federal Republic of Germany

Kaufmann, M. (p. 216)
Universitäts-Frauenklinik, Abt. Allgemeine Geburtshilfe und Gynäkologie, Voßstraße 9, 6900 Heidelberg, Federal Republic of Germany

Kirby, R. M. (p. 324)
The Liver Unit, Queen Elisabeth Hospital, Birmingham, UK

Klinga, K. (p. 216)
Universitäts-Frauenklinik, Abt. Gynäkologische Endokrinologie, Voßstraße 9, 6900 Heidelberg, Federal Republic of Germany

Knapp, W. (p. 32)
Immunologisches Institut der Universität Wien, Borschkegasse 8 a, 1090 Wien, Austria

Koelma, I. A. (p. 245)
Department of Pathology, State University of Leiden, P. O. Box 96 03, 2300 RC Leiden, The Netherlands

Kok, L. P. (p. 108)
Institute for Theoretical Physics, Groningen University, 9700 AV Groningen, The Netherlands

Kramer, A. E. J. L. (p. 340)
Department of Urology, University Hospital, Leiden, The Netherlands

Krapfl, E. (p. 216)
Universitäts-Frauenklinik, Abt. Allgemeine Geburtshilfe und Gynäkologie, Voßstraße 9, 6900 Heidelberg, Federal Republic of Germany

Kronenwett, M. (p. 90)
Patologiska Institutionen, Avd. för tumörpatologi, Karolinska Sjukhuset, 104 01 Stockholm 60, Sweden

Kübler, H.-Ch. (p. 216)
Universitäts-Frauenklinik, Abt. Gynäkologische Morphologie, Voßstraße 9, 6900 Heidelberg, Federal Republic of Germany

Kühn, W. (p. 216)
Universitäts-Frauenklinik, Abt. Gynäkologische Morphologie, Voßstraße 9, 6900 Heidelberg, Federal Republic of Germany

Kuenen-Boumeester, V. (pp. 108, *206*)
Dr. Daniel den Hoed Kliniek, Rotterdamsch Radio-Therapeutisch Instituut, Postbus 5201, 3008 AE Rotterdam, The Netherlands

Kyriakos, G. (p. 320)
Division of Endocrinology, Hippokration Hospital, Athens 11527, Greece

Lajous-Petter, A.-M. (p. 284)
Institut für Pathologie, Univ.-Klinikum Rudolf Virchow, Spandauer Damm 130, 1000 Berlin 19, Federal Republic of Germany

Lehmacher, W. (p. 408)
Institut für Medizinische Informatik und Systemforschung der GSF München, Ingolstädter Landstraße 1, 8042 Neuherberg, Federal Republic of Germany

Lellé, R. J. (p. 212)
Frauenklinik der MHH, Podbielskistraße 380, 3000 Hannover 51, Federal Republic of Germany

Lindholm, K. (p. 331)
Dept. of Clinical Cytology, Malmö General Hospital, 21401 Malmö, Sweden

Lipke, G. W. (p. 306)
Zentrum für Pathologie und Biophysik der Universität, Abt. für Zytopathologie, Moorenstraße 5, 4000 Düsseldorf, Federal Republic of Germany

Lubinski, J. (p. 178)
Department of Tumor Pathology, Medical Academy, 70-410 Szczecin, Poland

Lycklama à Nijeholt, A. A. B. (p. 340)
Department of Urology, University Hospital, Leiden, The Netherlands

Majdic, O. (p.32)
Immunologisches Institut der Universität Wien, Borschkegasse 8a, 1090 Wien, Austria

Malcolm, S. (p.22)
Mothercare Department of Paediatric Genetics, Institute of Child Health, 30 Guilford Street, London, WC1 N1EH, UK

Mareel, M.M. (p.27)
Laboratory of Experimental Cancerology, Department of Radiotherapy and Nuclear Medicine, University Hopital, 9000 Ghent, Belgium

Massari, F. (p.289)
I Tisiologic Clinic, University of Rome, Istituto Forlanini, Roma, Italy

Matzeu, M. (p.289)
I Tisiologic Clinic, University of Rome, Istituto Forlanini, Roma, Italy

Mechtersheimer, G. (p.226)
Pathologisches Institut der Universität, Im Neuenheimer Feld 220-221, 6900 Heidelberg, Federal Republic of Germany

Meinhardt, G. (p.71)
Universitätsfrauenklinik, Prittwitzstraße 43, 7900 Ulm, Federal Republic of Germany

Menaguale, L. (p.41)
Department of Internal Medicine, Università de L'Aquila, Italy

Messiaen, L. (p.27)
Laboratory of Experimental Cancerology, Department of Radiotherapy and Nuclear Medicine, University Hopital, 9000 Ghent, Belgium

Midena, E. (p.365)
Institute of Ophthalmology, University of Padova, Padova, Italy

Modesti, A. (pp.45, 289)
Dipartimento di Medicina Sperimentale, Università di Roma, "La Sapienza", Viale Regina Elena, 324, 00161 Roma, Italy

Möller, P. (p.226)
Pathologisches Institut der Universität, Im Neuenheimer Feld 220-221, 6900 Heidelberg, Federal Republic of Germany

Monsallier, M. (p. 350)
Urology Department, Hôpital Antiquaille, 69321 Lyon cedex 05, France

Moss, S. M. (pp. 399, 430)
DHSS Cancer Screening Evaluation Unit, Institute of Cancer Research, Section of Epidemiology, D Block, 15 Cotswold Road, Belmont, Sutton, Surrey SM2 5NG, U. K.

Mottolese, M. (pp. 233, *237, 359*)
Istituto Nazionale Tumori "Regina Elena", Servizio di Citopatologia, Viale Regina Elena, 291, 00161 Roma, Italy

Mouriquand, J. (p. 350)
Dept. of Cytopathology, Hôpital Les Sablons, Grenoble, France

Mouriquand, P. (p. 350)
Urology Department, Hôpital Antiquaille, 69321 Lyon cedex 05, France

Natali, P. G. (pp. 233, 237, 359)
Istituto Nazionale Tumori "Regina Elena", Servizio di Citopatologia, Laboratorio di Immunologia, Viale Regina Elena, 291, 00161 Roma, Italy

Nesland, J. M. (p. 257)
Department of Pathology, The Norwegian Radium Hospital and Institute for Cancer Research, Montebello, 0310 Oslo 3, Norway

Olszewski, W.
Department of Tumor Pathology, Medical Academy, 70-410 Szczecin, Poland

Ortisi, E. (p. 440)
Istituto di Medicina del Lavoro, Clinica del Lavoro "L. Devoto", Via S. Barnaba 8, 20122 Milano, Italy

Osborn, M. (p. 178)
Max-Planck Institute for Biophysical Chemistry, 3400 Göttingen, Federal Republic of Germany

Peen, U. (p. 421)
Departtment of Gynecology, Herlev Hospital, Herlev Ringvej, 2730 Herlev, Denmark

Perrin, P. (p.350)
Urology Department, Hôpital Antiquaille, 69321 Lyon cedex 05,
France

Perrone-Donnorso, R. (pp.50, 233, 237, 359)
Istituto Nazionale Tumori "Regina Elena", Servizio di
Citopatologia, Viale Regina Elena, 291, 00161 Roma, Italy

Piaton, E. (p.350)
Dept. of Cythopathology, Hôpital Les Sablons, Grenoble, France

Pfitzer, P. (p.306)
Zentrum für Pathologie und Biophysik, Abt. für Zytopathologie
der Universität, Moorenstraße 5, 4000 Düsseldorf,
Federal Republic of Germany

Quaglino, D. (p.41)
Department of Internal Medicine, Università de L'Aquila, Italy

Raap, A.K. (p.13)
Dept. Cytochemistry and Cytometry, Medical School University of
Leiden, 72 Wassenaarseweg, 2333 AL Leiden, The Netherlands

Ramaekers, F. (p.159)
Department of Pathology, University Hospital Nijmengen, Geert
Grooteplein Zuid 24, 6525 GA Nijmegen, The Netherlands

Rank, F. (p.421)
Institute of Pathology, Herlev Hospital, Herlev Ringvej,
2730 Herlev, Denmark

Rietveld, W.J. (p.425)
Department of Physiology, Leiden, The Netherlands

Rinaldi, M. (p.233, 359)
Istituto Nazionale Tumori "Regina Elena", Divisione di
Oncologia Medica II, Servizio di Citopatologia,
Viale Regina Elena, 291, 00161 Roma, Italy

Rivolta, G. (p.440)
Istituto di Medicina del Lavoro, Clinica del Lavoro "L. Devoto",
Via S. Barnaba 8, 20122 Milano, Italy

Robra, B.-P. (p.386)
MHH Epidemiologie und Sozialmedizin, Arbeitsgruppe II,
Postfach 610180, 3000 Hannover 61, Federal Republic of
Germany

Rodenacker, K. (p. 135)
GSF München, Ingolstädter Landstraße 1, 8042 Neuherberg, Federal Republic of Germany

Ruffing-Kullmann, B. (p. 408)
Institut für Klinische Zytologie der Technischen Universität München, Prinzregentenplatz 14, 8000 München 80, Federal Republic of Germany

Rummel, H. H. (p. 216)
Universitäts-Frauenklinik, Abt. Gynäkologische Morphologie, Voßstraße 9, 6900 Heidelberg, Federal Republic of Germany

Salet-van der Pol, M. (p. 159)
Department of Pathology, University Hospital Nijmegen, Geert Grooteplein Zuid 24, 6525 GA Nijmegen, The Netherlands

Sartori, C. (p. 355)
4 Divisione Chirurgica, Unità Sanitaria Locale, n° 10, Treviso, Italy

Scarpa, S. (pp. 45, *289*)
Dipartimento di Medicina Sperimentale, Università di Roma, "La Sapienza", Viale Regina Elena, 324, 00161 Roma, Italy

Schachter, A. (p. 404)
Cyto. Gynec. Unit, Beilinson Hospital, Petach Tiqva, Israel

Schenck, U. (pp. *135*, 404)
Institut für Klinische Zytologie, Technische Universität München, Prinzregentenplatz 14, 8000 München 80, Federal Republic of Germany

Schenck, U. B. (p. 135)
GSF München, Ingolstädter Landstraße 1, 8042 Neuherberg, Federal Republic of Germany

Schlienger, P. (p. 371)
Division of Radiotherapy, Institut Curie, 26, Rue d'Ulm, 75231 Paris Cedex 05, France

Schlotmann-Höller, E. (p. 306)
Zentrum für Pathologie und Biophysik der Universität, Abt. für Zytopathologie, Moorenstraße 5, 4000 Düsseldorf, Federal Republic of Germany

Schmauz, R. (p. 404)
Institut für Pathologie, Medizinische Universität Lübeck, 2400 Lübeck, Federal Republic of Germany

Schmid, U. (p. 223)
Departement Pathologie, Zytologisches und Immunhistochemisches Laboratorium, Kantonsspital, 9007 St. Gallen, Switzerland

Schneider, A. (p. 71)
Universitätsfrauenklinik, Prittwitzstraße 43, 7900 Ulm, Federal Republic of Germany

Scholman, H.-J. (p. 284)
Institut für Pathologie, Univ.-Klinikum Rudolf Virchow, Spandauer Damm 130, 1000 Berlin 19, Federal Republic of Germany

Schulte, E. (p. 376)
Anatomisches Institut der Universität, Abteilung Anatomie II, Albertstraße 17, 7800 Freiburg, Federal Republic of Germany

Schwarzmann, P. (p. 140)
Institut für Physikalische Elektronik der Universität, 7000 Stuttgart, Federal Republic of Germany

Schwechheimer, K. (p. 166)
Pathologisches Institut der Universität, Abt. Neuropathologie, Albertstraße 19, 7800 Freiburg, Federal Republic of Germany

Sciallero, S. (p. 104)
Endoscopic Service Gastrointestinal Unit of the National Cancer Institute (IST), Viale Benedetto XV, n. 10, 16132 Genoa, Italy

Soost, H.-J. (pp. 404, 408)
Institut für Klinische Zytologie, Technische Universität München, Prinzregentenplatz 14, 8000 München 80, Federal Republic of Germany

Spaander, P. J. (p. 108)
Rode Kruis Ziekenhuis, The Hague, The Netherlands

Spieler, P. (p. 223)
Departement Pathologie, Zytologisches und Immunhistochemisches Laboratorium, Kantonsspital, 9007 St. Gallen, Switzerland

Sprenger, E. (p. 3)
Abteilung Zytopathologie im Klinikum der Universität, Michaelistraße 11, 2300 Kiel 1, Federal Republic of Germany

Stauch, G. (p. 212)
Pathologisches Institut der MHH, Podbielskistraße 380,
3000 Hannover 51, Federal Republic of Germany

Sträßle, G. (p. 140)
Institut für Physikalische Elektronik der Universität, 7000 Stuttgart,
Federal Republic of Germany

Sulli, N. (p. 45)
Cattedra di Genetica Medica, Università di Roma, "La Sapienza",
00161 Roma, Italy

Tolis, G. (p. 320)
Division of Endocrinology, Hippocration Hospital, Athens 11527,
Greece

Us-Krašovec, M. (pp. 293, 301)
The Institute of Oncology, Zaloska 2, 61000 Ljubljana, Yugoslavia

Valkov, I. (p. 274)
Medical Academy, Department of Pathology, Central Laboratory
of Cytopathology, ul. "G. Sofiiski" N° 1, 1431 Sofia, Bulgaria

Van Aspert-van Erp, A. (p. 159)
Department of Pathology, University Hospital Nijmegen, Geert
Grooteplein Zuid 24, 6525 GA Nijmegen, The Netherlands

Van den Berghe, J. (p. 22)
Mothercare Department of Paediatric Genetics, Institute of
Child Health, 30 Guilford Street, London, WC1 N1EH, UK

Van der Kwast, T. H. (pp. 206, *265*)
Department of Pathology and Immunology, Erasmus University,
Burg. Oudln 50, Rotterdam, The Netherlands

Van der Meulen, E. A. (p. 108)
Riksuniversiteit, Subfaculteit Wiskunde en Informatica,
Department of Mathematics, P.O. Box 800, 9700 AV Groningen,
The Netherlands

Van der Ploeg, M. (p. 13)
Dept. Cytochemistry and Cytometry, Medical School University of
Leiden, 72 Wassenaarseweg, 2333 AL Leiden, The Netherlands

Van Heerde, P. (pp. 108, *313*)
Dept. of Pathology/Cytology, The Netherlands Cancer Institute,
Plesmanlaan 121, 1066 CX Amsterdam, The Netherlands

Van Krieken, H.J.H.M. (p. 340)
Department of Pathology, University Hospital, Leiden,
The Netherlands

Van Larebeke, N. (p. 27)
Laboratory of Experimental Cancerology, Department of
Radiotherapy and Nuclear Medicine, University Hopital,
9000 Ghent, Belgium

Van Nispen, A.C.M. (p. 108)
Dr. Daniel den Hoed Kliniek, Rotterdamsch Radio-Therapeutisch
Instituut, Postbus 5201, 3008 AE Rotterdam, The Netherlands

Van Roy, F.M. (p. 27)
Laboratory of Molecular Biology, State University of Gent,
9000 Ghent, Belgium

Van Steenis, G.J. (pp. 245, 340)
Department of Pathology, State University of Leiden,
P.O. Box 9603, 2300 RC Leiden, The Netherlands

Venturo, I. (pp. 50, 233, 237, 359)
Istituto Nazionale Tumori "Regina Elena" Divisione di
Oncologia Medica II, Servizio di Citopatologia,
Viale Regina Elena, 291, 00161 Roma, Italy

Verhest, A. (p. 37)
Departments of Cytogenetics, Dèpartement Pathologie, Hôpital
Erasme, 808 route de Lennik, 1070 Bruxelles, Belgium

Versnel, M.A. (p. 265)
Department of Pathology and Immunology, Erasmus University,
Burg. Oudln 50, Rotterdam, The Netherlands

Vielh, P. (p. 371)
Division of Clinical Cytopathology, Institut Curie, 26, Rue d'Ulm,
75231 Paris Cedex 05, France

Volkert, H. (p. 404)
Institut für Klinische Zytologie, Technische Universität München,
Prinzregentenplatz 14, 8000 München 80, Federal Republic of
Germany

Von Kleist, S. (p. 241)
Institut für Immunbiologie der Universität, Stefan-Meier-Straße 8,
7800 Freiburg, Federal Republic of Germany

Von Rosen, A. (p. 90)
Patologiska Institutionen, Avd. för tumörpatologi, Karolinska Sjukhuset, 10401 Stockholm 60, Sweden

Vooijs, G.P. (pp. 159, 412)
Department of Pathology, University Hospital Nijmegen, Geert Grooteplein Zuid 24, 6525 GA Nijmegen, The Netherlands

Vuzevski, V.D. (p. 265)
Department of Pathology and Immunology, Erasmus University, Burg. Oudln 50, Rotterdam, The Netherlands

Wagner, D. (p. 63)
Zytologisches Laboratorium am Evang. Diakonissen-Krankenhaus, Burgunderstraße 1, 7800 Freiburg, Federal Republic of Germany

Weber, K. (p. 178)
Max-Planck Institute for Biophysical Chemistry, 3400 Göttingen, Federal Republic of Germany

Wijsman-Grootendorst, A. (p. 108)
Leiden Cytology and Pathology Laboratory, P.O. Box 16084, 2301 GB Leiden, The Netherlands

Wikely, G. (p. 404)
Cytological Unit, Regional Laboratory Service, Durban, South-Africa

Wirnsberger, G. (p. 151)
Pathologisches Institut der Universität, Auenbruggerplatz 25, 8036 Graz, Austria

Witte, S. (p. 140)
Diakonissenkrankenhaus Karlsruhe, Abteilung Innere Medizin, 7500 Karlsruhe, Federal Republic of Germany

Wittekind, D. (p. 376)
Anatomisches Institut der Universität, Abteilung Anatomie II, Albertstraße 17, 7800 Freiburg, Federal Republic of Germany

Young, J.A. (p. 324)
Department of Pathology, The University of Birmingham Medical School, Birmingham B15 2TJ, UK

Zajdela, A. (p. 371)
Division of Clinical Cytopathology, Institut Curie, 26, Rue d'Ulm, 75231 Paris Cedex 05, France

Zillhardt, P. (p. 371)
Division of Clinical Cytopathology, Institut Curie, 26, Rue d'Ulm, 75231 Paris Cedex 05, France

I. Special Article

Cytology in Germany

E. Sprenger

The traditional way to begin this topic would be to trace the historical development of cytology in Germany. However, there is already a monograph on this subject, 'Klinische Zytologie in Deutschland - Die Anfangsjahre 1945-1957', which appeared in the *Frankfurter Beiträge* under the editorship of Horst Naujoks [7]. It is a superb documentation that absolves me from the obligation to provide a historical review, thus allowing me room to analyse the current situation in German cytology. Cytology is a diverse, complex structure that is in need of definition. The terms 'clinical cytology', 'diagnostic cytology', 'cytopathology', all closely related but still differentiated, may serve to mark off the sectors of cytology with which we shall concern ourselves in the following.

Every procedure, every method lives through the people who use it. In the field of cytology this is the group of clinicians with their special fields, the pathologists and the laboratory doctors. The spectrum of their activities involves the classical triad: medical care, teaching and research. Many aspects of the cytologists' activity are linked to auxiliary medical occupations, in this case the cytotechnician.

Medical Care: Cell Diagnosis

Cytology has attained a high degree of familiarity in Germany even among nonmedical people as a result of mass screening programms for cervical cancer. On July 1, 1971 [13], a law was introduced in the Federal Republic of Germany entitling every woman over 30 (and, since 1982 [1] every woman over 20) to take part in a screening programme involving an annual cervical smear. Approximately 30% of the women over 20 who are entitled to this examination actually to take advantage of it. This is approximately 5 million women annually. Fifteen years have passed since the introduction of mass screening programme for cervical cancer. Thus it should now be possible to attempt a preliminary appraisal of its effectiveness. Our main authority will be the results of a study by Soost et al. [12], based on 824402 persons examined during the period 1963-1976. The number of invasive cancers and precancerous lesions found decreased as participation in the screening programme increased. Soost's observation is not an isolated finding and represents the possible effectivity of the method under optimum conditions.

As it is so often the case, there is another side to the coin, and a study by Hilgarth and Schultz [3] has forcefully drawn our attention to it. All cytologically proven cases of a precancerous lesion, or of cervical cancer that were treated in the gynaecological clinic of the University of Freiburg between 1971 and 1979 were checked to ascertain whether the patient had previously participated in screening programmes for cervical cancer. For 11% of the previously screened patients with a precancerous lesion and 16% of the previously screened patients with cervical cancer, the screening results had been negative. One-third of the errors can be attributed to the cytologist, two-thirds to the technique used in taking the specimen and the method of preparing the smear. The author concludes, 'The effectivity of the entire programme for detecting early cancer of the cervix uteri lies between 74% and 84% for the material we studied.' The figure of 90% effectivity of mass screening programme, which stems from the leading cancer centres of the world and has found its way into the textbooks, has been given a counterpole.

The reasons for these lower figures are not easily quantified. In the past 15 years a transformation has taken place in the organization of mass screening programmes. Where as initially they were carried out by centralized institutions that were generally funded by non-profit organizations for the prevention of cancer, they are now becoming more and more decentralized, and this tendency has been increasing at a tremendous rate. Previously, laboratories with 100000 and more specimens were considered large, but nowadays laboratories with 20000 specimens can be regarded as big. Cytology is shifting more and more to the individual doctor's office or to a joint laboratory to which several doctors are affiliated. With the pressure due to the increasing number of medical practitioners, side income from cytology is becoming an important factor in the economic calculations of a doctor's office.

Non-gynaecological cytology is less prominent, but still the field of application is widening. A field with considerable tradition is the examination of effusions from body cavities for tumour cells. Sputum and urine cytology have also found a broad field of application.

A fundamental change in the field of cytological diagnosis has resulted from the development of the aspiration biopsy. Palpable nodules under the skin and nodules that can be visualized by X-ray, scintigraphy or sonography can be aspirated with a syringe and a needle with a diameter of 0.7 mm. The diagnostic reliability with which a malignant tumour up to 1 cm can be detected is around 90%–95%. Inflammatory lesions and benign nodules can also be analysed morphologically and followed up. An obstacle to the dissemination of this method lies in the discrepancy between the apparent simplicity of the method of obtaining material with this 'diagnostic acupuncture' and the difficulty of actually obtaining adequate material for the diagnosis. If the person doing microscopy performs the fine-needle aspiration himself, he can check on the success of the aspiration and reach an optimum effectivity, leading to a rate of less than 5% material that cannot be used for the diagnosis. When aspiration and diagnosis are separated, the rate of insufficient material increases to approximately 30%. This figure is an average that includes examples of good and not so good cooperation between collegues. On the whole, however, the constant, standardized diagnostic value of aspiration cytology still is in some places mistrusted. Cytopathological institutions that combine specimen collection and diagnosis are the rare exception in Germany.

A participation rate of only 30% in the annual mass screening programme for cervical cancer and the slow reduction in professional reservations about curative exfoliative and aspiration cytology indicate that cytological diagnosis has not yet achieved the full effectivity it could in promoting health care. This is a continual challenge to the German cytologist.

Teaching: Basic Training, Specialization, Continuing Education

Particularly in the transitional phase in which a technique is becoming a widely applied diagnostic procedure, teaching has a special significance as an importer of knowledge. In the catalogue of education goals for the *basic training of medical students,* diagnostic cytology finds only marginal acknowledgement. Among the clinical fields it is gynaecology and haemato-oncology that - marginally - orient students in the field of cytology. The natural home of cytological training should, however, be pathology, which fulfills this function to a varying degree from region to region. Until very recently, respected textbooks of pathology could get along with only a few references to cytology under the heading of general pathology and could restrict themselves to one representative illustration. From this it was quite easy for a student to gain the impression that cytology could be comfortably neglected. Following basic medial training, *specialization,* leading to the acquisition of a specialist qualification, is the next step in the training of a doctor. The specification for the training of pathologists includes an itemization of the number of cytology specimens and the subjects to be covered. In the field of gynaecology knowledge of cytology is expected, but no details are specified. Otherwise the practice of cytology is not established in the guidelines for specialized training, and it is pretty much left up to the individual to decide for her- or himself what non-prescribed certificates to procure.

In this situation, a Study Group for Cytology was brought into being at the end of the 1970s by the professional organization of pathologists, gynaecologists and laboratory doctors and the Deutsche Gesellschaft für Zytologie. This study group discussed the inclusion of cytology in the regulations on specialist training, but it has now discontinued its meetings, retaining the status quo. Thus, in Germany, there is still a vacuum of training regulations in vast areas of cytology.

Recognizing the incompleteness of the regulations for cytological practice, the federal organization of German medical doctors, whose responsibility this is, has taken several initiatives in the area of mass screening programmes for cervical cancer, but only in this area. Quality control regulations within the framework of these screening programmes went into effect in 1977 [4].

In 1981 [6] the regulations were revised and updated. Additionally, in 1980 [5] guidelines on the personal qualifications required for performing cytological screening within the screening programmes for cervical cancer (cytology guidelines) were set up.

For the field of curative cytology and aspiration cytology, however, there are no general regulations to date.

The third area of training is the *continuing education* of doctors with cytological training. For years the Deutsche Gesellschaft für Zytologie has awarded a certificate in gynaecological cytology following a practical test at the microscope. The success rate is around 30%. The certificate is something like a 'golden sports certificate' of diagnostic cytology. Besides this proof of one's own diagnostic qualification, this gives the opportunity for continued cytological training in microscopy courses under the auspices of the Deutsche Gesellschaft für Zytologie. An established system is offered that alternates annually between a big centralized programme in Munich at the Prof. Soost's Institute and a network of regional programmes set up so as to be available in all regions of Germany. The International Academy of Pathology also regularly runs continuing education programmes on cytological topics. In addition, individual medical councils and the Internatinoal Academy of Cytology offer courses.

Research: Conventional Diagnosis, Cell Markers, Automated Cell-Image Analysis

Research shows the imprint of creative individuality and thus resists being summarized in an overview. The organized basis for scientific presentations in the field of cytology has widened. The Deutsche Gesellschaft für Zytologie has been joined by the study group for cytopathology in the Deutsche Gesellschaft für Pathologie. Cooperation between the two institutions could become the driving force of scientific cytology.

In *conventional diagnostics* basic knowledge is well established. In some branches it is still necessary and possible to work on details.

There is a need for statistical evaluation of cytodiagnostic results, to answer questions regarding the efficacy of mass screening programmes in particular.

There are no sound cost-effect analyses available on mass screening programmes for cancer. According to Pflanz [8] such analyses can not be expected until individual data are available with details of the costs of the cytologically suspicious cases calculated against the cost of treatment of the cases in patients who were not screened at regular intervals.

Another object of research in the field of conventional cytodiagnostics is quality control of cytological work. We must distinguish between internal quality control and external quality control. The concept of internal quality control is based on test parameters made available to cytologists so that they can evaluate their own work and take the appropriate steps for their continuing education and/or their diagnostic decisions. An external quality control would be an examination brought into the laboratory that might possibly impose sanctions.

To date, the experience in the field of cytology with either internal or external quality control is insufficient. The Deutsche Gesellschaft für Zytologie has started a pilot study on external quality control, as has a regional group of cytologists. These activities will hopefully continue.

In addition to conventional staining procedures, *cytological markers* have begun to play a particularly important role in the diagnosis. In the discussion on the viral origin of cervical cancer the studies by zur Hausen [15] and Gissmann [2] have given important new impulses. The same holds true for nuclear DNA measurements, promoted by the German pathologist and cytologist Walter Sandritter [9, 10] in a unique way since the early 1950s.

The application of monoclonal antibodies has extended the spectrum of differential diagnoses to a degree that can hardly be assessed. The tumour cell that appears undifferentiated in conventional cytology can be given a therapeutically relevant differential diagnostic classification by demonstration of cytoskeletal proteins. In this field the expansion of the new diagnostic possibilities is still in full swing.

Automated cell-image analysis not only provides information additional to that in the conventional image but also arrives at independently formulated parameters that give a new diagnostic result. The cell-image diagnosis is based on quantifiable measurements in an automated system. This framework leads to standardization of diagnosis and is an important prerequisite for comprehensive quality control.

Such a system of standardized automated diagnosis comprises many components that together make up the total procedure. Specimen collection, fixation, preparation, staining, measurement, feature extraction, classification and documentation must be integrated. At present two alternative concepts are being developed under the sponsorship of the Federal Ministry of Research and Technology. These are the LEYTAS (Wasmund et al. [14]) project (Leyden Texture Analysis System) of the team in Leiden, Wetzlar and Fontainebleau and the FAZYTAN (Schwarzmann and Bloss [11]) project (*Früherkennung durch automatische Zytoanalyse:* early detection by automated cytoanalysis) of the Institute of Physical Electronics of the University of Stuttgart and the Kontron company in Munich.

First a prototype will have to be developed and guidelines for its use in practice will have to be established before a machine can be introduced to the market. The social environment of the diagnostic machine - the doctor and the cytotechnician - will be involved in the change that this technique could initiate. We must also except the structure of mass screening programme and curative cytology, including aspiration cytology, to be affected. The crucial event will probably be the transition from the qualitative parameters of conventional cytology to the quantitative parameters of automated cytology.

In the next few years we can expect important innovations that could be of significance not only for cytology but also for histology and cytogenetics.

Auxiliary Medical Professions: The Cytotechnician

An important prerequisite for the mass screening programmes for cervical cancer is the training of cytotechnicians at eight German schools of cytology. The training at these schools was at first geared to screening for cervical cancer. Soon, however, a change towards broad morphological training began. Starting with knowledge of anatomy and physiology, the necessary relations between these and pathological anatomy, cytogenetics and clinical medicine were imparted. Training in technical skills included not only cytological procedures but also those of histology, cell culture, cytogenetics and electron microscopy. As a formal expression of the increased breadth and depth of this training, the training period has been extended to 2 years. In addition to expanding and extending the training course for the cytotechnician to an auxiliary medical profession with comprehensive morphological training, government recognition of cytotechnician as designated professional occupation was pursued.

The range of the cytotechnician's practical work is not undisputed. A cytotechnician prepares the specimen and documents the material received for analysis. His or her function generally ends when the doctor begins his diagnosis. In special fields, e.g. cervical cytology, slides can be prescreened by a cytotechnician, who marks suspicious areas and thus prepares the diagnosis by the doctor. We all know, however, that this ideal cooperation between cytotechnician and cytologist can contrast with the sometimes very different reality of practice. Diagnoses are sometimes made by cytotechnicians, reports are even sometimes signed by technicians. This is an abuse of a cytotechnician's role that must be justified by the doctor who is his superior.

Cytology in Germany means: no revolutionary mood, but also not a closed chapter. For some perhaps already too much routine to stimulate enthusiasm, to attract young, competent people; for others, too little safeguarded, established structure to be attractive as a permanent life's work. Delegable monotony or a challenge to gifted morphologists with perfect sight. These are some pairs of concepts from discussions.

Cytology has not become a medical discipline of its own but lives at present as an appendix to clinical specialities and to pathology. To survive successfully within these basic disciplines cytology will have to be a powerful partner with influence on medical care and with an attractive image to the newcomers. This is the challenge to German cytologists at a moment when the star of cytology is losing its brightness.

References

1. Bundesausschuß der Ärzte und Krankenkassen (1982) Änderung von Richtlinien über die Früherkennung von Krebserkrankungen (Krebsfrüherkennungs-Richtlinien) vom 26. Februar 1982. Dtsch Ärztebl 79: 65-67
2. Gissmann L (1985) Kondylome - Hinweise für die Beteiligung der Papillomviren an der Entstehung des Zervixkarzinoms. Gynäkologe 18: 160-162
3. Hilgarth M, Schultz R (1981) Ursachen und Ausmaß falsch negativer Befunde in der gynäkologischen Krebsvorsorge. Frauenarzt 22: 324-328
4. Kassenärztliche Bundesvereinigung (1977) Durchführungsbestimmungen zur Qualitätssicherung zytologischer Untersuchungen im Rahmen der Früherkennung des Zervixkarzinoms. Dtsch Ärztebl 74: 1303-1304
5. Kassenärztliche Bundesvereinigung (1980) Richtlinien der Kassenärztlichen Bundesvereinigung über die Voraussetzungen zur Durchführung von zytologischen Untersuchungen im Rahmen der Krebsfrüherkennungsmaßnahmen bei Frauen (Zytologie-Richtlinien). Dtsch Ärztebl 77: 154
6. Kassenärztliche Bundesvereinigung (1981) Durchführungsbestimmungen zur Qualitätssicherung zytologischer Untersuchungen im Rahmen der Früherkennung des Zervixkarzinoms. Rundschreiben an die Kassenärztlichen Vereinigungen der Länder, February 1981, pp 1-3
7. Naujoks H (ed) (1986) Klinische Zytologie in Deutschland. Olms, Weidmann, Hildesheim (Frankfurter Beiträge zur Geschichte, Theorie und Ethik der Medizin, vol 2)
8. Pflanz M (1974) Kritik an der Zytologie als Massen-Screening-Verfahren. Fortschr Med 92: 351-352, 388
9. Sandritter W (1952) Über den Nukleinsäurestoffwechsel in Plattenepithel- und kleinzelligen Bronchialkarzinomen. Frankf Z Pathol 63: 387-422

10. Sandritter W (1952) Über den Nukleinsäuregehalt in verschiedenen Tumoren. Frankf Z Pathol 63: 423-446
11. Schwarzmann W, Bloss WH (1984) System FAZYTAN. In: Deutsche Forschungs- und Versuchsanstalt für Luft- und Raumfahrt e.V. (DFVLR) (eds) Automation der zytologischen Diagnostik. TÜV Rheinland, Cologne, pp 35-54
12. Soost HJ, Bockmühl B, Zock H (1979) Wie effektiv sind zytologische Krebsvorsorgeuntersuchungen bei der Bekämpfung des Zervixkarzinoms? Dtsch Med Wochenschr 104: 1331-1335
13. Stockhausen J (ed) (1971) Programmierte Krankheitsfrüherkennung - Vorsorgeuntersuchungen in der Krankenversicherung. Deutscher Ärzte-Verlag, Cologne
14. Wasmund H, Ploem JS, Meyer F (1984) Das LEITZ-System zur automatisierten zytologischen Krebsfrüherkennung. In: Deutsche Forschungs- und Versuchsanstalt für Luft- und Raumfahrt e.V. (DFVLR), (ed) Automation der zytologischen Diagnostik. TÜV Rheinland, Cologne, pp 17-33
15. zur Hausen H (1977) Human papillomaviruses and their possible role in squamous cell carcinomas. Curr Top Microbiol Immunol 78: 1-30

II. Incoming Methods: Cell Culture, In Situ Hybridization, Cytogenetics

In Situ Hybridization: An Overview

M. van der Ploeg and A. K. Raap

Introduction

In situ hybridization procedures allow the detection and localization of specific nucleic acid sequences in microscopic preparations. In combination with immunocytochemistry, they make it possible to study the topological relation between gene and gene products. The method stems from the fact that two strands in a DNA double helix or a DNA-RNA duplex can be separated by denaturation and then reannealed (hybridized) with labeled complementary strands under conditions which favor specific duplex formation. A labeled DNA or RNA sequence (probe) is used, to localize the gene sequences of interest (the target sequence).

The technique was originally developed by three groups independently: Pardue and Gall (1969), John et al. (1969), and Buongiorno-Nardelli and Amaldi (1970). Until recently, only radioactively labeled probes were applied, the specific molecular nucleic acid hybrids being detected by autoradiography. Radioactive in situ hybridization has at present reached a very high sensitivity, i.e., DNA sequences a few hundred base pairs in length can be localized in metaphase chromosomes with fairly short autoradiographic exposure times (Harper et al. 1981). Radioactive in situ detection of low copy number mRNA molecules in individual cells has also been described (Harper et al. 1986); the application of chemically synthesized oligodeoxyribonucleotides can be mentioned as a relatively recent development for the radioactive in situ localization of mRNA (for review see Coghlan et al. 1985). Small synthetic (18–30mer) oligonucleotide probes may have better accessibility to cellular mRNAs.

In spite of this sensitivity, application of radioactive hybridization procedures has been restricted to research laboratories, mainly by the inconveniences due to the radioactive labeling (safety measures; limited functional half-life of ^{32}P- or ^{35}S-labeled probes, which require frequent preparation and standardization; isotope disposal).

Nonisotopic Hybridocytochemistry

Since 1975; methods have been developed which aim at labeling probes or hybrids with a fluorochrome, a cytochemical enzyme endproduct, or - for electron microscopic evaluation - with gold particles. Rudkin and Stollar (1977) were the first to publish a fluorescent in situ hybridization procedure, employing antibodies specific for RNA-DNA hybrids. They described the localization of 5 *S* ribosomal DNA on polytene chromosomes of *Drosophila* by hybridization with a 5 *S* probe, followed by indirect immunocytochemistry. A few years later, Bauman et al. (1980) developed an end-labeling method of RNA probes with tetramethylrhodamine or fluorescein derivatives. The procedure was sensitive enough to allow direct fluorescence microscopic detection of virus DNA in infected cells, and also several genes on *Drosophila* polytene chromosomes (Bauman et al. 1981).

In this direct method, the reporter molecule is already bound to the probe before the hybridization, which implies that the hybrids can be visualized microscopically immediately after their formation in the preparation.

There are now also several indirect procedures available, in which the probe contains an element (introduced chemically or enzymatically) that enables the detection of the hybrid by affinity cytochemistry. For such methods, it is a prerequisite that the probe-reporter bond survives the hybridization and washing procedures, and also that the reporter molecule shall not, or only minimally, interfere with the hybridization reaction. The method of Bauman et al. (1980) and the labeling of nucleic acid probes with an enzyme, as described by Renz and Kurz (1984), fulfill these prerequisites. Four indirect methods will be described which also do so, because these are at present the ones most widely used.

Biotination of Probes

Biotinated nucleic acid probes can be obtained by following the procedure developed by Ward and co-workers (Langer et al. 1981) using biotin-containing nucleotide triphosphate analogues of cytidine (attached covalently to the C-5 position of the pyrimidine ring through an allylamine linker arm) and nick translation. The length of the linker arm is critical for the proper exposure of the biotin moiety in the hybrid (Brigati et al. 1983).

Recently, procedures have been developed to introduce the biotin by use of a photo-activatable analogue of biotin (Forster et al. 1985) or through a (chemical) transamination reaction (Viscidi et al. 1986). The detection of in situ hybridized probes after hybridization is mediated by avidin or anti-biotin-antibodies.

Acetylaminofluorene Labeling of Nucleic Acid Probes

The acetylaminofluorene (AAF) modification of nucleic acids is a chemical one. When RNA or DNA (single-stranded or double-stranded) is incubated with *N*-acetoxy-*N*-acetyl-2-aminofluorene at 37 °C for 1 h, this compound will be covalently bound to the C-8 position of the guanine residues in the probe (Landegent et al. 1984; Tchen et al. 1984). The relative concentrations of nucleic acid and activated AAF determine the extent of the haptenization of the probe. A 5%–6% substitution in most cases is optimal for hybridization purposes. When coupled to a protein carrier, AAF-guanosine is a strong immunogen, and poly- or monoclonal antibodies are easily obtainable.

Mercuration of Probes

This method differes from the previous ones by the fact that the hapten is introduced after the hybridization reaction. First the nucleic acid probe (RNA or DNA, double-stranded or single-stranded) is chemically mercurated at the C-5 position of the pyrimidines uracil and cytosine. Then the mercurated probe is hybridized, after which the hybrid is rendered immunochemically detectable by reacting it with a compound (ligand) that on one terminus carries the hapten and on the other a sulfhydryl group that will bind to the mercury atom in the probe (Hopman et al. 1986 a, b).

Sulfonation of Probes

A bisulfite-*O*-methylhydroxylamine mixture will, under mild conditions, cause the quantitative conversion of cytidine monomers into *N*4-methoxy-5,5-dihydrocytidine-6-sulfonate derivatives, the methylhydroxylamine serving to stabilize the sulfone. The reaction will also occur with cytidine residues in oligonucleotides and denatured DNA, resulting in modified probes (Sverdlov et al. 1974). Visualization (after the hybridization reaction) is carried out with a monoclonal antibody that binds specifically to the sulfone groups of the probe, followed by an enzyme or fluorescein or gold-conjugated anti-immunoglobulin antibody conjugate.

Multiple Simultaneous Hybridization

By combining several independent labeling procedures in different colors, it has become feasible to perform multiple simultaneous hybridization. Using the right probes and hybridization conditions, it has, for instance, been shown to be possible to visualize four different pairs of chromosomes in one metaphase spread or interphase nucleus (Nederlof et al. 1987). In this case, the probes were labeled with AAF, Hg, or with biotin. The receptor molecules were fluorescein isothiocyanate (FITC), tetramethylrhodamine isothiocyanate (TRITC), and aminomethylcoumarinacetic acid (AMCA; a recently developed blue fluorochrome); the fourth probe was labeled with AAF and with Hg, showing both red and green fluorescence emission when properly excited.

Sensitivity

In theory all four labeling methods mentioned are of equal sensitivity, because if more than 10% of the bases are modified, the number of antibody molecules that can be bound to the hybridized probe is limited by steric factors. Consequently, any actual differences in sensitivity between methods may be due to the quality of the antisera and the immunocytochemical detection or signal amplification systems used.

The complexity of a mammalian haploid genome is such that it has about 3×10^9 base pairs and consists of various classes of sequences. There are highly repetitive sequences which occur either clustered (e.g., mouse satellite DNA, alphoid sequences) or scattered throughout the genome (e.g., the Alu repeat); there are middle to low repetitive sequences (e.g., ribosomal DNA), and there are unique sequences. The nonradioactive hybridization procedures mentioned here can quite easily localize highly and middle repetitive nucleic acid sequences in metaphase spreads and interphase nuclei, in cell smears, or in tissue sections (Burns et al. 1985; Cremer et al. 1986; Durnam et al. 1985; Hopman et al. 1986b; Landegent et al. 1985a; Manuelidis and Ward 1984; Mitchell et al. 1986; Pinkel et al. 1986).

Unique sequences can also be visualized (the lower limit of detection at this moment is between 1 and 7 kb), which sometimes renders extensive experimentation, sophisticated microscopy, and statistical analysis necessary (Landegent et al. 1985a; Hopman et al. 1986 a, b; Ambros et al. 1987). However, procedures which amplify the signal obtained by precipitation of peroxidase-diaminobenzidine (DAB) using gold/silver-containing solutions may bring single copy gene detection into the realm of bright-field microscopy, provided background (control hybridization) signals can be clearly defined. A recent report by Garson et al. (1987) describes a protocol which allows chromosomal localization of small probes by using phase-contrast microscopy. Landegent et al. (1985a) were able, with AAF-labeled probes, to localize parts of the human thyroglobulin gene (±7 kb) in metaphase chromosomes; using biotinated probes, Albertson (1985) mapped muscle protein genes (±60 kb) to the X-chromosome of *Caenorhabditis elegans*. Most experience so far has been obtained with the detection and localization of DNA target sequences on glass slides and after hybridization with cell nuclei in suspension (Trask et al. 1988). Hybridization procedures to localize and quantitate specific mRNA sequences reliably are less commonly used so far and some are still under investigation (Singer and Ward 1982; McDougall et al. 1986; Dirks et al. 1988, unpublished results).

Applications in Cytopathology

The development of nonradioactive hybridochemical methods was a major advance, taking away most of the obstacles to wider application of this technique as a diagnostic tool. The hybridization approach may become a key diagnostic aid and the method of preference for detection of the expression of oncogenes and other cancer-related genes, although at present the detection of gene transcripts (mRNAs) with these methods is still largely unexplored.

Furthermore, in situ hybridization will be of value for the detection of viruses which for different reasons are difficult to detect otherwise; in vitro cultivation of parvo, human papillomavirus and Epstein-Barr virus, for instance, is not possible and cannot be used as a diagnostic tool; the cultivation of cytomegalovirus is difficult. Hybridocytochemistry thus may complement culture and immunopathology techniques. The special advantage of hybridization over immunocytochemistry is the fact that, whilst immunocytochemistry is able to indicate the presence of a protein, it is not definitive in regard to the function of the particular gene concerned. The protein may have arrived in those cells otherwise than by *de novo* synthesis, e.g., by pinocytotic or other transport processes. Hybridization with appropriate probes will definitely give "specific" results.

Cytogenetics and Cancer Cytogenetics

As stated above, detection of repetitive sequences in chromosomes but also in interphase nuclei is well within the limits of sensitivity of in situ hybridization. Consequently, the technique is applicable to the identification of species-specific chromosomes in somatic cell hybrids using either total genomic DNA as the probe, or probes for repetitive sequences (Durnam et al. 1985; Pinkel et al. 1986; Hopman et al. 1986b). Moyzis et al. (1987) used biotin-labeled recombinant DNA clones isolated from a human repetitive DNA library, which specifically hybridized to chromosome position 9qh and chromosome position 16qh. These locations were determined by in situ hybridization to metaphase chromosomes, using a fluorochrome as the reporter molecule.

By using double in situ hybridization and probes for both species of a somatic cell hybrid, subtle chromosomal translocations between chromosomes have been detected in minor fractions of the cells (Hopman et al. 1986b). Such phenomena are not so easily detected with single hybridization or other chromosome identification techniques.

Because karyotyping by classical means - depending on the tissue sample - can be cumbersome, a very interesting application of the manyfold nonradioactive in situ hybridization techniques in cancer cytogenetics (Cremer et al. 1986; Hopman et al. 1988), and also prenatal diagnosis (Julien et al. 1986), is the rapid identification of (parts of) chromosomes in interphase cells of solid tumors and chorion villi using chromosome-specific probes recognizing repetitive, clustered sequences (Burns et al. 1985; Manuelidis and Ward 1984; Cremer et al. 1986; Hopman et al. 1988).

The ability to use three differently labeled probes and three different immunofluorophores increases the versatility of "interphase karyotyping" considerably (Nederlof et al. 1987). The cultivation procedure is more cumbersome and carries the inherent danger that not all cell types will produce metaphase cells. The combination of DNA hybridization in suspension with the speed and the advantage of quantitative analyses provided by flow cytometry may offer a future diagnostic tool for quantification of hybridized chromosome-specific DNA probes, amplified genes and viral sequences in interphase nuclei.

In principle it should be possible to detect translocations in interphase nuclei by using double hybridizations with probes upstream and downstream of the specific chromosomal breakpoints: where there is translocation, the hybridization-linked colors will either merge or separate, dependent on the choice of probe. Clearly, deletions can in principle also be detected on the basis of the presence or absence of an in situ hybridization signal, but one should be absolutely sure that absence of signal is not caused by techical reasons. Furthermore, the techniques can be used to detect sites of gene amplification or the site of integration of transfected DNA (Boom et al. 1986; Smith et al. 1986). If for technical reasons, single-copy sequence detection cannot be performed on a routine basis, an elegant solution would be the use of a nested set of single copy sequences, as suggested by Cremer et al. (1986). It is also feasible to use a limited number of neighbouring or overlapping cosmid clones which can accommodate large stretches of foreign DNA (45 kb). These will most probably carry repetitive elements, which can in principle be eliminated from participation in the in situ hybridization by competition with excess unlabeled repetitive probe.

Viral Diagnosis

Viral diagnosis is an area in which the nonradioactive in situ hybridization techniques have penetrated quite rapidly, partly due to the fact that virus-infected cells in pathological tissues contain high copy numbers of that virus and thus are readily detectable by hybridocytochemical procedures. Some examples of human viruses and their associated diseases for which the technique has become operational in routine diagnostic settings in the fairly short time of existence of the methods are cytomegalovirus in immunodeficient (transplant) patients (Loning et al. 1986; Myerson et al. 1984 a, b; Raap et al. 1988; Jiwa et al. 1988, in preparation), herpes simplex viruses in herpes encephalitis (Burns et al. 1985; Boerman et al. 1988, in preparation), JC virus (a polyoma virus called after John Cunningham, the first patient from whom it was isolated) in progressive multifocal leukoencephalopathy (Aksamit et al. 1985; Nakamura et al. 1985; Boerman et al. 1988, in preparation), papilloma virus in various cancers, and parvo virus in fetal erythema infectiosum (Salimans and Raap 1988, in preparation). Sixbey et al. (1984) used to biotin method for the detection of Epstein-Barr virus in oropharyngeal epithelial cells. Hybridization procedures will permit analysis of genetic information already present in pathology tissue and cell archives for retrospective investigations when appropriate diagnostic probes become available (Jiwa et al. 1988, in preparation). The availability of several labeling techniques allows detection or screening for two or more viral genomes by using manyfold, simultaneous hybridiziation.

mRNA Localization

Numerous studies concerning the autoradiographic detection of mRNA sequences have been described. The list of successful nonradioactive studies is, however, still limited in spite of the importance of this approach for many areas of biomedical research. This restricted application may be partly due to technical difficulties. For example, after the first publication concerning fluorescent in situ actin mRNA detection (Singer and Ward 1982), a second paper appeared in which the technical difficulties are apparent (Lawrence and Singer 1985). Finally, a third article showed that these difficulties can be overcome (Lawrence and Singer 1986). Consequently, it appears that for each experimental set up, the various steps of the full in situ hybridization procedure have to be optimized against each other, which is also our experience with a biological model system that displays abundant expression (Dirks et al. 1988, in preparation).

Recently, a paper was published (Bresser and Evinger-Hodges 1987), which claims that, after careful examination of the various experimental variables, an extremely sensitive detection of in situ mRNA targets by nonisotopic means is possible.

Conclusions

Nonradioactive in situ hybridization procedures can be used as diagnostic and research tools. The methods allow the rapid, sensitive detection of specific nucleic acid sequences in morphologically preserved chromosome, cell, and tissue preparations.

The availability of such methods, in addition to other specific cytologic and cytochemical diagnostic procedures, improves the possibility of an early and accurate diagnosis of cytogenetic, viral, and oncological aberrations.

Further improvement in labeling techniques, assay formats, and enzymatic or fluorescent detection methods should allow for a wider application of nucleic acid hybridization reactions in diagnostic microbiology and basic research in infectious disease, cytogenetics, and oncology. The successful development of practical procedures for the detection of specific nucleic acids in clinical specimen then might markedly improve the care of patients with infectious diseases and allow a better understanding of the epidemiology and pathophysiology of such infections. The study of the connection between specific chromosome abnormalities and some forms of cancer or congenital defects may also benefit from these molecular diagnostic methods.

References

Aksamit AJ, Mourrain P, Sever JL, Mao EO (1985) Progressive multifocal leukoencephalopathy: investigation of three cases using in situ hybridization with JC virus biotinylated probes. Ann Neurol 18: 490-496

Albertson DG (1985) Mapping muscle protein genes by in situ hybridization using biotin-labeled probes. EMBO J 4: 2493-2498

Ambros PF, Bartram CR, Haas OA, Karlic H, Gadner H (1987) Non-isotopic in situ hybridization for mapping oncogenic sequences. In: Neth R, Gallo RC, Greaveo MF, Janka G (eds) Modern trends in human leukemia. Springer, Berlin Heidelberg New York

Bauman JGJ, Wiegant J, Borst P, Van Duijn P (1980) A new method for fluorescence microscopical localization of specific DNA sequences by in situ hybridization of fluorochrome labeled RNA. Exp Cell Res 128: 485-490

Bauman JGJ, Wiegant J, Van Duijn P (1981) Cytochemical hybridization with fluorochrome labeled RNA. II. Applications. J Histochem Cytochem 29: 238-246

Boerman RH, Raap AK, Peters ACB, Ter Schegget J, Van der Ploeg M (1988) Pre- and postmortem diagnosis of progressive multifocal leukoencephalopathy with a non-radioactive hybridization method (J clin Path in press)

Boom R, Geelen JL, Sol JC, Raap AK, Minnaar RP, Klaver BP, Van der Noordaa J (1986) Establishment of a rat cell line inducible for the expression of human cytomegalovirus immediate early gene products by protein synthesis inhibition. J Virol 58: 851-859

Bresser J, Evinger-Hodges M (1987) Comparison and optimization of in situ hybridization procedures yielding rapid, sensitive mRNA detections. Gene Anal Techn 4: 89-104

Brigati DJ, Myerson D, Leary JJ, Spalholz B, Travis SZ, Fong CKY, Hsiung GD, Ward DC (1983) Detection of viral genomes in cultured cells and paraffin-embedded tissue sections using biotin-labeled hybridization procedures. Virology 126: 32-50

Buongiorno-Nardelli N, Amaldi T (1970) Autoradiographic detection of molecular hybrids between RNA and DNA in tissue sections. Nature 225: 946-948

Burns J, Chan VTW, Jonasson JA, Fleming KA, Taylor S, Mc Gee JOD (1985) Sensitive system for visualizing biotinylated DNA probes hybridized in situ: rapid sex determination of intact cells. J Clin Pathol 38: 1085-1092

Coghlan JP, Aldred P, Haralambridis J, Niall HD, Penschow JD, Tregear GW (1985) Hybridization histochemistry. Ann Biochem 149: 1-28

Cremer T, Landegent J, Brueckner A, Scholl HP, Schardin M, Hager HD, Devilee P, Pearson P, Van der Ploeg M (1986) Detection of chromosome aberrations in the human interphase nucleus by visualization of specific target DNAs with radioactive and non-radioactive in situ hybridization techniques: diagnosis of trisomy 18 with probe L1.84. Hum Genet 74: 346-352
Durnam DM, Gelinas RE, Myerson D (1985) Detection of species chromosomes in somatic cell hybrids. Somatic Cell Mol Genet 11: 571-577
Forster AC, McInnes JL, Skingle DC, Symons RH (1985) Non-radioactive hybridization probes prepared by the chemical labeling of DNA and RNA with a novel reagent, photobiotin. Nucleic Acids Res 13: 745-761
Garson JA, Van den Berghe JA, Kemshead JT (1987) Novel non-isotopic in situ hybridization technique detects small (1 kb) unique sequences in routinely G-banded human chromosomes: fine mapping of N-myc and B-NFG genes. Nucleic Acids Res 15: 4761-4770
Harper ME, Ullrich A, Saunders GF (1981) Localization of the human insulin gene to the distal end of the short arm of chromosome 11. Chromosoma 83: 431-439
Harper ME, Marselle LM, Gallo RC, Wong-Stahl F (1986) Detection of lymphocytes expressing human T-lymphotropic virus type III in lymph nodes and peripheral blood from infected individuals by in situ hybridization. Proc Natl Acad Sci USA 83: 772-776
Hopman AHN, Wiegant J, Van Duijn P (1986a) A new hybridocytochemical method based on mercurated nucleic acid probes and sulfhydrylhapten ligands. I. Stability of the mercury-sulfhydryl bond and influence of the ligand structure on the immunochemical detection of the hapten. Histochemistry 84: 169-178
Hopman AHN, Wiegant J, Raap AK, Landegent JE, Van der Ploeg M, Van Duijn P (1986b) Bi-color detection of two target DNAs by non-radioactive in situ hybridization. Histochemistry 85: 1-4
Hopman AHN, Ramaekers FCS, Raap AK, Beck JLM, Devilee P, Van der Ploeg M, Vooijs GP (1988) In situ hybridization as a tool to study numerical chromosome aberrations in solid bladder tumors. Histochemistry (in press)
John H, Birnstiel ML, Jones KW (1969) RNA. DNA hybrids at the cytological level. Nature 223: 582-587
Julien C, Bazin A, Guyot B, Forestier F, Daffas F (1986) Rapid prenatal diagnosis of Down's syndrome with in situ hybridization of fluorescent DNA probes. Lancet II: 863-864
Landegent JE, Jansen in de Wal N, Baan RA, Hoeijmakers JHJ, Van der Ploeg M (1984) 2-Acetylaminofluorene-modified probes for indirect hybridochemical detection of specific nucleic acid sequences. Exp Cell Res 153: 61-72
Landegent JE, Jansen in de Wal N, Van Ommen G-J, Baas F, De Vijlder JJM, Van Duijn P, Van der Ploeg M (1985a) Chromosomal localization of a unique gene by non-autoradiographic in situ hybridization. Nature 317: 175-177
Landegent JE, Jansen in de Wal N, Ploem JS, Van der Ploeg M (1985b) Sensitive detection of hybridocytochemical results by means of reflection-contrast microscopy. J Histochem Cytochem 33: 1242-1246
Langer PR, Waldrop AA, Ward DC (1981) Enzymatic synthesis of biotin labeled polynucleotides: novel nucleic acid affinity probes. Proc Natl Acad Sci USA 78: 6633-6637
Lawrence JB, Singer RH (1985) Quantitative analysis of in situ hybridization methods for the detection of actin gene expression. Nucleic Acids Res 13: 1777-1799
Lawrence JB, Singer RH (1986) Intracellular localization of messenger RNAs for cytoskeletal proteins. Cell 45: 407-415
Loning T, Milde K, Foss HD (1986) In situ hybridization for the detection of cytomegalovirus (CMV) infection. Application of biotinylated CMV-DNA probes on paraffin embedded specimen. Virchows Arch 409: 777-790
Manuelidis L, Ward DC (1984) Chromosomal and nuclear distribution of the Hind III 1.9-Kb human DNA repeat segment. Chromosoma 91: 28-38
McDougall JK, Meyerson D, Beckman AM (1986) Detection of ciral DNA and RNA by in situ hybridization. J Histochem Cytochem 34: 33-38
Mitchell AR, Ambros P, Gosden JR, Porteous DJ (1986) Gene-mapping and physical arrangements of human chromatin in transformed, hybrid-cells: fluorescent and autoradiographic in situ hybridization compared. Somatic Cell Molec Genet 12: 313-324
Moyzis RK, Albright KL, Bartholdi MF, Cram LS, Deaven LL, Hildebrand CE, Joste NE, Long-

mire JL, Meyne J, Schwaracher-Robinson T (1987) Human chromosome-specific repetitive DNA sequences: novel marker for genetic analysis. Chromosoma 95: 375-386
Myerson D, Hackman RC, Nelson JA, Ward DC, McDougall JK (1984a) Widespread occurrence of histologically occult cytomegalovirus. Hum Pathol 15: 430-439
Myerson D, Hackman RC, Meyers JD (1984b) Diagnosis of cytomegalovirus pneumonia by in situ hybridization. J Infect Dis 150: 272-277
Nakamura S, Tourtelotte WW, Shapshak P, Darvish M (1985) Patient with progressive multifocal leukoencephalopathy by in situ hybridization using isotopic and biotinylated probes. No To Shinkei 37: 359-364
Nederlof PM, Raap AK, Robinson D, Wiegant J, Hopman AHN, Tanke HJ (1987) Simultaneous detection of three and more target DNA sequences by three color fluorescence in situ hybridization. Eur J Cell Biol [Suppl 20] 44: 97
Pardue ML, Gall JG (1969) Molecular hybridization of radioactive DNA to the DNA of cytological preparations. Proc Natl Acad USA 64: 600-604
Pinkel D, Straume T, Gray JW (1986) Cytogenetic analysis using quantitative, high sensitivity fluorescence hybridization. Proc Natl Acad Sci USA 83: 2934-2938
Raap AK, Geelen JL, Van der Meer JWM, Van de Rijke FM, Van den Boogaard P, Van der Ploeg M (1988) Non-radioactive in situ hybridization for the detection of cytomegalovirus infections. Histochemistry 88: 367-374
Renz M, Kurz C (1984) A colorimetric method for DNA hybridization. Nucleic Acids Res 12: 3435-3444
Rudkin GT, Stollar BD (1977) High resolution detection of DNA.RNA hybrids in situ by indirect immunofluorescence. Nature 265: 472-473
Singer JB, Ward DC (1982) Actin expression visualized in chicken muscle tissue culture by using in situ hybridization with a biotinated nucleotide analog. Proc Natl Acad Sci USA 79: 3331-3335
Sixbey JW, Nedrud JG, Raab-Traub N, Hanes RA, Pagano JS (1984) Epstein-Barr virus replication in oropharyngeal epithelial cells. N Engl J Med 310: 1225-1230
Smith GH, Doherty PJ, Stead RB, Gorman CM, Graham DE, Howard BH (1986) Detection of transcription and translation in situ with biotinylated molecular probes in cells transfected with recombinant DNA plasmids. Anal Biochem 156: 17-24
Sverdlov ED, Monastyrskaya GS, Guskova LI, Levitan TL, Scheichenko VI, Budowsky EI (1974) Modification of cytidine residues with a bisulfite-*O*-methylhydroxylamine mixture. Biochim Biophys Acta 340: 153-165
Tchen P, Fuchs RPP, Sage E, Leng M (1984) Chemically modified nucleic acids as immunodetectable probes in hybridization experiments. Proc Natl Acad Sci USA 81: 3466-3740
Trask B, Van den Engh G, Pinkel D, Mullikin J, Waldman F, Van Dekken H, Gray J (1988) Fluorescence in situ hybridization to interphase cell nuclei in suspension allows flow cytometric analysis of chromosome content and microscopic analysis of nuclear organization. Hum Genet 78: 251-259
Viscidi RP, Connelly CJ, Yolken RH (1986) Novel chemical method for the preparation of nucleic acids for non-isotopic hybridization. J Clin Microbiol 23: 311-317

In Situ Hybridization: Its Use as a Tool in Pathology Laboratories

S. Malcolm and J. van den Berghe

Introduction

In situ hybridization has not up to now been seen as a tool for the pathology laboratory, but rather as a specialist technique applied by geneticists and some other molecular biologists. As the power of the technique to answer important biological questions becomes more appreciated, and as the technology advances and becomes both more sensitive and more adaptable, this is likely to change. In the future it is likely that it will have some quite routine applications.

In situ hybridization involves the same techniques as nucleic acid hybridization in solution except that the nucleic acid is fixed in some sort of cytological preparation, either as DNA in chromosomes or nuclei or as RNA in cells. Clearly its importance has increased vastly over the last few years as so many genes of clinical interest have been cloned and the recombinant molecules resulting can be readily used as hybridization probes without further manipulation (Schmidtke and Cooper 1987).

The range of questions which can be tackled using this technique is wide.

1. Is a gene present?
2. Whereabouts in the genome does it occur?
3. Is it rearranged?
4. How much is there?
5. Is it expressed?

However, it is always likely that the experiments will be operating towards the limit of detection, as by necessity there is an extremely limited amount of target nucleic acid present at the single cell level. For example, the average loading of DNA on a gel in a Southern blot to detect a single copy gene is 5 μg. In contrast to this, a single cell contains only 6 pg DNA. Therefore the sensitivity of detection has to be a million times greater!

Technical Aspects

The techniques of probe preparation, labelling, hybridization and washing are all basically those of standard nucleic acid hybridization. The details as applied to chromosomal in situ hybridization are explained in two reviews (Buckle and Craig

1986; Malcolm 1986). In particular, these articles deal with the problem of maintaining chromosome morphology, so that the chromosomes can be identified after hybridization, whilst achieving sufficient denaturation of the two DNA strands to allow efficient hybridization. To a large extent these two aims are mutually incompatible and a compromise must be achieved. The exact method of banding for chromosome identification is fairly flexible and is most likely to reflect the previous expertise, apparatus and custom of the cytogeneticists in the laboratory. Among the banding techniques which have been successfully used are Lipsol G banding (Malcolm et al. 1981), trypsin banding (Garson et al. 1987a), Wright's Giemsa (Chandler and Yunis 1978), R banding (Bartram et al. 1983), Q banding (Kirsch et al. 1982) and BudR-induced replication banding (Buckle and Craig 1986).

One difference from solution hybridization is the necessity of building up a network of hybridized probe at the target site in order to increase the signal. Although the conditions for achieving this have not been rigorously studied, it is well established that labelling of the probe by nicktranslation and addition of dextran sulphate to the hybridization buffer aids the network or 'Christmas tree' formation.

The most exciting recent advances have involved the types of labelling and detection used. Previous experiments involved labelling with radioactive isotopes and detection using autoradiography, which was both time-consuming and fiddly to do. Unfortunately the use of ^{32}P is not possible in most experiments as, although it would be the most sensitive isotope to detect, because of its high initial particle energy the track length would be too long and therefore the resolution would be too poor. For most experiments tritium was the isotope of choice.

In the last couple of years many publications have appeared which use non-radioactive methods of detection of the hybridized sequences. Although some of the original studies described relatively insensitive experiments which were being used as controls to establish the system, it is clear that often the methods can be adapted to even the most sensitive uses. They have been based both on antibody detection of chemically modified nucleic acid and, very successfully, enzymatic detection of biotinylated nucleic acid. The original system for biotinylating DNA was developed in the laboratory of Dave Ward (Langer et al. 1981) and has been widely used since, all the necessary components for it are available commercially. More details of some of the non-radioactive methods available will be presented elsewhere in this volume.

Non-isotopic Method of In Situ Hybridization

A method has been developed which combines both high sensitivity with very fine resolution (Garson et al. 1987b). An outline of the method is presented in Fig. 1. The method involves labelling the probe with biotin by using biotin-11-dUTP as a substrate for DNA polymerase in a nick translation reaction. The protocol then exploits the specific and strong interaction between biotin and streptavidin to bind streptavidin-conjugated alkaline phosphatase to the hybridized sequences. The alkaline phosphatase is then detected with the chromogenic substrate mixture NBT/BCIP

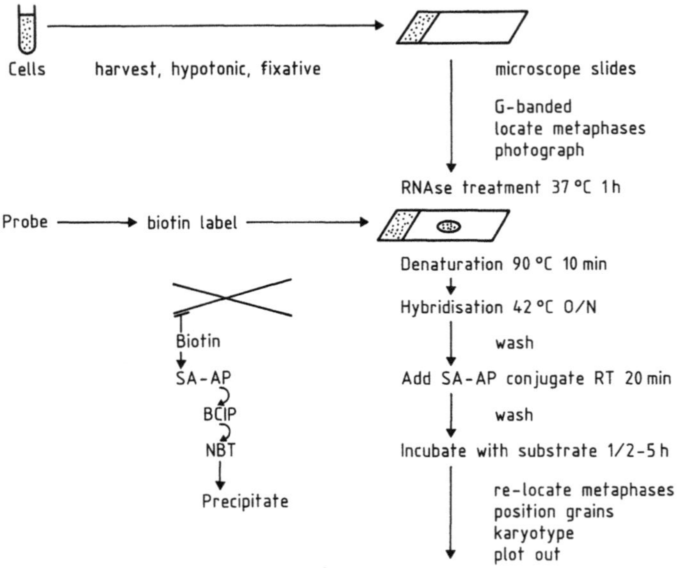

Fig. 1. Plan of in situ hybridization using biotinylated probe

(nitroblue tetrazolium and 5-bromo-4-chloro-3-indolyl phosphate). Previously located and photographed metaphases are then relocated and examined under phase contrast. The signal does not fade. Experimental details are as follows.

DNA probes were labelled with biotin-11-dUTP by nick translation (Bethesda Research Laboratories kit) and unincorporated nucleotides were removed by ethanol precipitation in the presence of 20 µg glycogen. The biotinylated probes can be stored at −20 °C for at least 9 months.

50–100 ng of biotinylated probe in 10 µl of hybridization buffer was applied to each slide under a sealed glass coverslip. The exact concentration of probe can be critical and pilot experiments may be necessary to determine the optimal concentration. This is the case if a first experiment shows apparently non-specific hybridization over many sites on the cell. Hybridization buffer contained 50% formamide (Sigma), 10% dextran sulphate (Sigma), 2 × SSC (1 × SSC = 0.15 *M* NaCl, 0.015 *M* Na citrate), 0.1 *M* EDTA, 0.05 m*M* Tris pH 7.5 and 100 µg/ml denatured sonicated salmon sperm DNA. The chromosomal DNA and probe were then denatured together at 80 °C for 10 min. Hybridization was performed in a humidified box at 42 °C for approximately 16 h and the slides subsequently washed in 2 × SSC for 30 min, 0.1 × SSC at 42 °C for 30 min, 2 × SSC for 15 min and finally in 0.1 *M* Tris pH 7.5, 0.1 *M* NaCl, 2 m*M* $MgCl_2$, 0.05% triton X-100, and 3% bovine serum albumin (TMN) for 15 min. Hybridized probe was detected by incubating slides for 20 min with 0.01 mg/ml streptavidin-conjugated alkaline phosphatase (BRL) in TMN. Excess conjugate was removed by washing for 3 × 5 min in TMN without the albumin, once in pH 9.5 buffer (0.1 *M* Tris 9.5, 0.1 *M* NaCl, 50 m*M* $MgCl_2$). Incubation with 150 µl of chromogenic substrate was performed in subdued light for 4–5 h. The substrate solution was made by adding 4.4 µl of

NBT at 75 mg/ml in 70% dimethyl formamide and 3.3 μl of BCIP at 50 mg/ml in dimethylformamide to 1 ml of pH 9.5 buffer. 1 m*M* levamisole (Sigma) was included to inhibit any residual alkaline phosphatase activity. Colour development was terminated by washing for 5 min in 20 m*M* Tris pH 7.5, 5 m*M* EDTA and the slides mounted in aqueous mountant (Dako).

Uses for In Situ Hybridization

One of the most important applications of in situ hybridization remains the mapping of single-copy human genes as part of the international effort to produce a full and accurate human gene map. It has the advantage over other techniques of using normal starting material, i.e. regular chromosome spreads, and until recently was the method with the greatest resolution as a gene was localised straight away to a band (about 0.5% of the total genome). Many genes have been localised by this method (Human gene mapping 8 1985). With the recent publication of a total human gene map based on genetic linkage using RFLPs (Donis-Keller et al. 1987), higher resolution mapping is now possible, but only once some detectable polymorphism is available for the gene or probe and only after extensive family studies. These localisation studies can be extended to study the rearrangements found in tumour cells, particularly leukaemias, and can throw some light on the involvement of genes in these rearrangements (Davis et al. 1984).

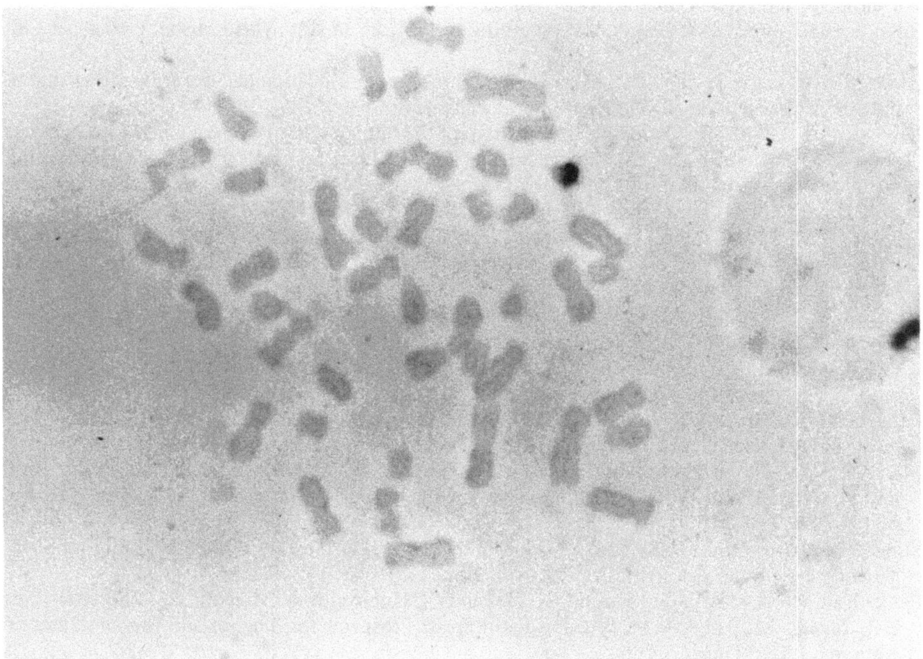

Fig. 2. In situ hybridization to a male metaphase spread and interphase nucleus using biotinylated probe pHY2.1 which is a highly repetitive sequence on the long arm of the Y chromosome

Although far from quantitative in general, the in situ technique can be usefully used to study highly amplified sequences. An example is the study of highly amplified N-*myc* genes in advanced stage neuroblastomas. Not only has it been shown that the site of the amplified genes is different from the original localisation of N-*myc* (Schwab et al. 1983; Garson et al. 1987b), but, using the biotinylated probe, it has been possible to demonstrate a periodicity in the structure of the amplified region which is by all other methods homogeneously staining (the HSR; Garson et al. 1987b). Another amplified sequence which lends itself to analysis by in situ hybridization is the highly repeated 2.1 kb sequence found abundantly on the fluorescent long arm of the Y chromosome. Although this sequence is not confined to the Y chromosome it is present on many more copies there and can be readily picked up in metaphase spreads and in interphase nuclei (see Fig. 2). This has great potential for rapid sexing both from chorionic villus sampling and potentially for embryos from the preimplantation stage (Penketh et al., in press).

Acknowledgements. Work in SM's laboratory has been generously supported by the Medical Research Council and Leukaemia Research Fund of Great Britain.

References

Bartram CR, de Klein A, Hagemeijer A, van Agthoven T, Guerts van Kessel A, Bootsma D, Grosveld G, Ferguson-Smith MA, Davies T, Stone M, Heisterkamp N, Stephenson J, Groffen J (1983) Translocation of c-abl oncogene correlates with the presence of a Philadelphia chromosome in chronic myelocytic leukaemia. Nature 306: 277-280

Buckle VJ, Craig IW (1986) In situ hybridization. In: Davies KE (ed) Human genetics: a practical approach. IRL Press, Oxford, pp 85-100

Chandler MF, Yunis JJ (1978) A novel banding method for in situ hybridization. Cytogenet Cell Genet 22: 352-356

Davis MB, Malcolm S, Rabbitts TH (1984) Chromosome translocation can occur on either side of the c-myc oncogene in Burkitt's lymphoma cells. Nature 308: 286-288

Donis-Keller H et al. (1987) Genetic linkage map of the human chromosomes. Cell 51: 319-337

Garson JA, van den Berghe JA, Kemshead JT (1987a) Novel non isotopic in situ hybridization technique detects small (1 kb) unique sequences in routinely G-banded human chromosomes: fine mapping of N-myc and BNGF genes. Nucleic Acids Res 15: 4761-4770

Garson JA, van den Berghe JA, Kemshead JT (1987b) High resolution in situ hybridization technique using biotinylated NMYC oncogene probe reveals periodic structure of HSRs in human neuroblastomas. Cytogenet Cell Genet 45: 10-15

Human Gene Mapping 8 Helsinki Conference (1985) 8. Internat. Workshop on human Gene Mapping. Cytogenet Cell Genet 40: Nos. 1-4

Kirsch IR, Morton CC, Nakahara K, Leder P (1982) Human immunoglobulin heavy chain genes map to a region of translocation in malignant B lymphocytes. Science 216: 301-303

Langer PR, Waldrop AA, Ward DC (1981) Enzymatic synthesis of biotin-labelled polynucleotides: novel nucleic acid affinity probes. Proc Natl Acad Sci USA 78: 6633-6637

Malcolm S (1986) Specialist techniques in research and diagnostic clinical cytogenetics. In: Rooney D, Czepulkowski B (eds) Human cytogenetics: a practical approach. IRL Press, Oxford, pp 197-226

Malcolm S, Barton P, Murphy C, Ferguson-Smith MA (1981) Chromosomal localisation of a single copy gene by in situ hybridization. Ann Hum Genet 45: 135-141

Penketh R, Monk M, van den Berghe JA, Delhanty J, Handyside A, Malcolm S, Middleton-Price HR, Tavares M, Pembrey M, Winston R (in press) Prospects for diagnosis in pre-implantation embryos. J Med Genet (in press)

Schwab M, Alitalo K, Klempnauer K-H, Varmus HE, Bishop JM, Gilbert F, Brodeur G, Goldstein M, Trent J (1983) Amplified DNA with limited homology to myc cellular oncogene is shared by human neuroblastoma cell lines and neuroblastoma tumours. Nature 305: 245-248

Correlations of Malignancy in Cultured Cells

M. M. Mareel, F. M. Van Roy, L. Messiaen, N. Van Larebeke,
and E. Boghaert

Introduction

Normal tissues are under strict growth control (G^- phenotype). Benign tumours differ from normal tissues by loss of growth control (G^+ phenotype), but they are not invasive (I^- phenotype) and not metastatic (M^- phenotype). Malignant tumours also display the G^+ phenotype, but at least part of their cell populations have acquired invasive (I^+ phenotype) and metastatic (M^+ phenotype) capabilities. We infer from these well-documented clinical and pathological observations that invasion and metastasis, more than loss of growth control, are the hallmarks of malignancy. Obviously, this does not mean that uncontrolled growth does not contribute to the progression of the disease.

It is our hypothesis that these three phenotypic traits are distinct and under separate genetic control (Mareel and Van Roy 1986). One argument in favour of this hypothesis is the natural occurrence of various combinations of these phenotypes: G^+, I^-, M^- in benign tumours; G^+, I^+, M^- in basocellular epitheliomas of the skin and in primary brain tumours; G^+, I^+, M^+ in many malignant tumours; G^-, I^+, M^- in cholesteatomas; G^-, I^+, M^+ in the trophoblast of some species, including humans. Another argument is that, at least in experimental systems, the mechanisms of invasion and growth can be separated; agents that arrest growth do not necessarily arrest invasion (Mareel et al. 1982; Distelmans et al. 1985), whereas invasion can be arrested without growth arrest (Mareel et al. 1984; Bracke et al. 1984; Storme et al. 1985).

Most, if not all, tumours are heterogeneous with regard to multiple biochemical and biological characteristics including the I and M phenotypes (review by Nicolson and Poste 1983). It is conceivable that the proportion of I^+ and M^+ cells and the degree to which these phenotypes are expressed will determine tumour progression. Furthermore, it is important to investigate how therapy influences I^+ and M^+ subpopulations (review by Poste and Greig 1982). Does non-curative chemotherapy result in more agressive or in less agressive tumours once relapses do occur? The importance of I^+ and M^+ subpopulations for prognosis and therapeutic strategy is a good reason to develop probes which recognize these subpopulations in biopsy specimens. We here review our work with cultured cells, illustrating that this task might well be feasible.

Experiments

Methods to test the I and M phenotypes of cultured cell populations and subpopulations are critical to the development of probes that are specific for these phenotypes. These methods have been discussed recently (Mareel et al. 1987). Briefly: for testing the I and M phenotypes in vivo, cell suspensions, cellular aggregates, or cells aggregated to a collagen sponge are implanted at various sites in syngeneic or in immunodeficient animals. Tumour-bearing animals are killed at fixed intervals or when moribund and examined for invasion and metastases both macroscopically and microscopically. Cells are considered to be I^+ or M^+ when they score positive in one of these tests; they are considered to be I^- or M^- when all tests are negative. For testing the I phenotype in vitro, cells are confronted with fragments of normal tissue in organotypic culture and confronting pairs are examined histologically after various periods of incubation. One purpose of the assay in vitro, which is relatively short (days) in comparison with the assays in vivo (weeks to months), is to see whether or not the I^+ phenotype might be acquired during expansion of the population inside the animal (for examples, see review by De Baetselier et al. 1984).

The hope that the molecular biology of oncogenes and proto-oncogenes would rapidly provide us with I^+ and/or M^+ specific probes has not been fulfilled. Two main strategies have been followed. First, I^- and/or M^- acceptor cell populations were transfected with plasmids containing specific gene sequences or with high molecular weight DNA from I^+ and/or M^+ sources. The major problems with these experiments are: (a) The I^+ and/or M^+ phenotypes could be conveyed upon acceptor cells through transfection with a variety of oncogenes and also non-oncogenes that encoded functionally unrelated molecules. (b) In some experiments the transfection procedure per se led to changes in the I and M phenotypes. (c) Acceptor cells frequently acquired the I^+ and/or M^+ phenotypes in an apparently spontaneous way after an extended passage in vitro. For example, we established cultures from C3H mouse eye lenses (MLE), because lens tumours never develop naturally and because such cultures consisted of epithelial cells only. MLE cells at passage 12 were found to be non-invasive in vitro and non-tumorigenic in vivo. The population was expanded up to passage 19 for transfection or cotransfection with plasmids encoding a mutated p21-*ras* protein and/or containing a bacterial neomycin resistance gene, and further selected or not for resistance to geneticin. Testing the phenotypes of interest necessitated further expansion of the cultures up to passage 21 and 23. When the tests were performed both in vivo and in vitro all cultures, controls, transfected, and mock-transfected ones, appeared to have acquired the I^+ and occasionally also the M^+ phenotype. It is quite clear that these and similar experiments do not instruct us about the role of *ras* genes in invasion and metastasis.

Second, antibodies against specific proto-oncogene or oncogene products were applied to tissue sections from carcinomas, from benign lesions, and from normal precursor tissues. So far, we do not have a probe that consistently distinguishes between normal tissues, benign lesions and non-invasive carcinoma on the one hand and invasive carcinoma on the other. For example: in biopsy specimens of human breasts with fibrocystic disease or with carcinoma, expression of p21-*ras* proteins at high level was found by immunohistochemistry with a group-specific monoclonal antibody (RAP-5, gift of J. Schlom, NCI, Bethesda, Md) in carcinomas, but also in hyperplastic lesions and in normal tissue during the post-ovulatory phase of the menstrual cycle (De Potter et al., to be published). These observations suggested that p21-*ras* expression is hormone-dependent in normal breast tissue, whereas it occurs constitutively in carcinomas.

In one family of cell lines established from a MMTV-induced Balb/c mouse mammary adenocarcinoma (Sonnenberg et al. 1986), the morphotype of cells in culture on tissue culture plastic substrate served as a marker for the I phenotype (Mareel et al. 1986). Subpopulations with three distinct morphotypes were cloned from the parental population (Fig. 1) and tested for invasiveness through confron-

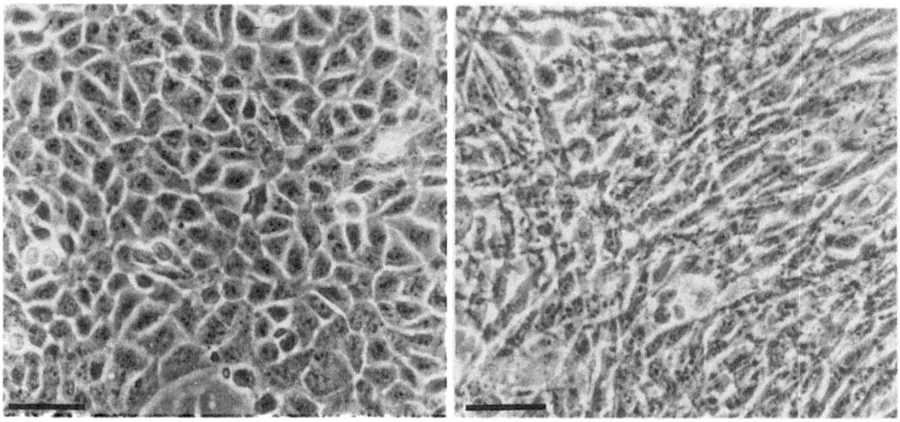

Fig. 1 a **b**

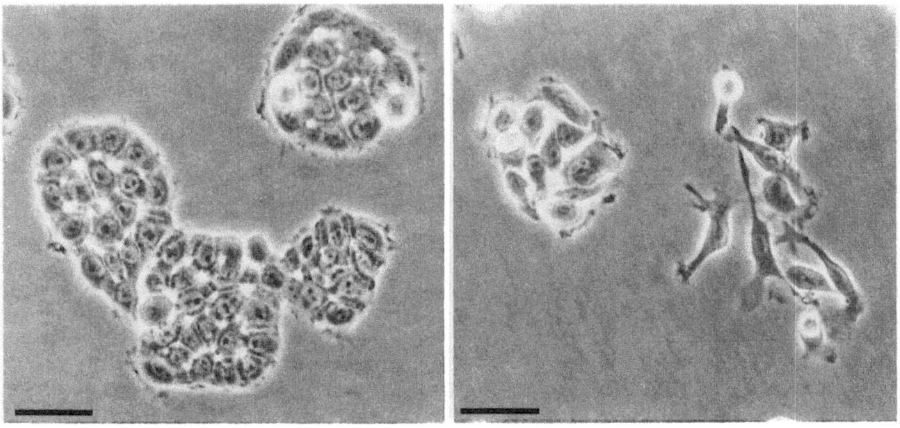

Fig. 2 a **b**

Fig. 1, 2. Photomicrographs of cells on tissue culture plastic substrata. **1** Polygonal epitheloid (**a**) and spindle-shaped fibroblastic (**b**) cells in cultures of subclones from a cell line that was derived from an MMTV-induced Balb/c mouse mammary adenocarcinoma. **2** Epitheloid (**a**) and fibroblastic (**b**) cells in cultures of subclones from the Nara rat bladder carcinoma NBTII cell line tation with fragments of embryonic chick heart in vitro and through subcutaneous injection as well as implantation under the renal capsula (kidney invasion test) in vivo. Populations of polygonal epitheloid cells displayed the I^- phenotype; it is, however, not certain that this phenotype is stable for a large number of passages. Populations of cuboidal cells appeared to be unstable inasmuch as they were morphotypically heterogeneous and scored both positive and negative in invasion tests, with a tendency to shift towards the I^+ phenotype at higher passage numbers. Populations of spindle-shaped fibroblastic cells were invasive and this I^+ phenotype can be expected to remain stable for a number of passages. So, in this particular cell family, transition of the epitheloid towards the fibroblastic phenotype served as a marker for the acquisition of the I^+ phenotype. This result cannot be extrapolated to all other epithelial cell lines. Epitheloid and fibroblastic subpopulations (Fig. 2) were also cloned from a parental Nara rat bladder carcinoma (NBTII) cell line, but both subpopulations were found to be invasive in vitro.

Transition from the epitheloid towards the fibroblastic morphotype marked the acquisition of the I^+ phenotype in a family of Madin-Darbey canine kidney cells (MDCK), consisting of parental cells and derivatives transformed by sarcoma viruses (Behrens et al., to be published). In the latter experiments the I phenotype was tested as mentioned above and also through seeding of cell suspensions on a matrix of collagen type I. The fibroblastic morphotype, in correlation with the I^+ phenotype, could be induced in two ways, namely in a constitutive way through virus transformation and epigenetically through treatment with a monoclonal antibody (anti-Arc-1) against a cellular adhesion molecule (L-CAM; uvomorulin). This suggests that deficient expression or aberrant function of uvomorulin is responsible for the transformation towards the I^+ phenotype. Comparison of I^- (epitheloid) with I^+ (fibroblastic) MDCK cell subclones by immunocytochemistry, immunoprecipitation and Western blotting showed absence of uvomorulin expression in I^+ populations. In this MDCK cell family, therefore, invasion could be explained through loss of a cellular adhesion molecule that normally keeps the cells within a well-organized epithelial monolayer. Work is now in progress to find out whether lack of uvomorulin expression correlates with the I^+ phenotype in other carcinomas.

Conclusions

1. Invasion and metastasis, more than loss of growth control, are hallmarks of tumour malignancy.
2. Tumours are heterogeneous, including with regard to the invasive and metastatic phenotypes, and minor subpopulations of the latter may determine prognosis.
3. In two families of cultured epithelial cells morphotypic and molecular markers for the invasive phenotype were found.

It is our opinion that the development of markers for invasive and metastatic cells applicable to specimens from human tumours will be invaluable for diagnosis and therapy planning, much to the benefit of the cancer patient.

Acknowledgements. Research in the authors' laboratories is supported by grants from the Ministerie van de Vlaamse Gemeenschap, Administratie van de Gezondheidszorg, the Fondation Philippe et Thérèse Lefèvre, the ASLK Kankerfonds, the FGWO (39.0009.83; 3.0005.84), and the Sport Vereniging tegen de Kanker, Brussels, Belgium. L.M. is a beneficiary of a PhD grant from the Instituut tot Aanmoediging van het Wetenschappelÿk Onderzoek in Nijverheid en Landbouw; F.M.V.R. is a Research Associate of the Nationaal Fonds voor Wetenschappelijk Onderzoek; E.B. is a beneficiary of a grant from the Belgisch Werk tegen Kanker, Brussels, Belgium.

References

Behrens J, Mareel MM, Van Roy FM, Birchmeier W. Essential steps in tumor cell invasion: epithelial cells acquire invasive properties following the loss of uvomorulin-mediated cell-cell adhesion (to be published)

Bracke ME, Van Cauwenberge RM-L, Mareel MM (1984) (+)-Catechin inhibits the invasion of malignant fibrosarcoma cells into chick heart in vitro. Clin Exp Metastasis 2: 161-170

De Baetselier P, Roos E, Brys L, Remels L, Feldman M (1984) Generation of invasive and metastatic variants of a non-metastatic T-cell lymphoma by in vivo fusion with normal host cells. Int J Cancer 34: 731-738

De Potter C, Pauwels C, Beghin C, Praet M, Mareel M, Roels H (to be published) The expression of p21-*ras* proteins in human neoplastic and non-neoplastic breast tissue. (to be published)

Distelmans W, Van Gynckel R, Vanherck W, De Brabander M (1985) The kidney invasion test: an assay allowing macroscopic quantification of malignant invasion in vivo. Invasion Metastasis 5: 170-184

Mareel MM, Van Roy FM (1986) Are oncogenes involved in invasion and metastasis? Anticancer Res 6: 419-436

Mareel MM, Bruyneel EA, De Bruyne GK, Dragonetti CH, Van Cauwenberge RM-L (1982) Growth and invasion: separate activities of malignant MO_4 cell populations in vitro. In: Galeotti T, Cittadini A, Neri G, Papa S (eds) Membranes in Tumour Growth. Elsevier, pp223-232

Mareel MM, Bruyneel EA, Dragonetti CH, De Bruyne GK, Van Cauwenberge RM-L, Smets LA, Van Rooy H (1984) Effect of temperature on invasion of MO_4 mouse fibrosarcoma cells in organ culture. Clin Exp Metastasiss 2: 107-125

Mareel MM, De Bruyne GK, Sonnenberg A, Hilgers J (1986) Experimental results on invasiveness of mouse mammary cells: clinical implications? In: Hollmann KH, Verley JM (eds) New frontiers in mammary pathology 1986. Nijhoff, Boston, pp351-362

Mareel MM, Van Roy FM, Messiaen LM, Boghaert ER, Bruyneel EA (1987) Qualitative and quantitative analysis of tumour invasion in vivo and in vitro. J Cell Sci Suppl 8: 141-163

Nicolson GL, Poste G (1983) Tumor cell diversity and host response in cancer metastasis - part II: host immune responses and therapy of metastases. Curr Probl Cancer 7: 3-42

Poste G, Greig R (1982) On the genesis and regulation of cellular heterogeneity in malignant tumors. Invasion Metastasis 2: 137-176

Sonnenberg A, Daams H, Calafat J, Hilgers J (1986) In vitro differentiation and progression of mouse mammary tumor cells. Cancer Res 46: 5913-5922

Storme GA, Berdel WE, van Blitterswijk WJ, Bruyneel EA, De Bruyne GK, Mareel MM (1985) Antiinvasive effect of racemic 1-0-octadecyl-2-0-methyl glycero-3-phosphocholine (ET-18-OCH_3) on MO_4 mouse fibrosarcoma cells in vitro. Cancer Res 45: 351-357

Cytological Analysis of Functional Alterations in Granulocytes

G. F. Fischer, O. Majdic, and W. Knapp

Introduction

Human granulocytes represent a population of resting cells. They have to be activated by an appropriate stimulus before they can function as effector cells in defense reactions, e.g., during bacterial infections. The activation process involves a series of metabolic events and results in degranulation and triggering of the respiratory burst, i. e., the production of bactericidal metabolic products of oxygen such as superoxide anion or hydrogen peroxide. Recently, the observation that certain recombinant cytokines seem to affect granulocyte activation [3-5, 8] has again stimulated interest in this biological process which may also explain some of the unpleasant side effects seen after therapeutic application of cytokines like TNF-α (TNF).

In order to gain better insight into the granulocyte activation process at single-cell level we performed flow cytometric studies with a series of monoclonal antibodies and a battery of granulocyte stimulants. We were particularly interested in changes in surface antigen expression upon stimulation of granulocytes with either complete secretagogues such as the phorbol ester phorbol myristate acetate (PMA) or the chemotactic peptide *N*-formyl methionyl-leucyl-phenylalanin (fMLP) together with the fungal product cytochalasin B (CyB) or with physiologically probably more relevant molecules such as tumor necrosis factor α (TNF), granulocyte-monocyte colony stimulating factor (CSF-GM) or interferon γ (IFN-γ). After stimulation cells were stained with monoclonal antibodies by indirect immunofluorescence and analyzed with a cytofluorometer. We further assessed the effects of these recombinant factors on the respiratory burst of granulocytes by measuring the production of hydrogen peroxide with the fluorescent dye dichlorodifluorescin diacetate (DCF-DA) at single-cell level.

Materials and Methods

Assessment of Activation-Associated Changes in Surface Antigen Expression

Granulocytes were isolated from heparinized blood by Ficoll-Hypaque density gradient centrifugation and dextran (2%) sedimentation. After being washed, granulocytes were suspended in Hank's balanced salt solution (HBSS) and stimulated with 10^{-7} *M* PMA (Sigma, St. Louis) for 20 min or pretreated for 30 min with 5 µg/ml CyB (Sigma, St. Louis) and subsequently stimulated for another 30 min with 10^{-7} *M* fMLP (Sigma). Incubations with recombinant CSF-GM (200 ng/ml), TNF (100 U/ml) and IFN-γ (100 U/ml) were performed for 2 h. As negative controls, cells were incubated in the absence of stimulants and cytokines for the same time period [2]. All incubation steps were performed at 37 °C. Subsequently cells were washed, fixed with 1% paraformaldehyde (Serva, Heidelberg) and then stained with the mAbs of the myeloid workshop panel submitted to the Third International Workshop on Human Leukocyte and Differentiation Antigens and FITC conjugated sheep F $(ab')_2$ anti mouse IgG + IgM antibodies (Grub, Scandic, Vienna). Membrane fluorescence intensities of single granulocytes were measured on a FACS 440 (Becton Dickinson, Sunny Vale, Calif.) flow cytometer in linear amplification mode. Mean fluorescence channel (MFC) values were determined and the changes upon stimulation were calculated as ratio of the MFC of stimulated cells to the MFC of control cells as described by Tetteroo et al. [6].

Assessment of Hydrogen Peroxide Production

Granulocytes were isolated as described above and incubated with recombinant TNF, CSF-GM and IFN-γ for 90 to 120 min at 37 °C. During the last 15 min 5 μM DCF-DA (Molecular Probes, Junction City, Kansas) was added in order to load the cells with this hydrogen peroxide-sensitive dye as described by Bass et al. [1]. Subsequently granulocytes were stimulated for another 30-min period with 10^{-8} *M* fMLP or medium and afterwards the fluorescence intensities were analyzed on a FACS 440. The ratio of the MFC of stimulated cells to the MFC of cells preincubated with medium only was calculated and the results are given as *x*-fold increase of the MFC values of stimulated cells.

Results and Discussion

Recombinant TNF and CSF-GM but not IFN-γ lead to a Respiratory Burst in Human Granulocytes

As shown in Table 1, preincubation with CSF-GM or TNF leads to a detectable increase in the hydrogen peroxide production of human granulocytes. The extent of this increase is similar to that seen upon stimulation with 10^{-8} *M* fMLP. Thus, with DCF-DA as a probe for the activation of the respiratory burst, we found a clear-cut activation of granulocytes by TNF, in agreement with a report of Klebanoff et al. [4] using a different technique. We also found that CSF-GM activates the respiratory burst by itself. This seems to be in contrast to reports using the less sensitive cytochrome C method for the respiratory burst [8]. Results presented in Table 1 confirm the findings of Weisbart et al. [8] that the preincubation of cells with CSF-GM enhances the respiratory burst in response to fMLP. In our hands the same phenomenon is observed if cells are incubated with TNF (Table 1). This effect is called "priming" of granulocytes [7]. Preincubation with IFN-γ for up to 2 h did not show any triggering of the respiratory burst nor any priming effect (Table 1).

Table 1. Activation of the respiratory burst in human granulocytes

Stimulation	Preincubation			
	Medium	CSF-GM	TNF	IFN
Medium	1	2.17 ± 0.32	1.96 ± 0.28	0.99 ± 0.04
fMLP	2.22 ± 0.34	3.09 ± 0.40	3.01 ± 0.32	2.23 ± 0.40

Values represent x-fold increases (mean ± SE, $n = 4$) in MFC of cells preincubated with medium or indicated factors for 90-120 min, loaded with DCF-DA and subsequently stimulated with fMLP or medium for additional 30 min and then analyzed on a FACS

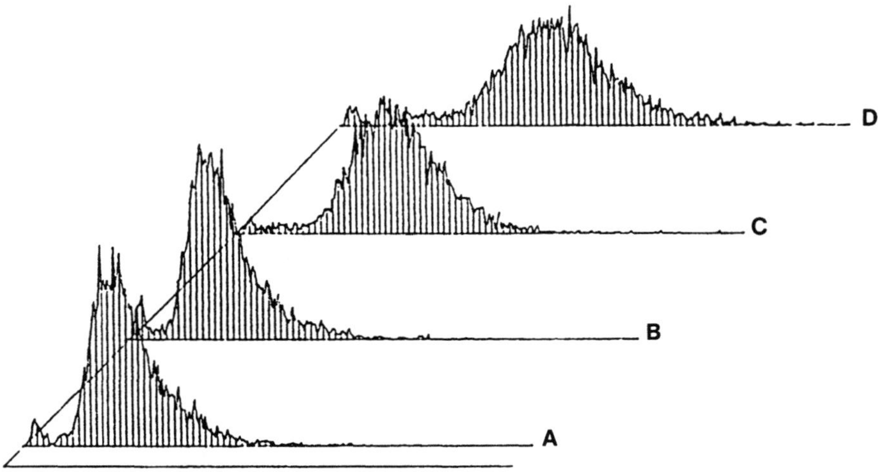

Fig. 1. FACS profiles of human granulocytes loaded with DCF-DA for 15 min and further incubated with *(A)* medium, *(B)* IFN-γ, *(C)* CSF-GM, *(D)* TNF at 37 °C for 30 min. Abscissa and ordinate represent fluorescence intensities and cell numbers respectively

The DCF-DA measurement of the respiratory burst at single-cell level also permits determination of the proportion of reactive cells. The possibility that certain activating agents only affect a subpopulation of granulocytes can therefore be checked. This could not be ruled out using global methods such as the measurement of the reduction of cytochrome C. So it was of interest to note that with all stimulants tested the whole population of granulocytes responded with an increase in fluorescence intensity. Figure 1 shows the FACS profiles of one representative experiment.

Activation-Associated Changes in Surface Antigen Expression

In order to learn more about the potential functional role of antibody-defined surface antigens, we tested the myeloid panel of monoclonal antibodies (mAbs) submitted to the Third International Workshop on Human Leukocyte and Differentiation Antigens for whether they detected antigens showing an altered pattern of expression on activated cells. Interestingly, this screening procedure revealed three

MAbs, namely 82H1, MOF11 and HL21 (Table 2), defining antigens which show two different cellular distribution patterns and are not detectable on resting granulocytes but can clearly be demonstrated by immunofluorescence staining on cells activated by PMA or fMLP/CyB. Other antigens already expressed on resting granulocytes which show an "up-" or "down-regulation" upon activation can further be divided into antigens the expression of which is influenced just by incubation with the cytokines TNF or CSF-GM or those which are expressed only upon stimulation with the complete secretagogues PMA or fMLP/CyB. The reaction pattern of the tested MAbs is summarized in Table 2. Only those antibodies of the myeloid panel are described which show a significant altered reaction pattern with activated cells. From the clustered mAbs for each cluster only one representative antibody is shown. The results with IFN-γ are not included in Table 2 since in no instance was a clear-cut alteration of the surface antigen expression observed. The potential of TNF and CSF-GM directly to activate granulocytes is underlined by our data indicating clear alterations in the expression of surface antigens upon incubation with TNF and CSF-GM as well as by the observation that both cytokines can trigger the respiratory burst. The stimulation of granulocytes with PMA or fMLP revealed an impressive amount of molecules which are expressed at higher or lower densities on activated cells; the function of most of these molecules is unknown so far. The question of whether the antigens which show different expression upon activation in vitro can serve as useful markers for monitoring an in vivo activation will remain a subject for future studies.

Table 2. Changes in surface antigen expression of granulocytes upon activation

Designation of MAb (CD)	Stimulus		
	fMLP/CyB	TNF	CSF-GM
Mo1 (CD11b)	2.8 ± 0.5*	1.9 ± 0.7	1.5 ± 0.2
G2G4 (CD16)	0.5 ± 0.3	0.9 ± 0.1	0.9 ± 0.2
T5A7 (CDw17)	0.5 ± 0.1	0.7 ± 0.2	0.7 ± 0.1
60.3 (CD18)	2.2 ± 0.6	2.1 ± 0.9	1.6 ± 0.5
TM3 (CD31)	0.4 ± 0.1	0.9 ± 0.1	0.8 ± 0.1
CF4	5.4 ± 2.1	1.3 ± 0.2	1.2 ± 0.1
VIM8	5.1 ± 2.6	1.3 ± 0.2	1.2 ± 0.1
GA11	7.0 ± 4.8	1.3 ± 0.3	1.2 ± 0.1
BY114A	8.3 ± 3.0	1.1 ± 0.2	1.1 ± 0.3
U1-58	2.1 ± 0.6	1.9 ± 0.6	1.3 ± 0.3
88H4	5.4 ± 1.9	2.0 ± 0.4	1.7 ± 0.7
G10F5	6.2 ± 0.1	2.1 ± 0.4	1.4 ± 0.2
HE1D	4.4 ± 2.0	1.2 ± 0.1	1.1 ± 0.1
88H7	3.2 ± 2.2	1.2 ± 0.1	1.0 ± 0.1
VIM2	2.8 ± 0.8	1.2 ± 0.1	1.1 ± 0.1
HL-21	8.5 ± 3.1	1.0 ± 0.1	1.0 ± 0.1
82H1	12.6 ± 8.4	1.0 ± 0.3	1.0 ± 0.1
MOF11	14.9 ± 3.5	1.0 ± 0.1	0.9 ± 0.1

* Values represent the ratio between the MFC value of stimulated cells and the MFC of resting granulocytes (mean ± SD of 3-4 experiments).

Summary

The activation process of human granulocytes was monitored at single-cell level. Cells were activated with the chemotactic peptide *N*-formyl-methionyl-leucyl-phenylalanin plus the fungal metabolite cytochalasin B or with the recombinant cytokines tumor necrosis factor α (TNF-α), granulocyte-monocyte colony stimulating factor (CSF-GM) or interferon γ (IFN-γ). Membrane antigens were analyzed by indirect immunofluorescence using a battery of myeloid-specific monoclonal antibodies. Granulocytes turned out to express some of their surface antigens at remarkably higher or lower levels upon activation with fLMP. Some antigens not expressed on resting cells were clearly seen on activated cells. While incubation with TNF-α and CSF-GM led to an altered expression of surface antigens, IFN-γ caused no changes. We further studied the potential of these cytokines to activate the respiratory burst, which is essential for the bactericidal function of granulocytes. The respiratory burst was assessed on the FACS using the hydrogen peroxide-sensitive, fluorescent probe dichlorodifluorescin diacetate. TNF-α and CSF-GM directly caused a production of hydrogen peroxide upon incubation with granulocytes. We did not find such an effect when testing IFN-γ. Thus, in our study, membrane antigens are described showing activation-associated changes in surface expression. Furthermore, our results indicate a direct activation of the respiratory burst by CSF-GM and TNF-α in human granulocytes.

References

1. Bass DA, Parce JW, DeChatelet LR, Szejda P, Seeds MC, Thomas M (1983) Flow cytometric studies of oxidative product formation by neutrophils: a graded response to membrane stimulation. J Immunol 130: 1910-1917
2. Fearon DT, Collins LA (1984) Increased expression of C3b receptors on polymorphonuclear leukocytes induced by chemotactic factors and by purification procedures. J Immunol 130: 370-375
3. Gamble JR, Harlan JM, Klebanoff SJ, Hubbard WC (1985) Stimulation of the adherence of neutrophils to umbilical vein endothelium by human recombinant tumor necrosis factor. Proc Natl Acad Sci USA 82: 8667-8671
4. Klebanoff SJ, Vadas MA, Harlan JM, Sparks LH, Gamble JR, Agosti JM, Waltersdorph AM (1986) Stimulation of neutrophils by tumor necrosis factor. J Immunol 136: 4220-4225
5. Shalaby MR, Aggarwal BB, Rinderknecht E, Svedersky P, Finkle BS, Palladino Jr MA (1985) Activation of human polymorphonuclear neutrophil functions by interferon-γ and tumor necrosis factor. J Immunol 135: 2069-2073
6. Tetteroo PAT, Bos MJE, Visser FJ, von dem Borne AEG (1986) Neutrophil activation detected by monoclonal antibodies. J Immunol 136: 3427-3432
7. Van Epps DE, Garcia ML (1980) Enhancement of neutrophil function as a result of prior exposure to chemotactic factor. J Clin Invest 66: 167-175
8. Weisbart RH, Golde DW, Clark SC, Wong GG, Gasson JC (1985) Human granulocyte - macrophage colony-stimulating factor is a neutrophil activator. Nature 314: 361-363

Perspectives of Cytogenetic Analysis of Human Solid Tumors

A. Verhest

Introduction

Since 1971, it has been well documented that chromosomal changes appear specific to certain types of leukemias (Sandberg 1980). More recent efforts to yield higher resolution banding (Yunis et al. 1981) have further suggested that these anomalies are not randomly distributed throughout the karyotype but affect a restricted number of breakpoints (Mitelman 1985). Apparently similar leukemias are associated with different karyotype profiles and also reveal discordant survivals and responses to treatment (Berger et al. 1987). This is the reason why, in the 1970s, cytogenetics gained a prominent role in the diagnosis and prognosis of leukemias and lymphomas (Third International Workshop on Chromosomes in Leukemia (1980) 1981; Fourth International Workshop on Chromosomes in Leukemia (1982) 1984). Their classification is now based on cytochemical morphololology, immunolabeling, and cytogenetics (the so-called MIC classification).

In the meantime, however, due to technical problems, our knowledge of solid tumors remained very poor.

Cytogenetics of the Hemopathies

In the early 1960s the Philadelphia chromosome (Ph^1) remained the sole consistent anomaly specifically associated with a well-defined type of leukemia. At present other anomalies have a diagnostic or a prognostic value: in the M2 subtype according to the FAB classification (Bennett et al. 1976), a t (8; 21) translocation indicates a better prognosis; a t (15; 17) is of diagnostic value in the M3 promyelocytic type, as is a chromosome 16 anomaly in cases of myelomonocytic leukemia (M4) with eosinophilia (Yunis et al. 1984).

In the lymphomas, t (8; 14) translocation characterizes the Burkitt and the small cleaved cell types, while t (11; 14) is encountered in the small cell lymphomas. A t (14; 18) translocation is remarkable in that, when it is recognized in fineneedle aspirated material, the follicular pattern of the lymphoma can be predicted! (Kristoffersson et al. 1985).

The specificity of these anomalies gives them a prognostic value comparable to that of the Kiel classification (Yunis et al. 1982).

Oncogenes, Fragile Sites, and Cancer Breakpoints

In the 1980s, independently of cytogenetics, rapid developments in molecular biology made it possible to assign an increasing number of oncogenes to precise loci of the genome. Viral oncogenes (v-*onc*) are known to induce experimental tumors and to have DNA sequences homologous to those in proto-oncogenes (p-*onc*), normally incorporated in the human genome (Bishop 1987). Activation of these p-*onc* in endogenous or cellular oncogenes (c-*onc*) alter or amplify their expression. Tumorigenesis, then is, due to a wrong triggering at a wrong time of a normal constituent of our genome. The compiling of the oncogene loci and the cancer breakpoints revealed a nonrandom distribution of both at the same sites, suggesting that some oncogene activations take advantage of those breakpoints (Pierce et al. 1986).

Two models have helped us to understand the molecular significance of some chromosome rearrangements. The specific translocations of Burkitt's lymphoma are t (8; 14) and, less often, t (2; 8) or t (8; 22) all resulting in the fusion of an oncogene, c-*myc*, with a cell type gene concerned with cell differentiation: c-*myc* translocates near the constant region of the heavy chain gene on chromosome 14 or the κ and λ light chain normally located on chromosomes 2 and 22 migrate in the vicinity of chromosome 8. These recombinations enhance the expression of c-*myc* involved in uncontrolled proliferation of B lymphocytes, with altered synthesis of the immunoglobulin corresponding to the involved locus (Mitchell et al. 1984).

The Ph^1 in chronic myeloid leukemia offers another model of deregulation of an oncogene by position effect. The classical Ph^1 translocation transfers the c-*abl* oncogene from chromosome 9 to a very narrow region of chromosome 22, the bcr (breakpoint cluster region). The resulting new messenger RNA orders the synthesis of a chimeric protein, with a tyrosine kinase activity stimulating the cellular proliferation. Even in cases of complex or masked Ph^1 translocations, this c-*abl*/bcr association is present and appears the constant event, mandatory for neoplastic transformation (De Klein et al. 1982).

Another stream of research on fragile sites recently converged with the fragile sites hypothesis of cancer (LeBeau 1986). Fragile sites are potential breakpoints, heritable and susceptible of rearrangements. They are observed in cultures as nonstaining gaps inducible under certain in vitro conditions (such as folate- and thymidine-deficient media). They coincide very often with cancer chromosome breakpoints. Therefore fragile sites have been hypothesized as preferred targets of mutagenic agents, connecting these directed injuries with predisposition to cancer chromosome rearrangements (Hecht and Hecht 1986; Yunis et al. 1987).

Malignant Solid Tumors

Recent improvements in culture techniques offer new insights in cytogenetic studies of solid tumors (Limon et al. 1986; Teyssier 1986). Preliminary results have already confirmed the nonrandom clonal, acquired character of chromosome rearrangements. Simple anomalies are the rule in sarcomas, which may explain why they have been more extensively analyzed than the far more common carcinomas. For example, synoviosarcoma is specifically associated with a t (X; 18). Liposarcomas can be subdivided in four histologic types; the myxoid variant with better prognosis is the only one bearing a t (12; 16). Alveolar rhabdomyosarcoma is authenticated by a t (2; 13).

The identification of consistent chromosome aberrations in sarcomas is a valuable aid to the pathologist in the difficulties encountered in the differential diagnosis of soft tissue tumors or undifferentiated metastases. In the diagnosis of small cell tumors of childhood, a t (11; 22) can accurately confirm a Ewing's tumor (Sandberg and Turc-Carel 1987).

The c-*sis* oncogene has been found similarly translocated in peripheral neuroepithelioma and in Ewing's tumor. It is, then, possible that both tumors share a common precursor cell type and histogenesis (Thiele et al. 1987).

The most common testicular cancers, despite their different histological patterns (seminoma, choriocarcinoma, embryonal carcinoma, teratoma), share an isochromosome 12 p marker indicating their identical germinal cell origins (Sandberg et al. 1987). Bladder cancers are uniformly urothelial; however, chromosome differences could reflect molecular heterogeneity, giving some clues on features such as grade, stage, response to therapy and survival (Sandberg 1986).

Benign Tumors

Some rearrangements, such as 12q- or trisomy 8 in adenomas and carcinomas of the colon, can satisfy our concept of precursor lesions in tumorigenesis. But what is the clinical meaning of monosomy 22 in meningioma (Al Saadi et al. 1987)?

What are we to think about those translocations in mixed salivary gland tumors (Mark et al. 1983), or in lipomas and leiomyomas? All examples of doubtless benign and often unexciting tumors to a pathologist's eyes!

No good explanation can be put forward, but at present aneuploidy cannot be taken as synonymous with malignant transformation.

Conclusion

Oncogenesis is a multiple step transformation brought about by pinpoint mutations and genic alterations at several locations of the genome; most of them are indetectable at the microscopic level and occur independently of visible chromosome translocations. The most probable hypothesis is that certain breakpoints play a permissive role in the subversive behavior of the cell on its way to increasing genic instability. These allow further genomic modifications and clonal chromosomic evolution, leading to increasing autonomy.

Systematic study of early chromosome alterations in solid tumors should add some knowledge concerning the control of cell proliferation and differentiation which will be of use in cancer detection and prevention.

References

Al Saadi A, Latimer F, Madercic M, Robbins J (1987) Cytogenetic studies of human brain tumors and their clinical significance. II. Meningioma. Cancer Genet Cytogenet 26: 127-141

Bennett JM, Catovsky D, Daniel MT, Flandrin G, Galton DAG, Gralnick HR, Sultan C (1976) Proposals for the classification of acute leukemias. Br J Haematol 33: 451-458

Berger R, Bernheim A, Ochoa-Noguera ME, Daniel MT, Valensi F, Sigaux F, Flandrin G, Boiron M (1987) Prognostic significance of chromosomal abnormalities in acute nonlymphocytic leukemia: a study of 343 patients. Cancer Genet Cytogenet 28: 293-299

Bishop JM (1987) The molecular genetics of cancer. Science 183: 305-311

De Klein A, van Kessel AG, Grosveld G, Bartram CR, Hagemeijer A, Bootsma D, Spurr NK, Heisterkamp N, Groffen J, Stephenson JR (1982) A cellular oncogene is translocated to the Philadelphia chromosome in chronic myelocytic leukemia. Nature 300: 765-767

Fourth International Workshop on Chromosomes in Leukemia, 1982 (1984) Clinical significance of chromosomal abnormalities in acute nonlymphocytic leukemia. Cancer Genet Cytogenet 11: 332-350

Hecht BK, Hecht F (1986) General correspondence of rare fragile sites to leukemia/lymphoma and common fragile sites to solid tumor chromosome breakpoints. In: Becher R, Sandberg AA, Schmidt CG (eds) Chromosomes in hematology. Zuckschwerdt, Munich p 69

Kristoffersson U, Olsson H, Åkerman M, Mitelman F (1985) Cytogenetic studies in non-Hodgkin lymphoma - results from fine needle aspiration samples. Hereditas 103: 63-76

LeBeau MM (1986) Chromosome fragile sites and cancer-specific rearrangements. Blood 67: 849-858

Limon J, Dal Cin P, Sandberg AA (1986) Application of long-term collagenase disaggregation for the cytogenetic analysis of human solid tumors. Cancer Genet Cytogenet 23: 305-312

Mark J, Dahlenfors R, Ekedahl C (1983) Specificity and implications of ring chromosomes and dicentrics in benign mixed salivary gland tumors. Acta Pathol Microbiol Immunol Scand [A] 91: 397-402

Mitchell KF, Battey J, Hollis GF, Moulding C, Taub R, Leder P (1984) The effect of translocations on the cellular *myc* gene in Burkitt lymphomas. J Cell Physiol [Suppl] 3: 171-177

Mitelman F (1985) A catalog of chromosome aberrations in cancer. Liss, New York

Pierce JH, Eva A, Aaronson SA (1986) Interactions of oncogenes with haematopoietic cells. Clin Haematol 15: 573-620

Sandberg AA (1980) The chromosomes in human cancer and leukemia. Elsevier North-Holland, New York

Sandberg AA (1986) Chromosome changes in bladder cancer: clinical and other correlations. Cancer Genet Cytogenet 19: 163-175

Sandberg AA, Turc-Carel C (1987) The cytogenetics of solid tumors: relation to diagnosis, classification and pathology. Cancer 59: 387-395

Teyssier JR (1986) Anomalies chromosomiques dans les tumeurs humaines Thesis, Faculty of Medicine, Reims

Thiele CJ, Whang-Peng J, Kao-Shan CS, Miser J, Israel MA (1987) Translocation of *c-sis* protooncogene in peripheral neuroepithelioma. Cancer Genet Cytogenet 24: 119-128

Third International Workshop on Chromosomes in Leukemia, 1980 (1981) Clinical significance of chromosomal abnormalities in acute lymphoblastic leukemia. Cancer Genet Cytogenet 4: 111-137

Yunis JJ, Bloomfield CD, Ensrud KM (1981) All patients with acute nonlymphocytic leukemia may have a chromosomal defect. N Engl J Med 305: 135-139

Yunis JJ, Oken MM, Kaplan ME, Ensrud KM, Howe RR, Theologides A (1982) Distinctive chromosomal abnormalities in histologic subtypes of non-Hodgkin's lymphoma. N Engl J Med 307: 1231-1237

Yunis JJ, Brunning RD, Howe RB, Lobell M (1984) High-resolution chromosomes as an independent prognostic indicator in adult acute nonlymphocytic leukemia. N Engl J Med 311: 812-818

Yunis JJ, Soreng AL, Bowe AE (1987) Fragile sites are targets of diverse mutagens and carcinogens. Oncogene 1: 59-69

Relationships Between Morphological Variants and Cytogenetic Findings in Acute Promyelocytic Leukaemia

C. Di Fusco, L. Menaguale, G. Del Porto, and D. Quaglino

Introduction

Acute promyelocytic leukaemia (APL) is a well-known variety of myeloid leukaemia, characterized by a typical karyotypic abnormality, t (15; 17).

Two different variants of APL have been recognized:

1. The hypergranular form, whose cells show numerous azurophilic granules and occasionally 'faggots' of Auer rods in the cytoplasm, more often associated with disseminated intravascular coagulation (DIC) and low or moderately raised leucocyte counts in the peripheral blood (Bennett et al. 1980)

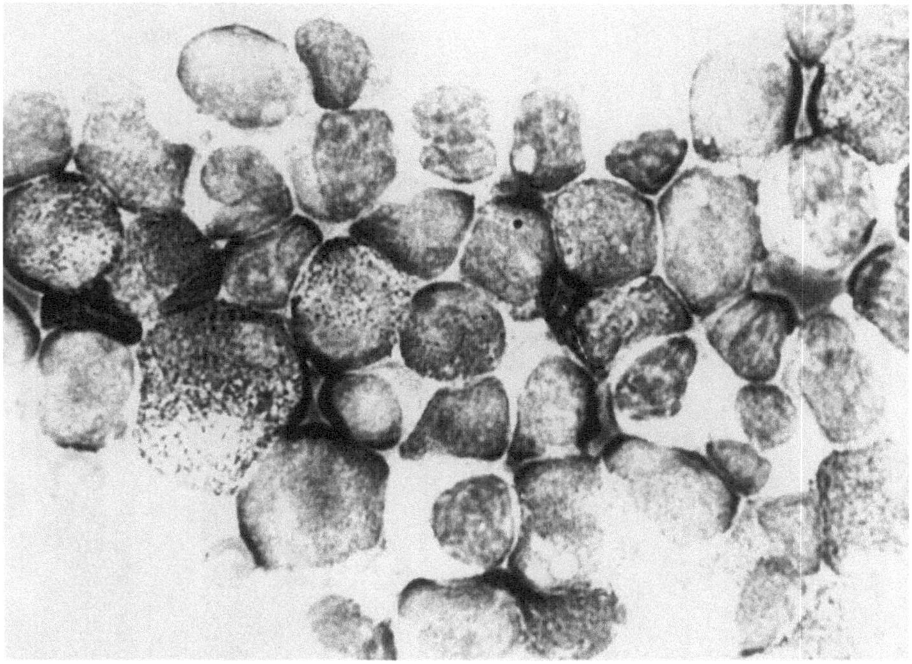

Fig. 1. Cytological appearance of the bone marrow at the onset of the disease. Most cells present fairly abundant azurophilic granules in their cytoplasm. May-Grünwald-Giemsa, ×960

2. The microgranular form, whose cells appear on light microscopy mostly devoid of cytoplasmic granules, characterized by especially high leucocyte counts and a lower frequence of DIC.

The above differences induced some groups of authors to define the two cytological variants of APL as distinct clinical entities (Golomb et al. 1980); the FAB classification also considers the microgranular variant as substantially different from hypergranular APL (Fourth International Workshop on Chromosomes in Leukemia 1984).

Case Report

A. P., a 60-year-old man, was first admitted to the hospital because of the recent onset of easy fatigue, pallor and ecchymoses. Laboratory findings were as follows: haemoglobin 7.2 g/dl, leucocytes $14 \times 10^9/l$ (85% abnormal promyelocytes), platelets $28 \times 10^9/l$, FDP more than 40 μg/ml, ethanol test positive.

An aspirate showed a markedly hypercellular bone marrow, almost completely represented by atypical promyelocytes, containing a variable but often high number of azurophilic granules (Fig. 1). Occasional faggots of Auer rods could be demonstrated by Sudan black B, myeloperoxidase and chloracetate esterase cytochem-

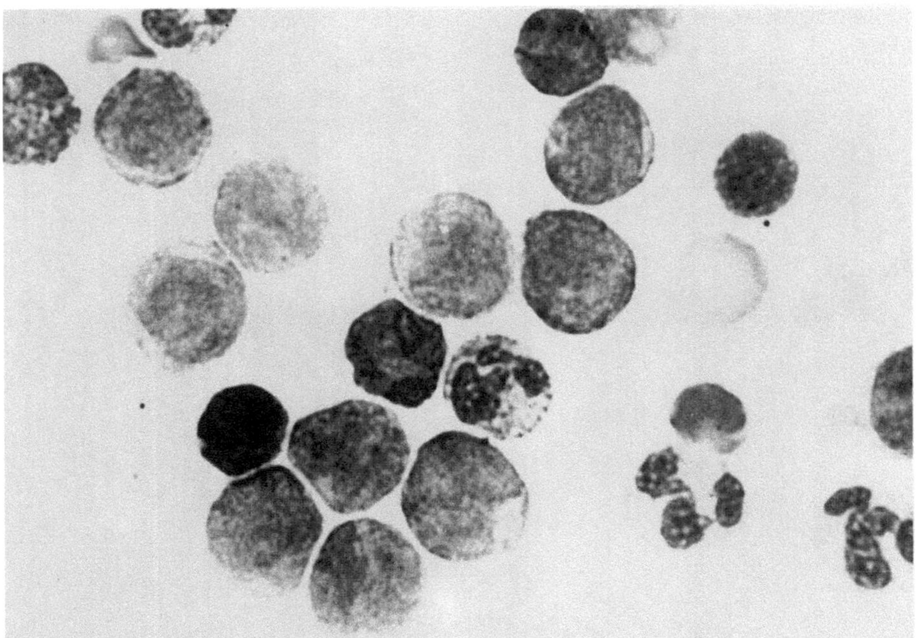

Fig. 2. Cytological appearance of the bone marrow at relapse. Along with normal-looking mature granulocytes, a large number of leukaemic blast cells is recognizable, without any evidence of cytoplasmic granules. May-Grünwald-Giemsa, ×960

ical reactions. Induction treatment with daunomycin and heparin followed by two consolidation courses of DAT (daunomycin, Ara-C, 6-thioguanine) was successful in achieving a complete clinical and haematological remission. A blood cell count showed: haemoglobin 14 g/dl, platelets $30 \times 10^9/l$, leucocytes $32.4 \times 10^9/l$ (80% atypical cells devoid of cytoplasmic granules). The bone marrow aspirate (Fig. 2) was heavily infiltrated by leukaemic cells, whose cytoplasm appeared completely agranular on light microscopy. At this stage of the disease, cytogenetic analysis of bone marrow cells, carried out employing both standard and GTG banding techniques, showed in all of the metaphases observed a 46, XY, t (15; 17) (q 22; 21) karyotype (Fig. 3).

Owing to the cumulative dose of daunomycin previously given to the patient, high-dose Ara-C treatment was attempted (3 g/m^2 every 12 h for 4 days, two courses). Five weeks after the previous control, a new aspirate showed a hypercellular bone marrow, almost completely made up of hypergranular promyelocytes.

A new therapeutic attempt, based on Ara-C and VP 16, subsequently proved capable of inducing a new remission, which is continuing the present time.

Discussion

The case reported contributes to the evaluation of the relationships existing between the two variants of APL. The patient's disease, diagnosed as APL on the ba-

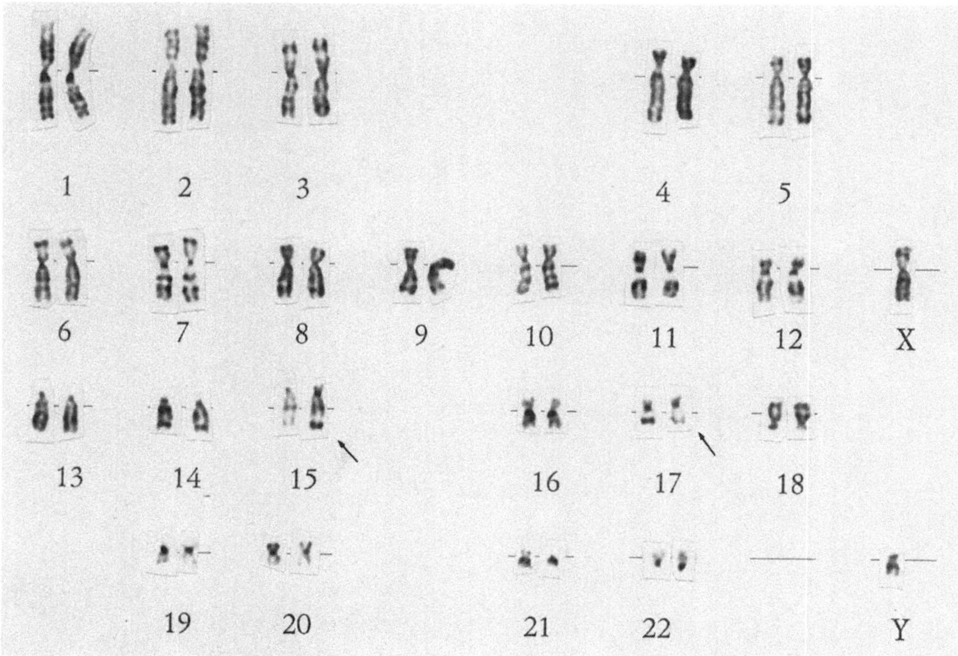

Fig. 3. Karyotype in complete clinical and haemotological remission: 46, XY, t (15; 17) (q 22; q 21)

sis of cytological, clinical and karyotypic evidence, presented in different phases both the hypergranular (at onset) and microgranular (at relapse) variants of APL. High-dose Ara-C treatment apparently reinduced the emergence of hypergranular cells.

On the basis of these findings, one may suggest that the hypergranular and microgranular variants of APL should not be regarded as distinct clinical entities, but rather represent different expressions of a single leukaemic disease. In particular, the changes in cell morphology observed in the present case suggest the hypothesis that microgranular and hypergranular promyelocytes represent different stages of maturation, which may also be induced by Ara-C treatment, in agreement with the hypothesis put forward by Hoffman et al. (1985).

References

Bennett JM, Catovsky D, Daniel MT, Flandrin G, Galton DAG, Gralnick HR, Sultan C (1980) A variant form of hypergranular promyelocytic leukaemia (M3). Br J Haematol 44: 169-170

Fourth International Workshop on Chromosomes in Leukaemia (1984) Chromosomes in acute promyelocytic leukemia. Cancer Genet Cytogenet 11: 288-293

Golomb HM, Rowley JD, Vardiman JW, Testa JR, Butler A (1980) Microgranular acute promyelocytic leukaemia: a distinct clinical, ultrastructural and cytogenetic entity. Blood 55: 253-259

Hoffman V, Berens ME, Moreillon MC, Cogoli M, Fruh U, Merchant RE, Kistler G (1985) Effects of 13-cis retinoic acid and Ara-C on differentiation and proliferation of non-promyelocytic acute myelogenous leukaemia. Leuk Res 9 (11): 1367-1371

Cytogenetic, Molecular, Immunologic, and Ultrastructural Study on an Established Cell Line of a Stage III Neuroblastoma

P. Grammatico, A. Modesti, S. Scarpa, S. Campo, C. Dominici, G. D'Orazi, N. Sulli, and G. Del Porto

Introduction

Neuroblastoma, representing about 7% of childhood malignant tumors, is the most common neoplasm in infancy and second in malignancy after lymphoma as a solid tumor outside the central nervous system in childhood. Its differential diagnosis from other tumors can be difficult and requires diagnostic approaches other than histological analysis.

The use of various markers has lately been considered for the identification of neuroblastoma, such as tumor-cell ultrastructural features, cell surface antigen immunoreactivity [5], chromosomal aberrations, and oncogene activation, among other approaches. Furthermore, the attempt to find new parameters with prognostic significance for this tumor has led to the study of N-*myc* amplification and also of chromosomal structural aberrations [8]. In particular, a great deal of cytogenetic work has been published on neuroblastoma.

Many authors report the preferential involvement of the short arm of chromosome N.1 in neuroblastoma [4, 7]. In addition, the presence of the homogeneously staining regions due to a gene amplification seems to be very significant in this tumor [7, 8]. Double minutes are also frequently described in the neuroblastoma cell lines analyzed [4, 7].

This report describes the establishment of a human continuous cell line, derived from neuroblastoma, and its characterization, including ultrastructural and immunochemical marker expression, and a cytogenetic study of the tumor cell line. This line may provide a valuable in vitro model for further investigation towards a better characterization of neuroblastoma.

Case Report

A 20-month-old girl was brought to us suffering from abdominal pain and enlargement. Physical examination revealed the presence of a hard, nontender mass in the right half of the abdomen, extending across the midline. X-rays of the abdomen showed intratumoral stippled calcifications, while skeletal survey and bone scintigraphy gave negative findings. Computerized tomography showed the presence of a tumor mass arising from the right adrenal gland and crossing the midline. In 24-h urine collection increased levels of catecholamines and vanillylmandelic acid were found. The girl underwent surgery and biopsy of the mass was performed. The histological diagnosis was neuroblastoma stage III according to Evans et al. [6].

Materials and Methods

The tumor specimen obtained was finely minced and placed in culture in RPMI-1640 medium with antibiotics and 20% fetal calf serum.

Trypsinization was continued for approximately 30 passages. For cytogenetic study cell culture was treated by hypotonic solution (KCl) and acetic acid-methanol (1:3).

Metaphases were analyzed after treatment with GTG and CBG banding techniques. For electron microscopy specimens were fixed in cacodylate-buffered 2.5% glutaraldehyde, postfixed in osmium and embedded in Epon 812.

N-*myc* amplification analysis was carried out by dot-blot analysis of genomic DNA extracted from the tumor cells.

For immunofluorescence cell layers were fixed in cold absolute ethanol, incubated with the antibody (1/100 dilution) and then with a second fluorescein-conjugated antibody (1/20 dilution).

Results

The cytogenetic analysis of the cell line revealed two different clones. The first polyploid presented many rearrangements and some markers. Three of them involved the chromosome N.1: M1 del(1)(p 32pter), M2 del(1)(q 32pter), and HSRq21. In addition, another marker was present in all metaphases of the clone M4 $3q^+$ and some double minute bodies (DM) (Fig. 1).

The second clone, with modal number 21, presented apparently the same chromosomes in all metaphases with many HRS but without DM (Fig. 2).

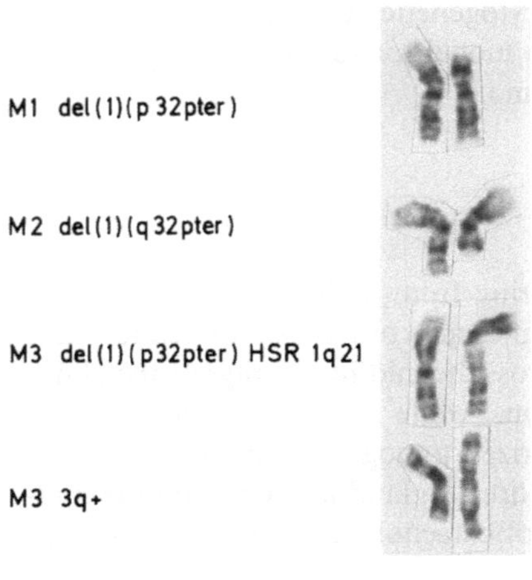

Fig. 1. Rearrangements and markers found in the first clone

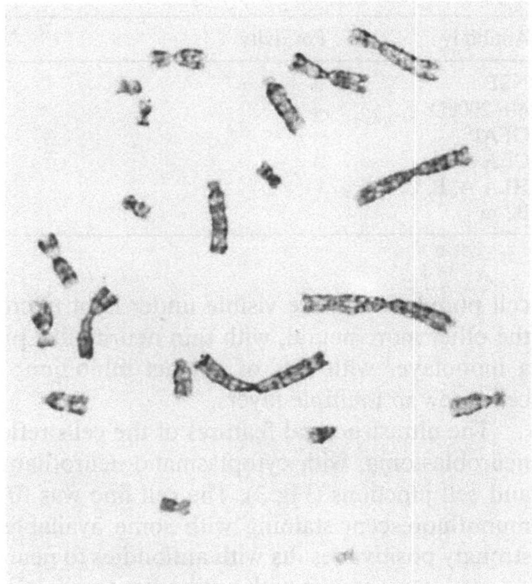

Fig. 2. Appearance of the second clone

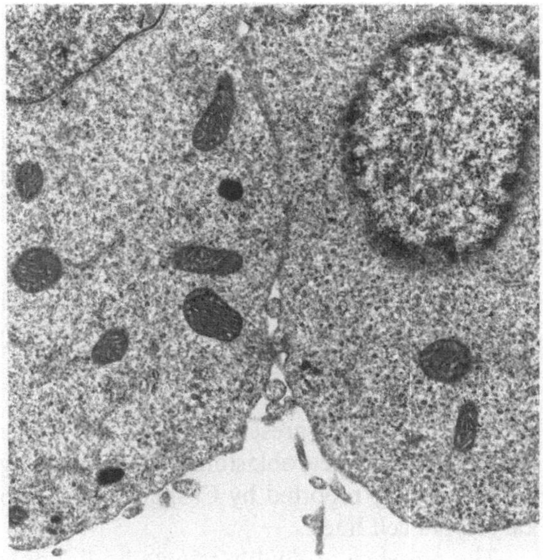

Fig. 3. Ultrastructure of the cells, characteristic of neuroblastoma

Flow cytometric analysis of monocellular suspension obtained from the biopsy sample demonstrated that DNA content was diploid (DI = 1).

Genomic DNA from the same tumor was analyzed by the dot-blot method and found to be amplified 63 times for N-*myc* oncogene compared to a normal control DNA. The biopsy tissue was then minced and plated in vitro. Within 4–5 days of plating, cells showed outgrowth and in about 10 days they were confluent. Two

Table 1. Results of immunostaining

Antibody	Positivity
NSE	++
NF-200kD	++
GFAP	+/−
CEA	+/−
HLA-A, B, C.	−
B2-m	−

cell populations were visible under light microscopy: one flat and rounded, and the other more neural, with thin neurite-like processes; cells multiplied rapidly as a monolayer with loss of contact inhibition; after confluence was reached, the cells grew in multiple layers.

The ultrastructural features of the cells reflected the distinctive morphology of neuroblastoma, with cytoplasmatic neurofilaments, neurotubules, neurogranules, and cell junctions (Fig. 3). The cell line was further partially characterized by immunofluorescent staining with some available antibodies: immunostaining gave strongly positive results with antibodies to neuron-specific enolase and to neurofilaments, weak results with antibodies to glial fibrillary acid protein and to carcinoembryonic antigen, and totally negative results with HLA-A, B, C, und B_2-microglobulin antibodies (Table 1).

Discussion

The aim of this paper is to describe the cytogenetic, molecular, immunologic, and ultrastructural features of a continuous cell line established from a human neuroblastoma in order to suggest new techniques, different from the routine diagnostic tools, which will allow certain diagnosis of undifferentiated and undiagnosed tumors.

As to cytogenetic findings, our results confirm the preferential involvement of chromosome N.1 and particularly, as recently reported in the literature [7], of breakpoint 1p32, where the BLYM1 oncogene is located and the highest frequency of HSR is found.

In our case, the presence of two cell lines expressing the same ultrastructural characteristic of neuroblastoma but morphologically different (by light microscopy), as already reported by Casper et al. [3], could be related to the two different karyotypic cell lines.

The presence of double minute bodies only in the polyploid line and of HSR mainly in the hypodiploid line is interesting and further studies could be done to investigate the correlation of this with N-*myc* amplification, which has a high prognostic value. The involvement of chromosome N.3, previously described as M3 by Biedler et al. [2], is also important.

Finally, N-*myc* amplification, ultrastructural, and immunological markers are constantly expressed either in the biopsy tissue or in the established cell line; this confirms that utilization of these parameters is important for diagnosis, prognosis, and treatment, especially in very difficult cases.

References

1. Balaban G, Gilbert F (1982) Homogeneously staining regions in direct preparations from human neuroblastomas. Cancer Res 42: 1838-1842
2. Biedler JL, Helson L, Spengler BA (1973) Morphology and growth, tumorigenicity, and cytogenetics of human neuroblastoma cells in continuous culture. Cancer Res 33: 2643-2652
3. Casper JT, Trent MJ, Harb J, Piaskowski V, Helmsworth M, Finlan J, Von Hoff DD (1983) Distinguishing characteristics of a new neuroblastoma cell line. Cancer Genet Cytogenet 10: 177-186
4. Christiansen H, Franke F, Bartram CR, Adolph S, Rudolph B, Harbott J, Reiter A, Lampert F (1987) Evolution of tumor cytogenetic aberrations and N-*myc* oncogene amplification in a case of disseminated neuroblastoma. Cancer Genet Cytogenet 26: 235-244
5. Donner L, Triche TJ, Israel MA, Seeger RC, Reynolds CP (1985) A panel of monoclonal antibodies which discriminated neuroblastoma from Ewing's sarcoma, rhabdomyosarcoma, neuroepithelioma and hematopoietic malignancies. In: Advances in neuroblastoma research. Liss, New York, pp 347-366
6. Evans AE, D'Angio GJ, Randolph J (1971) A proposed staging for children with neuroblastoma. Cancer 27: 374
7. Kaneko Y, Kanda N, Meseki N, Sakuray M, Tsuchida Y, Takda T, Okabe I, Sakurai M (1987) Different karyotypic patterns in early and advanced stage neuroblastomas. Cancer Res 47: 311-318
8. Seeger RC, Brodeur GM, Sather H (1985) Association of multiple copies of the N-myc oncogene with rapid progression of neuroblastoma. N Engl J Med 313: 111-1116.

Kaposi's Sarcoma: A Cytogenetic Study

P. Grammatico, A. M. Cianciulli, I. Venturo, S. Campo, R. Perrone Donnorso, and G. Del Porto

Introduction

There are in the literature only two specific studies about the cytogenetic analysis of Kaposi's Sarcoma (KS). The first [2] was based on 11 untreated KS patients. Lymphocyte cultures were made in all of these cases, fibroblast cultures additionally in five. The authors defined clonal evolution as the presence of at least two cells with the same numerical or structural aberration in the same culture, and their lymphocyte analysis showed clonal evolution in six of eleven cases. Additionally, in fibroblast cultures from two sarcomatous lesions they found seven cells in the first with 47, XXY and t (7; 13) in the second. The authors suggested that the chromosomal instability found in KS patients may be associated with an oncogenic virus.

In the second paper about cytogenetic study of KS [1], seven cell lines were analyzed and the authors concluded that the results suggested that there is no characteristic chromosomal aberration in KS, but that there is a chromosomal instability.

Case Report

We report a cytogenetic study on a 57-year-old male affected by sporadic KS.

In the 1973, at the age of 43 years, the patient consulted a physician for papulous lesions on the right leg. Similar lesions later also appeared on both arms. Only in 1983, after skin biopsy, was KS diagnosed.

In 1986 the patient was admitted in hospital with deep dyspnea, cough and hemoptysis. Cytological study of the effusion revealed the presence of red cells and many lymphocytes. The patient was then subjected to an alternate monthly combined therapy of two different cytostatic cocktails, the first composed of vincristine, cyclophosphamide, procarbazin, and prednisone, and the second of adriamycin, bleomycin, vincristine, and dacarbazine. Clinical remission of the pleural effusion was obtained after this therapy.

In February 1987, after an evacuative thoracocentesis, the patient was treated by administration of interferon (18 million units) in the pleural cavity, with positive results.

Materials and Methods

Lymphocyte culture was carried out using RPMI 1640 supplemented with 10% FCS, 1% L-glutamine, 0.1% antibiotics, and 2% PHA. Culture of the cells obtained by pleural effusion was carried out in RPMI 1640 supplemented with 20% FCS, 1% L-glutamine, and 0.2% antibiotics. Cytogenetic analysis was performed by treatment with KCl hypotonic solution and methanol-acetic acid (3:1) fixation. Chromosome preparations were made by GTG and CBG banding techniques. SCE analysis was preformed by the BrdU technique and was conducted in the sibship of the proband and in his two daughters.

Results

The results of the cytogenetic studies conducted on peripheral blood lymphocytes and cells from pleural effusion before therapy are shown in Table 1. Table 2 presents the results of the cytogenetic analysis of peripheral blood lymphocytes after interferon therapy.

SCE level is normal in the proband, his brother and two sisters, and his two daughters.

Discussion

As recently reported in the literature [1], we never found markers but some rearrangements.

After interferon therapy we found no differences in chromosomal analysis even though the rearrangements found were different before and after treatment.

Tabelle 1. Results of cytogenetic studies before chemotherapy

Lymphocytes
1) 45, X
2) 45, X
3) 46, XY
4) 46, XY
5) 46, XY
6) 46, XY
7) 45, XY, −8
8) 46, XY, −1, −6, −8, +?, +?, +?
Cells from pleural effusion
1) 39, XY, −2, −7, −8, −10, −12, −15, −20, −22, +7q+, +11p+, 8?
2) 89, XXYY, −8, −8, −20, −21, +?
3) 46, XY
4) 20, −1, −1, −2, −2, −7, −8, −8, −9, −10, −11, −11, −12, −14, −14, −15, −15, −16, −16, −18, −18, −19, −19, −21, −21, −X, −Y
5) 46, XY
6) 47, XY, del (1) (p36pter), +?
7) 46, XY
8) 46, XY
9) 46, XY, −1, −5, −6, −6, +?, +?, +?, +?

Tabelle 2. Results of cytogenetic studies (lymphocytes) after chemotherapy

1)	45, X
2)	43, X, −9, −18
3)	45, XX, −8
4)	47, XX, −1, +del (1) (p32pter), +del (1) (p32pter)
5)	40, XY, −1, −3, −4, −8, −14, −15
6)	46, XY
7)	46, XY
8)	46, XY
9)	46, XY
10)	46, XY
11)	46, XY
12)	46, XY
13)	46, XY
14)	46, XY
15)	46, XY
16)	46, XY
17)	46, XY
18)	46, XY
19)	46, XY
20)	44, XY, −2, −9
21)	45, Y, del (X) (q22qter), del (6) (q21qter), −13, −19, +t (19; 19) (q13; q13), +r (13)?
22)	48, XY, +?, +?
23)	42, XY, −4, −5, −10, −16
24)	46, XY

As to pleural effusion, we can affirm that interferon therapy was conducive to a good clinical result: in fact, the pleural effusion disappeared.

Our results do not accord with those of Scappaticci et al. [2], which emphasized preferential involvement of chromosomes N. 7, 10, 13, 15, 22, X, Y. In our case, only chromosomes N. 1 and N. 8 seem to be more involved.

We wish to mention the frequent loss of chromosome Y, which we think not significant, because it is often absent in neoplastic pathology and therefore cannot be considered as characteristic of KS. We also think that the hypothesis of polyclonality in the same skin lesion, as reported by Scappaticci et al. [2], is not sufficient to explain the chromosomal instability of KS. Analysis of SCE conducted in the sibship of the proband and his two daughters gave normal results, and this does not seem to fit with chromosomal instability.

Finally, we believe that analysis of a larger number of cases will be essential to achieving an explanation of chromosome instability or a more precise interpretation of the role of chromosomal aberrations in the pathogenesis of KS.

References

1. Dalli Bovi P, Donti E, Knowles II DM, Friedman-Kien A, Luciw PA, Dina D, Della Favera R, Basilico C (1986) Presence of chromosomal abnormalities and lack of AIDS retrovirus DNA sequences in AIDS-associated Kaposi's sarcoma. Cancer Res 46: 6333–6338
2. Scappaticci S, Cerimele D, Cottoni F, Pasquali F, Fraccaro M (1986) Chromosomal aberrations in lymphocytes and fibroblast cultures of patients with the sporadic type of Kaposi's sarcoma. Hum Genet 72: 311–317

III. Viral Infections

Human Papillomaviruses as an Essential Factor for the Development of Cervical Cancer

L. Gissmann

Introduction

It has been generally accepted that papillomaviruses are the cause not only of benign epithelial proliferations such as typical papillomas (verrucae vulgares, verrucae plantares, verrucae planae, laryngeal papillomas, condylomata acuminata) but also of atypical flat lesions at the skin and mucosa (e.g., cervical intraepithelial neoplasia, bowenoid papulosis, focal epithelial hyperplasia; for a description of these lesions see Syrjänen et al. 1987). In recent years evidence has accumulated suggesting that papillomaviruses are also involved in the development of malignant tumors, in particular of squamous cell carcinomas of the uterine cervix.

Papillomaviruses and Benign Warts

The infectious nature of human warts and the etiological role of papillomaviruses can be demonstrated by the following observations:

1. The natural history of warts clearly shows the epidemiological characteristics of an infectious disease (Oriel 1971; for review see Laurent and Kienzler 1985).
2. Injection of cell-free wart extracts into the skin of volunteers induces a papilloma at the site of inoculation (for review see Rowson and Mahy 1967).
3. Infection of human epithelial tissue in vitro with HPV 11 virus particles followed by heterotransplantation beneath the renal capsule of athymic mice leads to induction of the characteristic pathomorphological alterations also observed in HPV-associated lesions in vivo, i.e., koilocytosis and papillary growth. Moreover, virus particles can be recovered from such lesions and are active in a second round of infection (Kreider et al. 1985, 1986, 1987).
4. Transfection of molecularly cloned HPV 16 DNA into primary human foreskin keratinocytes leads to permanent growth in vitro but not to the fully transformed phenotype (tumorigenicity in vivo) of these cells (Dürst et al. 1987a; Pirisi et al. 1987).

The Link to Cervical Cancer

The association of an infectious agent with the development of cervical cancer was demonstrated by epidemiological studies initiated in the last century (for review see Doll 1986). The human papillomaviruses have been brought into the discussion as possible candidates for such an infectious agent (zur Hausen 1977), since the available evidence did not support the involvement of other agents such as herpes simplex virus type 2 (for review see zur Hausen 1983). The characterization of human papillomaviruses and their link to cervical cancer has been hampered for many years by two basic difficulties. Papillomaviruses cannot be propagated in cell culture. Therefore, material for biochemical and molecular biological analysis has to be purified directly from clinical specimens and thus is usually not available in sufficient quantities. Furthermore, there is a remarkable variety of individual human papillomavirus types (for review see Pfister and Fuchs 1987) that in many instances differ considerably from each other in the nucleotide sequence of their genomes. Therefore, molecular cloning of the DNAs of the different HPV types (for review see Gissmann and Schwarz 1985) was a prerequisite of screening genital tumors for the presence of these viruses. As demonstrated by DNA hybridization techniques in different laboratories, the papillomavirus type 6 (de Villiers et al. 1981) or 11 (Gissmann et al. 1982) is mostly found in benign genital warts, whereas HPV 16 (Dürst et al. 1983) or 18 (Boshart et al. 1984) is associated with approximately 70% of cervical cancer biopsies tested so far. In total, more than 90% of such tumor biopsies contain papillomavirus genomes, including other virus types such as HPV 31, 33, 35, or 39 (Lorincz et al. 1986, 1987a; Beaudenon et al. 1986, 1987).

Diagnosis of Genital Papillomavirus Infections

The well-defined precursor lesions of cervical cancer (cervical intraepithelial neoplasia: CIN) have been demonstrated to contain the different HPV types in approximately 10%-30% of cases (Gissmann et al. 1984; Lorincz et al. 1987b). Follow-up studies indicate that the risk of HPV 16-positive CIN lesions progressing to a higher grade of abnormality is significantly higher than that of a lesion containing HPV 6 or 11 (Campion et al. 1986). This implied that screening for papillomavirus infections might be applicable as a diagnostic tool to identify individuals at greater risk of tumor development (Wagner et al. 1984). By screening of a large number of cervical swabs by DNA hybridization it has been demonstrated, however, that the prevalence of subclinical genital HPV infections is one order of magnitude higher than expected from the frequency of clinically overt papillomavirus-associated lesions (de Villiers et al. 1987). In addition, testing of subsequent swabs obtained from the same women at monthly intervals gave, among all HPV infected individuals, a positive result only in a single specimen (L. Gissmann and A. Schneider, unpublished data). These observations support the hypothesis that papillomaviruses can persist in an silent state and and very often escape detection by the commonly used DNA hybridization techniques. Therefore, the mere identification of HPV 16 DNA does not significantly narrow down the number of people at risk of tumor development. On the other hand, and even more important, failure to detect papillomavirus DNA in a clinical specimen from a woman by no means excludes its presence, and thus does not permit discharge of the woman from the routine screening program. It is yet to be made clear, however, whether particular viral markers (individual viral proteins or a distinct immune response against them) are of any relevance in specifying the risk of tumor in any given person.

Causative Role of Papillomaviruses in Tumor Growth

Besides the regular presence of the viral DNA within the cancer biopsies, there are additional lines of evidence which strongly support a causative role of papillomaviruses in tumor growth (for review see Gissmann et al. 1987).

1. Papillomaviruses genomes are not only found in tumor specimens but also persist in cell lines established from such biopsies (e.g., HeLa cells; Schwarz et al. 1985). It is remarkable that, although these cell lines have been kept in culture for years and have undergone many chromosome rearrangements, certain genes of the persistent papillomavirus DNA have been conserved.
2. The viral DNA does not persist in a silent state but is actively transcribed within the tumor cells (Schwarz et al. 1985; Schneider-Gaedicke and Schwarz 1986), and nonstructural (early) papillomavirus proteins have been detected by different techniques such as Western blotting or immunoprecipitation (Smotkin and Wettstein 1986; Seedorf et al. 1987).
3. The HPV DNA replicates in benign lesions as extrachromosomal circular molecules but is found within cancer cells in an integrated state (Dürst et al. 1985). Interestingly, there is a specific region of integration within the viral genome (E1/E2 open reading frame; Schwarz et al. 1985; Baker et al. 1987; Awady et al. 1987), whereas integration within the human genome seems to occur at random (Mincheva et al. 1987). Therefore, activation of cellular genes by promotor insertion as a prerequisite for tumor development appears to be a rather unlikely mechanism, although in HeLa cells the HPV 18 DNA is integrated in the vicinity of the c-*myc* gene on chromosome 8 (Dürst et al. 1987b) and this gene has been shown to be overexpressed in a large number of cervical cancer biopsies (G. Riou, personal communication). On the other hand, the integration within the E1/E2 open reading frame of the viral genome certainly interferes with the regulation of viral gene expression. In fact, a transcriptional repressor function has been allocated to this region (Lambert et al. 1987) and it is tempting to speculate that the inactivation of this repressor ensures a constitutive expression of other early proteins such as E6 and E7 which seem to play a role in cell transformation in vitro (Matlashewski et al. 1987). At all events, the specificity of integration argues in favor of a causative role of the papillomavirus DNA in tumor development, rather than mere passenger status within the transformed cells.
4. By modulating the level of expression of the endogenous HPV DNA in cervical cancer cell lines, it can be demonstrated that the amount of early viral pro-

teins correlates with the growth efficiency of these cells in vitro (von Knebel Doeberitz et al. 1988; Rösl et al. 1988).

The Role of Cofactors

The following observations demonstrate that the formation of cervical cancer from a papillomavirus-infected cell is a very rare and slow process.

1. The latency period between primary infection with genital papillomaviruses (at the age of highest sexual activity) and the occurrence of tumors is usually several decades.
2. As mentioned before, epidemiological surveys indicate that the prevalence of infection with the (cancer-related) genital papillomaviruses is at least 10%-20% (de Villiers et al. 1987; Schneider et al. 1987), whereas in the Western countries the life-long risk for a woman of developing cervical cancer is in the order of 1% (Waterhouse et al. 1976).
3. The malignant tumors are monoclonal (Lehn et al. 1985), i.e., from a supposedly very large number of infected cells in the affected cervical epithelium, only one or a very few are able to escape normal cell growth control.

Hence it becomes evident that papillomavirus infection per se is not sufficient for the induction of cervical cancer, but that additional cofactors are required for a cell from normal cervical epithelium to develop into a malignant tumor. The nature of such cofactors is not well defined and their mode of action at present only a matter of speculation.

1. There is an increased risk of development of cervical cancer in women who smoke (Winkelstein 1977), and mutagenic substances can be isolated from their vaginal fluid (Holly et al. 1986). As discussed above, specific integration of papillomavirus DNA is a common feature of cervical cancer biopsies and established cell lines, and is assumed to be a prerequisite for malignant cell growth. As a response to mutational events, cellular DNA repair systems are induced which include excision and rejoining of DNA. Therefore, it seems likely that, at the same time, recombinational events are triggered (Sarasin 1985), thus favoring the integration of foreign DNA into the host genome. In addition, analysis of hybrids between malignant HeLa cells and normal fibroblasts or keratinocytes, has demonstrated that the transformed phenotype can be suppressed by a gene product present in normal cells (Stanbridge et al. 1982), and it has been recently postulated that this putative factor acts by down-regulation of the papillomavirus gene expression (zur Hausen 1986; Rösl et al. 1988). Therefore, the development of a malignant tumor from an HPV-infected cell is favored if this cell is constantly exposed to mutagens and the cellular gene becomes inactivated. Besides the mutagenic substances present in cigarette smoke, mutagens may also be produced by certain microorganisms which populate the lower genital tract of sexually active women (Harington et al. 1973) who are at particular risk of developing cervical cancer.

2. There are many anecdotal and a few published reports on the increased prevalence of genital warts during pregnancy (Woodruff and Peterson 1958; Garry and Jones 1985). In cervical swabs obtained from pregnant women papillomaviruses are detected more than twice as frequently as in material from nonpregnant controls (Schneider et al. 1987). In addition, multiple pregnancies and long term use of oral contraceptives are both considered to be risk factors for cervical cancer (for reviews see Rotkin 1973 and Doll 1986). The altered hormone status may have a direct impact on the papillomavirus expression, as, in the genomes of HPV 16 and HPV 18, several homologues to known glucocorticoid-responsive DNA sequences have been identified (M. von Knebel Doeberitz, personal communication), one of which was shown to become activated after hormone treatment (H. U. Bernhard, personal communication). Furthermore, an elevated rate of early virus gene expression was observed in an HPV 18-positive cervical cancer cell line upon dexamethasone treatment (M. von Knebel Doeberitz et al. 1988).
3. Alternatively or in addition to the hormonal influence, the temporary immunosuppression occurring during pregnancy may allow an HPV-positive cell to escape from the regular immune surveillance. The increased prevalence of benign and malignant HPV-associated tumors in immunosuppressed patients (for review see Spradbrow 1987) clearly demonstrates that the immune system actually plays a critical role in tumor development.

Summary

Human papillomaviruses (HPV) represent a very heterogeneous group of epitheliotropic agents and to date cannot be replicated in tissue culture. For these reasons the investigation of their role in human malignancies has been hampered, although the idea has been discussed for more than a decade that they are an important factor in the development of squamous cell cancer of the uterine cervix. Chiefly due to the advances of recombinant DNA techniques, it could be demonstrated that certain HPV types are associated with cervical cancer biopsies as well as with the precursor lesions of these tumors (cervical intraepithelial neoplasia). In addition, molecular biological analysis of cell lines established from cervical cancer biopsies (e.g., the HeLa cells) suggests that papillomaviruses are not only important for the induction of tumor formation but are also essential for the maintenance of the malignant phenotype. On the other hand epidemiological surveys indicate that papillomaviruses are not the only factor required for tumor development. The nature of the putative "cocarcinogens" and their possible mode of action is discussed.

Acknowledgements. I am grateful to Dr. H. zur Hausen for stimulating discussions and to Dr. M. Pawlita for critical reading of the manuscript.

References

Awady MK, Kaplan JB, O'Brien SJ, Burk RD (1987) Molecular analysis of integrated papillomavirus 16 sequences in the cervical cancer cell line SiHa. Virology 159: 389-398

Baker CC, Phelbs WC, Lindgren V, Braun MJ, Gonda MA, Howley PM (1987) Structural and transcriptional analysis of human papillomavirus type 16 sequences in cervical carcinoma cell lines. J Virol 61: 962-971

Beaudenon S, Kremsdorf D, Croissant O, Jablonska S, Wain-Hobson S, Orth G (1986) A novel type of human papillomavirus associated with genital neoplasias. Nature 321: 246-249

Beaudenon S, Kremsdorf D, Obalek S, Jablonska S, Pehau-Arnaudet G, Croissant O, Orth G (1987) Plurality of genital human papillomaviruses: characterization of two new types with distinct biological properties. Virology 161: 374-384

Boshart M, Gissmann L, Ikenberg H, Kleinheinz A, Scheurlen W, zur Hausen H (1984) A new type of papillomavirus DNA: its presence in cervical cancer biopsies and in cell lines derived from cervical cancer. EMBO J 3: 1151-1157

Campion MJ, Cuzick J, McCance DJ, Singer A (1986) Progressive potential of mild cervical atypia: prospective cytological, colposcopic, and virological study. Lancet II: 237-240

de Villiers E-M, Gissmann L, zur Hausen H (1981) Molecular cloning of viral DNA from human genital warts. J Virol 40: 932-935

de Villiers E-M, Wagner D, Schneider A, Wesch H, Miklaw H, Wahrendorf J, zur Hausen H (1987) Human papillomavirus infections with and without abnormal cervical cytology. Lancet II: 703-706

Doll R (1986) Implications of epidemiological evidence for future progress. In: Peto R, zur Hausen H (eds) Viral etiology of cervical cancer. Banbury report, vol 21. Cold Spring Harbor, New York, pp 321-326

Dürst M, Gissmann L, Ikenberg H, zur Hausen H (1983) A papillomavirus DNA from a cervical carcinoma and its prevalence in cancer biopsies from different geographic regions. Proc Natl Acad Sci USA 80: 3812-3815

Dürst M, Kleinheinz A, Hotz M, Gissmann L (1985) The physical state of human papillomavirus type 16 DNA in benign and malignant tumors. J Gen Virol 66: 1515-1522

Dürst M, Dzarlieva-Petrusevska RT, Boukamp P, Fusenik NE, Gissmann L (1987a) Molecular and cytogenetic analysis of immortalized human primary keratinocytes obtained after transfection with human papillomavirus type 16 DNA. Oncogene 1: 251-256

Dürst M, Croce CM, Gissmann L, Schwarz E, Huebner K (1987b) Papillomavirus sequences integrate near cellular oncogenes in some cervical carcinomas. Proc Natl Acad Sci USA 84: 1070-1074

Garry R, Jones R (1985) Relationship between cervical condylomata, pregnancy and sub-clinical papillomavirus infection. J Reprod Med 30: 393-399

Gissmann L, Schwarz E (1985) Cloning of papillomavirus DNA. In: Becker Y (ed) Recombinant DNA research and viruses. Nijhoff, Boston, pp 173-190

Gissmann L, Diehl V, Schultz-Coulon HJ, Schnürch HG, zur Hausen H (1982) Molecular cloning and characterization of human papillomavirus DNA derived from a laryngeal papilloma. J Virol 44: 393-400

Gissmann L, Boshart M, Dürst M, Ikenberg H, Wagner D, zur Hausen H (1984) Presence of human papillomavirus (HPV) DNA in genital tumors. J Invest Dermatol 83: 26s-28s

Gissmann L, Dürst M, Oltersdorf T, von Knebel Doeberitz M (1987) Human papillomaviruses and cervical cancer. In: Steinberg BM, Brandsma JL, Taichman LB (eds) Papillomaviruses. Cold Spring Harbor, New York, pp 275-280 (Cancer cells, vol 5)

Harington JS, Nunn JR, Irwig L (1973) Dimethylnitrosamine in the human vaginal vault. Nature 241: 49-51

Holly EA, Petrakis NL, Fried NF, Sarles DL, Lee RE, Flander LB (1986) Mutagenic mucus in the cervix of smokers. JNCI 76: 983-986

Kreider JW, Howett MK, Wolfe SA, Bartlett GL, Zaino RJ, Sedlacek TV, Mortel R (1985) Morphological transformation of human uterine cervix with papillomavirus from condylomata acuminata. Nature 317: 639-641

Kreider JW, Howett MK, Lill NL, Bartlett GL, Zaino RJ, Sedlacek TV, Mortel R (1986) In vivo transformation of human skin with human papillomavirus type 11 from condylomata acuminata. J Virol 59: 369-376

Kreider JW, Howett MK, Stoler MH, Jaino RJ, Welsh P (1987) Susceptibility of various human tissues to transformation in vivo with human papillomavirus type 11. Int J Cancer 39: 459-465

Lambert PF, Spalholz BA, Howley PM (1987) A transcriptional repressor encoded by BPV 1 shares a common carboxy-terminal domain with the E2 transactivator. Cell 50: 69-78

Laurent R, Kienzler J-L (1985) Epidemiology of HPV infections. In: Jablonska S, Orth G (eds) Warts human papillomaviruses. Lippincott, Philadelphia, pp 64-70 (Clinics in dermatology, vol 3)

Lehn H, Krieg P, Sauer G (1985) Papillomavirus genomes in human cervical tumors; analysis of their transcriptional activity. Proc Natl Acad Sci USA 82: 5540-5544

Lorincz AT, Lancaster WD, Temple G (1986) Cloning and characterization of the DNA of a new human papillomavirus from women with cervical dysplasia of the uterine cervix. J Virol 58: 225-229

Lorincz AT, Quinn AP, Lancaster WD, Temple GF (1987a) A new type of papillomavirus associated with cancer of the uterine cervix. Virology 159: 187-190

Lorincz AT, Temple GF, Kurman RJ, Jenson AB, Lancaster WD (1987b) Oncogenic association of specific human papillomavirus types with cervical neoplasia. JNCI 79: 671-677

Matlashewski G, Schneider J, Banks L, Jones N, Murray A, Crawford L (1987) Human papillomavirus type 16 DNA cooperates with activated *ras* in transforming primary cells. EMBO J 6: 1741-1746

Mincheva A, Gissmann L, zur Hausen H (1987) Chromosomal integration sites of human papillomavirus DNA in three cervical cancer cell lines mapped by in situ hybridization. Med Microbiol Immunol (Berl) 176: 25-256

Oriel JD (1971) Natural history of genital warts. Br J Vener Dis 47: 1-13

Pfister H, Fuchs PG (1987) Papillomaviruses: particles, genome organization and proteins. In: Syrjänen K, Gissmann L, Koss LG (eds) Papillomaviruses and human disease. Springer, Berlin Heidelberg New York, pp 1-18

Pirisi L, Yasumoto S, Feller M, Doninger J, DiPaolo JA (1987) Transformation of human fibroblasts and keratinocytes with human papillomavirus type 16 DNA. J Virol 61: 1061-1066

Rösl F, Dürst M, zur Hausen H (1988) Selective suppression of human papillomavirus transcription in non-tumorigenic cells by 5-azacytidine. EMBO J (in press)

Rotkin ID (1973) A comparison review of key epidemiological studies in cervical cancer related to current searches for transmissible agents. Cancer Res 33: 1353-1367

Rowson KEK, Mahy BWJ (1967) Human papova (wart) virus. Bacteriol Rev 31: 110-131

Sarasin A (1985) SOS response in mammalian cells. Cancer Invest 3: 163-174

Schneider A, Hotz M, Gissmann L (1987) Increased prevalence of human papillomaviruses in the lower genital tract of pregnant women. Int J Cancer 40: 198-201

Schneider-Gaedicke A, Schwarz E (1986) Different human cervical carcinoma cell lines show similar transcription patterns of human papillomavirus type 17 early genes. EMBO J 5: 2285-2292

Schwarz E, Freese UK, Gissmann L, Mayer W, Roggenbuck B, Stremlau A, zur Hausen H (1985) Structure and transcription of human papillomavirus sequences in cervical carcinoma cells. Nature 314: 111-114

Seedorf K, Oltersdorf T, Krämmer G, Röwekamp W (1987) Identification of early proteins of the human papillomaviruses type 16 and type 18 in cervical cancer cells. EMBO J 6: 139-144

Smotkin D, Wettstein FD (1986) Transcription of human papillomavirus type 16 early genes in a cervical cancer and a cancer-derived cell line and identification of the E7 protein. Proc Natl Acad Sci USA 83: 4680-4684

Spradbrow PB (1987) Immune response to papillomavirus infection. In: Syrjänen K, Gissmann L, Koss LG (eds) Papillomaviruses and human disease. Springer, Berlin Heidelberg New York, pp 334-370

Stanbridge EJ, Der CD, Doersen CJ, Nishimi RY, Peehl DM, Weissman BE, Wilkinson J (1982) Human cell hybrids: analysis of transformation and tumorigenicity. Science 215: 252-259

Syrjänen K, Gissmann L, Koss LG (eds) (1987) Papillomaviruses and human disease. Springer, Berlin Heidelberg New York

von Knebel Doeberitz M, Oltersdorf T, Schwarz E, Gissmann L (1988) Modulation of human papillomavirus gene expression in C4-1 cervical carcinoma cells is correlated with altered growth properties. Cancer Res (in press)

Wagner D, Ikenberg H, Böhm N, Gissmann L (1984) Identification of human papillomavirus in cervical smears by DNA in situ hybridization. Obstet. Gynecol 64: 767-772
Waterhouse J, Muir C, Correa P, Powell J (eds) (1976) Cancer incidence in five continents, vol III. IARC, Lyon (IARC scientific publication no 15)
Winkelstein WJ (1977) Smoking and cancer of the uterine cervix: hypothesis. Am J Epidemiol 106: 257-259
Woodruff JD, Peterson WF (1958) Condylomata acuminata of the uterine cervix. Am J Obstet Gynecol 75: 1354-1362
zur Hausen H (1977) Human papillomaviruses and their possible role in squamous cell carcinomas. Current Topics in Microbiology and Immunology, vol 78. Springer, Berlin Heidelberg New York, pp 1-30
zur Hausen H (1983) Herpes simplex virus in human genital cancer. Int Rev Exp Pathol 25: 307-326
zur Hausen H (1986) Intracellular surveillance of persisting viral infections. Lancet II: 489-491

HPV Infection of the Cervix Uteri: Prevalence and Age Distribution of a Sexually Transmissible Disease in a Cancer Screening Program

D. Wagner and G. Gross

Introduction

There is growing evidence from both epidemiological and clinical studies that infection with human papillomavirus (HPV) as a sexually transmitted disease may play a role in the genesis and development of cervical and penile cancer (zur Hausen 1977; Meisels and Fortin 1976; Meisels and Morin 1981; Gissmann 1984; Gissmann et al. 1984).

Recent advances in microbiology and virology have led to the introduction of newly developed HPV DNA Cloning- and hybridization techniques into clinical medicine (Gissmann et al. 1983). Different HPV DNA types in cervical cells obtained from women routinely screened for cervical cancer were detected by the in situ filter hybridization technique (Wagner et al. 1984). This was the rationale for applying the method in a large scale screening program aimed at establishing the prevalence rate of HPV infections in clinically healthy woman and in woman with abnormal clinical findings (de Villiers et al. 1987). Further, by using these new techniques in addition to cytology and colposcopy, the gynecologist is now equipped to detect and to judge minute cellular changes in the cervical smear and in apparently benign colposcopic lesions of the uterine transformation zone as being related to an HPV infection. The combination of HPV hybridization techniques with cervical cytology und colposcopy has significantly improved the sensitivity of these widely used clinical methods of investigation, especially with regard to HPV-induced precancerous lesions of the uterine cervix (Schneider et al. 1987).

Studies of HPV-related disease in the female population have steered the attention of the dermatologist to HPV-induced alterations in the sexual partners of women presenting with genital condylomata or precancerous lesions of the uterine cervix. Minute penile lesions detected only by means of a colposcope ("peniscope") have been recently described as Bowenoid papulosis (Gross et al. 1985) and have also been related to an HPV infection, probably transmitted by sexual intercourse (Gross et al. 1986).

The following report confirms the significance of genital HPV infections with regard to a sexually transmitted disease which may have a considerable impact on the planning and realization of forthcoming screening programs for genital cancer.

Prevalence and Age Distribution of HPV Infections in a Female Population

From 1985 to 1987, 4394 women were screened for HPV infections in addition to precursor lesions of cervical cancer in the Gynecology Department and the outpatient clinic of the Freiburg Diaconate Hospital. The objective was to improve our cancer screening program by early diagnosis of clinically silent HPV infections, routinely combining colposcopy and cytology with in situ filter hybridization. We acknowledge the close cooperation with the German Cancer Research Center, Heidelberg (E. M. de Villiers, L. Gissmann and H. zur Hausen), in carrying out the filter hybridization.

The results are shown in Table 1: specimens from 781 out of 4394 women showed various types of HPV, i.e., types 6, 11, 16, and 18. The prevalence rate in the total population screened was 17.8%. The relatively high rate is due to the fact that patients in our colposcopy clinic were included in the study.

The prevalence rate of clinically "healthy" woman with negative cytology and colposcopy but positive HPV findings was 15.4% (626 out of 4058 women). In the group of women with positive cytology and/or colposcopy findings, positive HPV findings were obtained in 155 of 336 women, increasing the prevalence rate in this group to 46.1%. It was interesting that only 155 out of 781 HPV-positive women revealed positive cytologic or colposcopic findings. This indicates that only about 20% of HPV-infected women can be diagnosed by clinical examination, and about 80% of cases remain clinically inapparent.

The clinical diagnosis may be significantly improved by the integration of the minimal HPV-induced morphological changes in the cytologic smear and in the colposcopic picture (for more details see below).

The age distribution of the 781 HPV-positive women is shown in Fig. 1. The highest prevalence rate, 25.1%, was found in the age group 21-30 years, closely followed by 24.4% in the age group 31-40 years. However, even in the teenage group the prevalence rate amounted to 20.7%. A significant decrease in the prevalence rate, to 9.9%, occurred in the age group 51-60 years, with a further drop to 4.4% in the group of women over 70 years. This may indicate an important effect of sexual activity on the HPV infection rate, which supports the hypothesis of virus spreading by sexual transmission.

The age distribution of the 781 HPV-positive women classified into groups with positive and negative colposcopic and/or cytological results is shown in

Table 1. Prevalence of HPV infection of the uterine cervix

Ü	*n*	HPV-positive	%	HPV 6 or 11	%	HPV 16 or 18	%	HPV 6, 11, 16, 18	%
Total of patients screened	4394	781	17.8	147	3.4	295	6.7	339	7.7
Colposcopy/cytology negative	4058	626	15.4	127	3.1	234	5.8	265	6.5
Colposcopy/cytology positive	336	155	46.1	20	6.0	61	18.1	74	22.0

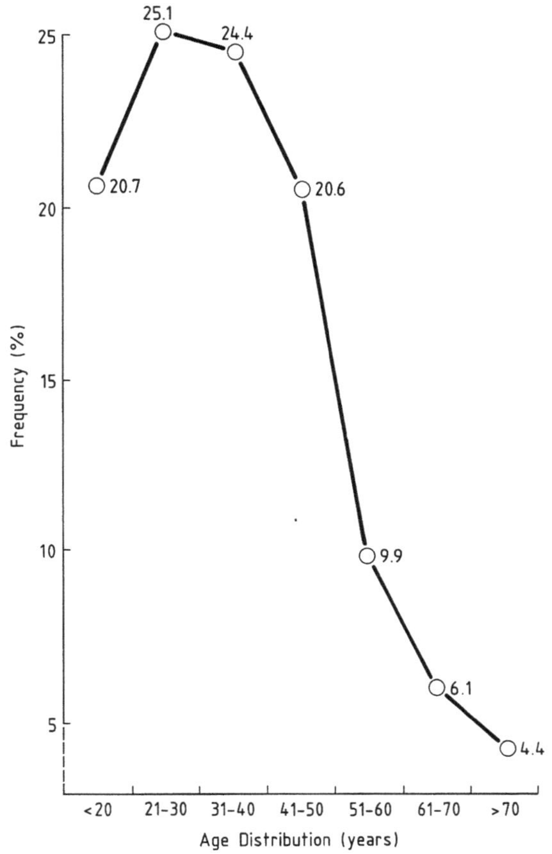

Fig. 1. Age distribution of 781 HPV-positive women

Fig. 2. The highest rate of HPV infection in women with positive cytology/colposcopy was found in the age group 20-30 years and reached 29% of all cases, steadily decreasing to 14% in the age group 41-50 years and to 11% and 5% in the older age groups. In contrast, the frequency of positive HPV findings and negative cytological colposcopic findings was highest among the women of age group 41-50 years. That means that efforts towards the clinical detection of HPV infection should be focussed on women under 40 years of age.

The age distribution of 336 women with positive cytological and/or colposcopic results is shown in Fig. 3. Out of 336 women, 155 were HPV-positive, and in every age group HPV-positivity was determined in about 50% of the patients. Thus, the gynecologist can assume an HPV infection in every second woman with positive cytological and/or colposcopic results.

Minimal Morphological Changes in HPV-Positive Women

Using colposcopy and cytology in combination with hybridization techniques, the gynecologist can now become aware of minimal morphological changes which

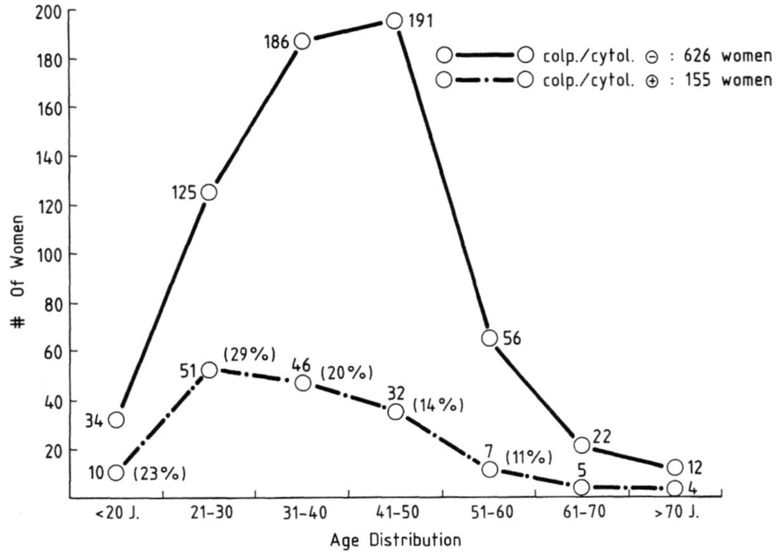

Fig. 2. Age distribution of 781 HPV-positive women. O—O Colposcopy/cytology negative (626 women); O – · – O Colposcopy/cytology positive (155 woman)

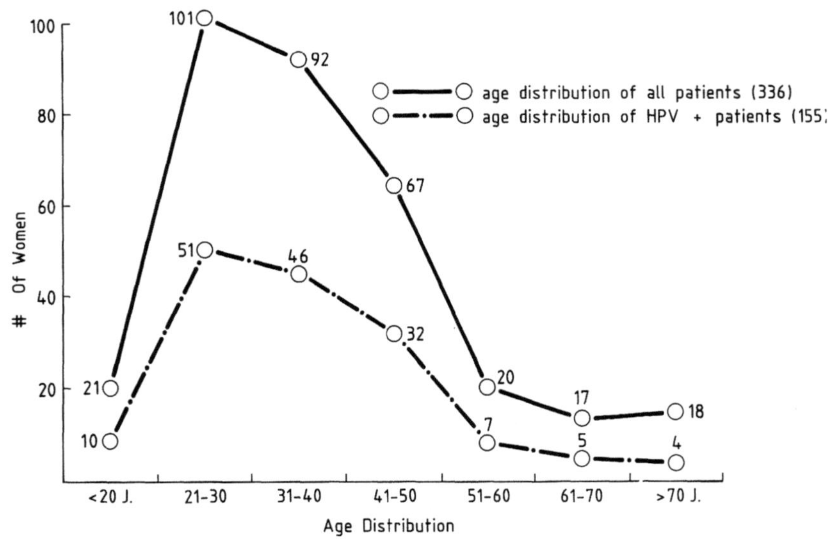

Fig. 3. Rate of HPV infections in 336 woman with positive colposcopic and/or cytological results. O—O Age distribution of all patients (336); O – · – O Age distribution of HPV-positive patients (155)

used to be considered an "unspecific inflammatory reaction," but which have now to be related to an HPV infection.

In cytology the following, mostly cytoplasmic changes of the differentiated squamous epithelial cell have been observed:

1. Perinuclear halo without nuclear atypia
2. Dyskeratotic cells (very small superficial cells with strong eosinophilic staining)
3. Giant superficial cells
4. Amphophilic cytoplasm staining
5. Cracked cytoplasm and basophilic granules
6. Hyperchromatosis of double- and multinucleated superficial cells

In a correlation study of 112 patients showing these minimal changes, 64 (57%) were HPV-positive, compared with 11% HPV-positive women in a control group showing normal cytology. (For more detail see Schneider et al., this volume).

In colposcopy the gynecologist now concentrates on very small, nonproliferating, acetic-white and iodine-negative lesions within the transformation zone of the uterine cervix. These had been previously defined as metaplasia following unspecific inflammation. Out of 66 women presenting with such lesions, 28 (42%) were HPV-positive according to the in situ filter hybridization technique, which has a relatively low sensitivity. When the more sensitive Southern blot technique was used on cervical biopsies taken from those lesions, HPV-positive findings ranged up to about 70%-80%.

Sexual Transmission of HPV Infection

If we consider an HPV infection as being sexually transmissible, we must expect positive HPV-findings in the sexual partners of patients with and without clinical signs of HPV-infection. To verify this hypothesis we examined 121 couples of whom one of the partners presented clinical signs of an HPV disease, i.e., condylomata acuminata, cervical intraepithelial neoplasia, or Bowenoid papulosis of the male or female genital tract.

The following methods for detection of an HPV infection were used: clinical examination for visible lesions, colposcopy ("peniscopy"), smears from various locations of the genital tract taken from either partner for cytology and in situ filter hybridization, tissue biopsies for histology, and HPV typing by Southern blot analysis.

The following results were obtained: If one of the partners of the 121 couples examined had a clinically or colposcopically visible lesion, the partner also revealed a macroscopic or microscopic visible lesion, which could be related to an HPV infection in 43 cases (35.5%). However, in 72 cases (59.5%) the HPV-findings were positive in the partner but no visible lesion could be demonstrated at the time of examination, indicating that the partner was a virus carrier but had no disease.

In 46 of the 121 couples examined, one of the partners presented with a precancerous lesion, i.e., cervical, vaginal, or vulvar intraepithelial neoplasia in the woman or Bowenoid papulosis of the penis (penile intraepithelial neoplasia) in the man (Fig. 4). When we subdivided these 46 couples according whether the first precancerous lesion was detected in the woman (28 cases) or in the man (18 cases), the following results for the partners were obtained: for the 28 women, 4 partners (14.4%) had clinically visible precancerous lesion, but in 13 partners (46.6%) only the HPV findings were positive and no clinical lesion was visible. This indicates that the male partner harbored the virus and might have induced the precancerous lesion in his female partner.

Even more striking were the results in the group of 18 couples, when the precancerous lesion was detected first in the male partner. Of the partners of 18 men with Bowenoid papulosis of the penis, seven women (38.8%) presented with a macroscopically or colposcopically visible precancerous lesion, mostly of the uterine cervix (cervical intraepithelial neoplasia), and in 13 women (72.2%) only the HPV findings were positive without any visible lesion. Since most of the precancerous lesions in men and women are strongly related to HPV types 16 and 18, the

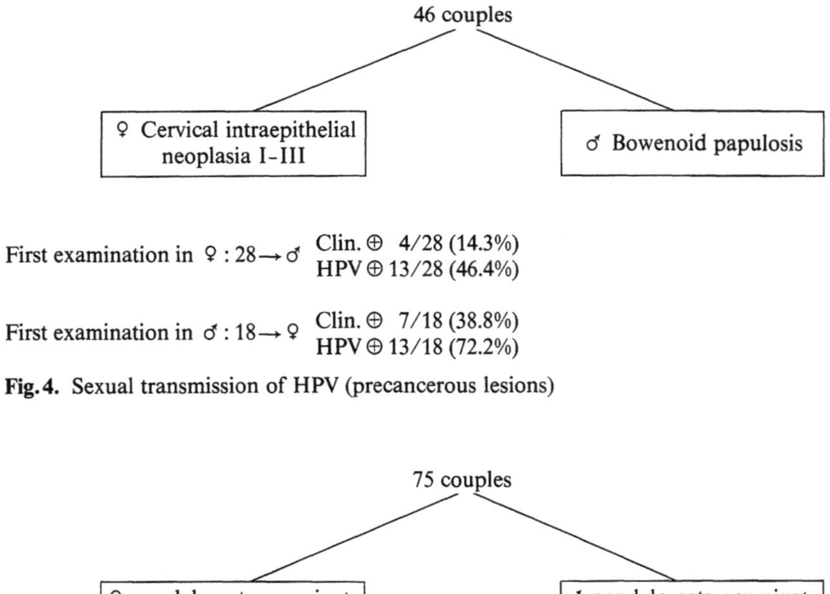

Fig. 4. Sexual transmission of HPV (precancerous lesions)

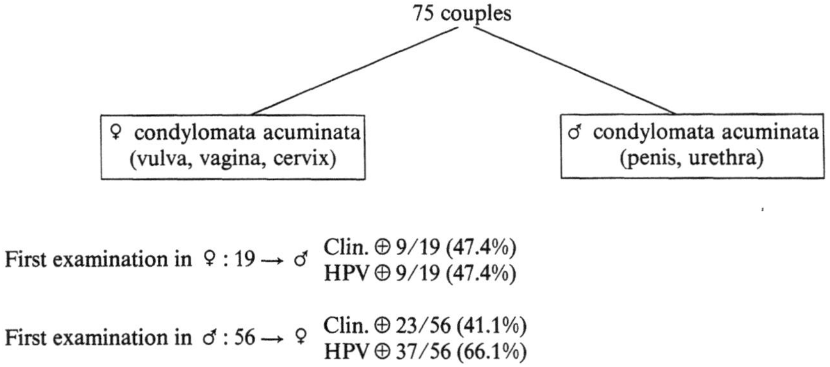

Fig. 5. Sexual transmission of HPV (condylomatous lesions)

high rate of HPV-positive women in couples the male partner of which has a penile Bowenoid papulosis may indicate a considerable risk for his sexual partner of developing a precancerous lesion in the near future. The consequence of this observation should be close follow-up of those women, using all available methods for detection of an HPV infection.

Similar results were obtained in 75 couples of whom one of the partners demonstrated condylomata acuminata of different genital regions (Fig. 5). In a group of 19 couples condylomata acuminata were first detected in the female partner. In nine men (47.4%), condylomata of the penis or urethra could be clinically demonstrated and the same incidence of positive HPV findings was observed. In the group of 56 couples with first detection of condylomata acuminata in the male partners, corresponding lesions were clinically demonstrated in 23 women (41.1%) and 37 woman (66.1%) were HPV-positive. Again, these results give clear evidence that the female partner of a man with condylomata acuminata is at high risk of developing an HPV infection, particularly with types 6 and 11, even if there was no visible lesion in the woman at the time of first examination.

Summary and Conclusion

The results of this study clearly show that papillomavirus infection of the female genital tract (vulva, vagina, and cervix), with a prevalence rate of 15.4% of clinically healthy woman, has to be considered as one of the most frequent genital infections. The true prevalence rate cannot be determined by clinical diagnosis only, but requires application of the newly developed HPV DNA hybridization techniques. Only by complementary use of the cytological, colposcopic, histological, and virological findings obtained by those new techniques will the gynecologist be able to detect the minute morphologic changes in lesions which appear to be benign, but which in fact are related to an HPV infection.

In a screening program provided for the early detection of cervical cancer, a woman presenting with an HPV-positive lesion should be assigned to a higher risk group for development of the disease. She should then be followed up at least once a year. For the large group of HPV-negative women, however, a lengthening of the screening intervals might be justified for reasons of cost-effectiveness.

The results of this study confirm that HPV infections are sexually transmitted and that, to be able to treat a patient with an HPV-related disease effectively, one has to examine his or her sexual partner. This requires the close cooperation of gynecologists and dermatologists, who should be experts in this field of common interest.

Cancer prevention today should not only be confined to new diagnostic procedures in addition to well-established screening methods. Aspects such as sexual hygiene and even sex education must be made an integral part of any cancer prevention program in order to ensure long-term effectiveness.

References

de Villiers EM, Wagner D, Schneider A, Wesch H, Miklaw H, Wahrendorf J, Papendick U, zur Hausen H (1987) Human papillomavirus infections in women with and without abnormal cervical cytology. Lancet: 703-705

Gissmann L (1984) Papillomaviruses and their association with cancer in animals and in man. Cancer Surv 3 (1): 1961-1981

Gissmann L, Wolnik L, Ikenberg HI, Koldovsky U, Schnürch HG, zur Hausen H (1983) Human papillomaviruses types 6 and 11 DNA sequences in genital and laryngeal papillomas and in some cervical cancers. Proc Natl Acad Sci USA 80: 560-563

Gissmann L, Boshart M, Dürst M, Ikenberg H, Wagner D, zur Hausen H (1984) Presence of human papillomavirus (HPV) DNA in genital tumors. J Invest Dermatol 83 (Suppl 1): 265-285

Gross G, Wagner D, Hauser B, Ikenberg H, Gissmann L (1985) Bowenoide Papulose und Carcinoma in situ der Cervix uteri bei Sexualpartnern. Hautarzt 36: 465-469

Gross G, Ikenberg H, de Villiers EM, Schneider A, Wagner D, Gissmann L (1986) Bowenoid papulosis: a venereally transmissible disease as reservoir for HPV 16. In: zur Hausen H, Peto R (eds) Origins of female genital cancer: virological and epidemiological aspects. Banbury report. Cold Spring Harbor, New York, pp 149-165

Meisels A, Fortin R (1976) Condylomatous lesions of the cervix and vagina. I. Cytologic patterns. Acta Cytol (Baltimore) 20: 505-509

Meisels A, Morin C (1981) Human papillomavirus and cancer of the uterine cervix. Gynecol Oncol 12: 111-123

Schneider A, Meinhard G, de Villiers EM, Gissmann L (1987) Sensitivity of the cytologic diagnosis of cervical condyloma in comparison with HPV DNA-hybridization studies. Diagn Cytol III (3): 250-255

Wagner D, Ikenberg H, Böhm N, Gissmann L (1984) Type specific identification of human papillomavirus in cervical smears by DNA in situ hybridization. Obstet Gynecol 6: 767-772

zur Hausen H (1977) Human papillomaviruses and their possible role in squamous cell carcinoma. Current Topics in Microbiology and Immunology, vol 78. 1-30 Springer, Berlin Heidelberg New York

Conventional Cytology is Insensitive for the Detection of Genital HPV Infection*

A. Schneider, G. Meinhardt, and E.-M. De Villiers

Introduction

Cytological examination and colposcopy are indispensable techniques for the diagnosis of cervical intraepithelial neoplasia. Cytological examination is regarded as a superior screening method, whereas with colposcopy lesions can be localized and graded accurately [1, 2]. Recent studies show that cervical infection with human papillomavirus (HPV) precedes development of cervical intraepithelial neoplasia [3-5]. Thus, prevention of the latter should focus on the detection and treatment of cervical HPV infection not yet associated with intraepithelial neoplasia. Highly sensitive molecular biological techniques show a surprisingly high prevalence rate of HPV infection of between 10% and 38% in asymptomatic women [6-11]. The cytological smear shows "classic" signs of HPV infection - koilocytosis and dyskeratocytosis - in only 1%-2% [12, 13]. In order to improve the sensitivity of cytological examination we tested a panel of "nonclassic" signs. In a second study we critically evaluated the sensitivity of conventional cytology and colposcopy for the detection of HPV infection.

"Classic" and "Nonclassic" Cytological Signs of Condyloma [14]

We selected Papanicolaou-stained smears from 100 patients, who were positive for HPV DNA on parallel smears with filter in situ hybridization, for rescreening. Another 100 smears obtained from patients negative for HPV DNA and matched for age were used for comparison. Cells with nuclear atypia, defined as enlargement of the nucleus combined with a irregular chromatin pattern, had to be absent in all these smears in order to be part of this study. These smears (with known HPV DNA result) were rescreened for the presence of the classic cytomorphological signs of koilocytosis and dyskeratocytosis. Nine nonclassic cytological signs were evaluated additionally (Fig. 1):

1. "Mild koilocytosis", which in contrast to classic koilocytosis, shows no nuclear atypia and a faint border of the central cavity.

* This work was supported by the Deutsche Forschungsgemeinschaft (SFB 31: Tumor genesis and development).

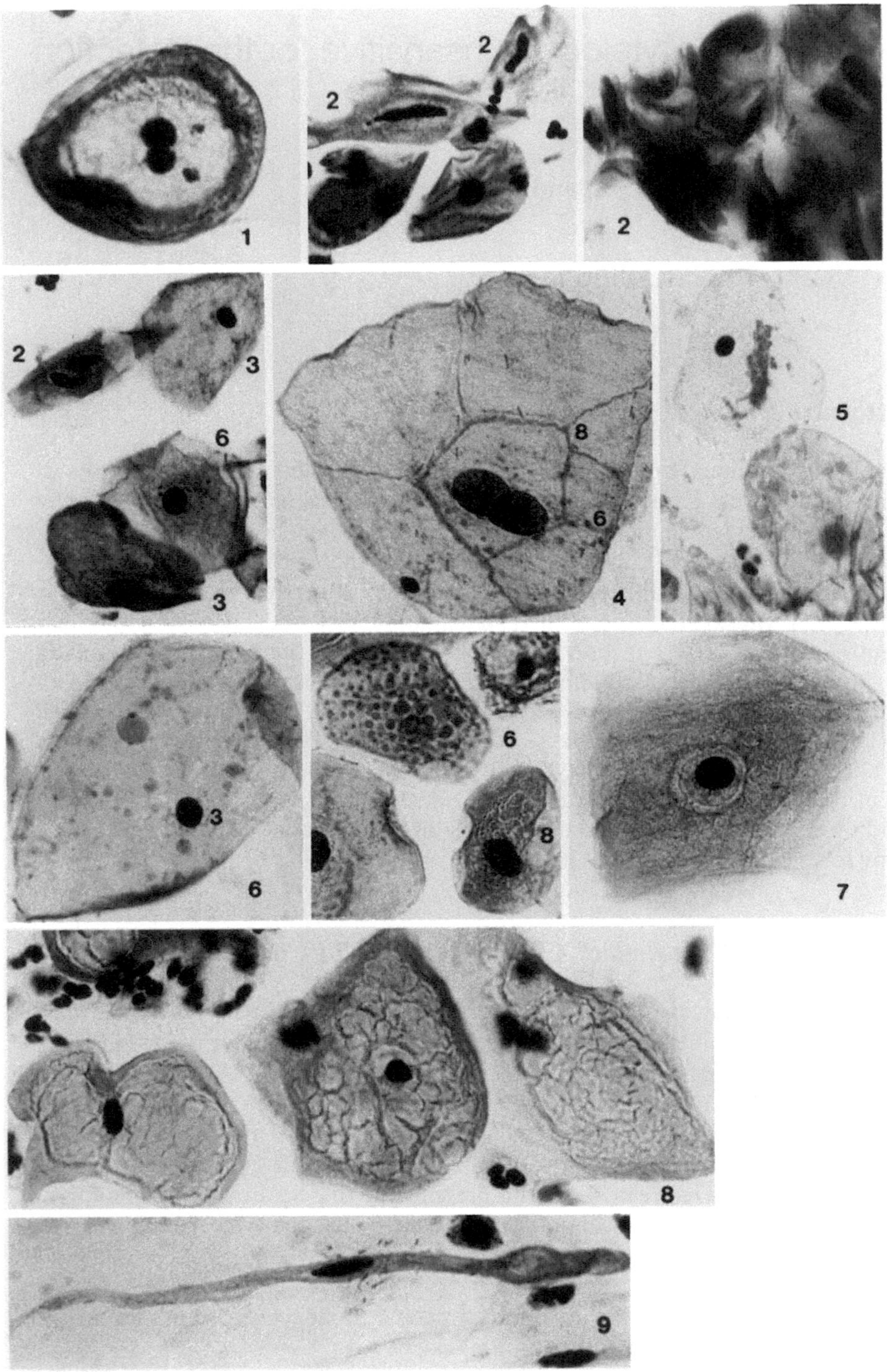

Fig. 1. The nine "nonclassic" cytologic signs of HPV infection. *1* "Mild koilocytosis"; *2* "mild dyskeratocytosis"; *3* nuclear hyperchromatism; *4* bi- or multinucleation; *5* "cleared cytoplasm"; *6* keratohyalin or keratohyalin-like granules, "measled cells"; *7* perinuclear halos; *8* condensation of filaments; *9* spindle cells (×340–510)

2. "Mild dyskeratocytosis", which differs from classic dyskeratocytosis in having less pronounced nuclear changes. These cells are frequently arranged three-dimensionally.
3. Nuclear hyperchromatism, distinguished from nuclear atypia by increased stainability in the absence of irregularities of the chromatin and nuclear membrane.
4. Bi- or multinucleation.
5. "Cleared cytoplasm", defined as an apparent lack of stainable matter between nucleus and cell border.
6. Keratohyalin and keratohyalin-like granules, which are basophilic or eosinophilic condensations of varying size in the cytoplasm. The nucleus may be missing in these cells. When the entire cytoplasmic matter has condensed into granules we call them "measle cells".
7. Perinuclear halos.
8. Condensed filaments, appearing as crack-like cytoplasmic fissures.
9. Spindle cells, distinguished from similar cells found in keratinizing squamous carcinoma by their regular chromatin pattern.

Statistical analysis for each nonclassic sign was done according to the Bonferroni inequality test and a combination of the most valuable signs was identified by discriminant analysis.

The classic cytomorphological signs, koilocytosis and/or dyskeratocytosis, were found in 15 out of 100 HPV-positive smears and three out of 100 HPV-negative samples. Statistical analysis confirmed a significant correlation between the detection of HPV DNA and each nonclassic cytomorphological sign, with the exception of spindle cells, due to the overall low prevalence of this criterion. Mild koilocytosis, mild dyskeratocytosis, hyperchromatic nuclei, bi- or multinucleation and cleared cytoplasm, correlated most significantly with HPV infection. Using these five signs, 84% of the HPV-positive cases and 92% of the HPV-negative smears were identified correctly. A mean of 4.8 nonclassic signs were found in each smear of the HPV-positive group, whereas a mean of only 2.0 signs were detected in the HPV-negative group. There was significant correlation with the HPV DNA detection when four or more nonclassic signs were present in the HPV-positive group, whereas about 90% of the cases with less than three signs were negative for HPV DNA.

In summary, the classic cytological signs koilocytosis and dyskeratocytosis are not very sensitive for the detection of cervical condyloma. A panel of nonclassic signs can be detected in gynecological smears in women with subclinical HPV-infection of the lower genital tract. None of these nonclassic signs is pathognomonic by itself, and several have to be present in one smear to render a definite diagnosis.

Conventional Cytology and Colposcopy in Condyloma [15]

Two hundred and fifty women shown to be positive for HPV by filter in situ hybridization during the previous 3 months were reexamined colposcopically and cytologically. Sixty-four women had previously had a hysterectomy. None of these patients had an abnormal cytological smear in their history, and none of the gynecological smears showed any signs of intraepithelial or invasive neoplasia. The cytological diagnosis of condyloma was made in the presence of the classic signs. An atypical transformation zone was identified by colposcopy, when white focal lesions (acetowhite epithelium or spikes), punctuation, and/or mosaicism were found [16]. Punch biopsies were taken in 25 patients. For statistical analysis the confidence intervals were calculated according to Clopper and Pearson [17].

The complete transformation zone could be evaluated in 150 patients (Table 1). Colposcopic findings were atypical in 104 patients, whereas cytology revealed HPV infection in 23 cases only. This difference between colposcopic and cytological findings was statistically significant. In patients in whom colposcopy was incorrectly performed ($n = 36$), five patients had positive cytological findings, whereas 31 patients had negative findings. Flat acetowhite epithelium was the most frequent colposcopic finding (29%), punctuation was seen in 18%, acetowhite spikes in 17%, and mosaicism in 9% (Figs. 2, 3).

Of patients who had had a hysterectomy, evaluation of the vagina showed colposcopic abnormalities in 38, whereas cytological examination showed signs of

Table 1. Correlation between cervical cytological ("classic" signs of condyloma) and colposcopic findings in 150 HPV-positive women (no atypical cells present in the cytological smears)

Colposcopy	Cytology					
	Positive		Negative		Total	
	n	(%)	*n*	(%)	*n*	(%)
Positive	15	(10)	89	(60)	104	(70)
Negative	8	(5)	38	(25)	46	(30)
Total	23	(15)	127	(85)	150	(100)

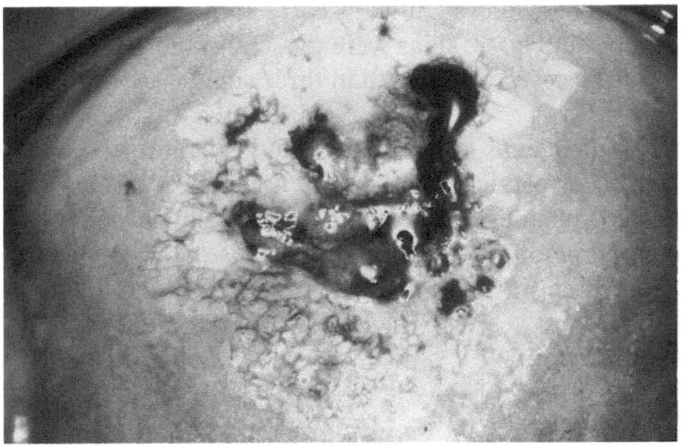

Fig. 2. This circular atypical transformation zone shows acetowhite epithelium with mosaicism. Cytological findings were negative and histological examination showed elongated rete pegs, lack of glycogenation, and acanthosis. Filter in situ hybridization was positive for HPV 6/11 and 16/18 (×6)

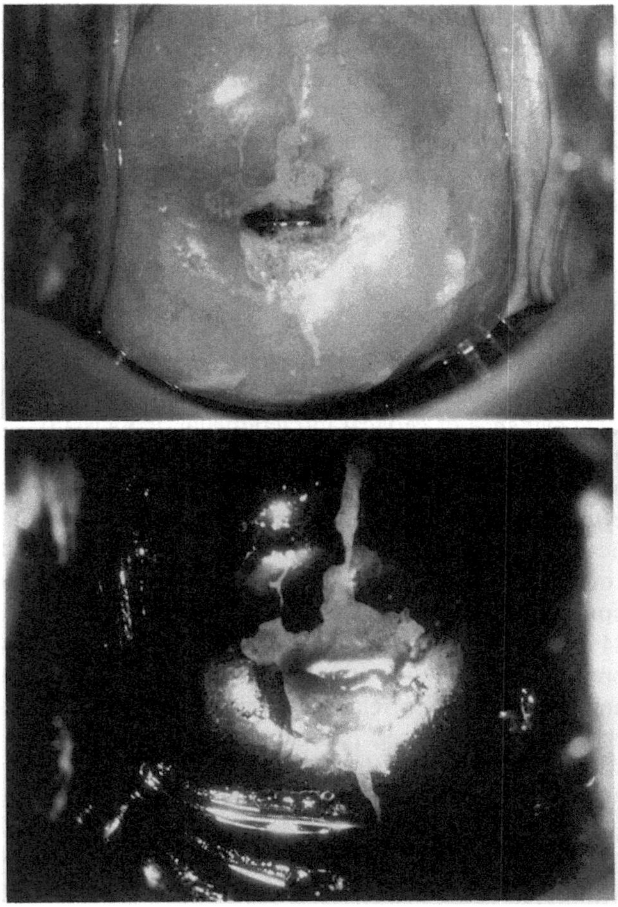

a

b

Fig. 3. **a** Within this atypical transformation zone, tongues of acetowhite epithelium extend to the outer area of the ectocervix. Mosaicism and punctuation are present. Cytological findings were negative, and histological examination showed elongated rete pegs and acanthosis. Filter in situ hybdridization was positive for HPV 16/18 (×6). **b** Application of Lugol's solution reveals partial iodine uptake. The lesions can be identified more clearly. Same patient as in **a** (×6)

HPV infection in only six. This difference between the two methods was statistically significant. Most commonly, the colposcopic feature acetowhite spikes was found (52%), whereas acetowhite epithelium and punctuation were observed rather rarely (5% and 3%, respectively).

Histologically, 15 cervical and 10 vaginal biopsies showed only minor changes such as lack of glycogenation, elongation of rete pegs, and acanthosis. Classic signs of condyloma were observed in cases with positive cytology only.

Thus, colposcopic signs of an HPV infection were present in 66% of cases, whereas cytological examination revealed an HPV infection in only 14%. The colposcopic signs are those summarized under the well-established term "abnormal transformation zone." In up to 80% of biopsies from abnormal transformation zones histological examination did not reveal any significant changes [18]. Similarly, we found signs of classic condyloma only in about 20% of cases, and in the majority of sections only minor changes were observed. It appears from our data that the abnormal transformation zone correlates in most cases with an HPV infection and may represent a precursor lesion of the so-called flat condyloma. These lesions are more easily recognized by colposcopy than by conventional cytological or histological examination.

Summary

The DNA filter in situ hybridization technique was used to document the presence of human papillomavirus (HPV) DNA in cervical swabs where there was no cytologic evidence of cervical intraepithelial neoplasia (CIN). Empirically established cytological signs of HPV infection were critically evaluated and compared with a panel of nine "nonclassic" signs in a comparative retrospective study. "Classic" cytological signs such as koilocytosis and dyskeratocytosis enabled correct identification of only 15% of the HPV-DNA-positive cases. Of the nonclassic signs, mild koilocytosis, mild dyskeratocytosis, hyperchromatic nuclei, bi- and multinucleation, and clear cytoplasm were the most valuable and enabled identification of 84% of the HPV-positive smears. Four or more of the nonclassic signs had to be present in the HPV-positive smears and more than seven signs absent in the HPV-negative cases to correlate significantly with HPV positivity. None of these nonclassic signs is by itself pathognomonic.

In another study classic cytologic signs were compared with colposcopy. In 150 HPV-DNA-positive patients the entire transformation zone could be visualized and HPV infection was diagnosed colposcopically in 70% of these patients. Cytological examination showed evidence of HPV infection in only 15%. Acetowhite epithelium (40%), acetowhite spikes (24%), punctuation (19%), mosaicism (12%), and papillary structures (5%) were the colposcopic features found. Vaginal colposcopy revealed HPV infection in 59% of 64 HPV-DNA-positive patients who had had a hysterectomy. Vaginal cytological examination showed signs of HPV infection in only 9%. Acetowhite spikes (87%), acetowhite epithelium (8%), and punctuation (5%) were the colposcopic signs identified.

We conclude that nonclassic cytological signs and diligent colposcopic examination are obligatory prerequisites for the detection of non-CIN-related HPV infection of the lower genital tract.

References

1. Stafl A (1976) Colposcopy. Cancer 38 (Suppl): 432-436
2. Talebian F, Shayan A, Krumholz BA, Palladino VS, Mann LI (1976) Colposcopic evaluation of patients with abnormal cervical cytology. Obstet Gynecol 49: 670-674
3. Mitchell H, Drake M, Medley G (1986) Prospective evaluation of risk of cervical cancer after cytological evidence of human papillomavirus infection. Lancet 1: 573-575
4. Mittal KR, Miller HK, Lowell DM (1987) Koilocytosis preceding squamous cell carcinoma in situ of uterine cervix. Am J Clin Pathol 87: 243-245

5. Nash JD, Burke TW, Hoskins WJ (1987) Biologic course of cervical human papillomavirus infection. Obstet Gynecol 69: 160-162
6. Wagner D, Ikenberg H, Boehm H, Gissmann L (1984) Identification of human papillomavirus in cervical swabs by deoxyribonucleic acid in situ hybridization. Obstet Gynecol 64: 767-772
7. Wickenden C, Steele A, Malcolm AD, Coleman DV (1985) Screening for wart virus infection in normal and abnormal cervices by DNA hybridization of cervical scrapes. Lancet 1: 65-67
8. McCance DJ, Campion MJ, Clarkson PK, Chesters PM, Jenkins D, Singer A (1985) Prevalence of human papillomavirus type 16 DNA sequences in cervical intraepithelial neoplasia and invasive carcinoma of the cervix. Br J Obstet Gynaecol 92: 1101-1105
9. Burk RD, Kadish AS, Calderin S, Romney SL (1986) Human papillomavirus infection of the cervix detection by cervicovaginal lavage and molecular hybridization: correlation with biopsy results and Papanicolaou smear. Am J Obstet Gynecol 154: 982-989
10. Cox MF, Meanwell CA, Maitland NJ, Blackledge G, Scully C, Jordan JA (1986) Human papillomavirus type-16 homologous DNA in normal human ectocervix. Lancet 2: 157-158
11. De Villiers E-M, Wagner D, Schneider A, Wesch H, Miklaw H, Wahrendorf J, Papendick U, zur Hausen H (1987) Human papillomavirus infections in women with and without abnormal cervical cytology. Lancet 2: 703-706
12. Schneider A, Kraus H, Schuhmann R, Gissmann L (1985) Papillomavirus infection of the lower genital tract: detection of viral DNA in gynecological swabs. Int J Cancer 35: 443-448
13. Meisels A, Morin C (1981) Human papillomavirus and cancer of the uterine cervix. Gynecol Oncol 12: 111-123
14. Schneider A, Meinhardt G, De Villiers E-M, Gissmann L (1987) Sensitivity of the cytologic diagnosis of cervical condyloma in comparison with HPV-DNA hybridization studies. Diagn Cytopathol 3: 250-255
15. Schneider A, Sterzik K, Buck G, De Villiers E-M (1988) Colposcopy is superior to cytology for the detection of early genital human papillomavirus infection. Obstet Gynecol 71: 236-241
16. Kolstad P, Stafl A (1982) Atlas of colposcopy. Universitetsforlaget, Oslo
17. Documenta Geigy (1975) Wissenschaftliche Tabellen. Thieme, Stuttgart, pp 185-188
18. Wespi HJ (1986) Colposcopic-histologic correlations in the benign acanthotic nonglycogenated squamous epithelium. Colp Gynecol Laser Surg 2: 147-158

Alteration in Cell Morphology of a Human Neuroblastoma Line Under the Influence of Persistent Viral Infection

M. Halbach

Introduction

In recent years studies of changes in functional properties of cultured cells of different origin under the influence of persistent viral infection have been attracting growing interest, especially in the field of neurobiology. Several investigators have managed to establish persistent infection with a variety of different kinds of viruses of cells of different origin and character. Focussing, for example, on specialized cellular functions playing an important role in central nervous intercellular communication, an impairment of a variety of highly specialized cell functions has been reported. Thus synthesis of neuron specific enzymes like cholineacetyltransferase (CAT) or acetylcholineesterase (AChE) has been found to be impaired in persistently cytomegalovirus-infected murine neuroblastoma cells (Oldstone et al. 1977); membrane receptors for neurotransmitters and receptor-dependent membrane functions have been found to be altered under the influence of viruses from the paramyxo (Halbach and Koschel 1979) and rhabdo groups (Koschel and Halbach 1979; Koschel and Münzel 1984), and influence on the functions of ion channel mechanisms (modifying bioelectric membrane potentials) has been attributed to persistent infection with mumps virus in murine pheochromocytoma cells (Ziegler and Stauffer 1987). Although in all cases severe alteration of cellular functional properties could be observed in basal cell functions as reflected by cell morphology, growth rate in culture remained unaffected.

Material, Methods, and Results

In order to study the virus-induced alteration of membrane receptor function, formerly observed in a murine hybrid cell line with neuronal properties (Koschel et al. 1986), we tried to establish a persistent infection with the rabies virus hep flury strain in a different cell line with similar neuronal features, the human neuroblastoma cell line IMR_{32} (Tumilowicz et al. 1970). This cell line - similar to other neuronal cell populations from different sources - has, besides other features, membrane receptors for prostaglandins, catecholamines, acetylcholine, and opiates which are functionally coupled to the adenylatecyclase system. Under standard culture conditions these cells have a typical morphology characteristic for neuronal cells (Fig. 1), forming in early culture phases rather spindle-like, then - with

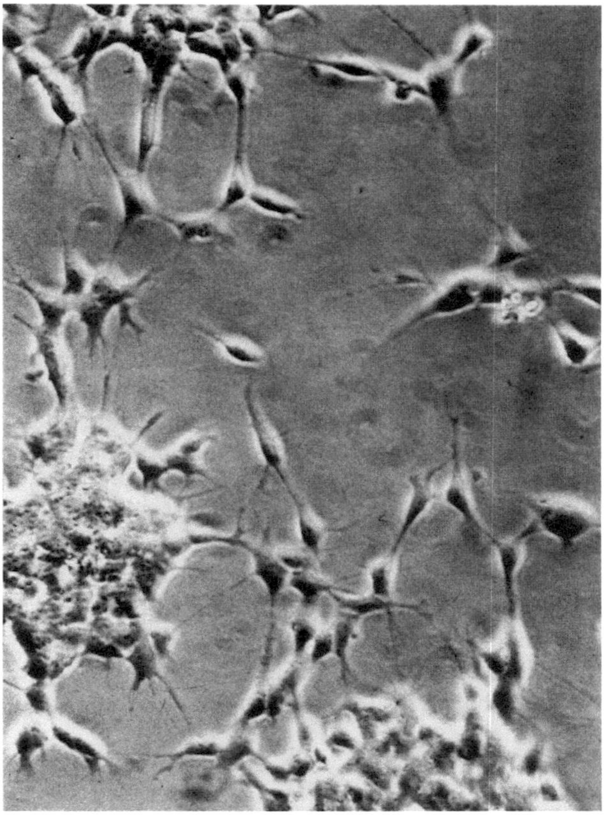

Fig. 1. Native IMR_{32} human neuroblastoma cells. Phase contrast optics, magnification ×216

growing density of the cell layer - rounded cell bodies and small, later enlarging cell processes which somewhat resemble neurites.

Rabies virus, added to the culture medium in appropriate amounts of infectious particles, easily enters neuronal cells and following extensive reproduction normally destroys the host cells by inducing typical morphological alterations well known to virologists as a cytopathogenic effect (CPE). Under suitable experimental conditions, however, which cannot be described here in detail, the rabies virus infection of the cell culture can be regulated in a way that at least a few cells, although carrying virus, survive the general CPE. These cells can be kept in culture and may be raised to a viable cell population, proliferating with a normal growth rate but constantly bearing infectious viral particles, thus representing a persistently infected cell culture.

When IMR_{32} cells were incubated with rabies virus hep flury strain under conditions normally used to establish a persistent infection, at first the expected situation could be observed: approximately 48 h after virus inoculation, more than 98% of all culture cells showed severe morphological signs of cell damage, with final cell destruction and complete structural disintegration several hours later. At this time all cells which could still be harvested contained viral antigen, as detected by immunofluorescence. When these surviving cells were transfered to new culture flasks and fed with fresh medium, they attached to the bottom and after a few hours showed a quite normal-looking shape, absolutely comparable to the cell morphology just before viral infection. Under optimum culture conditions, regularly fed with fresh medium, the infected cultures remained stable for more than 2 months but no significant propagation of cells could be observed.

When we had almost decided to finish the experiment and to discard the infected cultures we made an unexpected observation (Fig. 2): within a few days the sparse layer of single cells on the bottom of the flasks had almost completely disappeared and had been replaced by a dense, confluent monolayer of large, fibroblast-like, regularly orientated cells, looking completely different to the IMR_{32} cells infected more than 8 weeks ago. The formation of this cell type took place nearly simultaneously in several different culture flasks. In all cells harvested, viral antigen could be found by immunofluorescence; infectious virus particles could not be detected in the culture medium. There was almost no cell growth in cultures beyond confluency; attempts to further propagate this cell type by passaging failed because of a very poor plating efficiency.

We were able to follow up this state of rabies virus-infected neuroblastoma cell culture for a period of almost 3 months; cells were regularly fed with fresh medium, remained in good condition and showed no noticeable changes in morphology. Quite suddenly, without any recognizable cause, within a few days another dramatic change in cell morphology occurred. Starting simultaneously at several

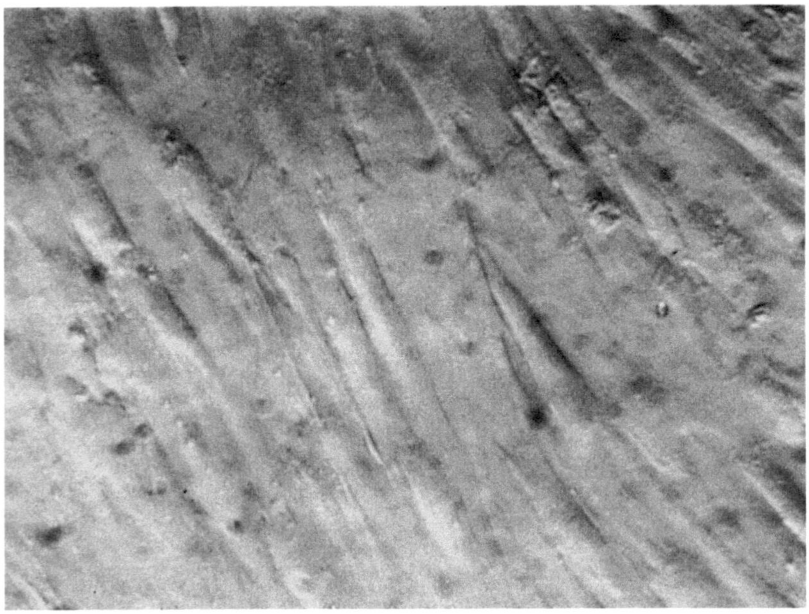

Fig. 2. IMR_{32} cells infected with rabies hep flury virus, 10 weeks after virus inoculation. Phase contrast optics, magnification ×216

separate regions of the same monolayer, and with only a few days variation in different flasks, a morphologically completely distinct cell population emerged out of the stationary cell layer and rapidly grew up (Fig. 3). After approximately 48 h a new population of cells, looking just like the IMR_{32} neuroblastoma cells before infection, had overgrown the fibroblast-shaped cells. This newly emerged population could easily be propagated by passaging the cells to new flasks; in almost all culture cells, which showed approximately the same growth rate as uninfected controls, virus antigen could be found. No infectious virus was detected in the culture medium.

Since then we have maintained this rabies virus-infected IMR_{32} cell population in culture for a period of more than 2 years, performing a series of experiments. During this time the proportion of infected cells in culture varied in a characteristic way well known for other persistently rabies virus-infected neuroblastoma cell lines. Basal growth parameters remained unaltered. We never again observed a change of cell morphology like the one described above. Attempts to reproduce this special morphological alteration of cells by fresh virus inoculation to uninfect-

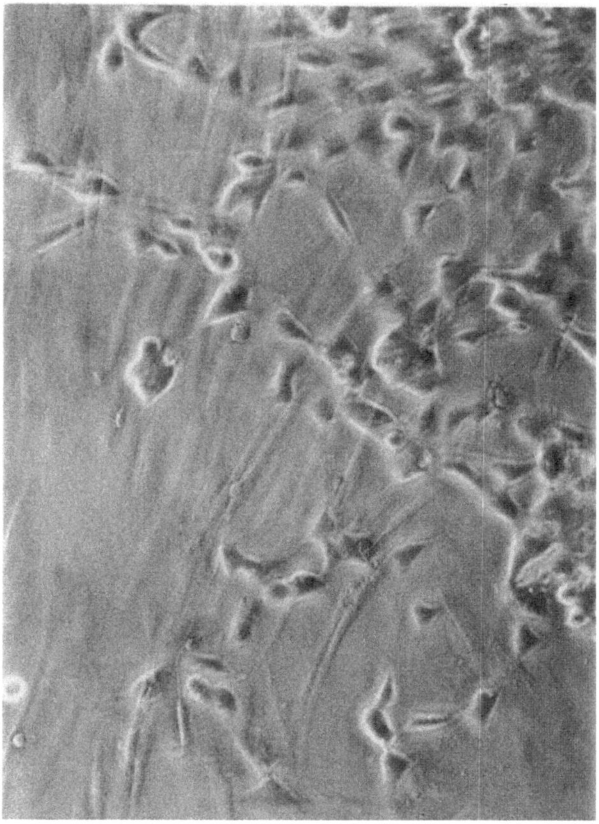

Fig. 3. IMR_{32} cells infected with rabies hep flury virus, 22 weeks after virus inoculation. Phase contrast optics, magnification ×216

ed cultures failed, as did experiments in which a superinfection of cells preinfected with rabies or other viruses was studied. Owing to the experimental limitations due to the inability to propagate the cells in the culture phase of altered cell morphology, reflections on the mechanisms of this phenomenon, a complete change in cell morphology induced by a virus, can only be hypothetical.

Discussion

Although it is well known that viruses from the rhabdo virus group in infected cell populations can induce the formation of syncytia, which are not able to further multiply, the transformation of all cells to syncytia of a completely uniform shape, forming a regularly orientated, tissue-like monolayer, is highly improbable. After persistent infections with rabies virus are established in murine neuroblastoma cells, only single syncytia of highly variable shape and size can be observed. Neither could this assumption explain the complete reversion of the cell morphology after a stationary period of 3 months. However, this possibility cannot completely be ruled out, as we were unable to harvest enough cells to carry out suitable experiments.

The possibility that the cultures might have been contaminated with cells of different origin can be excluded: all biochemical tests of specialized neuronal cell functions of the persistently rabies-infected cells showed the general existence of IMR_{32}-specific features, although in part affected by viral attack.

Tumilowicz and coworkers (1970) originally described two morphologically distinct forms of cells constituting the IMR_{32} population. It could therefore be possible that a morphological subtype, normally hidden by a predominant morphological variant, might become prevalent under viral influence for some period. This could be explained by a difference in the vulnerability of the two cell forms to viral attack. The reversion of general cell morphology in culture could then be caused by a virus-dependent or cell-dependent adaptive process.

A direct viral influence on cell functions constituting cellular morphology (i.e., the cytoskeleton) can also be assumed. A highly selective viral attack on different specialized cell functions has been reported for several experimental virus/cell systems (Koschel et al. 1986). Unfortunately there is no possibility of further elucidation of the mechanisms underlying the phenomenon described above; all attempts so far to reestablish this fibroblast-like culture state again with reasonable amounts of cells have failed.

Summary

In order to establish a persistent infection of a human neuronal cell line with neurotropic virus, the neuroblastoma line IMR_{32} was inoculated with the rabies virus strain hep flury. As in other, comparable murine neuroblastoma cell lines freshly infected with the same rabies virus strain, an excessive cell destruction in culture took place at first, leaving only a small number of viable cells behind. These surviving cells had been maintained in culture for 2 months when a morphologically different cell species suddenly emerged from the culture, forming a completely new cell population. Only 3 months later, a complete reversion of cell morphology to the original state was observed. In all phases of culture, intracellular viral antigen could be detected.

The mechanisms of the virus-induced alteration in cell morphology, which has never been observed in comparable neuroblastoma lines infected with the same virus, remain obscure; some possibilities with respect to the IMR_{32} line are discussed.

References

Halbach M, Koschel K (1979) Impairment of hormone dependent signal transfer by chronic SSPE virus infection. J Gen Virol 42: 615-619

Koschel K, Halbach M (1979) Rabies virus infection selectively impairs membrane receptor functions in neuronal model cells. J Gen Virol 42: 627-632

Koschel K, Münzel P (1984) Inhibition of opiate receptor-mediated signal transmission by rabies virus in persistently infected NG-108-15 mouse neuroblastoma-rat glioma hybrid cells. Proc Natl Acad Sci USA 81: 950-954

Koschel K, Münzel P, Halbach, Metzner R (1986) In: Levi-Montalcini R, Calissano P, Kandel ER, Maggi A (eds) Molecular aspects of neurobiology. Springer, Berlin Heidelberg New York, pp 158-161

Oldstone MBA, Holmstoen J, Welsh RM (1977) Alterations of acetylcholine enzymes in neuroblastoma cells persistently infected with lymphocytic choriomeningitis virus. J Cell Physiol 91: 459-472

Tumilowicz JJ, Nichols WW, Cholon JJ, Greene AE (1970) Definition of a continuous human cell line derived from neuroblastoma. Cancer Res 30: 2110-2118

Ziegler RJ, Stauffer EK (1987) Effects of persistent mumps virus infection on the excitability of PC12 and TE671 cells. Proceedings of the 7th International Congress of Virology, 9-14 Aug 1987, Edmonton, Canada. Abstract R4.23, p 73

IV. Quantitative Methods

Introduction

K. Goerttler

In the field of both cytologic and histopathologic diagnostics, standardization of complex statements is essential. Only by this means can results from different laboratories be compared. Staging and grading of tumors have proved very helpful for statements concerning prognosis and therapy.

The cell nucleus and the DNA contained in it play a crucial role in oncological diagnosis. The introduction of stoichiometrically binding dyes (Feulgen and Rossenbeck 1924) and examination under monochromatic light (Caspersson 1936) led to objective, quantitative measurements of the DNA content of the cell nucleus. Sandritter (1952) employed these methods for tumor diagnosis and was able to show that variations in the mean values determined for a diploid stem line could be used as criteria for malignancy. At the beginning of these investigations, measurements of single cells under the microscope were time-consuming. Subsequently, however, highly sophisticated methods of image analysis with automated measurement and evaluation of numerous single cells were developed which led to the possibility of reexamining under the microscope any given cell evaluated as atypical in the cytoautomat. Moreover, methods of three-dimensional morphometry were added to these image-analytical methods.

A dichotomy of quantitative methods resulted from the introduction of flow cytometry, enabling one to register the DNA in more than 1000 cells per second and set up the results as histograms which also allowed conclusions to be drawn about proliferation behaviours by determining the numbers of cells in S and G2 + M phases. This further development seemed to open new avenues of automated cytology.

But there were setbacks. Scientists learned that the information gained about DNA needed interpretation, e.g., in the differential diagnosis of high proliferative activity in regenerating systems or in the case of inflammatory processes. A critical discovery was that malignant tumors display diploid DNA stem lines with varying percentages from organ to organ. This destroyed the hope that a reliable statement on malignant and benign samples could be made by virtue of the detection of aneuploidy alone. In the case of a malignant tumor, the determination of a diploid stem line could lead to a false-negative result and would therefore discredit the method of automated tumor diagnosis, and in fact this uncertainty has considerably hindered the introduction of flow cytometry and the development of the measuring equipment belonging to it. Image analytical methods in connection with microscopical analysis were less affected.

Further development led to the employment of additional parameters in both methods. In the case of image analysis these were the simultaneous determination of the volumes of nuclei, the ratio of nucleus to cytoplasm, and the two-dimensional measurement based on this relationship. Besides nucleus size, flow cytometry encompassed the cell volume and the nucleus-related proliferation parameter Ki-67, for when analyzing solid tumors only, the nucleus remains as a source of information, the cytoplasm is being removed and more or less destroyed during preparation. This does not, however, hold true for measurements of intact cell suspensions, e.g., blood, effusions, urine, or gynecological smears. During measurements of the latter, we used two-parameter flow cytometry (measurement of cell nucleus DNA and cell protein) and were thus able to reduce the number of false-positive results from 30% to 15% and the number of false-negative results from 15% to 8%. It can be expected that a third parameter will lead to even lower figures. Flow cytometry and image analysis are therefore both able to achieve or reduce the margin of error presently achieved by experts in diagnosis by biopsy.

In conventional histopathologic and cytologic diagnosis the proliferative activity of a sample is not assessed sufficiently. Counting mitoses is not an adequate method. The in vivo employment of radioactive precursors (^{3}Hthymidine) is not acceptable for ethical reasons. Since the use of bromodeoxyuridine (BrdU) as proliferation marker has yet not been proved to be devoid of risk, its employment in conventional diagnosis does not seem possible. Image analysis and flow cytometry, however, enable a more precise determination of the kinetics of proliferation. Treatment studies with case control have shown that the feature "proliferation" in addition to the ploidy index (DNA index) ranks higher than conventional cytologic and histologic diagnosis.

Automated methods have therefore developed far beyond the standard reached by cytologists and pathologists, as far as the analysis of cell kinetics is concerned. A new dimension of therapeutically as well as prognostically relevant diagnosis appears. Will it therefore supersede conventional cytology and histology? Let us refer to the experience gained with the introduction of aspiration cytology. Only with the help of histopathological control did this method gain the particular authority it possesses today. At the same time, the mutual control of those two methods resulted in an efficient division of labor between cytology and histology. In introducing automated cytology, we will have to ascertain over a sufficiently long period of time whether the percentage of false-negative and false-positive results is less than that in results determined by conventional cytological and histological examination. If this can be achieved, automated measuring methods will certainly be used by the experts, i.e. cytologists and pathologists, who are able to assess its value. They are more likely to recognize possible improvements and to appreciate them.

References

Caspersson T (1936) Über den chemischen Aufbau der Strukturen des Zellkerns. Skand Arch Physiol 73 [Suppl 8]: 1-151

Feulgen R, Rossenbeck H (1924) Der mikroskopisch-chemische Nachweis einer Nukleinsäure vom Typ der Thymusnukleinsäure und darauf beruhende elektive Färbung von Zellkernen in mikroskopischen Präparaten. Z Physiol Chem 135: 203-248

Sandritter W (1952) Über den Nukleinsäurestoffwechsel in verschiedenen Tumoren (Adenocarcinome, Nieren-, Schilddrüsen-, Lebercarcinome, und Sarkom). Frankf Z Path 63: 423-446

DNA Measurement: Its Value for Diagnosis and Prognosis

G. Auer, M. Kronenwett, and A. von Rosen

Introduction

The use of quantitative DNA analysis in human tumor material has increased markedly during the last years. Considering that benign tumors in general exhibit an euploid DNA distribution pattern, while malignant tumors are predominantly aneuploid, DNA measurements have been utilized as a complement to cytological or histopathological diagnosis in cases where additional information has been needed in order to distinguish between benign and malignant lesions. The main reason for the growing interest in DNA analysis is, however, the great number of publications reporting a relationship between malignancy potential and DNA ploidy in various tumors, such as mammary, prostatic, ovarian, endometrial, and thyroid carcinomas, where morphological malignancy grading is inadequate. These studies strongly indicate that nuclear DNA content is of important prognostic value.

Representative tumor material and reliable techniques of cell preparation and analysis are prerequisites for the application of the diagnostic and/or prognostic information given by DNA analysis, as well as for adequate interpretation of the DNA histograms.

Value of DNA Analysis in Clinical Practice

Diagnosis

DNA analysis contributes diagnostic information in addition to that obtained by conventional methods. Experimental investigations of the carcinogenic process in bronchial and cervical mucosa in laboratory animals [10, 15] have shown that invasive carcinoma is always preceded by aneuploid transformation of morphologically indistinguishable intraepithelial cellular structures. Retrospective studies of archive material of human bronchial secretion or cervical smears [13, 14] have demonstrated aneuploid cell populations in patients who subsequently developed squamous cell carcinomas in the bronchial system or cervix uteri, whereas cells in patients with bronchitis or cervicitis were euploid. DNA analysis could therefore be an useful complementary technique, in patients with morphologically low or intermediate dysplasia, in distinguishing between those at high risk and those at low risk of developing an invasive carcinoma.

DNA analysis also gives additional information in tumors where the morphological criteria for malignancy are insufficient, aneuploidy being strongly indicative of malignancy. On the other hand, a morphologically well differentiated and euploid tumor may still be a carcinoma.

Prognosis

To judge from hitherto published results, the nuclear DNA content of tumor cells is closely related to the course of the disease. The prognostic significance of DNA content is not equal in all tumor types: DNA analysis provides no prognostic information beyond that obtained by conventional means such as histopathological diagnosis or clinical stage in, for example, bronchial squamous cell carcinoma, small-cell lung carcinoma [17], and osteosarcoma [5], all of which are usually aneuploid tumors, indicating high malignancy potential. DNA analysis is, consequently, also less valuable in predominantly euploid tumors of low malignancy potential, such as Kaposi's sarcoma or carcinoid tumors [16], while the prognostic value of DNA analysis in adenocarcinomas of the mammary gland [2], prostate [18], thyroid gland [4], ovary [6], and endometrium [12], and in chondrosarcoma [11], is significant. In patients with these tumors DNA measurements have been shown to provide additional prognostic information over and above that obtained from conventional predictors. One important observation is that the additional prognostic information given by DNA ploidy is greater in the early stages of the disease and frequently not significant in patients with advanced disease [8]. DNA profile as single prognostic predictor is more important in, for example, mammary and thyroid carcinomas than all other traditional variables in combination [3, 8].

Cytometric Instruments

Two main and basically different measuring techniques can be distinguished: (a) DNA analysis of morphologically identified cells on slide preparations (slide technique) and (b) DNA analysis of cellular suspensions where morphological identification of the individual cells at present is impossible (flow technique). In the slide method, the amount of DNA-bound dye in each nucleus is measured directly by absorption (after, e.g., Feulgen staining) or fluorometric (after, e.g., ethidium bromide or propidium iodide staining) techniques. The advantage of the slide method is that it allows morphological identification of the cells analyzed, which at present is the only way of obtaining tumor cell-specific DNA data. This technique is relatively time-consuming with respect to simple so-called interactive measuring procedures. New measuring instruments are being constructed that will drastically shorten the time needed for analysis. Thus, in the near future, the whole procedure, from preparation of tissue to presentation of DNA data, will take no longer than a simple flow-cytometric analysis.

With the flow cytometric technique, a great number of cells can be analyzed within a short time. In isolated tumor cell populations (e.g., tumor cells cultured in vitro), the DNA content can be determined with high precision, allowing the demonstration of subtle quantitative differences in cellular DNA content as well as cell kinetic analyses, e.g., estimation of fractions of cells in the S phase region. In a clinical tumor specimen with unknown proportions of normal cells and tumor cells, well preserved or damaged, the flow cytometric curves are often difficult to interpret, thus limiting the usefulness of the flow technique in routine clinical practice. The main application of flow cytometry in solid tumors is in the determination of the mean value of the DNA content of the tumor stemline(s) (DNA index), provided there is a significant deviation from the DNA value of the control cells and that the representativity and quality of the cell sample has been assessed morphologically prior to flow cytometric analysis. The fraction of cells in S phase seems to be more easily assessed in cellular suspensions of lymphomas, where the background count is usually low. However, this material also requires additional methods for assessment of the proportion of tumor cells, e.g., morphological examination or analysis by immunocytochemistry.

The results indicate that DNA measurement of slide preparations is at present the best method for evaluating the tumor malignancy potential, whereas flow cytometry should preferably be used for determination of the DNA index of the tumor stemline(s). This flow cytometric information is, above all, of diagnostic value, but it is in certain tumors of prognostic value when combined with assessment of the S phase fraction.

Tumor Material

The lack of tumor-specific markers makes morphological study of clinical tumor material necessary. In order to obtain reliable DNA histograms, tumor cells must be distinguished from nontumor cells in preparations with varying proportions of malignant cells. This distinction is at present only possible in cytological or histological material on slides or in imprints of fresh surgical biopsy specimens. Disintegration of fresh or frozen biopsy specimes for flow cytometric analysis includes mechanical procedures and enzymatic treatment which are difficult to control. Studies of the influence of these methods on the disintegration of tissues composed of cells with varying intercellular adhesion, differences in proliferative activity, and different sensitivity to enzymatic treatment are still lacking. When disintegrating paraffin-embedded material, according to the method described by Hedley [9], these problems become even more pronounced. Even so, this method could be of great importance in the future for retrospective analyses of paraffin-embedded tumor material, provided extensive methodological improvements are made.

DNA analysis of histological sections by the slide technique has the advantage of visual selection of tumor cells, which gives a high degree of tumor cell specificity. The technique is highly important, since the available tumor material in most cases consists of fresh or paraffin-embedded surgical biopsies. The drawback of DNA analysis on tissue sections is that some nuclei are cut, usually giving high coefficients of variation and wide modal peaks. Such patterns can be misinterpreted as representing aneuploid tumors. By restricting the information extracted from the DNA histograms to the percentage of cells with DNA values exceeding P 90 (i.e., values not exceeded by 90% of the control cells) or 2 × P 90, DNA analyses performed on histological sections have given important prognostic information in, among others, chondrosarcomas, ovarian tumors and mammary carcinomas [6, 7, 11].

Interpretation of Results from DNA Measurements

In the early studies of the relationship between DNA content and clinical course in breast cancer patients, a possible relation between patient prognosis and distribution of DNA values was explored without presuming that DNA content would provide prognostic information. A statistically significant correlation between the modal DNA value of the tumor cells and the clinical course was shown by Atkin [1], tumors with modal values corresponding to those of normal diploid cells being associated with a favorable prognosis, whereas increased modal values were indicative of high malignancy. A similar approach was used in studies at the Karolinska Hospital [2]. The four different types of DNA distribution patterns identified in breast cancer patients, classified as I (diploid), II (polyploid), and III and IV (aneuploid) have been found to correlate strongly to the course of the disease.

Both methods of interpretation of DNA histograms (modal value and histogram type) require specific techniques of cell preparation, fixation, staining, and analysis of DNA data. However, the same interpretations have later been applied to data obtained by totally different techniques. The problem of adequate interpretation has been further emphasized by flow cytometric data. This is due to the varying proportions of tumor cells with different sensitivity to enzymatic disintegration in the analyzed tumor samples, which produce DNA curves that are often not representative of the tumor cell population. In addition, the high resolution in the flow technique with very low coefficients of variation (CV) detects tumor stemlines with minimal deviation from the DNA index of the diploid control cells and defines them as aneuploid. Such stemlines are likely to be identified as diploid by the slide technique, where the CV is considerably higher. This aneuploid DNA distribution, as judged by flow cytometric analysis, will then be interpreted as indicative of high malignancy potential, which is an assumption that has not been preceded by careful analysis of controlled patient populations.

Conclusion

DNA analysis has been shown to provide diagnostic and prognostic information in addition to that obtained by conventional clinical and morphological methods in a great number of human tumors.

The additional information given by DNA measurement is at present evaluated at certain hospitals in order to explore the contribution of DNA to diagnostic or therapeutic decisions. There is so far no general agreement on the value of DNA analysis, since the slide and the flow techniques often give conflicting results.

Comprehensive analyses of carefully controlled patient material are still required for optimal utilization of DNA data, in addition to further methodological studies of preparation of tumor material, identification of tumor cells, measuring techniques, and interpretation of results.

References

1. Atkin NB (1972) Modal deoxyribonucleic acid value and survival in carcinoma of the breast. Br Med J 86: 271-272
2. Auer G, Caspersson T, Wallgren A (1980) DNA content and survival in mammary carcinoma. Anal Quant Cytol 3: 161-165
3. Bäckdahl M (1985) Nuclear DNA content and prognosis in papillary, follicular and medullary carcinomas of the thyroid. Thesis, Karolinska Hospital, Stockholm
4. Bäckdahl M, Auer G, Forsslund G et al (1986) Prognostic value of nuclear DNA content in follicular thyroid tumours. Acta Chir Scand 152: 1-7
5. Bauer H (1987) DNA cytometry of osteosarcoma. Thesis, Karolinska Hospital, Stockholm
6. Erhardt K, Auer G, Björkholm E, Forsslund G, Moberger B, Silfverswärd C, Wicksell G, Zetterberg A (1984) Prognostic significance of nuclear DNA content in serous ovarian tumors. Cancer Res 44: 2198-2202
7. Erhardt K, Auer G, Folin A (1987) Mammary adenocarcinoma: nuclear DNA content in non-invasive and invasive parts of the same tumor. Anal Quant Cytol Histol 9: 263-267
8. Fallenius A (1986) DNA content and prognosis in breast cancer. Thesis, Karolinska Hospital, Stockholm
9. Hedley D, Friedlander M, Taylor I, Rugg C, Musgrove E (1983) Method for analysis of cellular DNA content of paraffin-embedded pathological material using flow cytometry. J Histochem Cytochem 31: 1333-1339
10. Konaka C, Auer G, Nasiell M, Hayata Y, Kato H, Caspersson T, Ono J (1982) Sequential cytomorphological and cytochemical changes during development of bronchial carcinoma in beagle dogs exposed to 20-methylcholanthrene. Acta Histochem Cytochem 15: 779-797
11. Kreicbergs A, Zetterberg A (1980) Cytophotometric DNA measurements of chondrosarcoma: methodological aspects of measurements in tissue sections from old paraffin embedded specimens. Anal Quant Cytol 2: 84-92
12. Moberger B, Auer G, Forsslund G, Moberger G (1984) The prognostic significance of DNA measurements in endometrial carcinoma. Cytometry 5: 430-436
13. Nasiell M, Kato H, Auer G, Zetterberg A, Roger V, Karlén L (1978) Cytomorphological grading and Feulgen DNA-analysis of metaplastic and neoplastic bronchial cells. Cancer 41: 1511-1521
14. Nasiell K, Näslund I, Auer G (1984) Cytomorphologic and cytochemical analysis in the differential diagnosis of cervical epithelial lesions. Anal Quant Cytol 3: 196-200
15. Näslund I, Rubio C, Auer G (1987) Nuclear DNA changes during pathogenesis of squamous cervical carcinoma in 3,4-benzopyrene-treated mice. Anal Quant Cytol Histol 9: 411-418
16. Nobin AP, Erhardt K, Auer G, Falkmer S, Mårtensson H (1987) Nuclear DNA patterns and survival in metastasizing ileal carcinoid. World J Surg 11: 372-377
17. Yoneyama K, Erhardt K, Stenkvist B, Wiman G, Kato H, Auer G (1987) Nuclear DNA content in squamous carcinoma, oat cell carcinoma and alveolar carcinoma of the lung. Am J Clin Oncol 10 (4): 330-334
18. Zetterberg A, Esposti PL (1980) Prognostic significance of nuclear DNA levels in prostatic carcinoma. Scand J Urol Nephrol [Suppl] 55: 53-58

DNA Flow Cytometry in Diagnostic Cytopathology: Methodologic Considerations and Current State of Clinical Applications

G. E. Feichter

Introduction

The recent progress of flow cytometry was made possible by technical improvement of the flow cytometers, by developments in computer techniques, with better possibilities of data storage and evaluation, and by increased experience with the methods of staining and preparation.

The main motives for the first developments of flow cytometric techniques were the expectations of clinical applicability of the method, especially for the detection of tumor cells. Although the theoretical basis existed, these attempts were not successful because of the limitations of the method at that time.

The rapid development of the applications of flow cytometry started as soon as it became evident that DNA ploidy is one of the most effective indicators of malignancy in human tumors. Accordingly, the clinical applications of flow cytometry are now focused entirely on its implications for tumor prognosis, especially in the predictability of recurrence-free and overall survival times.

The number of publications on the prognostic relevance of either the DNA index or S-phase fractions has increased continuously during recent years. It became a real problem to keep pace with the literature on flow cytometry. Some extensive review articles have contributed significantly to define the actual position of this technique in tumor analysis (Coon et al. 1987; Merkel et al. 1987)..

However, successful applications of DNA flow cytometry are limited by (a) admixtures of different cell populations, especially lymphocytes, in samples from complex tumors, and (b) the presence of unspecific background fluorescence resulting from cell debris. These problems have not been solved satisfactorily so far, giving rise to criticism to the method as a whole. Some clinical studies have been started *before* solutions to some of the most important methodological problems of flow cytometry have been found.

Admixtures of Different Cell Populations

Malignant tumors contain variable amounts of lymphocytes. Their pulses will appear together with the pulses resulting from diploid tumor cells in the same peak. Therefore the total number of tumor cells cannot be determined exactly in diploid tumors. But this is essential for the calculation of the cell cycle phases. An apparent increase in the number of cells, caused by admixtures of nonneoplastic cells, will result in correspondingly lower S-phase fractions.

In aneuploid tumors, the pulses of lymphocytes and tumor cells in the G0/1 phase will appear in different peaks. In some particular cases, such as lymph node metastases with relatively low amounts of tumor cells, the aneuploid peaks may be too small for cell cycle analysis. In cases with extremely high numbers of lymphocytes, an apparently purely "diploid" histogram can be obtained, which can lead to serious misinterpretations.

The problem of cellular admixtures in flow cytometry can be handled basically by two approaches, consisting in (a) the isolation of epithelial cells, using monoclonal antibodies directed against epithelial cell markers (Ramaekers et al. 1986), or (b) the separation of lymphocytes by centrifugation on a lymphocyte separation medium (LSM), which is used in our laboratory. In our experience, LSM centrifugation, compared with flow cytometry without lymphocyte separation, shows:

1. Significantly higher S-phase fractions in diploid tumors
2. A significant increase in the amount of aneuploid cells in the sample
3. No change in the S-phase fractions in suspensions from aneuploid tumors after separation of lymphocytes
4. The appearance of aneuploid peaks in some specimens which were initially classified as diploid.

Conclusion: the separation of lymphocytes from cellular suspensions is an indispensable precondition for correct flow cytometric analyses, especially for the calculation of the S-phase fraction in diploid tumors.

Procedure for Background Correction

DNA distributions invariably contain pulses resulting from cellular debris, producing unspecific background fluorescence. The background pulses overlay different parts of the histogram. Calculations of the cell cycle phases will include these unspecific pulses, and the reflected percentages of the cell cycle phases, including S-phase, will be considerably higher than the real percentages in the tumor.

Although models for appropriate correction procedures have been described in the past (Haag et al. 1987), they have not been applied to the evaluation of histograms for a long time. The S-phase fractions published in such studies, mainly in the 1970s and early 1980s, were therefore much higher than the corresponding tritium-labeling indices obtained by autoradiography. Simultaneous determination of LI and of S-phase fractions with and without correction for background fluorescence clearly demonstrated that calculation of the S-phase fraction gives reliable results only when a procedure for background correction is applied (Czech 1987).

Conclusion: correction procedures for unspecific background fluorescence are an important prerequisite for correct calculation of S-phase fraction in flow cytometry in both diploid and aneuploid tumors.

Clinical Applications

The correlation of flow cytometry results to well-established prognostic factors and to clinical course will be presented below separately for different tumor types, on the basis of data from the literature and our own results.

Breast Cancer

Breast cancer is probably the tumor which has been studied most by flow cytometry. Both DNA index and S-phase fraction are significantly associated to the histologic grade and to the estrogen receptor content (Horsfall et al. 1986; Owainati et al. 1987; Feichter et al. 1988).

Patients with diploid carcinomas have a significantly better prognosis, i.e., longer recurrence-free and overall survival times, than those with aneuploid ones (Cornelisse et al. 1987; Hedley et al. 1987; Kallioniemi et al. 1987; Kaufmann et al. 1988).

In addition to the reports on the prognostic significance of the thymidine-labeling index in breast cancer (McGuire et al. 1985), we found a significant correlation between the length of recurrence-free and overall survival times using fresh material (Kaufmann et al. 1988) as well as paraffin-embedded carcinomas (Leidinger 1987; Fig. 1).

Conclusion: flow cytometrically determined DNA ploidy (DNA index) and S-phase fraction correlate significantly to some of the traditional prognostic factors and to the clinical outcome in breast cancer. In the special case of breast cancer, flow cytometry is ready for clinical application, provided that the method is practiced carefully, as mentioned above.

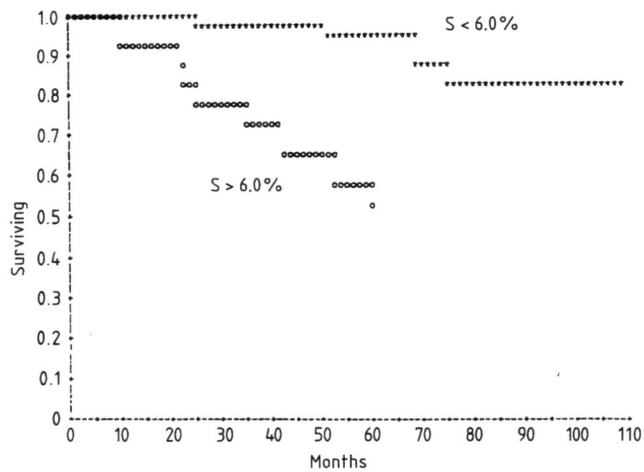

Fig. 1. Survival of 82 patients with breast cancer with different amounts of cells in the S phase (Kaplan-Meier). $p=0.0014$. *, less than 6.0% of cells in S phase (1.0%–6.0%); O, more than 6.0% of cells in S phase (6.1%–14.6%)

Large-Bowel Carcinoma

The relative DNA content of colorectal carcinomas has been the subject of many flow cytometric examinations (see Giaretti et al. 1988; Scott et al. 1987). Tumor heterogeneity is a particular problem in this tumor (Quirke et al. 1986; Hiddemann et al. 1986).

An adverse effect of an increase in the S-phase fraction on prognosis was recently reported by Bauer et al. (1987), who stated that S-phase fraction was a more powerful prognostic factor than the DNA index. Patients with diploid carcinomas had a better clinical course than those with aneuploid ones in the studies of Emdin et al. (1987) and Scott et al. (1987).

An important clinical application of flow cytometry could be the examination of colorectal adenomas (Giaretti et al. 1988), with the possibility of identifying adenomas at high risk of transforming into invasive carcinomas.

Conclusion: whereas the DNA index seems to be of prognostic value in colorectal carcinomas, more data on the involvement of S-phase fraction in the prognosis of this tumor are still needed. Clinical application of flow cytometry to colorectal carcinomas is rendered more difficult because of the heterogeneity, especially in S-phase fractions, of this tumor.

Ovarian Carcinoma

Early publications on the prognostic value of the DNA index (see Friedlander et al. 1984) were later confirmed by Volm et al. (1985 a), Rodenburg et al. (1987), and Rutgers et al. (1987). In addition to this, Rutgers et al. (1987) and Volm et al. (1985 a) reported possible connections between cell cycle data and clinical outcome in ovarian carcinoma. The data were confirmed by Iversen and Skaarland (1987).

In our own studies we have found a significant correlation of both DNA index and S-phase fraction to histologic grade (Feichter et al. 1985) and to the length of recurrence-free and overall survival times (Kühn et al. 1987; Figs. 2, 3).

Conclusion: the DNA index is one of the strongest prognostic factors in ovarian carcinoma. Numerous clinical data justify its clinical application. S-phase fractions have at least the same predictive value as the DNA index in our material; however, more clinical studies are needed before information about S-phase fractions can be put to clinical use.

Endometrial Carcinoma

Only a few flow cytometric studies have been carried out on endometrial carcinoma. We found in our material a significant correlation between the histologic grade and the flow cytometric data (DNA index and S-phase fraction; Feichter et al. 1982). Similar results have been reported by Geisinger et al. (1986) and Lindahl et al. (1987). Oud et al. (1986) could not confirm these results. Results of clinical studies on the correlations between either DNA index or S-phase fraction and follow-up of patients with endometrial cancer are still not available.

Conclusion: The correlation between flow cytometric data and histologic grade could be regarded as a hint towards clinical applicability, but until clinical data are not available, this is not a subject for immediate further discussion.

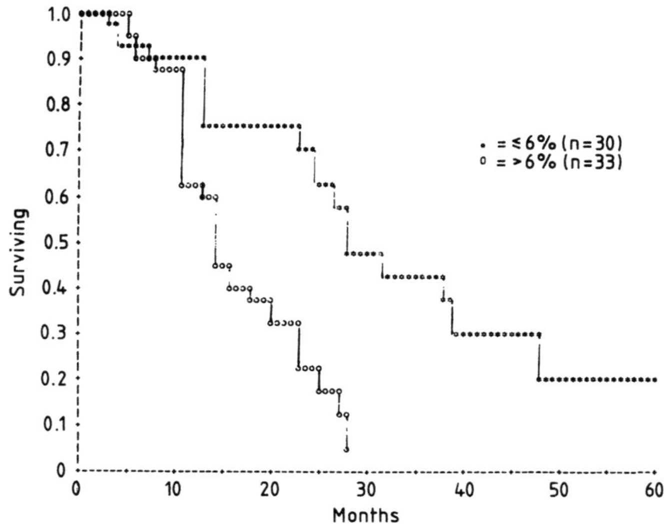

Fig. 2. Survival of 63 patients with ovarian carcinomas FIGO stages III/IV as a function of the DNA index (Kaplan-Meier). $p = 0.016$

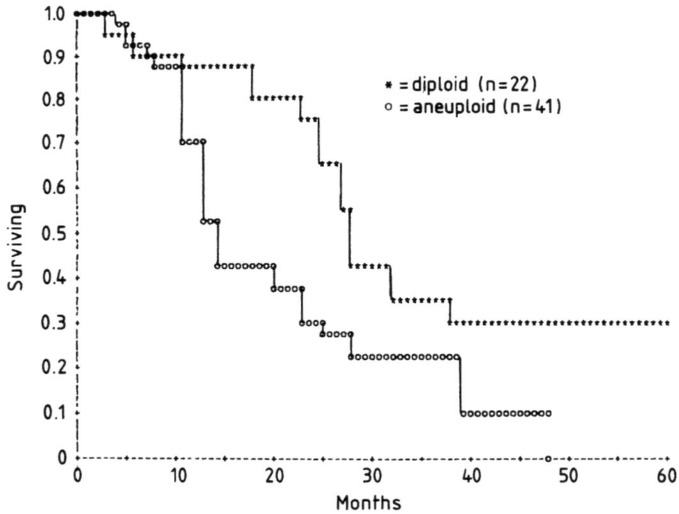

Fig. 3. Survival of 63 patients with ovarian carcinomas FIGO stages III/IV as a function of the percentage of cells in the S phase (Kaplan-Meier). $p = 0.001$

Oropharyngeal Squamous Epithelium Carcinoma

Johnson et al. (1985) found that patients who have both aneuploid distributions and high S-phase fractions have a five-fold worse prognosis than those who have neither. In the study of Tytor et al. (1987) patients with diploid tumors had a better prognosis, but the differences between the survival intervals were not significant.

In our material (Feichter et al. 1987) the frequency of aneuploid distributions and the S-phase fractions correlated significantly with the histologic grade. Furthermore, we have found correlations between S-phase fractions and response to chemotherapy in this tumor. Due to the numerous subgroups, statistical analysis of the overall survival time could not be carried out. To overcome this problem, which is encountered regularly in clinical practice, we have developed a prognostic index with an extremely high predictive value (Feichter et al. 1987).

Conclusion: there are clear correlations between DNA index, S-phase fraction and other prognostic factors of this tumor, but due to a lack of follow-up data the method cannot yet be given clinical application.

Other Solid Tumors

Extensive work on the flow cytometry of bronchial carcinoma was done by Volm et al. (1985b), demonstrating a significantly better prognosis for patients whose "non-small cell carcinomas" were either diploid or had low S-phase fractions. Tirindelli-Danesi et al. (1987) found a significant correlation between multiclonality and clinical outcome, but Bunn et al. (1983) consider one-parameter flow cytometry to be impractical for use in automated cytology of this tumor. So far, only the data published by Volm et al. (1985b) hints towards clinical applicability of flow cytometry in bronchial carcinoma, and more results from clinical studies are needed before such application can be considered.

Similar correlations between DNA index and clinicopathologic data have been reported by Frankfurt et al. (1985) and Lundberg et al. (1987). An up-to-date opinion was given by Murphy (1987). In the case of carcinomas of the urinary bladder and the prostate, as well as for renal cell carcinoma, it has basically been proved that DNA ploidy and clinical outcome are significantly associated, but it is still unclear how the data furnished by flow cytometry could be applied practically to the individual case (Murphy 1987; Tribukait 1987).

In malignant non-Hodgkin lymphomas, S-phase fractions but not DNA indices proved to be of major prognostic value when compared to other prognostic factors (Bauer et al. 1986; Lenner et al. 1987). In this tumor group as well as in soft tissue sarcomas (Kreicbergs et al. 1987; Berndt and Feichter 1987) clinical application is not yet in sight because of lacking follow-up data. The same is true for uterine cervical carcinoma with few exceptions (Strang et al. 1987).

Conclusions

The evolution of the technique of flow cytometry has made much progress, but improvements are still necessary. Separation of different cell populations, especially lymphocytes, should be practised on a larger scale than at present. The methods of cell separation need further improvement, especially when multiparameter analyses requiring an intact cytoplasm are planned. For the correct calculation of S-phase frations from DNA histograms, procedures for background correction are essential. It is probably less important which model for background correction is applied than that one is used at all.

As far as clinical applications of DNA flow cytometry are concerned, only a few tumor types have been studied extensively enough for prognostic purposes: at present these are breast cancer and ovarian carcinoma. Although numerous flow cytometric studies of solid tumors have been published, there is still a lack of sufficient clinical follow-up data. Furthermore, it is still not clear whether the data furnished by flow cytometric DNA analysis should be integrated into a complex prognostic score, as we did (Feichter et al. 1987), or if some of the less effective classic prognostic parameters should be replaced by either DNA index or S-phase fraction.

Thus, flow cytometry is on the way to becoming a most effective method for the analysis of tumor aggressiveness, with the real prospect of clinical application on large scale. However, improvements in the method and more results from clinical studies are still needed for the majority of malignant solid tumors.

Summary

Flow cytometry has made astonishing progress in both methods and clinical applications. Successful flow cytometry is only possible if some preconditions of the method are fulfilled, such as elimination of background pulses and separation of different cell populations from complex tumors. Both DNA ploidy (DNA index) and S-phase fractions, the main data furnished by flow cytometry, have been proved to be of prognostic relevance in the assessement of a number of tumors such as breast cancer, carcinoma of the ovary, bronchi, urinary bladder, prostate, oral cavity and malignant lymphomas, in which aneuploidy and a high S-phase fraction are associated with an unfavorable clinical outcome. It is to be expected that developments in the method, including multiparametric analyses, and more results from clinical studies will lead to further broadening of the clinical application of flow cytometry in the near future.

Acknowledgements. The major part of the flow cytometric examinations were the excellent work of Christine Koch.

References

Bauer KD, Merkel DE, Winter JN, Marder RJ, Hauck WW, Wallemark CB, Williams TJ, Variakojis D (1986) Prognostic implications of ploidy and proliferative activity in diffuse large cell lymphomas. Cancer Res 46: 3173-3178

Bauer KD, Lincoln ST, Vera-Roman JM, Wallemark CB, Chmiel JS, Madurski ML, Murad T, Scarpelli DG (1987) Prognostic implications of proliferative activity and DNA aneuploidy in colonic adenocarcinomas. Lab Invest 57: 329-335

Berndt R, Feichter G (1987) DNS-Durchflußzytophotometrie, histologisches Typing und Grading bei Weichteilsarkomen. Verh Dtsch Ges Path 71: 350

Bunn PA, Carney DN, Gazdar AF, Whang-Peng J, Matthews MJ (1983) Diagnostic and biological implications of flow cytometric DNA content analysis in lung cancer. Cancer Res 43. 5026-5032

Coon JS, Landay AL, Weinstein RS (1987) Advances in flow cytometry for diagnostic pathology. Lab Invest 57: 453-479

Cornelisse CJ, van de Velde CJH, Caspers RJC, Moolenaar AJ, Hermans J (1987) DNA ploidy and survival in breast cancer. Cytometry 8: 225-234
Czech W (1987) Vergleich von Labeling-Index nach ^{3}H-Thymidin-Markierung und Zellzyklusanalyse durch DNA-Impulszytophotometrie zur Quantifizierungg DNA-synthetisierender Zellen. Thesis, University of Heidelberg
Emdin StO, Stenling R, Roos G (1987) Prognostic value of DNA content in colorectal carcinoma. A flow cytometric study with some methodologic aspects. Cancer 60: 1282-1287
Feichter GE, Höffken H, Heep J, Haag D, Heberling D, Brandt H, Rummel H, Goerttler K (1982) DNA-flow-cytometric measurements on the normal, atrophic, hyperplastic and neoplastic human endometrium. Virchows Arch Pathol Anat 398: 53-65
Feichter GE, Kühn W, Czernobilsky B, Müller A, Heep J, Abel U, Haag D, Kaufmann M, Rummel HH, Kubli F, Goerttler K (1985) DNA flow cytometry of ovarian tumors with correlation to histopathology. Int J Gynecol Pathol 4: 336-345
Feichter GE, Maier H, Adler D, Born IA, Abel U, Haag D, Goerttler K (1987) S-phase fractions and DNA-ploidy of oropharyngeal squamous epithelium carcinomas. Acta Otolaryngol 104: 377-384
Feichter GE, Müller A, Kaufmann M, Haag D, Born IA, Abel U, Klinga K, Kubli F, Goerttler K (1988) Correlation of DNA flow cytometric results and other prognostic factors in breast cancer. Int J Cancer (in press)
Frankfurt OS, Chin JL, Englander LS, Greco WR, Pontes JE, Rustum YM (1985) Relationship between DNA ploidy, glandular differentiation, and tumor spread in human prostate cancer. Cancer Res 45: 1418-1423
Friedlander ML, Hedley DW, Taylor IW, Russel P, Coated AS, Tattersall (1984) Influence of cellular DNA content on survival in advanced ovarian cancer. Cancer Res 44: 397-400
Geisinger KR, Kute TE, Marshall RB, Homesley HD, Morgan TM (1986) Analysis of the relationships of the ploidy and cell cycle kinetics to differentiation and the female sex steroid receptors in adenocarcinoma of the endometrium. Am J Clin Pathol 85: 397-400
Giaretti W, Sciallero S, Bruno S, Geido E, Aste H, DFi Vinci A (1988) DNA flow cytometry of endoscopically examined colorectal adenomas and adenocarcinomas. Cytometry 9: 238-244
Haag D, Feichter GE, Goerttler K, Kaufmann M (1987) Influence of systematic errors on the evaluation of the S-phase portions from DNA distributions of solid tumors as shown for 328 breast carcinomas. Cytometry 8: 377-385
Hedley DW, Rugg CA,, Gelber RD (1987) Association of DNA index and S-phase fraction with prognosis of N+ early breast cancer. Cancer 47: 4729-4735
Horsfall DJ, Tilley WD, Orell SR, Marshall VR, Cant ELMcK (1986) Relationship between ploidy and steroid hormone receptors in primary invasive breast cancer. Br J Cancer 53: 23-28
Iversen OE, Skaarland E (1987) Ploidy assessment of benign and malignant ovarian tumors by flow cytometry. Cancer 60: 82-87
Johnson TS, Williamson KD, Cramer MM, Peters LJ (1985) Flow cytometric analysis of head and neck carcinoma DNA Index and S-fraction from paraffin-embedded sections: comparison with malignancy grading. Cytometry 6: 461-470
Kallioniemi OP, Blanco G, Alavaikko M, Hietanen T, Mattile J, Lauslahti K, Koivula T (1987) Tumor DNA ploidy as an independent prognostic factor in breast cancer. Br J Cancer 56: 637-642
Kaufmann M, Feichter GE, Nhila A, Klinga K, Abel U (1988) Flow-zytophotometrische Parameter, Hormonrezeptoren und axillärer Lymphknotenstatus als Prognosefaktoren beim Mammakarzinom. Geburtsh u Frauenheilk (in press)
Kreicbergs A, Tribukait B, Willems J, Bauer HCF (1987) DNA flow analysis of soft tissue tumors. Cancer 59: 128-133
Kühn W, Feichter GE, Hanke J, Rummel HH, Kaufmann M, Schmid H (1987) Klinischer Verlauf des Ovarialkarzinoms in Abhängigkeit von morphologischen Prognosefaktoren und zellkinetischen Parametern. Geburtsh u Frauenheilk 47: 446-451
Leidinger J (1987) Beziehungen zwischen DNS-durchflußzytophotometrisch ermittelten S-Phasen, mammographisch bestimmter Tumorverdoppelungszeit und Überlebenszeit beim Mammakarzinom. Thesis, University of Heidelberg
Lenner P, Johansson H, Lindh J, Dige U (1987) Non-Hodgkin-Lymphoma. Multivariate analysis of prognostic factors including fraction of S-phase cells. Acta Oncol 26: 179-183

Lindahl B, Alm P, Fernö M, Killander D, Langström E, Norgren A, Trope C (1987) Prognostic value of flow cytometrical DNA measurements in stage I-II endometrial carcinoma. Anticancer Res 7: 791-798

Lundberg S, Carstensen J, Rundquist I (1987) DNA flow cytometry and histopathological grading of paraffin-embedded prostate biopsy specimens in a survival study. Cancer Res 47: 1973-1977

McDivitt RW, Stone KR, Craig RB, Palmer JO, Meyer JS, Bauer WC (1986) A proposed classification of breast cancer on kinetic information derived from a comparison of risk factors in 168 primary operable breast cancers. Cancer 57: 269-276

Merkel DE, Dressler LG, McGuire WL (1987) Flow cytometry, cellular DNA content, and prognosis in human malignancy. J Clin Oncol 5: 1690-1703

Moran RE, Melamed MR (1984) Flow cytometric analysis of human lung cancer. Correlation with histologic type and stage. Anal Quant Cytol 6: 99-104

Murphy WM (1987) DNA flow cytometry in diagnostic pathology of the urinary tract. Human Pathol 18: 317-319

Oud PS, Reubsaet-Veldhuizen JAM, Beck HLM, Pahlplatz MMM, Hesselmanns GHFM, Hermkens HG, Tas J, James J, Vooijs GP (1986) DNA and nuclear protein measurement in columnar epithelial cells of human endometrium. Cytometry 7: 325-330

Owainati AAR, Robins RA, Hinton C, Ellis IO, Dowle CS, Ferry B, Elston CW, Blamey RE, Baldwin RW (1987) Tumour aneuploidy, prognostic parameters and survival in primary breast cancer. Br J Cancer 55: 449-454

Quirke P, Fozard JBJ, Dixon MF, Dyson JED, Giles GR, Bird CC (1986) DNA aneuploidy in colorectal adenomas. Br J Cancer 53: 477-481

Ramaekers CS, Beck HLM, Feitz WFJ, Oud PS, Debruyne FMJ, Vooijs GP, Herman ChJ (1986) Application of antibodies to intermediate filament proteins as tissuespecific probes in flow cytometric analysis of complex tumors. Anal Quant Cytol 8: 271-280

Rodenburg CJ, Cornelisse CJ, Heintz PAM, Hermans J, Fleuren GJ (1987) Tumor ploidy as a major prognostic factor in advanced ovarian cancer. Cancer 59: 317-323

Rutgers DH, Wils IS, Schaap AH, von Lindert ACM (1987) DNA flow cytometry, histological grade, stage, and age as prognostic factors in human epithelial ovarian carcinomas. Pathol Res Pract 182: 207-213

Scott NA, Wieand HS, Moertel ChG, Cha StS, Beart RW, Lieber MM (1987) Colorectal cancer: Duke's stage, tumor site, preoperative plasma CEA level, and patient prognosis related to DNA ploidy pattern. Arch Surg 122: 1375-1379

Strang P, Eklund G, Stendahl U, Frankendahl B (1987) S-phase rate as a predictor of early recurrences in carcinoma of the uterine cervix. Anticancer Res 7: 807-810

Tirindelli-Danesi D, Teodori L, Mauro F, Modini C, Botti C, Ciocconetti F, Stipa S (1987) Prognostic significance of flow cytometry in lung cancer. A 5-year study. Cancer 60: 844-851

Tribukait B (1987) Flow cytometry in assessing the clinical aggressiveness of genito-urinary neoplasms. World J Urol 5: 108-122

Tytor M, Franzèn G, Olofsson J, Brunk U, Nordenskjöld B (1987) DNA content, malignancy grading and prognosis in T1 and T2 oral cavity carcinomas. Br J Cancer 56: 647-662

Volm M, Brüggemann A, Günther M, Kleine W, Pfleiderer A, Vogt-Schaden M (1985a) Prognostic relevance of ploidy, proliferation, and resistance-predictive tests in ovarian carcinoma. Cancer Res 45: 5180-5185

Volm M, Drings P, Mattern J, Sonka J, Vogt-Moykopf I, Wayss K (1985b) Prognostic significance of DNA patterns and resistance-predictive tests in non-small cell lung carcinoma. Cancer 56: 1396-1403

Flow Cytometry of DNA in Colorectal Adenomas and Adenocarcinomas

W. Giaretti, S. Sciallero, S. Bruno, E. Geido, H. Aste, and A. Di Vinci

Introduction

Histopathological, clinical, and epidemiological data strongly indicate the majority of cases of colorectal carcinomas develop from preexisting adenomas [4].

In recent years it has become accepted policy to remove all polyps visible during endoscopy. Polyps, however, tend to reappear and, in some cases, a colorectal cancer may develop. Follow-up of patients after the removal of a colorectal adenoma represents an important clinical and economical problem and still no clear guide lines have been established.

DNA aneuploidy in colorectal adenomas may prove to be of use in predicting the risk of recurrences and/or of malignant transformation and therefore contribute to the identification of high-risk subgroups of patients.

The purpose of this study is to assess the incidence and degree of DNA aneuploidy of colorectal adenomas and adenocarcinomas, to compare their DNA indices, and to study the correlation of DNA ploidy with histopathological and clinical features.

Clinical Material, Flow Cytometry, and Data Analysis

The material for the study was taken from 64 colorectal adenomas and 49 adenocarcinomas of unselected patients who underwent colonoscopy. The specimens (three forceps biopsies from different sites on the head of the polyps and on the adenocarcinomas and two biopsies from normal mucosa) were treated according to Thornthwaite's method [7]. The measurements were performed using an arc lamp flow cytometer (model ICP22A; Ortho Diagnostic Systems, Westwood, Mass., USA) with suitable filters for the DAPI excitation (window between 300 and 400 nm) and for the DAPI fluorescence emission (window between 450 and 490 nm). The fluorescence signals were input to signal processing electronics using 1024 channels for subsequent storage and analysis in a DEC PDP-11-23 computer (Digital Equipment Co., Maynard, Mass., USA). DNA histograms obtained from mixed populations of normal mucosa and cells from the lesion were used to evaluate the DNA index [3]. Trout red blood cells (TRBC) were used as an internal DNA standard [3].

The coefficient of variation (CV) of every peak was calculated as width at half maximum of the peak divided by the peak mean channel and the factor 2.35. The correlation between DNA ploidy and histological and clinical features was evaluated by means of contingency tables. The adenomas were histologically classified according to Morson et al. [4] in the Department of Pathology of our Institute.

Results and Discussion

Control mucosa taken from 105 patients showed in all cases a single G0/G1 peak. Adenomas and adenocarcinomas showed instead at least two G0/G1 DNA peaks in 20 out of 64 adenomas (31%) and in 36 out of 49 adenocarcinomas (74%).

The frequency of the DNA index values of these 20 DNA aneuploid adenomas and 36 DNA aneuploid adenocarcinomas is shown in Fig. 1. The DNA index values of the six cases of adenomas with foci of adenocarcinomas were respectively 0.91, 1.89, 1.17, 1.53, 1.33, 1.70. The first three cases were adenomas with foci of intramucous adenocarcinoma. The last ones had foci of adenocarcinoma infiltrating the stalk. The incidence of DNA aneuploidy in the DNA index range 0.80-1.20 was 45% for the adenomas and 8% for the adenocarcinomas. Among these, adenomas showed in most cases only small changes in DNA content (of the order of 5%-10% above or below DNA index = 1.0). The detection of such cases requires coefficients of variation of the G0/G1 peaks not higher than 2%-3%, as may be seen in Fig. 2 (given as an example of the standard measurements obtained for every lesion).

These findings suggest that the incidence and degree of DNA aneuploidy in colorectal premalignant and malignant lesions are a function of the tumor progression and therefore substantiate the adenoma-carcinoma sequence by DNA flow cytometry. The significance of the small changes in DNA content in 45% of DNA aneuploid colorectal adenomas is at present only a matter for speculation. It also remains to be proven that patients with adenomas having DNA index values

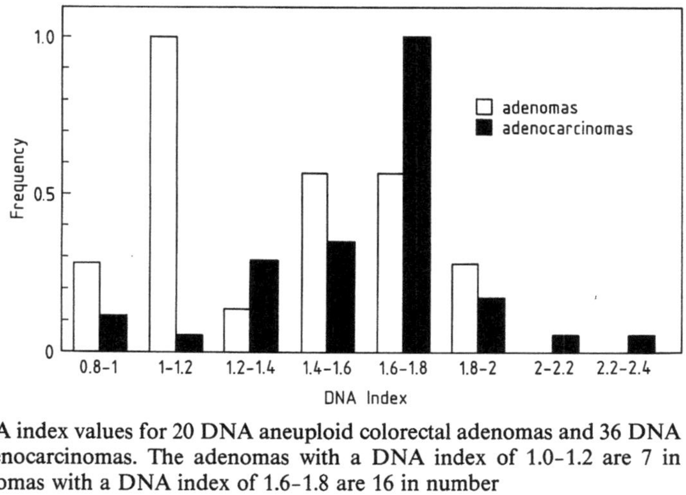

Fig. 1. Frequency of DNA index values for 20 DNA aneuploid colorectal adenomas and 36 DNA aneuploid colorectal adenocarcinomas. The adenomas with a DNA index of 1.0-1.2 are 7 in number, the adenocarcinomas with a DNA index of 1.6-1.8 are 16 in number

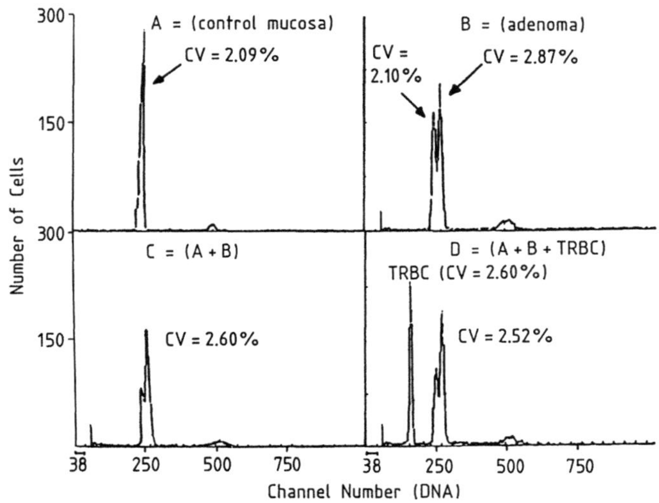

Fig. 2 A–D. Standard set of DNA histograms obtained for every lesion. **A** DNA histogram of the normal mucosa with a single G0/G1 peak; **B** DNA histogram of one adenoma showing two separate G0/G1 peaks; **C** DNA histogram obtained after mixing control mucosa (**A**) and adenoma (**B**), showing an increase in the percentage of cells for the second peak; **D** DNA histogram obtained after mixing trout red blood cells (TRBC) to the previous sample (**C**)

similar to the most frequent DNA index values for adenocarcinomas are at higher risk of further adenomas and/or of cancer.

DNA aneuploidy was present in none out of 8 adenomas (0%) smaller than 1 cm in size, in 9 out of 35 (25%) 1-2 cm in size, and in 11 out of 21 (52%) larger than 2 cm in size. The correlation of DNA ploidy to size of adenomas was statistically significant (χ^2 value = 8.5; $p = 0.02$). This result is in agreement with previous work [2, 5, 8]. DNA aneuploidy was present in 7 out of 42 adenomas (16%) with mild/moderate dysplasia and in 13 out of 22 (59%) with severe dysplasia. The correlation was statistically significant (χ^2 value = 12.09; $p < 0.01$). This finding was previously reported by Goh and Jass [2]. An interesting correlation was found between DNA ploidy of adenomas and family history of colorectal cancer, as previously reported [6]. DNA aneuploidy has been studied in relation to histological type, number, and site of the adenomas, and age of patients, but no significant correlation was found with these features [1]. Similarly, no statistically significant correlation was found, among patients with adenocarcinomas, between DNA ploidy and Dukes' stage, site of the lesion, and age of the subjects [1]. A follow-up study is in progress to elucidate the relationship of DNA ploidy of both adenomas and adenocarcinomas to clinical outcome.

References

1. Giaretti W, Sciallero S, Bruno S, Geido E, Aste H, Di Vinci A (1988) DNA flow cytometry of endoscopically examined colorectal adenomas and adenocarcinomas. Cytometry 9: 238-244
2. Goh HS, Jass JR (1986) DNA content and the adenoma-carcinoma sequence in the colorectum. J Clin Pathol 39: 387-392

3. Hiddemann W, Schumann J, Andreeff M, Barlogie B, Herman CJ, Leif RC, Mayall BH, Murphy RF, Sandberg AA (1984) Convention on nomenclature for DNA cytometry. Cytometry 5: 445-446
4. Morson BC, Dawson IMP (1979) Adenomas and the adenoma-carcinoma sequence. In: Morson BC, Dawson IMP (eds) Gastrointestinal pathology II. Blackwell Scientific, Oxford, pp 615-647
5. Quirke P, Fozard JBJ, Dixon MF, Dyson JED, Giles GR, Bird CC (1986) DNA aneuploidy in colorectal adenomas. Br J Cancer 53: 477-481
6. Sciallero S, Bruno S, Di Vinci A, Geido E, Aste H, Giaretti W (1988) Flow cytometric DNA ploidy in colorectal adenomas and family history of colorectal cancer. Cancer 61: 114-120
7. Thornthwaite JT, Sugarbaker EV, Temple WJ (1980) Preparation of tissue for DNA flow cytometric analysis. Cytometry 1: 229-237
8. Van den Ingh HF, Griffioen G, Cornelisse CJ (1985) Flow cytometric detection of aneuploidy in colorectal adenomas. Cancer Res 45: 3392-3397

Sense and Nonsense of Cytomorphometry and Cytophotometry in Predicting Survival

E. A. van der Meulen, M. E. Boon, A. Böcking, V. Kuenen-Boumeester, P. J. Spaander, P. van Heerde, A. C. M. van Nispen, A. Wijsman-Grootendorst, and L. P. Kok

Introduction

In a study in which several European institutes participated, it was investigated whether cytomorphometry and cytophotometry can be of value for predicting survival in breast cancer, follicle center cell lymphoma, and prostate cancer. The study was especially designed to be presented at the 15th European Congress of Cytology in Baden-Baden, FRG, in 1987.

Breast Cancer

There is a wide variation in the biological behavior of breast carcinomas. Several histologic characteristics, established by objective or subjective methods, have proved to be of value (Bloom and Richardson 1957; Baak et al. 1982, 1985; McGuire 1987). Wallgren et al. (1976) and Zajdela et al. (1979) used cytologic smears for their morphometric studies: they found that prognosis correlates to nuclear size and variability. The involvement of axillary lymph nodes is the most important prognostic factor (McGuire 1987), but additional predictors are needed for a multivariate analysis to predict early relapse and survival (Baak et al. 1985).

In earlier studies, we found that cytomorphometry can be used to obtain objective cytologic data for survival studies (Kuenen-Boumeester et al. 1984). In accordance with other investigations, we found that nuclear size and variation are important. As for cytophotometry, we found that DNA malignancy grading was successful in predicting survival (Böcking and Auffermann 1986), which is in keeping with other DNA studies performed on cytological material (Auer et al. 1984). In the present study, both approaches are used.

Follicle Center Cell Lymphoma

The cytomorphological diagnosis of follicle center cell lymphoma with a mixed, mainly small-cell population, may be one of the most difficult ones in cytology. The diagnostic problems that may arise with regard to neoplastic and nonneoplastic proliferations of follicle center cells are reflected in the relatively large number of false-positive, false-negative, and doubtful conclusions, even in histological specimens. A number of subjective criteria are used, giving rise to rather poor reproducibility (Jones et al. 1977; Velez-Garcia et al. 1983). In spite of this, therapeutic management is largely based on subjective criteria such as cell type and growth pattern (NCI 1982). Fortunately, immunological marker typing reduces differential diagnostic problems. Marker typing, however, is not routinely practised everywhere. Concerning correlations between cell type, cell size, and clinical parameters, the literature is rather contradictory (Foucar et al. 1983; Lennert et al. 1978). In some reports a relation between cell type and survival is described (NCI 1982). Some cytohistologic characteristics of non-Hodgkin lymphomas seem to be promising predictors (Look et al. 1985; Morgan et al. 1986; Böcking et al. 1986). To the best of our knowledge cytomorphometry has not yet been used to predict survival of patients with follicle center cell lymphoma.

Prostate Cancer

DNA cytophotometry of prostate cancer has prognostic value (Böcking et al. 1988). Since nuclear area and variability correlate with degree of differentiation of the tumor (Böcking et al. 1984b), it is anticipated that cytomorphometry on Giemsa-stained smears can be used to estimate prognosis.

The clinical importance of cytologic grading of smear preparations from prostatic carcinomas has been repeatedly stressed (Esposti 1971), and it has been shown that cytological grading correlates well with nuclear size (Böcking et al. 1984b). In an earlier pilot study, we showed that measurement of selected cells in smear preparations can be used for grading prostate carcinoma (Spaander et al. 1982). In the present study we measured random carcinoma cells in the smears to investigate whether this cytomorphometric approach is valuable for estimating prognosis.

Statistical Analysis

To clarify the involvement of statistical inferences, two methods were used, the first one serving a somewhat different purpose than the second:

1. A C-Year survival model. In this method it is possible to distinguish characteristics which are important for short-term survival from those which are good predictors of long-term survival. As will become clear, these two may differ. This finding is important for further patient management.
2. The Cox proportional hazards model, widely used in survival studies, is used to test the significance of the information contained in the characteristics considered and to test whether a certain set of characteristics provides significant additional information with regard to characteristics incorporated earlier. In addition, a procedure will be given for estimating the probability of survival of cancer (together with its statistical uncertainty) of any individual patient in dependence on his/her prognostic characteristics (Kardaun 1986).

The main aims of this study were to investigate whether cytomorphometry and cytophotometry provide additional prognostic information to the "usual" predictors and to estimate the probabilities of survival of cancer of any individual patient (representative for the population on which the statistical analysis is based), using quantitative data. Moreover, this novel approach makes it possible to express the statistical uncertainties.

Data

Patients

Breast Cancer

Sixty patients were treated for breast cancer in the period 1973-1978 in the Daniel den Hoed Cancer Center in Rotterdam. The following clinical data were recorded: age, menopausal status, tumor status (T), and nodal status (N, indicating the presence or absence of metastatic tumor in the axillary lymph nodes in the surgical specimen; Table 1). Patients were included in the analysis if the tumor was confirmed histologically ($T>0$) and distant metastases other than metastatic tumor of the axillary lymph nodes could not be found ($M=0$). Only patients with complete follow-up were included.

Patients with medial tumors and those with lateral tumors combined with metastatic tumor in the axillary lymph nodes ($N=1$) underwent postoperative radiotherapy.

Patients were followed up for a minimum of 8 years or until death. They were clinically reexamined at regular intervals. The period between diagnosis (about the time of operation) and first relapse will be referred to as the disease-free period. Local or distant recurrences were treated with conventional methods. Twenty-six patients died of breast carcinoma and eight of other causes.

Follicle Center Cell Lymphoma

In the Netherlands Cancer Institute a larger number of lymphoma patients are treated according to international protocols. Fine-needle aspirations are performed in most patients at the first visit. From 1974 to 1985 about 8000 lymph node aspirations were performed, of which 2600 revealed lymphomas (mostly B-cell lymphomas).

Thirty-nine patients suffering from follicular center cell lymphomas (centroblastic/centrocytic; Lennert et al. 1978) with a follicular and/or diffuse growth pattern, a long follow-up ranging from 5 to 10 years, treated in one protocol (EORTC: European Organisation for Research on Teatment of Cancer), with cytological slides of high quality and histological confirmation, were selected. The patients were clinically reexamined at regular intervals. The disease-free period was recorded. All patients who died, died of lymphoma. See Table 1 for a summary of the data.

Prostate Cancer

Fine-needle aspirations were available for 38 patients treated in the Leiden University Hospital in the period 1974-1976 and with complete follow-up data. The patients were treated by surgery, radiotherapy, or orchidectomy. Histologic specimens were available for all patients. Of the 38 patients, 33 were dead at the time of

Table 1. Summary of data recorded

	Notation	Breast cancer	Lymphoma	Prostate cancer
Patients				
Age	Age	*	*	*
Sex	Sex	♀	*	♂
Menopausal status	Meno	*	–	–
Stage of disease	Stage	–	*	–
Tumor status	T	*	–	*
Nodal status	N	*	–	–
Cytology/Histology				
Centroblastic/centrocytic cells	CbCc	–	*	–
Lymphocytic/histiocytic cells	LcHc	–	*	–
Size of cells	Size	–	*	–
Cleavage	Cleav	–	*	–
Tumor grade	TumGr	–	*	*
Follicular/diffuse tumor tissue	Arch	–	*	–
Cytomorphometry				
Number of nuclei measured		100	100	50
Area of nuclei	see pp. 112–113	*	*	*
Perimeter of nuclei	see pp. 112–113	*	*	*
Max. diameter of nuclei	see pp. 112–113	*	*	*
Axis of smallest ellipsis around the nucleus	see pp. 112–113	–	*	–
Cytophotometry				
DNA malignancy grade	DNAMG	*	–	–

*, recorded; –, not recorded

evaluation. In this group of very old patients it was not possible to establish with certainty whether they died of cancer.

Cytological and Histological Data for Lymphoma and Prostate Cancer

The non-Hodgkin's lymphomas were classified according to the Kiel classification (Lennert et al. 1978). For the histological slides from excised lymph nodes the classification systems of Rappaport (1966), Lennert et al. (1978), and the Working Formulation (NCI 1982) were used. Tumors were classified as centrocytic or centroblastic according to Lennert et al. (1978), as lymphocytic or histiocytic according to Rappaport (1966), and as small or large cell size and cleaved or noncleaved according to the Working Formulation (NCI 1982). In addition, the tumor grade was recorded and the tumor was (histologically) designated as follicular or diffuse (Table 1). Histological specimens of the tumors removed from all prostate cancer patients were graded (Spaander et al. 1982).

Smear Preparation

Fine-needle aspirations were performed by the cytologists and pathologists of the Daniel den Hoed Cancer Center and Netherlands Cancer Institute for the breast carcinoma patients and the lymphoma patients, respectively. The prostate aspirations were made by the urologist H.J. de Voogt, working in close cooperation with the Department of Pathology of Leiden University Hospital (Spaander et al. 1982). All smears were air-dried and stained with the Giemsa method. For DNA cytophotometry, the smears were destained and restained according to the Feulgen method (Böcking and Auffermann 1986).

Cytophotometrical Data

The Feulgen-stained breast carcinoma smears were measured on a TV image analysis system, Leitz TAS Plus, combined with a Leitz automatic microscope and plotter, and the scalar DNA malignancy grade was determined (Auffermann et al. 1984; Böcking and Auffermann 1986).

Cytomorphometrical Data

Using a camera lucida and a graphic tablet (MOP Videoplan, Kontron) nuclei were measured in the Giemsa-stained smears. The breast carcinoma smears were measured using a magnification of 400. The smears were mounted with a grid measuring 20×40 mm^2 containing 8×625 squares. At random, 10 squares were selected at low magnification and 10 cancer cells in each square measured. Only nonoverlapping nuclei were used. The lymphomas were measured at a magnification of 1000. Some smears contained mature lymphocytes: these were not measured. Nuclear overlapping was not an obstacle to measurement in these smears.

The prostate smears were measured at a magnification of 400. The smears contained exclusively cancer cells, which were randomly measured. Only nonoverlapping nuclei were used.

The area (μm^2), perimeter (μm), and maximum diameter (μm) of each nucleus were measured. In lymphoma specimens the lengths of the main axes of the smallest ellipse outlining the nucleus was also recorded.

Data Reduction

Of course an enormous number of characteristics can discussed on the basis of the cytomorphometrical measurements. In these case studies, however, attention will be drawn to the mean (MnAr) and standard deviation (SdAr) of the nuclear area.

For lymphoma the mean (MnStr) and standard deviation (SdStr) of the nuclear stretch and the mean (MnSh) and standard deviation (SdSh) of the nuclear shape are taken into consideration as well. The nuclear stretch is defined as the ratio of the greatest to the smallest length of the main axes of the smallest ellipse outlining the nucleus, while the shape of the nucleus is defined as the ratio of its perimeter to its maximum diameter. The reasons for introducing this shape and stretch factor are obvious: the nuclear shape is a measure of the unevenness of the outline of the nucleus, while the nuclear stretch is a measure of oblong form. Moreover, both nuclear shape and stretch do not correlate well with nuclear area (see Tables 3 and 4), suggesting that it is not a redundant choice. MnSh and SdSh are also recorded for prostate cancer, but, since the nuclei were similarly shaped, not for breast cancer.

Table 2. Correlation matrix of the cytomorphometrical features measured in breast cancer

	MnAr	SdAr	Range
MnAr	1.000	0.8734	0.8683
SdAr	0.8734	1.000	0.9750
Range	0.8683	0.9750	1.000

Table 3. Correlation matrix of the cytomorphometrical features measured in lymphoma

	MnAr	SdAr	MnSh	SdSh	MnStr	SdStr
MnAr	1.000	0.8386	0.0606	0.5014	0.4513	0.3878
SdAr	0.8386	1.000	−0.0410	0.3871	0.4848	0.4335
MnSh	0.0606	−0.0410	1.000	0.4124	−0.6546	−0.4745
SdSh	0.5014	0.3871	0.4124	1.000	0.2931	0.3792
MnStr	0.4513	0.4848	−0.6546	0.2931	1.000	0.8617
SdStr	0.3878	0.4335	−0.4745	0.3792	0.8617	1.000

Table 4. Correlation matrix of the cytomorphometrical features measured in prostate cancer

	MnAr	SdAr	Range	MnSh	SdSh
MnAr	1.000	0.8482	0.9158	−0.0271	0.2609
SdAr	0.8482	1.000	0.9444	0.0656	0.4311
Range	0.9158	0.9444	1.000	0.0125	0.2906
MnSh	−0.0271	0.0656	0.0125	1.000	0.1515
SdSh	0.2609	0.4311	0.2906	0.1515	1.000

For breast and prostate cancer the 80% interpercentile range (Range) of the nuclear areas is added. It can be seen from Tables 2 and 4 that considering both SdAr and Range is rather unnecessary. Nevertheless Range is added, since this statistic was studied in an earlier paper. Since, for lymphoma, we already have six features to consider, Range is not included here.

See Tables 2-4 for correlations between the cytomorphometrical features considered.

Methods

Testing Strategy

The general strategy applied for testing whether covariables provide prognostic information is as follows. First of all the information contained in each single covariable is tested. Then the attention is drawn to any two covariables which mutually contain additional prognostic information. Thereupon a third covariable is added to any such pair, to be tested for its additional prognostic value, and so on. The joint information in the sets of covariables thus obtained is expressed too, perhaps somewhat redundantly.

Reduced Sample *c*-Year Survival Model

To apply an honest *c*-year survival analysis, all cases which entered the study after *c* years before the termination date have to be left out. Moreover, no other reasons for leaving out patients can be justified, since then the population is manipulated. This implies that censoring (i. e., only knowing that 'failure' will occur or would have occurred beyond a certain time) before *c* years is not allowed. Defining failure time T by the period between diagnosis and death by any cause, solves most of this problem, since then the only possibility of a case being censored is emigration or being lost in some way, which occurred very rarely.

Dichotomizing failure time T by defining $D = I_{(T<c)}$, we employ the logit model

$$P(D=1|Z=z) = \frac{\exp(\alpha+\beta' z)}{1+\exp(\alpha+\beta' z)} = p_\theta(z), \tag{1}$$

where $Z=(Z_1,\ldots,Z_p)'$ is the vector of explanatory variables under consideration and $\theta=(\alpha, \beta)$. Of course, other models like the probit model could have been considered here.

With regard to the data set (d_i, z_i), $i=1,\ldots,m$, where m is the reduced sample size, the joint conditional likelihood function of θ on $d_i|z_i$, $i=1,\ldots,m$, is given by

$$l\,(\theta) = \prod_{i=1}^{m} p_\theta(z_i)^{d_i}(1-p_\theta\,(z_i))^{1-d_i} \tag{2}$$

The covariance matrix of the maximum likelihood estimate $\hat{\theta}$ can, because of its asymptotic efficiency, be suitably estimated by $I^{-1}(\hat{\theta})$, where $I(\theta)$ is Fisher's information matrix, i. e., $I(\theta)_{i,j} = -E_\theta(\partial^2/\partial\theta_i 7\partial\theta_j \log l\,(\theta))$. Frequently the observed Fisher information matrix, i. e., $I(\hat{\theta})_{i,j} = -\partial^2/\partial\theta_i\partial\theta_j \log l\,(\theta)|_{\theta=\hat{\theta}}$, is used (Efron and Hinkley 1978).

To test the significance of the additional information in a covariable Z_i, one can focus on $\hat{\theta}_{i+1}/(I(\hat{\theta})_{i+1,i27}+1|)^{1/2}$. A more general way to test whether a set of covariables carries additional prognostic information with regard to the set already incorporated, is to apply the likelihood ratio test. This states that the hypothesis H_0: $\beta_{r+1}=\ldots=\beta_p=0$ is to be rejected if

$$-2\log\frac{\max_{\theta\in\Theta_0} l(\theta)}{\max_{\theta} l(\theta)} > \chi^2_{p-r;\alpha}, \tag{3}$$

where $\Theta_0=\{(\alpha, \beta_1,\ldots,\beta_p)|\beta_{r+1}=\ldots=\beta_p=0\}$.

For the practical application of this logit model the statistical package GLIM was used. The results are given on pp. 117 ff.

Cox Proportional Hazards Model

Model Assumption and Theory

Consider the data (t_i', δ_i, z_i), $i=1,\ldots,n$, where $t_i'=\min(t_i, c_i)$, $\delta_i=I(t_i<c_i)$, indicating whether or not case i is censored, and $z_i=(z_1,\ldots,z_p)_i^t$ the vector of scores of case i. Let failure time T have distribution function $F(t)$, survival function $S(t)=1-F(t)$, and density function $f(t)$. It will be assumed that T is distributed independently of the censoring time C. The hazard or failure rate function

$$\lambda(t) = f(t)/S(t) = \frac{d}{dt}(-\log S(t)) \tag{4}$$

will play a prominent part in the model formulation. Note that $\lambda(t)$ is the infinitesimal probability of dying at t, given that one has lived up to t. Therefore it is an obvious measure of the risk of failure. For example, a patient randomly chosen from a population for which failure time is exponentially distributed, i.e., $S_\alpha(t)==e^{-\alpha t}$, is constantly at the same risk, since $\lambda(t)=\alpha$. In other words, a patient from this population will always have the same prospects. Note that in general $S(t)=\exp\{-\Lambda(t)\}$, where

$$\Lambda(t)=\int_0^t \lambda(u)\,du \tag{5}$$

is the cumulative hazard rate function.

The proportional hazards model is based on the assumption (Cox 1972)

$$\lambda(t|Z=z) = e^{\beta^t z}\lambda_0(t), \tag{6}$$

where $\lambda_0(t)$ is an arbitrary and unspecified baseline hazard function and $\beta=(\beta_1,\ldots,\beta_p)^t$ the vector of regression parameters. Note that $\lambda_0(t)$ is a nuisance function.

In order to find an estimate for β, Cox (Cox 1972; Cox and Oates 1984) suggested consideration of the 'likelihood'

$$l(\beta) = \prod_{i=1}^{n}\left(\frac{\exp(\beta^t z_i)}{\sum_{k\in R(t_i')}\exp(\beta^t z_k)}\right)^{\delta_i}, \tag{7}$$

where $R(t) = \{i | t_i' \geq t\}$ denotes the set of patients still at risk just prior to t. An intuitive interpretation of $l(\beta)$ is as follows. Given the risk set $R(t)$, the probability of patient $k \in R(t)$ failing is proportional to $\exp(\beta' z_k)$. Given that someone of the risk set $R(t_i')$ fails at t_i', the probability that patient $j \in R(t_i')$ is the victim is

$$\exp(\beta' z_j) / \sum_{k \in R(t_i')} \exp(\beta' z_k).$$

It should be noted that $l(\beta)$ is not the full conditional likelihood on (t_i', δ_i, z_i), $i = 1, \ldots, n$, since it lacks, e.g., the information provided by the order of censored observations between two successive failures. Nevertheless, Cox argues, and others have proven more rigorously, that the estimate $\hat{\beta}$ obtained by maximizing the partial likelihood $l(\beta)$ is asymptotically efficient. Hence the covariance matrix of $\hat{\beta}$ can be estimated by $I^{-1}(\hat{\beta})$, where $I(\beta)$ is Fisher's information matrix, i.e., $I(\beta)_{i,j} = -E_\beta(\partial^2/\partial\beta_i\partial\beta_j \log l(\beta))$. Note that $\lambda_0(t)$ remains completely unspecified and the only statistics required from the data are the rank vector of the ordered life times $t_{(1)}' < \ldots < t_{(n)}'$, together, of course, with $z_{(1)}, \ldots, z_{(n)}$ and $\delta_{(1)}, \ldots, \delta_{(n)}$.

The cumulative baseline hazard function $\Lambda_0(t)$ can be estimated consistently by (see, e.g., Sec. 4.8.2 in Kalbfleisch and Prentice 1980, or Sec. 7.8 in Cox and Oates 1984)

$$\hat{\Lambda}_0(t) = \sum_{\{i | t_i' < t\}} \frac{\delta_i}{\sum_{k \in R(t_i')} \exp(\hat{\beta}' z_k)}. \tag{8}$$

As $\Lambda(t|z) = \exp(\beta' z)\,\Lambda_0(t)$, the survival curve of an arbitrarily chosen patient with vector of covariables $Z = z$ is estimated by

$$\hat{S}(t|z) = \exp(-\hat{\Lambda}_0(t)\exp(\hat{\beta}' z)). \tag{9}$$

To obtain instantaneous $1-\alpha$ confidence bands for $\hat{S}(t|z)$ one needs the asymptotic joint distribution of $\hat{\beta}$ and $\hat{\Lambda}_0$. This distribution has been derived under somewhat different conditions and model assumptions by Bailey (1979, 1983) and Anderson and Gill (1982). Here we mention only that this joint distribution is asymptotically normal with mean zero and a covariance matrix which is rather complex but which can be estimated consistently. Therefore, since $\text{var}(\log(\hat{\Lambda}(t|z))) = \text{var}(\log(\hat{\Lambda}_0(t)) + \hat{\beta}' z)$, a two-sided $1-\alpha$ instantaneous confidence band for $\log(\hat{\Lambda}(t|z))$ and thus, by transforming back, also for $\hat{S}(t|z)$, can be constructed.

Testing the Proportional Hazards Assumption

One way of testing whether the proportional hazards assumption (Eq. 6) holds at any time t with respect to a certain covariable Z is as follows. Assume that

$$\lambda(t | Z = z) = e^{\beta_1 z + \beta_2 \log(t) \cdot z} \lambda_0(t) \tag{10}$$

and test whether the regression coefficient β_2 corresponding with the interaction term $\log(t) \cdot z$ differs significantly from zero. In fact, the regression coefficient belonging to the interaction of Z and any nonconstant function of t should not differ significantly from zero, if Eq. 6 holds. For simplicity however, we only focus on $\log(t)$.

For information regarding time-dependent covariables, see, e.g., Kalbfleish and Prentice (1980) and Cox and Oates (1984).

Practical Implementation

For the practical implementation of this analysis, the statistical special purpose package Concern, written by Kardaun and Kiel, is available at the Groningen University. It enables the user to calculate the regression parameters along with their standard errors and the survival curve together with a two-sided $1-\alpha$ confidence band.

Since Concern did not provide the maximum of the likelihood function, the likelihood ratio test can not be applied. But an asymptotically equivalent method, introduced by Wald, is as follows. Consider the model where all p covariables are incorporated. Then the hypothesis H_0: $\beta_{r+1} = \ldots = \beta_p = 0$ is to be rejected if

$$\hat{\beta}^{*t} (C I^{-1} (\hat{\beta})\, C^t)^{-1} \hat{\beta}^* > \chi^2_{p-r;\alpha}, \tag{11}$$

where $\hat{\beta}^* = (\hat{\beta}_{r+1}, \ldots, \hat{\beta}_p)^t$ and $C = [0; I_{p-r}] \in \mathbb{R}^{(p-r) \times p}$. See for instance Sec. 4.4.4, p. 156, in Serfling (1980). For more information on the statistical package Concern the reader is referred to Kardaun (1986).

Since Concern cannot deal with time-dependent covariables, the statistical package BMDP was used to test the model assumption (Eq. 6).

Results

Analysis Based on *c*-Year Survival Time

Survey

In this section, the results based on the method described above (p. 114–115), and concerned with the true survival probabilities in regard to a certain fixed period of c years, will be given and discussed. In each case study we will focus on several values of c, namely $c=3$, $c=4$, $c=5$, and for breast cancer, additionally, $c=8$. The significance of the single, additional, and joint information in a (set of) covariable(s) is expressed by means of p values based on the likelihood ratio test (see p. 114–115). The results for breast cancer, lymphoma, and prostate cancer are listed in Tables 5, 6, and 7, respectively.

Results for Breast Cancer

The most significant covariables for $c=3$ and $c=4$ appear to be N, SdAr, Range, and DNAMG (for abbreviations, see Table 1 and section 'Data Reduction'), whereas interesting pairs of covariables are N/Range and N/DNAMG, since each covariable contains significant additional information with regard to the oth-

Table 5. *p* Values for breast cancer based on *c*-year survival time

Single information in:	$c=3$	$c=4$	$c=5$	$c=8$
Age	0.89	0.33	0.32	0.37
Meno	0.29	0.66	0.61	0.50
N	0.0048 *	0.059	0.11	0.22
T	0.75	0.64	0.29	0.24
MnAr	0.096	0.15	0.42	0.35
SdAr	0.010 *	0.024 *	0.093	0.11
Range	0.0065 *	0.037 *	0.11	0.12
DNAMG	0.058	0.011 *	0.033 *	0.098

Additional information in:	with regard to:	$c=3$	$c=4$	$c=5$	$c=8$
MnAr	N	0.026 *	0.10	0.35	0.31
N	MnAr	0.0014 *	0.041 *	0.091	0.20
Range	N	0.00028*	0.017 *	0.070	0.095
N	Range	0.00020*	0.028 *	0.069	0.17
SdAr	MnAr	0.018 *	0.037 *	0.044 *	0.13
MnAr	SdAr	0.19	0.26	0.17	0.39
MnAr	Range	0.12	0.45	0.23	0.44
Range	MnAr	0.0076 *	0.089	0.065	0.14
DNAMG	N	0.079	0.017 *	0.051	0.14
N	DNAMG	0.00031*	0.039 *	0.082	0.19
DNAMG	Range	0.58	0.064	0.088	0.27
Range	DNAMG	0.14	0.86	0.89	0.81
DNAMG	N, Range	0.65	0.15	0.18	0.43
Range	N, DNAMG	0.0015 *	0.54	0.84	0.65

Joint information in:	$c=3$	$c=4$	$c=5$	$c=8$
N, DNAMG	0.00025 *	0.0046 *	0.023 *	0.11
N, Range	0.000024*	0.010 *	0.052	0.12

* Significant at the 5% level

er. Using all three of DNAMG, N, and Range, instead of either pair mentioned, does not significantly improve the prognosis (except for $c=3$, where Range does provide significant additional information). Remarkable is the fact that all covariables are insignificant when $c=8$. Note that, since follow-up was sufficiently long, data reduction was not needed here.

Results for Lymphoma

The most significant feature for each value of *c* is MnAr, whereas Arch, TumGr, SdAr, SdSh, MnStr, and SdStr are also significant for some values of *c*. The only covariable which contains significant additional information with regard to any of

Table 6. *p* Values for lymphoma based on *c*-year survival time

Single information in:	$c=3$ ($n=38$)	$c=4$ ($n=38$)	$c=5$ ($n=35$)
Sex	0.48	0.71	0.60
Age	0.75	0.47	0.23
Stage	0.16	0.17	0.76
TumGr	0.033 *	0.0053 *	0.12
Arch	0.024 *	0.0030 *	0.040 *
LcHc	0.74	0.36	0.82
CbCc	0.067	0.11	0.24
Size	0.74	0.36	0.82
Cleav	0.82	0.99	0.55
MnAr	0.00044 *	0.00010 *	0.0075 *
SdAr	0.047 *	0.0046 *	0.087
MnSh	0.34	0.51	0.35
SdSh	0.021 *	0.0067 *	0.063
MnStr	0.15	0.054	0.022 *
SdStr	0.15	0.037 *	0.038 *

Additional information of:	with regard to:	$c=3$ ($n=38$)	$c=4$ ($n=38$)	$c=5$ ($n=35$)
Arch	TumGr	0.45	0.31	0.094
TumGr	Arch	0.99	0.99	0.30
Arch	MnAr	0.92	0.57	0.66
MnAr	Arch	0.0071 *	0.010 *	0.077
TumGr	MnAr	0.89	0.64	0.99
MnAr	TumGr	0.0051 *	0.0060 *	0.031 *
SdAr	MnAr	0.10	0.66	0.40
SdSh	MnAr	0.42	0.26	0.49
MnStr	MnAr	0.82	0.92	0.22
SdStr	MnStr	0.65	0.39	0.82

* Significant at the 5% level

the others is MnAr, implying that using another feature besides MnAr does not significantly improve the accuracy of prognosis. Note that for $c=5$, 3 patients are omitted from the study, one because of emigration, the other two because of their recent entry.

Results for prostate cancer

All cytomorphometrical features appeared to be insignificant for each value of *c*. Only Age turned out to be significant for $c=5$. Of course, this is not very surprising, since most of the patients are over the age of 65. Note that one patient had to be omitted from the study because of his recent entry.

Table 7. p Values for prostate cancer based on c-year survival time

Single information in:	$c=3$ ($n=36$)	$c=4$ ($n=36$)	$c=5$ ($n=36$)
Age	0.22	0.12	0.018 *
T	0.89	0.70	0.92
TumGr	0.82	0.50	0.58
MnAr	0.89	0.84	0.78
SdAr	0.64	0.38	0.58
MnSh	0.61	0.92	0.89
SdSh	0.72	0.20	0.60
Range	0.69	0.57	0.79

* Significant at the 5% level

Analysis Based on the Cox Proportional Hazards Model

Survey

In this section the results based on the Cox proportional hazards model, described above (pp. 115 ff), will be given. Whereas in the previous section failure is defined as death by any cause, here, it will be defined as either death by the cancer under consideration or relapse (disease-free period). Note that by doing so the independence of failure time and censoring time becomes somewhat questionable.

Since the program Concern cannot (yet) deal with data sets for which the number of noncensored cases exceeds 30, the statistical package BMDP is used for prostate cancer.

Instead of the likelihood ratio statistic, the Wald statistic, briefly described on p. 117, is used to test the significance of the single, additional, and joint information in a (set of) covariable(s).

Survival plots are displayed for some interesting sets of covariables. Here also a 95% confidence band is displayed, which is a useful instrument for expressing the statistical uncertainty in the estimated survival probabilities.

Results for Breast Cancer

Since the results of the c-year survival analysis indicated that for some covariables significances decreased with increasing time (see Table 5), some doubts arise regarding the validity of the proportional hazards assumption (Eq. 6). In the Cox proportional hazards model it is assumed that the significance of a covariable remains constant over time.

Restricting our attention to the most striking covariables in this respect, namely N, Range, and, to a somewhat lesser extent, MnAr, and DNAMG, it was found that, using the statistical package BMDP and the testing procedure described above (p. 116–117), the model assumption (Eq. 6) can be used. It should be noted however that regarding N the evidence against Eq. 6 was close to being significant ($p=0.10$).

The results concerning death by cancer, displayed in Table 8, correspond

Table 8. *p* Values for breast cancer based on the Cox proportional hazards model

Single information in:	Death by cancer		Disease - free period	
Age	0.42		0.092	
Meno	0.061		0.032	*
N	0.042	*	0.046	*
T	0.22		0.30	
MnAr	0.085		0.13	
SdAr	0.027	*	0.068	
Range	0.029	*	0.070	
DNAMG	0.029	*	0.080	

Additional information in:	with regard to:	Death by cancer		Disease - free period	
N	Range	0.021	*	0.026	*
Range	N	0.013	*	0.030	*
N	Meno	0.045	*	0.050	
Meno	N	0.068		0.035	*
N	MnAr	0.032	*	0.032	*
MnAr	N	0.050	*	0.058	
N	DNAMG	0.034	*	0.036	*
DNAMG	N	0.033	*	0.087	
DNAMG	N, Range	0.21		0.49	
Range	N, DNAMG	0.074		0.095	
Meno	N, Range	0.033	*	0.026	*
N	Meno, Range	0.022	*	0.030	*
Range	N, Meno	0.0080	*	0.030	*
Meno	N, DNAMG	0.089		0.042	*
N	Meno, DNAMG	0.041	*	0.044	*
DNAMG	N, Meno	0.039	*	0.11	

Joint information in:	Death by cancer		Disease - free period
N, Range	0.0031	*	0.020
N, Meno	0.030	*	0.011
N, DNAMG	0.0061	*	0.029
N, Meno, Range	0.00087	*	0.0016
N, Meno, DNAMG	0.0065	*	0.0078

* Significant at the 5% level

for the greater part with the results of the previous analysis. Again, N/Range, N/MnAr, and N/DNAMG are the most interesting pairs of predictors, whereas N/Meno is interesting as well. An interesting triplet of covariables appears to be N/Meno/Range, since each covariable contains significant additional information with regard to the other two.

With respect to the disease-free period, Meno and N turned out to be significant features, while SdAr, SdSh, Range, and DNAMG seem still of some interest. Significant pairs appeared to be N/Range, N/Meno, N/MnAr, and N/DNAMG, while the most significant triplet is again N/Meno/Range.

With regard to both death by cancer and disease-free period, incorporating more covariables did not significantly improve the accuracy of prognosis.

The survival (from cancer) curve, displayed in Fig. 1, corresponds to patients with N = 0 and Range = 115 μm² (≈ mean (Range)). For other patients the survival curve is obtained by shifting the curve upwards or downwards according to the arrows. One arrow length corresponds to N = 1 (lymph nodes positive) and a Range difference of + 50 μm² (≈ SD (Range)), respectively. A similar survival plot is displayed in Fig. 2, where instead of Range, DNAMG is used. The curve displayed corresponds to patients with N = 0 and DNAMG = 1.0 (≈ mean (DNAMG)). One arrow length corresponds to N = 1 and a DNAMG difference of + 0.6 (≈ SD (DNAMG)), respectively. The uncertainty in this shift is expressed by means of the interval $[\hat{\beta} - SD(\hat{\beta}), \hat{\beta} + SD(\hat{\beta})]$ next to the arrow. It is noted that the width of the confidence band also varies with the value of the covariable. However, this variation is too complicated to express in Figs. 1 and 2 (see Kardaun 1986).

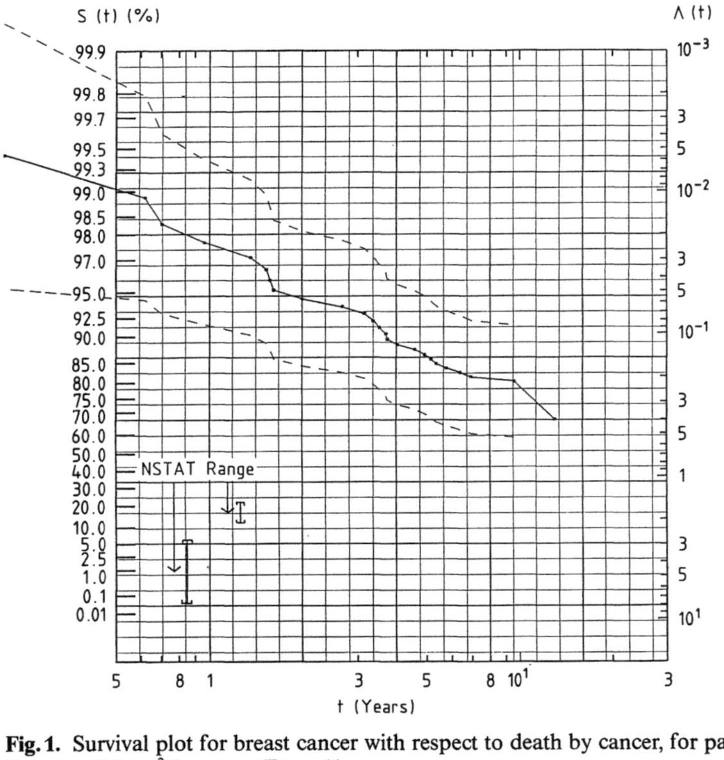

Fig. 1. Survival plot for breast cancer with respect to death by cancer, for patients with N = 0 and Range = 115 μm² (≈ mean (Range))

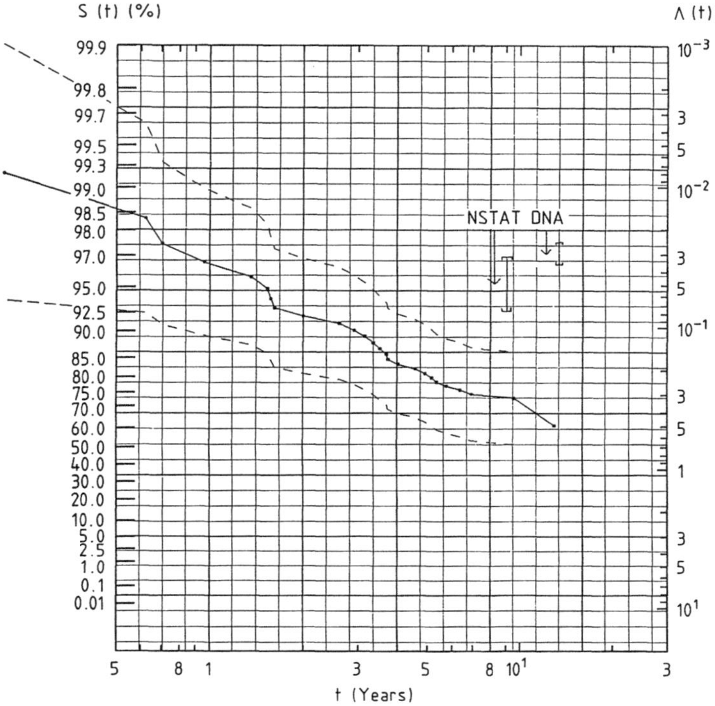

Fig. 2. Survival curve for breast cancer with respect to death by cancer, for patients with N=0 and DNAMG=1.0

Results for Lymphoma

The results, displayed in Table 9, show much similarity with the results of the previous analysis. With respect to death by cancer, MnAr again turned out to be most significant, while Arch, TumGr, MnStr, and SdStr are also significant or close to significant. Again, MnAr is the only feature containing significant additional information with regard to any other feature.

With regard to the disease-free period CbCc is just significant, while Size and LcHc are close to significant.

The survival (from cancer) curve, displayed in Fig. 3, corresponds to a patient with MnAr = 85 μm^2 ($\approx$ mean (MnAr)). One arrow length corresponds to +30 μm^2 ($\approx$ SD (MnAr)).

Results for Prostate Cancer

Since for nine patients the moment of first occurrence of distant metastases was not recorded and since for eleven patients distant metastases were established at the diagnosis, the statistical analysis has been restricted to studying the regression of time to death by cancer on the covariables considered only. The results, listed in Table 10, are that none of the characteristics are significant, although being very old is almost significantly "favorable".

Table 9. *p* Values for lymphoma based on the Cox proportional hazards model

Single information in:		Death by cancer	Disease - free period
Stage		0.51	0.53
Sex		0.40	0.13
Age		0.071	0.45
TumGr		0.041 *	0.16
Arch		0.028 *	0.099
LcHc		0.43	0.069
CbCc		0.099	0.046 *
Size		0.43	0.069
Cleav		0.38	0.14
MnAr		0.0074 *	0.097
SdAr		0.036 *	0.47
MnSh		0.30	0.53
SdSh		0.028 *	0.080
MnStr		0.049 *	0.43
SdStr		0.051	0.15
Additional information		**Death by cancer**	**Disease - free period**
in:	**with regard to:**		
Arch	MnAr	0.35	0.25
MnAr	Arch	0.027 *	0.27
TumGr	MnAr	0.75	n.c.
MnAr	TumGr	0.020 *	n.c.
MnStr	MnAr	0.39	n.c.
MnAr	MnStr	0.014 *	n.c.
SdAr	MnAr	0.19	n.c.
MnAr	SdAr	0.015 *	n.c.
SdSh	MnAr	0.16	0.15
MnAr	SdSh	0.015 *	0.21
CbCc	MnAr	0.19	0.060
MnAr	CbCc	0.0086 *	0.14
TumGr	Arch	0.27	0.27
Arch	TumGr	0.086	0.13

* Significant at the 5% level
n.c. not calculated

Discussion

In these analyses it is shown that quantification is valuable for the prediction of survival of patients with breast cancer and follicle center cell lymphoma, but it was not successful in predicting the outcome of prostate cancer. It is important to note that the aspirations were performed by cytopathologists and pathologists with long experience in aspiration cytology: it is a well-known fact that the quality of

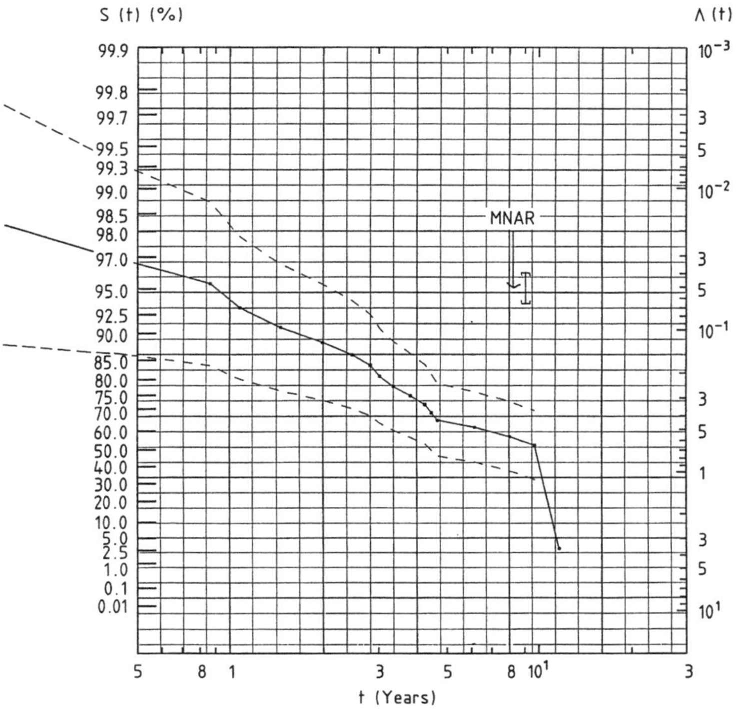

Fig. 3. Survival curve for lymphoma with respect to death by cancer

the smear depends on the aspirator, and that nuclear size can be influenced by the technique of preparing the smear (Boon and Kok 1987).

Breast Cancer

For this group of patients with breast cancer, the cytomorphometric characteristic Range proved to have prognostic value for short-term survival (Table 5) and time till death from cancer (Table 8), and additional information with respect to the characteristic N. The importance of this characteristic is in keeping with our earlier findings (Kuenen-Boumeester et al. 1984) and of other cytologic studies (Wallgren et al. 1976; Zajdela et al. 1979). It is of interest that menopausal state was important in prediction of the disease-free period, and had additional information with regard to N and Range (Table 8). The survival probabilities for any individual patient (with $T>0$ and $M=0$) can be estimated using cytomorphometry (Fig. 1).

In our breast cancer patients, estrogen receptor status was not established (the cases date from the 1970s, when this procedure was not routinely performed). Mossler et al. (1982) showed that there is a negative correlation between estrogen receptor values and nuclear area. A similar approach to that applied in this paper should be used for a more recent group of patients, in whom estrogen receptor status is established, to investigate the additional information provided by this measurement with regard to cytomorphometry.

Table 10. *p* Values for prostate cancer based on the Cox proportional hazards model

Single information in:	Death by cancer
Age	0.070
T	0.13
TumGr	0.20
MnAr	0.77
SdAr	0.40
MnSh	0.87
SdSh	0.25
Range	0.61

* Significant at the 5% level

The value of nuclear DNA content for prediction of short- and long-term survival is a matter of debate (Taylor et al. 1983; Stuart-Harris et al. 1985; McGuire 1987). DNA diploid tumors were found to be associated with small-size nuclei with limited size variations, whereas DNA aneuploid tumors were associated with large nuclei with wide size variations (Owainati et al. 1987). In addition, mitotic activity correlates with nuclear size (Owainati et al. 1987) and variance (Stenkvist et al. 1981). In the present study, using other statistical methods than in earlier studies concerning DNAMG, it was shown again that it is an useful predictor. The characteristic DNAMG contained information concerning death by cancer, and had additional information to that provided by N. For an individual patient, the survival probabilities can be estimated with the aid of Fig. 2.

The cytomorphometeric characteristic Range did not contain additional information to that provided by N and DNAMG, nor did DNAMG to that provided by N and Range (Table 8). This is in keeping with our earlier findings (Boon et al. 1984).

It becomes clear that many different factors in concert have an impact on predictions of survival and relapse (McGuire 1987). A longer follow-up of a larger group of patients is needed to establish more precisely the predictive value of cytomorphometry and cytophotometry also for patients receiving adjuvant therapy. It should be noted that our patients, treated in the 1970s, did not receive such therapy. It might be useful to include the characteristics Range and DNAMG in a prospective study of patients treated according to current protocols.

Follicle Center Cell Lymphoma

For these patients with follicle center cell lymphoma, the short- and medium-term survival (Table 6) and time till from cancer death (Table 9) are most influenced by the cytomorphometric characteristic MnAr. Although the histologic characteristics Arch and TumG are significant predictors, they do not contain additional information. For an individual patient, prognosis can be estimated with the aid of Fig. 3.

As for these lymphomas, future studies should focus on the question of whether antigen expression (using monoclonal antibodies) adds to morphometry. Until we have obtained this information, we can use the curves published here to estimate the prognosis of the individual patient.

Prostate Cancer

The group of prostate carcinoma patients was limited in size, and most patients were very old, as is often the case in prostate carcinomas. Both factors complicated our survival study. In addition, cell selection in poorly differentiating prostate carcinoma poses problemes (Spaander et al. 1982), further adding to the difficulties in establishing the value of cytomorphometry in predicting survival of these patients. For grading prostate cancer, it was obviously very important to measure bare nuclei: these are present in poorly differentiated prostate carcinoma smears but absent in smears from well-differentiated carcinoma (Spaander et al. 1982). In addition to these bare nuclei, smears from poorly differentiated carcinomas contain cell groupings with intact cytoplasm. In the study mentioned it was shown that when the smear is measured by cytotechnicians, only the cells with intact nuclei are measured, and the bare nuclei are thought to be unfit for quantification. It seems likely that in other morphometric projects also these bare nuclei are not measured, to look at the published nuclear values of poorly differentiated carcinomas, which are much lower than when bare nuclei are included (Spaander et al. 1982; Stöber and Schmidt 1980). The measurements published in the present paper were performed by a cytotechnician, and here too the diagnostically important bare nuclei were judged too "degenerated" for morphometry. We anticipate that if the instruction is given to measure these bare nuclei, better results can be obtained.

Comments Concerning the Statistical Approach

Regarding the statistical approach it is noted - perhaps stating the obvious - that it is useful not to restrict oneself to one method. In our case, the results of the *c*-year analysis gave rise to doubts concerning the validity of the assumption on which the proportional hazards model is based. That is why the problem was investigated and in fact solved.

It is recommended that in estimating survival probabilities the statistical uncertainties are expressed. Normal statistical packages cannot be used for this. A special-purpose package like Concern has been especially devised to deal with this problem and has been used for survival analysis using clinical data (Kardaun 1986) and in the present study for cytophotometric and cytomorphometric data.

References

Anderson PK, Gill RD (1982) Cox's regression model for counting processes: a large sample study. Ann Stat 10: 1100-1121

Auer GU, Eriksson E, Azavedo E, Caspersson T, Walgren A (1984) Prognostic significance of nuclear DNA content in mammary adenocarcinomas in humans. Cancer Res 44: 394-396

Auffermann W, Repges R, Böcking A (1984) Rapid diagnostic DNA cytometry with an automatic microscope and a TV image-analysis system. Anal Quant Cytol Histol 6: 179-188

Baak JPA, Kurver PHJ, de Snoo-Nieuwlaat AJE, de Graef S, Makkink B, Boon ME (1982) Prognostic indicators in breast cancer: morphometric methods. Histopathology 6: 327-339

Baak JPA, van Dop H, Kurver PHJ, Hermans J (1985) The value of morphometry to classic prognosticators in breast cancer. Cancer 56: 374-382
Bailey KR (1979) The general maximum likelihood approach to the Cox's regression model. Thesis, University of Chicago, Chicago, Ill
Bailey KR (1983) The asymptotic joint distribution of regression and survival estimates in the Cox's regression model. Ann Stat 11: 39-48
Bloom HJG, Richardson WW (1957) Histological grading and prognosis in breast cancer: a study of 1409 cases of which 359 have been followed for 15 years. Br J Cancer 11: 359-377
Böcking A, Auffermann W (1986) Algorithm for DNA-cytophotometric diagnosis and grading of malignancy. Anal Quant Cytol Histol 8: 363
Böcking A, Adler C-P, Common HH, Hilgarth M, Granzen B, Auffermann W (1984a) Algorithm for a DNA-cytophotometric diagnosis and grading of malignancy. Anal Quant Cytol Histol 6: 1-8
Böcking A, Auffermann W, Schwarz H, Bammert J, Dorrjer G, Vucicuja S (1984b) Cytology of prostatic carcinoma. Anal Quant Cytol Histol 6: 74-88
Böcking A, Chatelain R, Auffermann W, Krüger GRF, Asmus B, Wohltmann D, Schuster C (1986) DNA-grading of malignant lymphomas. I. Prognostic significance, reproducibility, and comparison with other classifications. Anticancer Res 6: 1205-1216
Böcking A, Chatelain R, Orthen U, Gien G, Kalckreuth GV, Jocham D, Wohltmann D (1988) DNA-grading of prostatic carcinoma: prognostic validity and reproducibility. Anticancer Res 8: 1-7
Boon ME, Kok LP (1987) An explanation for the reported variability of nuclear areas in air-dried Romanowsky-Giemsa stained smears of follicular tumors of the thyroid. Acta Cytol (Baltimore) 31: 527-530
Boon ME, Auer GU, van Kaam H, Schwinghammer H (1984) Classifying breast carcinomas with DNA measurements and morphometry. Cytometry 5: 469-472
Cox DR (1972) Regression models and life tables (with discussion). J R Stat Soc [B] 34: 187-220
Cox DR, Oates D (1984) Analysis of survival data. Monographs on statistics and applied probability. Chapman and Hall, London
Efron B, Hinkley DV (1978) Assessing the accuracy of the maximum likelihood estimator: observed versus expected Fisher information (with discussion). Biometrika 65: 457-489
Esposti PL (1971) Cytologic malignancy grading of prostatic carcinoma by transrectal aspiration biopsy. Scand J Urol Nephrol 5: 199-209
Foucar K, Armitage JO, Dick FR (1983) Malignant lymphoma, diffuse mixed small and large cell. A clinicopathologic study of 47 cases. Cancer 51: 2090-2099
Jones SE, Butler JJ, Byrne GE, Coltman CA, Moon TE (1977) Histopathologic review of lymphoma cases from Southwest Oncology Group. Cancer 39: 1071-1076
Kalbfleisch JD, Prentice RL (1980) The statistical analysis of failure time data. Wiley, New York
Kardaun OJWF (1986) On statistical survival analysis and its application in medicine. Wolters-Noordhoff, Groningen
Kuenen-Boumeester V, Hop WCJ, Blonk DI, Boon ME (1984) Prognostic scoring using cytomorphometry and lymph node status of patients with breast carcinoma. Eur J Cancer Clin Oncol 20: 337-345
Lennert K (1978) Malignant lymphomas other than Hodgkin's disease. Springer, Berlin Heidelberg New York (Handbuch der speziellen pathologischen Anatomie und Histologie, Vol 1, part 3 B)
Look AT, Roberson PC, Williams DL, Rivera G, Bowman WP, Pui CH, Ochs J, Abromowitch N, Kalwinsky D, Dahl GV, George S, Murphy SB (1985) Prognostic importance of blast cell DNA content in childhood acute lymphoblastic leukaemia. Blood 65: 1079-1086
McGuire WL (1987) Prognostic factors for recurrence and survival in human breast cancer. Breast Cancer Res Treat 10: 5-9
Morgan DR, Williamson JMS, Quirke P, Clayden AD, Smith MEF, O'Brien CJ, Alison DL, Child JA, Bird CC (1986) DNA content and prognosis of non-Hodgkin's lymphoma. Br J Cancer 54: 643-649
Mossler JA, McCarthy KS, Woodard BH, Mitchener LM, Johnston WW (1982) Correlation of mean nuclear area with oestrogen receptor content in aspiration cytology of breast carcinoma. Acta Cytol (Baltimore) 26: 417-422

NCI (1982) National Cancer Institute sponsored study of classifications of non-Hodgkin's lymphomas; summary and description of a working formulation for clinical usage. Cancer 49: 2112-2135

Owainati AAR, Robins RA, Hinton C, Ellis ID, Dowle CS, Ferry B, Elston CW, Blamey RW, Baldwin RW (1987) Tumour aneuploidy, prognostic parameters and survival in primary breast cancer. Br J Cancer 55: 449-454

Rappaport H (1966) Tumors of the hematopoietic system. In: Atlas of tumor pathology, sect 3, part 8. Armed Forces Institute of Pathology, Washington, DC

Serfling RJ (1980) Approximation theorems of mathematical statistics. Wiley, New York

Spaander PJ, Ruiter DJ, Hermans J, de Voogt HJ, Brussee JAM, Boon ME (1982) The implications of subjective recognition of malignant cells in aspirations for the grading of prostatic cancer using cell image analysis. Anal Quant Cytol 4: 123-127

Stenkvist B, Bengtsson E, Eriksson O, Jarkrans T, Nordin B, Westman-Naeser S (1981) Correlation between cytometric features and mitotic frequency in human breast carcinoma. Cytometry 1: 287-291

Stöber U, Schmidt U (1980) Zur Klinik des Prostatakarzinoms unter Berücksichtigung zyto- und histomorphologischer Befunde. Urol Int 35: 233-239

Stuart-Harris R, Hedley DW, Taylor IW, Levene AL, Smith IE (1985) Tumour ploidy, response and survival in patients receiving endocrine therapy for advanced breast cancer. Br J Cancer 51: 573-576

Taylor IW, Musgrove EA, Friedlander ML, Foo MS, Hedley DW (1983) The influence of age on the DNA ploidy levels of breast tumours. Eur J Cancer Clin Oncol 19: 623-628

Velez-Garcia E, Durant J, Gams R, Bartolucci A (1983) Results of a uniform histopathologic review system of lymphoma cases; a ten-years study from the Southeastern Cancer Study Group. Cancer 52: 675-679

Wallgren A, Silversward C, Zajicek J (1976) Evaluation of needle aspirates and tissue sections as prognostic factors in mamma carcinomas. Acta Cytol (Baltimore) 20: 313-318

Zajdela A, Saravis de la Riva L, Ghossein NA (1979) The relation of prognosis to nuclear diameter of breast cancer cells obtained by cytologic aspiration. Acta Cytol (Baltimore) 23: 75-81

Prognostic Validity of the DNA Malignancy Grade in Breast Cancer

A. Böcking, C. Goecke, and S. Biesterfeld

Introduction

Grading of tumor malignancy should provide an index for the malignant potential of a tumor, independently of its histogenetic classification. Clinically valid grading systems should fulfill the following requirements:

1. High prognostic validity, which means correlation with survival probability of the individual patient and with other clinically relevant features like tumor stage, lymph node involvement, metastases, relapse-free interval, or response to therapy.
2. High representativity, which means that samples from different areas of the same tumor should provide similar results.
3. High reliability, which means inter- and intraobserver reproducibility.

Conventional grading systems are based on the subjective evaluation of histo- or cytomorphological patterns and differentiate between only two to four groups of malignancy grades. Their main disadvantage is their bad interobserver reproducibility, which was found to be only 69% for the Bloom and Richardson grading for breast cancer, tested by Champion and Wallace (1971). Stenkvist et al. (1979) found a 65% interobserver reproducibility for that grading system which was adopted by the WHO.

To overcome these problems attempts were made to objectively quantitate prognostically relevant morphological features using morphometrical methods. So far, however, histomorphological patterns, relevant for grading tumor malignancy, cannot be analyzed automatically by image analysis systems. Not even with manual or semiautomated systems has it been possible, up to now, to grade histomorphological patterns. Thus, attempts to measure morphological features for grading purposes have so far been concentrated on less complex elements like cells or nuclei. This is the main reason, why attempts towards malignancy grading by morphometry and image analysis were first carried out on cytologic material.

Material and Methods

Tissue specimens were investigated from 104 breast cancer patients with a known TNM stage, treated by mastectomy and axillary lymphadenectomy, and with a follow-up period up to 10 years. The following variables were tested for their correlation with survival time and with each other: TNM staging factors according to the UICC (Spiessl et al. 1982), histopathologic grading according to Bloom and Richardson (1957), mean nuclear area and variability, DNA histogram classification according to Auer et al. (1980), and DNA malignancy grade according to Böcking et al. (1984), and Böcking and Auffermann (1986). DNA malignancy grading was additionally tested for its interobserver reproducibility.

Monolayer smears of cancer cells were prepared from paraffin-embedded, formalin-fixed tissue derived from mastectomy specimens, using mechanical and enzymatic techniques according to Delgado et al. (1984) and van Driel-Kulker et al. (1985). Measurements of nuclear DNA and area were performed semiautomatically using a TV image analysis system combined with an automated microscope (Fig. 1; Auffermann et al. 1984). Feulgen staining was performed automatically with pararosaniline using a modified Shandon Varistain 24-3 (Chatelain et al. 1988). Only tumor cells were measured and lymphocytes were used as reference cells (correction factor × 1.19).

Results

All variables under consideration, besides the age of the patient, had a significant overall influence on survival time according to the Cox multivariate regression model. The best prognostic indices were given by the TNM staging system. The Breslow test revealed highly significant differences in survival probabilities between groups of different T, N, and M categories (besides T3 vs. T4).

The histopathologic grading according to Bloom and Richardson (1957) was

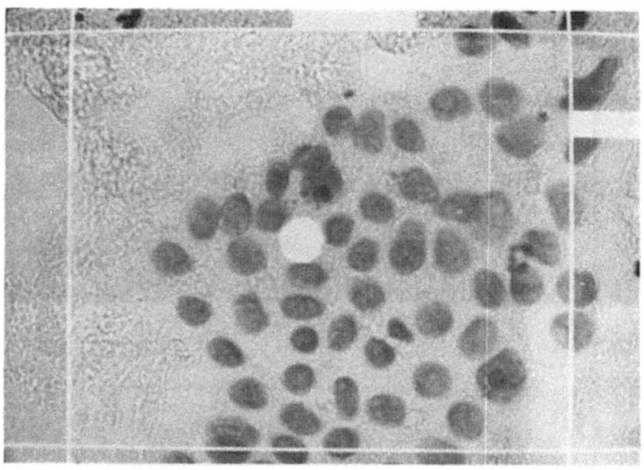

Fig. 1. Interactive monitor of the TV image analysis system TAS Plus (Leitz, Wetzlar, FRG) for rapid nuclear DNA measurements. The nucleus to be measured is covered by a mask individually adapted to its size and shape

also found to be of great prognostic significance; the Breslow test showed highly significant differences in survival time between the three grades of malignancy.

For the statistical evaluation of the DNA malignancy grade and the nuclear area the patients were grouped according to a 25%/50%/25% quantilization model, as the values for these measurements are continuous without preformed grouping. The Breslow test revealed significant differences in survival probability for the three groups with different DNA malignancy grades (Fig. 2). Measurements of nuclear area showed significant differences only between groups 2 and 3. The same holds for the variability of nuclear areas. Auer DNA histogram classification, which differentiates between four prognostic groups, revealed significant differences only between groups 3 and 4, not between groups 1 and 2 or 2 and 3.

These variables were also tested for comparison of their different prognostic validities, using the Cox multivariate stepwise regression model. The TNM system had the highest prognostic validity. Histopathologic grading had second place for prognostic validity, meaning that it provided prognostic information additional and complementary to that given by the staging system. If these variables were known, no other index added significant prognostic information. If the Bloom and Richardson grading was removed from the Cox model, because of its bad reproducibility, the DNA malignancy grade was accepted as the second best prognostic index after the TNM system and gave similar additional prognostic information to that given by the subjective histological grading. Again, no other variable could add further significant prognostic information once staging and DNA grading were known, not even nuclear area. If the various grading systems were introduced into the Cox model, the histopathologic grading revealed the best prognostic validity, followed by DNA malignancy grade, while the Auer classification and mean nuclear area added no further prognostic information.

A strong correlation of DNA malignancy grade with lymph node status was found. Eleven percent of the patients in the DNA malignancy grade group 1 and 58% in the other groups revealed lymph node metastases.

For the final decision on the clinical applicability of these grading indices, their different reproducibilities were determined. A 57% interobserver reproduc-

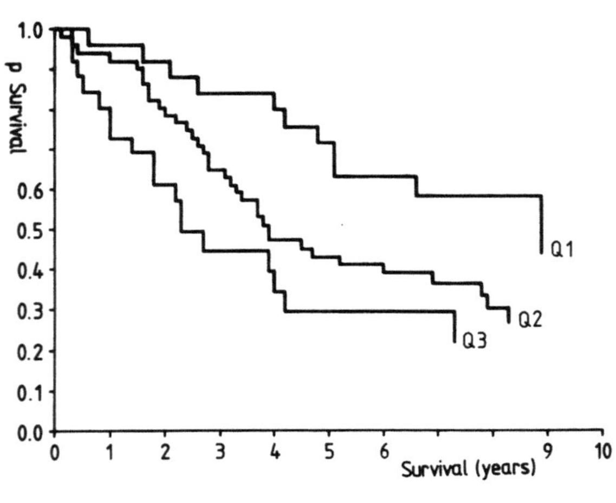

Fig. 2. Kaplan-Meyer survival curves for three groups of 104 breast cancer patients (25%/50%/25% quantilization) according to different DNA malignancy grades

ibility was found for the subjective histopathologic grading and a 62% reproducibility for the histogram classification according to Auer. An 82% interobserver reproducibility was found for the DNA malignancy grading when this scalar parameter was artificially divided in three groups. A more adequate measure for this index is that the 95% confidence limits lie within ± 0.3 DNA malignancy grade, i.e., that in 95% of cases a second measurement by another observer gives the same result ± 0.3 DNA malignancy grade.

Discussion

Up to now at least four major publications have reported on the prognostic significance of the nuclear DNA distribution of breast cancer. Atkin (1972) reported on 67 patients with a follow up of 8 years. Auer et al. (1980) reported on 112 patients with a follow up of 10 years. Krug and Schönfelder (1984) reported on 62 patients with 5-year follow up. Klintenberg et al. (1985) reported on 150 patients with 5-year follow up. Yet Opfermann et al. (1987) found insufficient correlation between DNA measurements and survival time for individual patients. These authors measured 200 epithelial but not definitely cancer cells per smear. In all of these studies DNA histograms were subjectively typed in two to four groups, resulting in insufficient reproducibility. Opfermann et al. (1987) reported a 59% interobserver reproducibility for certain patterns in the Auer classification.

In order to arrive at an objective and reproducible interpretation of nuclear DNA distribution in human tumors, we proposed in 1984 and 1986 a DNA malignancy grade. This is based solely on the variance of the nuclear DNA distribution around the normal, diploid value. This is the prognostically most relevant variable according to our own investigations. The biologic background for this assertion is the observation in tumor cytogenetics that the range of chromosomal aneuploidy is a prognostically highly relevant variable in many tumors, in addition to the degree of chromosomal aberration (Pauwels 1987). This variability of chromosomal aneuploidy and its extent corresponds with the variance of the nuclear DNA distribution around the normal 2c value in cytophotometry. This is for example quantified by our 2c deviation index (2cDI, Böcking and Auffermann 1986). As the absolute figures of variance will not be understood as a prognostic index by most physicians, we perform a logarithmic transformation to arrive at a continuous scale of DNA malignancy grade, ranging from 0.01 to 3.0, meeting the proposals of the UICC.

Summing up the results described above, TNM staging was shown to have the strongest prognostic validity. Besides that, the histopathologic grading system of Bloom and Richardson had significant additional prognostic validity. Leaving this subjective grading, because of its insufficient reproducibility, DNA malignancy grade revealed similarly good prognostic validity but with a much better reproducibility than all subjective grading systems tested. The nuclear area was found to be less prognostically significant than the DNA malignancy grade.

References

Atkin NB (1972) Modal deoxyribonucleic acid value and survival in carcinoma of the breast. Br Med J 1: 271-272

Auer G, Caspersson TO, Wallgren AS (1980) DNA content and survival in mammary carcinoma. Anal Quant Cytol 2: 161-165

Auffermann W, Repges R, Böcking A (1984) Rapid diagnostic DNA cytometry with an automatic microscope and a TV image analysis system. Anal Quant Cytol 6: 179-188

Bloom HJG, Richardson WH (1957) Histological grading and prognosis in breast cancer. Br J Cancer 11: 359-377

Böcking A, Auffermann W (1986) Algorithm for DNA-cytophotometric diagnosis and grading of malignancy (letter to the editor). Anal Quant Cytol Histol 6: 363

Böcking A, Adler CP, Common HH, Hilgarth M, Granzen B, Auffermann W (1984) Algorithm for a DNA-cytophotometric diagnosis and grading of malignancy. Anal Quant Cytol 6: 1-8

Champion HR, Wallace IWJ (1971) Breast cancer grading. Cancer 25: 441-448

Chatelain R, Willms A, Biesterfeld S, Auffermann W, Böcking A (1988) Automated Feulgen staining with a temperature controlled machine. Anal Quant Cytol Histol (submitted for publication)

Delgado R, Mikuz G, Hofstädter F (1984) DNA-Feulgen-cytometric analysis of single cells from paraffin embedded tissue. Pathol Res Pract 179: 92-94

Klintenberg C, Wallgren A, Bjelkenkrantz K, Castenssen J, Humla S, Nordenskoeld B, Shoog L (1985) DNA distribution, cytosol estrogen receptors and axillary nodes as prognostic predictors in breast carcinoma. Acta Radiol Oncol 24: 253-258

Krug H, Schönfelder M (1984) Impulszytophotometrie and Katamnese beim Mammakarzinom. Acta Histochem [Suppl] (Jena) 30: 117-127

Opfermann M, Brugal G, Vassilakos P (1987) Cytometry of breast carcinoma. Significance of ploidy balance and proliferation index. Cytometry 8: 217-224

Pauwels RPE, Smeets AWGB, Skapers RFM, Geraedts JPM, Debruyne FMY (1987) Grading of superficial bladder cancer. 2. Cytometric classification. Br J Urol (in press)

Spiessl B, Scheibe O, Wagner G (eds) (1982) TNM Atlas. Springer, Berlin Heidelberg New York

Stenkvist B, Westman-Naeser S, Vegelius J, Holmquist J, Nardin N, Bengtson F, Eriksson O (1979) Analysis of reproducibility of subjective grading systems for breast carcinoma. J Clin Pathol 32: 979-983

Van Driel-Kulker AMJ, Mesker WF, v Velzen J, Tanke HJ, Feichtinger J, Ploem JS (1985) Preparation of monolayer smears from paraffin-embedded tissue for image cytometry. Cytometry 6: 268-272

Correlations of Cytomorphology, Hormone-Receptor Status, and Age in Patients with Breast Carcinoma*

U. Schenck, W. Eiermann, G. Burger, P. Gais, U. Jütting, K. Rodenacker, and U. B. Schenck

Introduction

Breast carcinomas with negative results for hormone receptor tests show more striking criteria of malignancy than hormone receptor-positive ones. By means of high resolution image cytometry we were able to separate hormone receptor-positive carcinomas from hormone receptor-negative carcinomas better than by visual or planimetric examination (Burger et al. 1986; Schenck et al. 1985, 1986). In the present study of 69 cases we tried by means of image cytometry to reclassify and correlate the cases for age, visual nuclear grade, and different methods of hormone receptor evaluation. The main questions were: Which method of hormone receptor evaluation gives results which correlate best with those of cytomorphology? Is image cytometry able to show differences between cells from breast carcinomas of younger and older patients which might be anticipated from the different age distributions of receptor-positives and receptor-negative carcinomas? Can image analysis be used to reclassify cases with regard to visual nuclear grade? How closely do case evaluations by different classifiers correlate with each other?

Material and Methods

The 69 cases are from a continuous series of breast tumors sent for hormone receptor analysis but restricted to those cases having adequate slides. Biopsy smears were taken from the deep frozen tumors, air dried, and stained according to Pappenheim for visual microscopical analysis. Smears for image cytometry were prepared in the same way, but stained according to Feulgen (hydrolysis data: 5 *N* HCl, 28°, 20 min). All specimens were analyzed visually at a maximum magnification of 500. The cases were assigned to seven diagnostic groups:

Group 1: only small nuclei, possibly benign
Group 2: only or almost only small nuclei
Group 3: predominantly small nuclei
Group 4: predominantly large nuclei

* Work supported by the Deutsche Forschungsgemeinschaft, Sonderforschungsbereich 324.

Group 5: nucleoli small and/or infrequent
Group 6: prominent nucleoli
Group 7: multiple prominent nucleoli

Intermediate diagnostic groups (e.g., 4,7) are given so that the slides are attributed to the group which characterizes the case better and a continuous grading system arises.

Estrogen receptor (ER) status was evaluated by the dextran-coated charcoal method (DCC), by estrogen receptor enzyme immunoassay (ER-EIA, Abbott), and by estrogen receptor immunocytochemical assay (ER-ICA, Abbott) applied to biopsy smears. The progesteron receptor (PR) status was determined by the DCC method.

Images of about 100 nuclei per case were taken by TV camera (objective magnification of 100, oil, narrow band optical filter of 550 nm). Each cell image was preprocessed and more than 50 planimetric and chromatin features were measured. All nuclei from all specimens belonging to one class (e.g., ER-DCC positive) were pooled, and linear discriminant analysis was applied to reclassify the single cells. In this reclassification all nuclear features were considered and selected stepwise on the basis of F statistics. Depending on the F values set for feature rejection, up to 15 features were used. From the single cell reclassification, a value was obtained for the a posteriori probabilities (APOPs) of each cell belonging to either one of the two classes. All cells of a specimen were used to obtain an APOP distribution. This represents the position of the specimen in the multidimensional feature space. The mean value of the APOP distribution was used to rank the specimens. In order to reclassify the specimens, an optimal threshold was introduced. Specimens for which the distance of the mean APOP to the threshold is less than the standard error of the mean or two times this error were classified as unclear.

Simplified, the APOP value of a case can be understood as representing its image cytometrical evaluation. It indicates the probability of a case belonging to one of two groups, the nuclei of which have been pooled together. When the nuclei are pooled according to different case characteristics (e.g., ER-DCC negative vs. ER-DCC positive; cases with nucleoli vs. cases with no nucleoli; patients younger than 60 years vs. patients older than 60 years) different APOPs result from the different classifiers. In the APOP/APOP correlation the evaluation of the cases using one classifier is correlated to the evaluation of the cases using another classifier.

Results and Discussion

Reclassification of the positive and negative results gained by all the different methods of hormone receptor evaluation was possible in a majority of cases (Table 1). Especially if the estrogen receptor was determined by monoclonal antibodies, cases with negative and positive results were clearly separated. The best classification results were obtained if only the 29 cases with concordant cytosolic receptor evaluations were regarded (16 cases: ER(DCC)-, PR(DCC)-, and ER(EIA)-; and 13 cases: ER(DCC)+, PR(DCC)+, and ER(EIA)+).

Table 1. Image cytometrical reclassification of results gained by different methods of hormone receptor evaluation

Receptor evaluation			Reclassification			
	Positive	Negative	Correct		Discrepant	Unclear
			positive	negative		
ER-EIA	47	22	36	16	10	7
ER-ICA	32	33	23	26	12	4
ER-DCC	38	31	23	19	9	18
PR-DCC	22	47	12	28	16	13
Concordant cases	13	16	12	11	2	4

The case evaluations correlate as to ER-EIA APOP and ER-DCC APOP ($r=0.80$). This indicates that both methods give essentially the same information, although not necessarily for any given individual case, of course. Since the best reclassification results are based on hormone receptor evaluation with monoclonal antibodies, and since the image analytical ranking predominantly follows the ER-EIA result when hormone receptor evaluations disagree, ER-EIA seems to be more closely correlated to morphology than ER-DCC.

Our material (Oncology Laboratory, Frauenklinik Großhadern) confirms that there are different age distributions for hormone receptor-positive and hormone receptor-negative breast carcinomas (Fig. 1). The age distribution of all 1698 cases shows a double-peaked curve. The peak of the receptor-negative carcinomas is in the group of patients under 50 years; the peak of the receptor-positive carcinomas is in the group over 60 years.

The age distribution of the 69 cases which underwent reclassification was similar to that of the large series. It was possible to classify patients by age (Table 2).

The case evaluation correlates in terms of age-APOPs (thresholds 50 and 60 years) and ER-EIA APOPs ($r=0.76$ and 0.86 respectively). It remains unclear

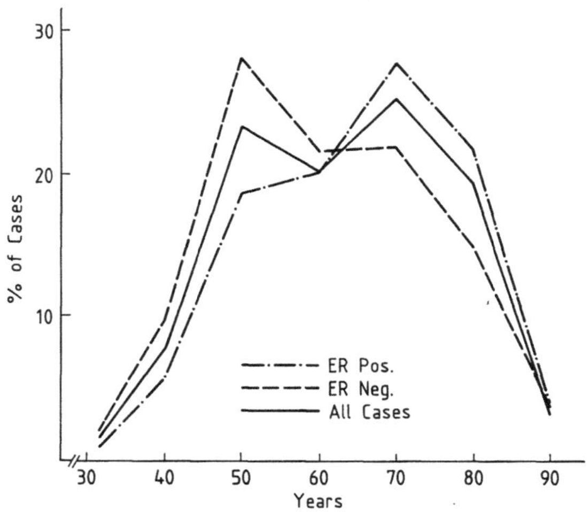

Fig. 1. Age distribution of 1698 patients with breast carcinoma

whether the classification by age mainly reflects the hormone receptor content or whether other influences like hormone levels play a major role.

When the 69 tumors were subdivided in groups without nucleoli (visual nuclear grade below 4.5) and with nucleoli (visual nuclear grade above 4.4), the reclassification based on visual nuclear grade showed only seven discrepancies (Table 3).

The evaluation based on ER-EIA APOP and that based on the visual nuclear grade APOP correlate closely (Fig. 2). This means that a classifier based on the visual nuclear grade can predict the hormone receptor status (ER-EIA) reasonably well, and vice versa, i.e., the ER-EIA can be used as a basis for an image analytical grading system. The average ER content of 75 fmol/mg was lower among the

Table 2. Image cytometrical reclassification by patient age

AGE (years)	Reclassification				AGE (years)	Reclassification			
	<50	≥50	Unclear	Total		<60	≥60	Unclear	Total
<50	15	4	–	19	<60	21	8	5	34
≥50	11	28	11	50	≥60	7	24	4	35

Table 3. Image cytometrical reclassification by visual nuclear grade

Visual nucleoli	Reclassification			
	Without	With	Unclear	Total
Without	22	3	1	26
With	4	37	2	43

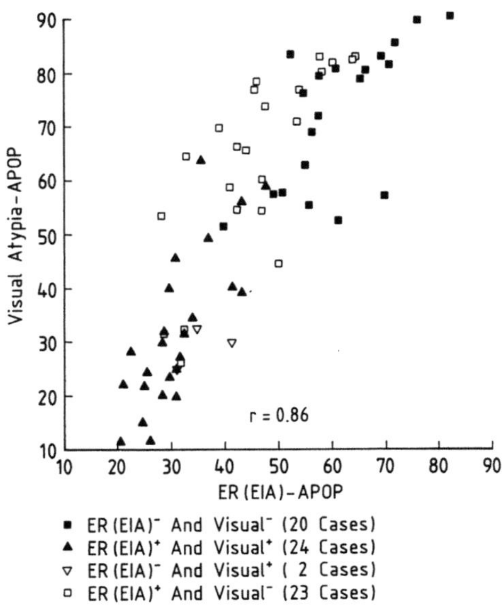

Fig. 2. Correlation of ER-EIA a posteriori probabilities *(APOP)* and visual atypia APOP

24 receptor-positive cases with nucleoli (visual –) as compared to the 150 fmol/mg of the 23 cases without nucleoli (visual +). Only two of the ER-EIA negative cases showed no nucleoli, which again indicates that visual cytology already gives useful information concerning hormone receptor status.

Summary

High resolution image cytometry was applied to 69 Feulgen-stained biopsy smears of primary breast carcinomas. The results were related to the estrogen receptor content of their tissue as determined by different methods including those with monoclonal antibodies, to the visual nuclear grade of Pappenheim-stained specimens, and to the age of the patients.

We were able to show that by multivariate discriminant analysis based on the feature distributions of the cells examined, a substantial reclassification of the cases with respect to their hormone receptor status, to their visual grade, and to the age of the patients was possible.

References

Burger G, Jütting U, Gais P, Rodenacker K, Schenck U, Giaretti W, Wittekind D (1986) The role of chromatin pattern in automated cancer cytology. In: Mary JY, Rigaut JP (eds) Quantitative image analysis in cancer cytology and histology. Elsevier, Amsterdam, pp 91-103

Schenck U, Burger G, Jütting U, Peters-Welte C, Eiermann W (1985) Punktionszytologie der Mamma: Korrelation visueller und bildanalytischer Untersuchungen mit dem Hormonrezeptorstatus. Geburtsh Frauenheilk 45: 17-21

Schenck U, Burger G, Jütting U, Eiermann W (1986) Zytomorphologisches Grading und Hormonrezeptorstatus beim Mammakarzinom - visuelle Untersuchungen an Ausstrichpräparaten. Dtsch Med Wschr 111: 1949-1953

Quantitative Ultraviolet Microscopy*

S. Witte, W. H. Bloss, P. Schwarzmann, and G. Sträßle

Introduction

Quantitative description of the "chromatin texture" of cell nuclei is the main method of high resolution cytometry. It is usually performed on Papanicolaou-stained smears [10]. Other authors have analyzed fluorescent-stained cells which also show a similar nuclear texture [4]. A nuclear texture is apparent even after the rather vigorous hydrolytic procedure of the Feulgen pretreatment with subsequent fluorescent staining, and this nuclear texture can be described quantitatively [8]. The question arises as to the extent to which the nuclear structure is influenced by the fixation of the cell or by the staining procedures themselves.

In this context, it would be of advantage to be able to analyze a nuclear texture in unfixed and even unstained cells, thus avoiding artifacts which may result from fixation and staining. The method of choice for microscopy of unfixed wet cell smears is phase contrast microscopy. However, this does not reveal any clear nuclear structure at all. Even using the more sophisticated method of differential interference contrast microscopy, we cannot see more details within the nucleus, in fact even fewer than with the phase contrast method [14].

These poor results with unfixed unstained material contrast sharply with those obtained using the monochromatic absorption method with ultraviolet illumination. Nucleic acids with an absorbance peak at a wavelength of 263 nm can be visualized very clearly by means of the ultraviolet microscope and the ultraviolet-sensitive television camera (Fig. 1). This method was first introduced by Caspersson [1] and further developed by his laboratory [2] and by Sandritter [5].

The aim of our study was to investigate whether ultraviolet absorbance microscopy can be used for quantitative evaluation of the nuclear texture of unfixed and unstained cellular material derived from routinely obtained abrasive cell smears.

* Supported by the Deutsche Forschungsgemeinschaft.

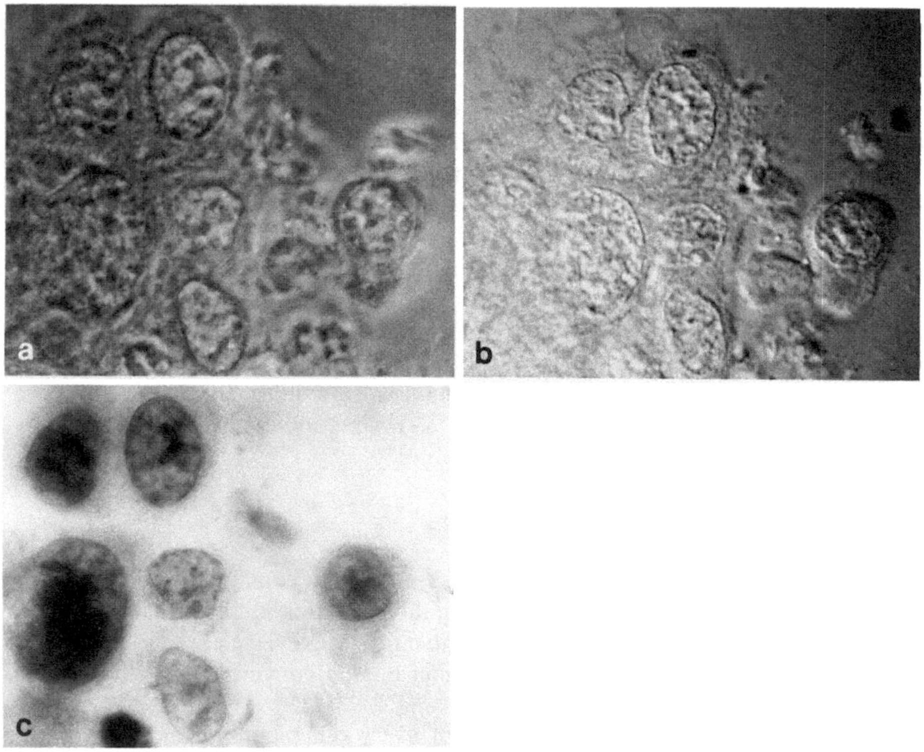

Fig. 1a–c. Tumor cells of gastric carcinoma, obtained by gastroscopic brushing. Wet unfixed preparation, ×560. **a** Phase contrast. **b** Differential interference contrast. **c** Ultraviolet monochromatic absorbance $\lambda = 263$ nm

Methods

Cellular Material

We investigated abrasion material from gastroenterological investigations obtained from the stomach and colon with the endoscopically guided brush abrasion technique in the Endoscopy Department of our hospital [12]. It is routinely available to the cytology laboratory for conventional cytodiagnostics, which we carry out with the phase contrast native method and Papanicolaou staining. The diagnoses were confirmed by forceps biopsies taken from the same area of the mucosa at the same endoscopy and examined histologically. The cells collected by brushing the suspect mucosal area were rinsed from the brush with physiological saline solution and centrifuged, and the sediment was investigated immediately after its collection as wet unfixed smear preparations on clean slides under cover glass. The time which elapsed between endoscopic brushing and microscopy was around 10–20 min.

Microscopy

The ultraviolet microscope (Carl Zeiss, FRG) was used [11]. The following components are arranged on an optical bench: (a) xenon high-pressure burner XB0450 with stabilized superposed appliance, (b) heat protection cuvette with blue filter, (c) monochromator with variable diaphragm slit width, (d) deflecting mirror in the foot of the microscope, which can be switched for monochromator beam projection or for a built-in low-voltage lamp and collector with diaphragm element for bright field or phase rings, (e) microscope with quartz optical systems: Ultrafluar condenser 0.8, Ultrafluar 32:1, n. A. 0.4 as objective (these objectives are used optionally with and without phase ring), (f) beam divider in the microscope tube for attachable camera, look-in eyepiece with electronic image transducer or TV camera, (g) UV-sensitive TV camera in combination with monitor, (h) the microscope table has a motor-driven scanning facility in X-Y direction controlled with a joystick.

Taking Pictures

In the cytology laboratory, the cell image pick-up system consists of the following components: video camera with analog digital converter, digital image storage 512 × 256 × 8 bit, host computer, and magnetic tape station. The stored pictures are rendered visible on a TV monitor via a digital analog converter and checked in this way.

A series of six single images are taken of every cell picture scene. These are monochromatic absorption pictures with the wavelengths 263, 270, 280, 290, and 313 nm as well as the control picture in phase contrast (Figs. 2, 3). The monochromator slit was adjusted to optimal sensitivity of the TV camera and registered.

Image Processing System

The FAZYTAN system has been used [3]. The preprocessing includes a shading correction, noise reduction, the segmentation of the nuclear images, and a gray level histogram process.

During the registration of a spectral cell picture sequence, slight locational alterations of the unfixed and suspended cells may occur in the visual field of the microscope. An algorithm was developed to compensate for these object move-

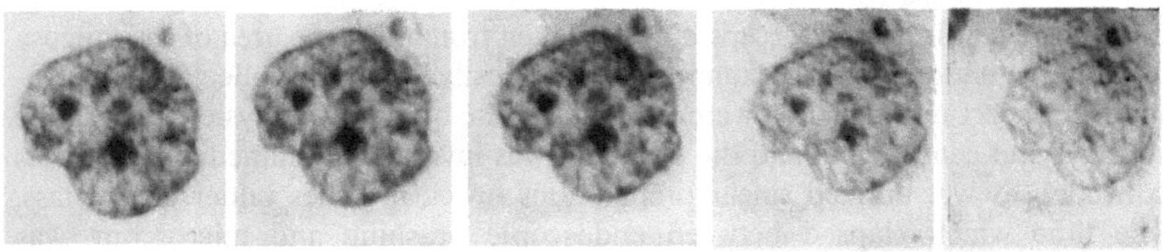

Fig. 2. Spectral cell image sequence, *left to right* 263, 270, 280, 290, and 313 nm. Gastric tumor cell, wet unfixed preparation, × 800

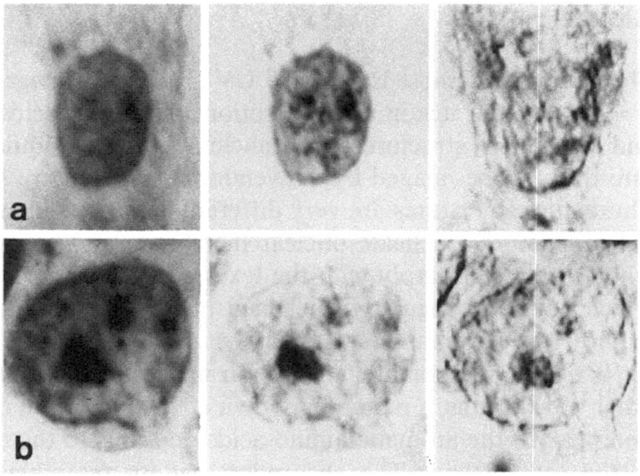

Fig. 3 a, b. Spectral cell images, 263 nm, 280 nm, 313 nm *(from left to right)*. Wet unfixed preparations, × 560. **a** Case of gastric ulcer. **b** Case of gastric adenocarcinoma. Cells collected by gastroscopic brushing

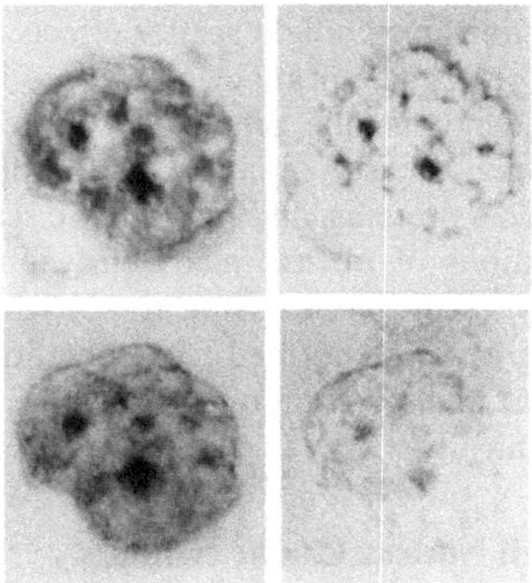

Fig. 4. Evaluated nucleic acid specific and nonhiston protein specific nuclear distribution from the recorded monochromatic cell image sequence. Gastric tumor cell, wet unfixed preparation (same cell as in Fig. 2). *Upper row,* without motion compensation; *lower row,* motion compensated

ments; it insures exact superposition of the monochromatic component pictures (see Fig. 4) [7].

Segmentation of the cell nucleus is necessary in order to be able to measure the nuclear characteristics. Because of the weak contrast of the ultraviolet absorption images of the cell nuclei from the cytoplasm, a texture-evaluating segmentation algorithm was developed which is based on a pixel classification. The calculation of the nuclear characteristics largely follows the FAZYTAN-IPS strategy. Combinations of the linear independent Minkowsky dimensions, area, size, and connectivity in sectional images are calculated from the gray-shade topography of original and nonlinear filtered cell nucleus pictures (erosion, dilatation).

Results

The subjectively most impressive UV absorption images result at the wavelength $\lambda = 263$ nm, the maximum absorption of nucleic acids. The nuclear membrane and an internal structure of the nucleus, which is similar to the chromatin structure that can be imaged by conventional staining, are shown with high contrast. These nuclear features are very different in various kinds of cells. Normal epithelial cells show a delicate nuclear membrane and a pale, fine reticular structure which contrasts sharply with the texture of the tumor cells from the same organ, which have an irregularly thickened nuclear membrane and a coarse trabecular, clumpy nuclear texture (see Fig. 3).

If the monochromatic absorption images of the same cells are investigated at other wavelengths, a pattern is shown at 280 nm which corresponds to the absorption peak of the aromatic amino-acids and thus to the nonhistone nucleoproteins in the cell nucleus. The absorption pattern resembles the picture obtained at 263 nm. The reason for this is the major overlap in the absorption characteristics of nucleic acids and proteins [9]. The extinction coefficient of proteins is much smaller than that of nucleic acids. At increasing wavelengths, the absorption images increasingly lose contrast, until a picture results at $\lambda = 313$ nm which is merely nonspecific light scatter effects and differences in refractive index.

In order to attain a separate image of the local distribution of nucleic acids and nonhistone proteins in the cell nucleus, a sandwich model was developed which is based on a linear superposition of the absorption components of nucleic acids, proteins, and nonspecific scattering in each image point [6]. The local distributions of nucleic acids and nonhistone proteins calculated in this way reveal marked differences in texture in the same cell (Fig. 4). It was thus possible to attain a gray-value normed representation of the specific texture of the two most important nuclear components in nonfixed cells of a cytodiagnostic material obtained clinically.

Cell Nucleus Characteristics

In a first series of classification experiments, two different feature sets were calculated which are based on a gray-value normed or texture-transformed representation of the cell nucleus. In contrast to the sequence filtering introduced (erosion, median, and dilatation), the central element of an image window is not assigned the gray-*value* window, but the *rank of the* gray *value* of the central element with regard to all window elements. The texture transformation (gray-value rank sequence transformation, GRT) thus emphasizes local differences in texture and suppresses global interrelationships with gray values (extinction).

The feature set UVM 120xxx contains 120 measurement values per spectral cell picture. The gray-value peak of the cell nucleus in the original picture, in the eroded picture (Ero 2 = erosion filtering with radius 2), and in the dilated picture (Dil 2 = dilatation filtering with radius 2), and the three linearly independent Minkowski dimensions area (A), size (S), and connectivity (C) are calculated from equidistant sectional images. These values constitute the measurement parameters A, S, S*S/A, C/A. The measurements are applied on cell nucleus pictures at

263 nm and 280 nm. Two procedures were chosen for preprocessing: UVM 120 norm is gray value norming, UVM 120 GRT is gray value rank sequence transformation.

Classification Experiments

The native data bank created comprised 711 cell picture scene sequences. These were divided into cell classes (Fig. 5) which were laid down by the investigator in microscopy of the original preparations. These cell classes are:

1 Normal cylindrical form of the epithelia ($n = 137$)
2 Round cell form of the epithelia ($n = 51$)
3 Epithelial cells with swelling of the nuclei ($n = 62$)
4 Epithelial cells with slight to moderate nuclear atypia and solitary enlarged nucleoli ("ulcer cells") ($n = 150$)
5 Cells with severe nuclear atypia ($n = 174$)
6 Tumor cells ($n = 137$)

The cell classifiers were calculated by means of a linear polynomial set. The six cell classes were grouped into the two classes N (for normal, 1–3) and P (for pathological, 4–6).

In a first experiment, the feature set UVM 120 norm was used. The reclassification of the learning data set revealed a classification error of 15% (Table 1) in the two-class case. The test data set of 461 cells revealed a classification error of

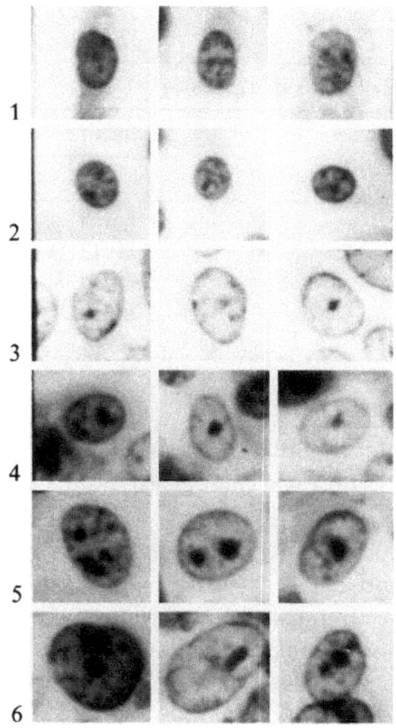

Fig. 5. Cytological cell nuclei classes of benign and malignant cells, obtained by gastrointestinal endoscopic brushing (see text). Wet unfixed preparations, ×256. Monochromatic absorbance, $\lambda = 263$ nm

Table 1. Classification experiment I

Learning set		Machine		Total
		N	P	
Observer	N	222	28	250
	P	47	203	250

Misclassified: 15%

Test set		Machine		Total
		N	P	
Observer				
N	1	127	10	250
	2	44	7	
	3	51	11	
P	4	74	76	461
	5	35	139	
	6	8	129	

Misclassified: 20%

Classifier type: polynomial; feature set: UVM 120 norm

Table 2. Classification experiment II

Learning set		Machine		Total
		N	P	
Observer	N	227 (232)	23 (18)	250
	P	37 (21)	213 (229)	250

Misclassified: 12% (8%)

Test set		Machine		Total
		N	P	
Observer				
N	1	129 (131)	8 (6)	250
	2	43 (44)	8 (7)	
	3	55 (57)	7 (5)	
P	4	67 –	83 –	461
	5	26 (34)	148 (140)	
	6	11 (5)	126 (132)	(311)

Misclassified: 17% (10%)

Classifier type: polynomial; feature set: UVM 120 GRT

20%. Cell class 4 showed an equivalent assignment of the two selective classes N and P.

In a second experiment, the feature set UVM 120 GRT was used. A classification error of 12% now resulted from the same cell material in the reclassification experiment (Table 2). The more extensive test data set revealed a classification error of 17%.

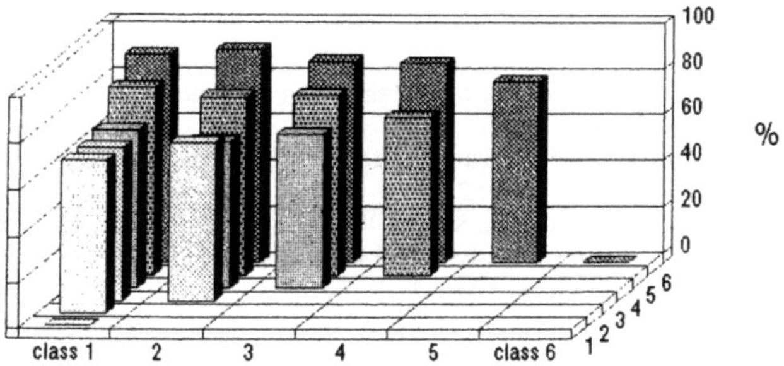

Fig. 6. Separation efficiency of cytological classes 1-6 (see Fig. 5 and text), expressed by results of classification experiment 2. Recognition rates in percent

According to the results of the first experiment with regard to cell class 4, we have repeated the classification experiments after omission of this cell class (numbers in brackets). In this way, the classification performance was improved from 12% to 8% classification errors in the reclassification of the learning data set, and from 17% to 10% classification errors in the test data set.

Accordingly, these first classification experiments show that the nuclear textures obtained allow calculation of cell-class-specific differences (Fig. 6). In relation to a two-class case (normal and pathological), cell class 4 does not show any specific group affiliation. The cells of class 4 belong to the clinical picture of gastric ulcer [13], a malignant tumor being ruled out histologically. This cell class thus forms a group of its own which differs from the groups of the normal to slightly reactively altered cell nuclei on the one hand and the highly atypical cells or the tumor cells on the other hand. Furthermore, the atypical classification experiments showed that diagnostic capacity is raised by application of texture-related gray-value rank transformation (GRT) compared to a gray value normed preprocessing.

Summary

By ultraviolet microscopy of unfixed unstained wet cell smears a nuclear texture resembling chromatin structure can be demonstrated. It is mainly caused by the spectral absorbances of the cell constituents. Using multispectral images, detection of and compensation for movement of cells between consecutive images, and image separation procedures, material-specific textures of nucleic acids and nonhiston proteins are obtained simultaneously for automated segmented cell nuclei. High resolution nuclear texture analysis is based on the FAZYTAN-IPS strategy. For classification experiments we used cellular material obtained by routine endoscopic abrasive techniques of the gastrointestinal tract. Using subjective cytodiagnostic criteria, cells are classified into six classes and collected as learning set data base. Classification tests using a two-class case (normal cylindrical cells and reactive cellular changes versus tumor cells and severe atypias) gave the following results: Learning set of 500 cells misclassified (false positives and false negatives) 8%, test set of 711 cells misclassified 10%.

References

1. Caspersson T (1936) Über den chemischen Aufbau der Strukturen des Zellkernes. Skand Arch Physiol 73, Suppl 8
2. Caspersson TO (1987) History of the developement of cytophotometry from 1935 to the present. Anal Quant Cytol 9: 2-6
3. Erhardt R, Reinhardt ER, Schlipf W, Bloss WH (1980) FAZYTAN: a system for fast cell segmentation, cell image analysis and feature extraction based on TV-image pickup and parallel processing. Anal Quant Cytol 2: 25-40
4. Gledhill BL, Campbell GL (1983) Microfluorimetric comparison of chromatin during cytodifferentiation. In: Thaer AA, Sernetz M (ed) Fluorescence techniques in cell biology. Springer, Berlin Heidelberg New York, pp 151-162
5. Sandritter W (1958) UV-Mikrospektrophotometrie. In: Graumann W, Neumann I (eds) Handbuch der Histochemie, vol 1/1. Fischer, Stuttgart, pp 220-338
6. Sträßle G, Reinhardt ER, Witte S (1983) Texturanalyse an nativem Zellmaterial mit Hilfe der Ultraviolettmikroskopie. Fifth DAGM Symposion, Karlsruhe 1983. VDE Fachberichte 35: 131-136
7. Sträßle G, Straub B (1985) Determination of motion vectors applied to cell image sequences. In: Suetens P, Young IT (eds) Medical image processing. SPIE Proc 593: 64-71
8. Tanke HJ, von Ingen EM, Ploem JS (1979) Acriflavin-Feulgen-Stilbene (AFS) staining: a procedure for automated cervical cytology with a television based system (LEYTAS). J Histochem Cytochem 27: 84-86
9. Thiessen G, Thiessen H (1977) Microspectrophotometric cell analysis. Prog Histochem Cytochem 9/4
10. Wied GL, Bartels PH, Bibbo M, Sychra JJ (1980) Cytomorphometric markers for uterine cancer in intermediate cells. Anal Quant Cytol 2: 257-263
11. Witte S (1968) Moderne Entwicklungen der Ultraviolettmikroskopie in der klinischen Zytologie. Verh Dtsch Ges Inn Med 74: 617-619
12. Witte S (1970) Gastroenterologische Zytologie. In: Ottenjann R, Elster K, Witte S (eds) Gastroenterologische Endoskopie, Biopsie und Zytologie. Thieme, Stuttgart, pp 123-142
13. Witte S, Bressel D (1965) Die zytologische Diagnose des Ulcus ventriculi. Dtsch Med Wschr 90: 1100-1104
14. Witte S, Ruch F (eds) (1979) Moderne Untersuchungsmethoden in der Zytologie, 2nd edn. Witzstrock, Baden-Baden

V. Immunocytochemistry – Survey Articles

Immunocytochemistry in Cytology

H. Höfler, G. Wirnsberger, and H. P. Dinges

In several reviews (Nadji 1980; Walts and Said 1983; Koss 1984; Chess and Hajdu 1986) the considerable value of the application of immunocytochemical methods in cytology (ICC) has been indicated. However, despite the high number of reports focused on the application of ICC as a diagnostic tool it did not have the enormous practical impact of immunohistochemistry. This may be due to methodological difficulties and the existence of a vast number of different protocols for fixation and storage and the different immunocytochemical techniques used for cytological preparations. Because of inaccuracies in the descriptions of fixation and processing of cell preparations in many papers on ICC, a comparison of data obtained by different authors is impossible or at least questionable.

The goal of this study was comparatively to evaluate ICC procedures. We studied the applicability of antibodies directed to three different groups of antigens (intermediate filament proteins, other cytoplasmic antigens, and cell surface antigens) for ICC and compared the results obtained with different antibodies after several methods of fixation and storage.

Material and Methods

To allow a comparison of different procedures for fixation and cell processing, the reactivity of individual antibodies was studied using large series of identically prepared smears. To obtain this material we used (a) imprint preparations of benign breast lesions (fibrocystic disease, $n=5$) and invasive ductal breast carcinomas ($n=2$); (b) imprint preparations of hyperplastic tonsils ($n=6$); and (c) bronchoalveolar lavages (BALs, $n=12$). From each case 70–100 cell preparations were prepared and processed in different ways, as described below. For each different procedure at least two identical preparations were processed.

Preparation of Glass Slides and Fixation. Imprints of fresh tissue were immersion-fixed in different fixatives (Table 1) after or without air drying prior to fixation. Air-dried smears were stored for 1 day at room temperature prior to fixation. After fixation and immersion in phosphate-buffered saline (PBS; 0.1 *M*, pH 7.2) the preparations were either processed immediately for ICC (without air drying) or 2 months later after storage under special conditions (see below). Spray fixation (Merckofix spray) was performed according to the specifications of the manufac-

Table 1. Fixatives and times of fixation (at room temperature)

Fixative	Time
Ethanol 70%	5 min
Ethanol 95%	5 min
Acetone 100%	5 min
Methanol 100%	5 min
Acetone 100%/methanol 100% 1:1	5 min
Ethanol 95%/acetic acid 5%	5 min
Delaunay's fixative, modified[a]	5 min
Formalin 10%, buffered in PBS pH 7.2	1 min
Paraformaldehyde 4%, buffered in PBS pH 7.2	1 min
Formalin 10%/acetone 100% 1:1	1 min
Formalin 10%/methanol 100% 1:1	1 min
Glutaraldehyde 0.05%, buffered in Tris pH 7.2	1 min
Merckofix spray	

[a] Delaunay's fixative, modified: To 900 ml acetone was added 100 ml ethanol and 0.5 ml trichlor-acetic acid.

turer (Merck, FRG), but the imprints were not allowed to dry completely after the application of the spray.

Cell preparations of BALs were prepared by gentle centrifugation at 500 *g* for 10 min. The supernatant was discarded and the cell pellet was resuspended with 300 μl Tyrode working solution.[1] Smears were prepared from the cell suspension and processed as described above.

Antibodies. In order to cover a broad spectrum of antigens useful for diagnostic cytology we performed ICC with antibodies directed to three different types of antigens. The first group of antibodies was directed to the intermediate filament proteins cytokeratin (monoclonal antibody Lu5, 1:500; Hoffmann-LaRoche, Basel, Switzerland) and vimentin (monoclonal antibody V9, meat; Boehringer Mannheim, FRG). The second group of antibodies recognized the cytoplasmic proteins leukocyte common antigen (m-h LCA, neat; Amersham, UK) and lysozyme (ra-h lysozyme 1:500; Dakopatts, Glostrup, Denmark). The third group was represented by antibodies to cell membrane-bound antigens (monoclonal antibodies Leu 1, 2a, 4, and 14, 1:500; Becton-Dickinson, Mountain View, CA, USA).

Immunocytochemistry. Immunocytochemical reactions were performed either using the avidin-biotin-complex (ABC) technique (Hsu et al. 1981; Gower et al. 1985) or the alkaline phosphate-anti-alkaline phosphatase technique (APAAP; Cordell et al. 1984). For the ABC technique endogenous peroxidase was blocked by 3% H_2O_2 diluted in dH_2O for 5 min at room temperature. Slides were incubat-

[1] *Tyrode working solution.* Stock I: dissolve 160 g NaCl, 4 g KCl, 5.3 g $CaCl_2 \cdot 2H_2O$, 4.26 g $MgCl_2 \cdot 5H_2O$ in dH_2O and bring to a total volume of 1000 ml with dH_2O. Stock II: dissolve 20 g $NaHCO_3$, 1.3 g $NaH_2PO_4 \cdot 2H_2O$ in 1000 ml dH_2O. To prepare the working solution add 50 ml of stock I to 750 ml dH_2O, add 50 ml of stock II, and bring to a total volume of 1000 ml with dH_2O. Finally, add 1.1 g glucose.

ed with the primary antibody containing 1% bovine serum albumin in a humid chamber for 60 min at room temperature, followed by incubation with biotinylated IgG (BioMacor; Rehovot, Israel) diluted in PBS (1:20) for 30 min at room temperature. Vectastain (Vector, Burlingame, USA) was applied for 30 min at room temperature. The site of reaction was visualized with 3,3-diaminobenzidinetetrahydrochloride (DAB, 0,05% w/v) and H_2O_2 (0,01% v/v). Cell preparations were counterstained with Mayer's hematoxylin for 30 s. The preparations were dehydrated in graded alcohols and mounted with Eukitt (Kindler, Freiburg i. Br., FRG).

In those experiments in which the APAAP technique was employed, rabbit anti-mouse (pan)Ig (Dakopatts; Glostrup, Denmark) was used as second antibody (1:30) after incubation with primary antibodies. For polyclonal primary antibodies generated in rabbits, an affinity-purified mouse anti-rabbit IgG (1:500, Dakopatts) was used as linking antibody. The APAAP complex (1:100 in PBS) was applied for 30 min and the color reaction was performed with fast red.[2] The cell preparations were counterstained with Mayer's hematoxylin for 30 s and mounted in glycerol-gelatin.

Storage of Cell Preparations. Identically prepared and processed imprints were used to enable the comparison of several different storage conditions. Unfixed and fixed cell preparations were stored air dried at room temperature, −20 °C, or −70 °C. Undried preparations were transferred immediately after the PBS step following the fixation into 100% methanol and were stored at 4 °C. In other experiments the cell preparations were kept in a solution of sucrose, magnesium chloride, and glycerol[3] at −20 °C. For all experiments in which the influence of prolonged storage was studied, smears were stored for 2-8 months prior to the immunoreactions.

Immunocytochemistry after Papanicolaou (PN) or May-Grünwald-Giemsa (MGG) Staining. Fresh, undried imprint preparations of breast lesions and tonsils were spray-fixed and stained according to PN; MGG staining was performed after air drying. After dehydration coverslips were placed over the preparations. After at least 1 week the coverslips were removed with xylene and the cells destained with a solution of acidic ethanol (3% HCl/70% ethanol); ICC was then performed as described above.

Controls. As a positive control for each experiment and each antibody an adequate tissue section was included in the experiments and processed identically.

---

[2] *Fast red TR working solution.* 10 mg naphthol-AS-MX-phosphate (Serva, FRG), was disolved in 0.5 ml *N,N*-dimethylformamide (Serva) and 50 ml veronal acetate buffer, pH 9.2. 50 mg fast red TR (Serva, FRG) and 10 mg levamisol (Sigma, St. Louis, USA) were added. The solution was used after filtration and adjustment of the pH to 9.2-9.8 for 30 min with 2 *N* NaOH at room temperature.

[3] *Sucrose-magnesium chloride-glycerol solution.* 42.8 g sucrose (Sigma, Westbury, USA) and 0.33 g $MgCl_2.6H_2O$ (Merck) was disolved in 250 ml glycerol (Sigma) and was finally brought to a total volume of 500 ml with PBS.

Replacement of the primary antibody with normal serum and omission of H_2O_2 from the DAB color reaction served as negative controls.

Results

Generally, the results of the different experiments varied significantly with regard to cell preservation and immunoreactivity. Air drying of the cells prior to fixation significantly influenced the quality of immunoreactivity. Besides a complete loss

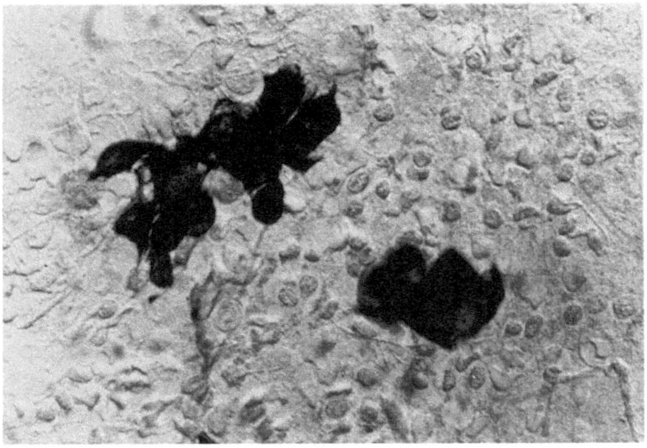

Fig. 1. Demonstration of strong cytokeratin immunoreactivity (antibody Lu5) in squamous cells after fixation with glutaraldehyde (0.05%) and storage in sucrose-magnesium chloride-glycerol solution at −20 °C for 2 months. Lymphoid cells are unstained. Imprint preparation of tonsil; ABC technique, no counterstain, interference contrast optics, ×410

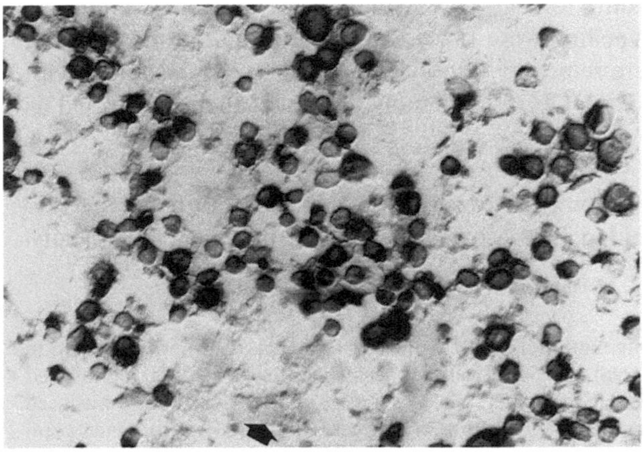

Fig. 2. Vimentin immunoreactivity (antibody V9) in lymphoid cells; epithelial cells remain unstained *(arrow)*. Tonsil, imprint; spray fixation, followed by immunocytochemistry. ABC technique, no counterstain; interference contrast optics, ×250

of immunoreactivity, false-positive staining occurred in some cases, particularly when antibodies to cytoskeletal proteins were used. Similarly, air drying of cell preparations after fixation prior to the immunohistochemical procedures was associated with inconsistent results. To obtain reliable and reproducible immunocytochemical reactions and to preserve morphological details, fixation was required. The suitability of the fixatives depended on the antibodies used and the antigen to be determined. Generally, the best results with antibodies to all three groups of antigens were obtained after fixation with 0.05% glutaraldehyde for 1 min or spray fixation (Figs. 1-3). Other fixatives such as ethanol, acetone, or formalin/acetone mixtures resulted in fair results, and unbuffered acidic fixatives (ethanol/acetic acid), formalin, and paraformaldehyde were found to be inadequate because of poor preservation of cell morphology and antigenicity. The effect of the most widely used fixatives on the immunohistochemical reactions is listed in Table 2.

In imprint preparations in which larger cell complexes were attached to the glass slides, these complexes were specifically stained only in the periphery,

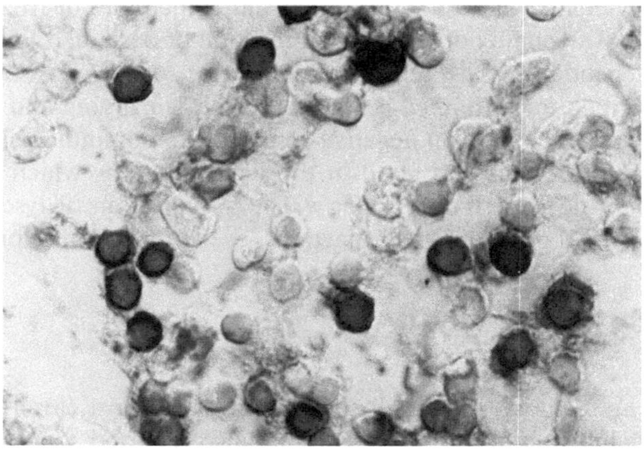

Fig. 3. Immunocytochemical visualization of Leu 1 epitopes on lymphoid cells on imprints of the tonsil after spray fixation and subsequent immunocytochemical examination. APAAP technique, no counterstain; interference contrast optics, ×410

Table 2. Results of immunocytochemistry after using different fixatives and different antibodies. Group-I antibodies are directed to intermediate filament proteins, antibodies of group II to soluble cytoplasmic antigens, and antibodies of group III to cell membrane-bound antigens (compare Table 2)

Fixative	I	II	III
Glutaraldehyde 0,05%	+++	+++	+++
Merckofix spray	+++	+++	++
Acetone 100%	+++	++	++
Methanol 100%	+++	++	++
Formalin 10%	+++	++	−/+
Unfixed	++	+	−/+

whereas the central portions of the cell complex remained nonreactive. This phenomenon was noticed with antibodies to cytoskeletal proteins and was not influenced by the addition of detergents to the antibody or washing solutions. After formalin or paraformaldehyde fixation of tonsil imprints, however, only the central parts of the multilayered cell complexes were reactive with antibodies directed to LCA, probably due to a decrease of fixative concentrations in the central parts. Because of the inconsistent results in multilayered cell complexes, the evaluation of the quality of immunoreactions was restricted to areas with single cell layers.

In those experiments in which the influence of prolonged storage of cell preparations was studied, we observed that storage of air-dried slides - independent of storage temperatures - was associated with a significant reduction of antigenicity. The best results with excellent preservation of antigenicity and of morphological details were obtained when fixed cell preparations were kept in sucrose-magnesium chloride-glycerol solutions at −20 °C (Fig. 1). Furthermore, storage in sucrose-magnesium chloride-gelatin solutions helped to reduce nonspecific binding of antibodies in preparations rich in blood, lymphatic fluid, or mucus. Storage of smears in methanol at 4 °C also allowed good preservation of morphology and antigenicity, whereas storage at room temperature or frozen led to poor results.

The quality of the immunoreaction of destained PN- or MGG-stained material depended on the cell processing prior to PN or MGG staining. As already described for unstained cell preparations, air drying and the use of inadequate fixatives caused (false) negative results. The immunolocalization of intermediate filament proteins was good in those preparations which were never allowed to dry out and were fixed properly. The visualization of lysozymes and LCA, however, resulted in inconsistent reactions, even after optimal processing.

Discussion

The methods and results of our study are most closely comparable with those of Dalquen et al. (1986). This group also reported some inconsistencies of immunocytochemical results, e.g., larger cell groups remaining unstained whereas individual cells or smaller cell groups were more uniformly immunoreactive. This phenomenon is probably caused by impaired accessibility of the antigen in the center of larger cell complexes. This is one of the main differences between ICC and immunohistochemistry on tissue sections, where accessibility usually does not pose a significant problem. The observation that LCA immunoreactivity was restricted to the center of larger cell complexes on imprints of tonsils after formalin and paraformaldehyde fixation may be explained by incomplete fixation of the central parts which prevents the epitopes from destruction by an inadequate fixative. We agree with Judd and Britton (1982), Leenen et al. (1985), and Dalquen et al. (1986) that air drying and prolonged storage of fixed or unfixed smears in dry conditions cause inconsistent results. Drying of fixed or unfixed cells may lead to changes in the cell membrane, causing nonspecific binding of antibodies or a decreased accessibility of antibodies to cytoplasmic antigens. In contrast, however, several authors did not find a reduction of immunoreactivity after air drying and storage at temperatures between +20 °C and −70 °C (DeWaele et al. 1983; Erber et al. 1984;

Li et al. 1984; Ramaekers et al. 1984; Springall 1986). A major disagreement between our study and the paper by Dalquen et al. (1986) is that spray fixation yielded good results in our hands if prolonged air drying after fixation was avoided. Judd and Britton (1982) obtained the best results after fixation in acetone, ethanol, or acetone/ethanol mixtures, whereas glutaraldehyde (0.01% and 0.1%) was found to be inadequate for both preservation of cell morphology and immunoreactivity of antibodies to cytoplasmic and surface proteins. In contrast to our fixation time (1 min), these authors used rather long fixations (15 min). Our results are in accordance with those of van Ewijk et al. (1984) and Leenen et al. (1985) who reported excellent preservation of (cytoplasmic) mononuclear phagocyte differentiation antigens with 0.05% glutaraldehyde as fixative. Nadji (1980) and Szpak et al. (1984) used higher concentrations of antibodies for ICC than for immunohistochemistry. In our experiments an increase of the antibody concentration was not necessary for ICC.

Several authors reported the possibility of ICC succeeding PN staining (Faravelli et al. 1984; Geddie et al. 1984; Keshgegian and Kline 1984; Kobayashi et al. 1985; Chess and Hajdu 1986; Dalquen et al. 1986; Travis and Wold 1987). In our hands, and in line with Dalquen et al. (1986) and Travis and Wold (1987), ICC succeeding PN or MGG staining is associated with inconsistent results. For some antigens, however, particularly the rather stable intermediate filament proteins and some soluble cytoplasmic antigens, reliable results may be obtained.

Generally, nonspecific reactions with monospecific antibodies may be caused by: (a) (nonspecific) ionic binding of antibodies and reagents, (b) cross-reacting natural antibodies, and (c) the existence of the same or similar epitopes on different antigens. The problem of nonspecific reactions in ICC and hints for their prevention were addressed by Gosselin et al. (1985), Springall (1986), and Denk (1987).

We conclude that immunocytochemistry may represent a valuable tool in diagnostic cytology. ICC, however, cannot replace the experience and skill of the cytopathologist, as diagnostic errors resulting from uncritical use of the techniques and methodical difficulties will arise. The interpretation of immunocytochemical results, therefore, must be guided by subtle conventional cytological evaluation and should be used only in line with appropriate controls.

Acknowledgements. The authors whish to thank Mrs. R. Zeller for laboratory work, Mrs. H. Schleich for secretarial help, and Mr. I. Georgiev for the photographic reproductions of the figures.

References

Chess Q, Hajdu SI (1986) The role of immunoperoxidase staining in diagnostic cytology. Acta Cytol (Baltimore) 30: 1-7

Cordell JL, Falini B, Erber WN, Ghosh AK, Abdulaziz Z, MacDonalds S, Pulfords KAF, Mason DY (1984) Immunoenzymatic labeling of monoclonal antibodies using complexes of alkaline phosphatase and monoclonal anti-alkaline phosphatase (APAAP complexes). J Histochem Cytochem 32: 219-229

Dalquen P, Bittel D, Gudat F, Overbeck J, Heitz PH (1986) Combined immunoreaction and Papanicolaou's stain on cytological smears. Pathol Res Pract 181: 50-54

Denk H (1987) Immunohistochemical methods for the demonstration of tumor markers. In: Seifert G (ed) Morphological tumor markers. Springer, Berlin Heidelberg New York, pp 47-69 (Current topics in pathology, vol 77)

DeWaele M, DeMey J, Moeremans M, DeBrabander M, van Camp B (1983) Immunogold staining method for the light microscopic detection of leukocyte cell surface antigens with monoclonal antibodies: its application to the enumeration of lymphocyte subpopulations. J Histochem Cytochem 31: 376-381

Erber WN, Pinching AJ, Mason DY (1984) Immunocytochemical detection of T and B cell populations in routine blood smears. Lancet I: 1042-1045

Faravelli A, Sironi M, Villa E, Radice F (1984) Immunoperoxidase study of cytologic smears in oral pemphigus. Acta Cytol (Baltimore) 28: 414-418

Geddie WR, Bedard YC, Strawbridge HTG (1984) Medullary carcinoma of the thyroid in fine needle aspiration biopsies. Am J Clin Pathol 82: 552-558

Gosselin EJ, Dennett JC, Sorenson GD, Pettengill OS, Cate CC (1985) Immunocytochemical staining of cytocentrifuge prepared cultured cells: nonspecific staining and its elimination. Histochem J 17: 847-858

Gower WR, McSweeney EM, Dyben TE, Fabri PJ (1985) Identification of gastrinproducing cells in cell cultures and smears: an avidin-biotin-peroxidase complex (ABC) technique. J Histochem Cytochem 33: 1087-1089

Hsu SM, Raine L, Fanger H (1981) The use of avidin-biotin-peroxidase complex in immunperoxidase techniques. Am J Clin Pathol 75: 816-861

Judd MA, Britten KJM (1982) Tissue preparation for the demonstration of surface antigens by immunoperoxidase techniques. Histochem J 14: 747-753

Keshgegian AA, Kline TS (1984) Immunoperoxidase demonstration of prostatic acid phosphatase in aspiration biopsy cytology (ABC). Am J Clin Pathol 82: 586-589

Kobayashi TK, Mizuhara S, Arai M, Sawaragi I (1985) Application of immunoperoxidase staining in the cytodiagnosis of herpes simplex keratitis. Diagn Cytopathol 1: 317-321

Koss LG (1984) Cytochemistry (Editorial). Acta Cytol (Baltimore) 28: 353-355

Leenen PJM, Jansen AMA, van Ewijk W (1985) Fixation parameters for immunohistochemistry. The effect of glutaraldehyde or paraformaldehyde fixation on the preservation of mononuclear phagocyte differentiation antigens. In: Bullock GR, Petrusz P (eds) Techniques in immunohistochemistry, vol III. Academic, NY, pp 1-24

Li CY, Ziesmer SC, Yam LT, English MC, Janckila AJ (1984) Practical immunocytochemical identification of human blood cells. Am J Clin Pathol 81: 204-212

Nadji M (1980) The potential value of immunoperoxidase techniques in diagnostic cytology. Acta Cytol (Baltimore) 24: 442-447

Ramaekers F, Haag D, Jap P, Vooijs PG (1984) Immunocytochemical demonstration of keratin and vimentin in cytologic aspirates. Acta Cytol (Baltimore) 28: 385-392

Springall DR (1986) Immunocytochemistry in diagnostic cytology. In: Polak JM, van Noorden S (eds) Immunocytochemistry. Modern method and applications. Wright, Bristol, pp 547-567

Szpak CA, Johnston WW, Lottich SC, Kufe D, Thor A, Schlom J (1984) Problems of reactivity of four novel monoclonal antibodies (B72.3, DF3, B1.1 and B6.2) with cells in human malignant and benign effusions. Acta Cytol (Baltimore) 28: 356-367

Travis WD, Wold LE (1987) Immunoperoxidase staining of fine needle aspiration specimens previously stained by papanicolaou technique. Acta Cytol (Baltimore) 31: 517-520

Van Ewijk W, van Soest PL, Verkerk A, Jongkind JF (1984) Loss of antibody to prefixed cells: fixation parameters for immunohistochemistry. Histochem J 16: 179-193

Walts AE, Said JW (1983) Specific tumor markers in diagnostic cytology: immunoperoxidase studies of carcinoembryonic antigen, lysozymes and other tissue antigens in effusions, washes and expirates. Acta Cytol (Baltimore) 27: 408-416

Monoclonal Antibodies in Diagnostic Cytopathology

G. P. Vooijs, A. Huijsmans, M. Salet-van der Pol, A. van Aspert-van Erp, and F. Ramaekers

Introduction

Especially in cytopathologic diagnosis, immunohistochemistry may be a particularly strong tool because fine-needle aspiration biopsy smears, sputum smears, or cytologic preparations from body cavity effusions are often partially or entirely devoid of any morphologic information. Since the diagnosis of pathologic changes by cytologic examination of exfoliated or aspirated cellular material is becoming an increasingly important part of diagnostic pathology, considerable effort has been put into the development of immunohistochemical methods for use in cytology during the past few years [3, 10].

Antibodies to Intermediate Filament Proteins and Tumor Diagnosis

Intermediate filament proteins (IFP) form a cytoskeletal structure composed of 10-nm filaments in most mammalian cells and have been shown to be tissue specific to a certain extent. Five different types of IFP have been distinguished so far, and the distribution of these different proteins has been shown to follow well-known histologic principles [6].

Monoclonal and polyclonal antibodies to the different types of IFP have been prepared by several groups working in the field (see for example [6, 9]) and many of them are now commercially available. In general, antibodies to keratin react only with epithelial tissues, antibodies to desmin only with muscle tissues, glial fibrillary acidic protein (GFAP) antibodies only with glial cells, and neurofilament antibodies with neural cells. Antibodies to vimentin normally react only with cells and tissues of mesenchymal origin, but for this type of IFP several exceptions are known. For example, mesothelial cells have been suggested to coexpress keratins and vimentin [7].

Antibodies to IFP have been tested extensively for use in tissue characterization in the past few years. From the data obtained in these studies it is obvious that IFP can be used as powerful markers in tumor typing (for reviews see [6, 13]). Most important, it has been shown that tumor cells generally retain their original IFP, with the exception of a few types of neoplasms. Therefore, carcinomas can be stained with antibodies to keratins, while mesenchymal tumors such as lymphomas [11], most soft tissue sarcomas, melanomas, seminomas, and Schwannomas are negative for keratins but positive for vimentin. Tumors derived from striated or smooth muscle tissues, i.e., rhabdo- and leiomyosarcomas, can be differentiated from other soft tissue tumors because of their reactivity with desmin antibodies [1]. Moreover, astrocytomas are strongly positive for GFAP, while ganglioneuroblastomas and some neuroblastomas are positive with neurofilament antisera. However, some neuroendocrine neoplasms such as Merkel cell tumors and small cell lung cancers [2] have also been described as containing neurofilament proteins.

In contrast to normal adult human tissues, certain neoplastic tissues may coexpress different types of IFP [1, 2, 4]. When tumor cells contain more than one IFP type, vimentin is usually one of them. In the case of epithelial tumors, vimentin IFP have been found next to keratin IFP in, for example, pleomorphic adenomas of the parotid gland, adenoid cystic carcinomas of the salivary gland, and renal carcinomas and nephroblastomas, as well as in some adenocarcinomas of the lung, mesotheliomas, and thyroid carcinomas [4]. Coexpression of vimentin and desmin is observed in most if not all rhabdomyosarcomas [1] and in leiomyosarcomas. GFAP and vimentin are found simultaneously in astrocytomas. Coexpression of keratins and neurofilament proteins has also been observed. This is the case in some neuroendocrine tumors such as Merkel cell tumors and lung carcinoids [2]. Very recently it has been suggested that smooth muscle cells and smooth muscle cell tumors may be positive for certain keratin antibodies as well as reacting with desmin antisera [13]. The fact that some tumor types coexpress different types of IFP can be of help in the diagnosis of such malignancies.

Cytokeratin Distribution in Normal and Malignant Epithelial Tissues

The keratins are a family of IFP characterized by a remarkable biochemical diversity; they are represented in human epithelial tissues by at least 19 different keratin polypeptides ranging in molecular weight from 40 to 68 KDa. These polypeptides are not expressed randomly throughout epithelia but occur in cell type specific combinations [5]. The polypeptide composition of epithelial keratin filaments varies with the state of differentiation. The various types of carcinomas also differ in their keratin polypeptide content. Several monoclonal antibodies to different keratins have been prepared and different reactivities of such antibodies with different subtypes of epithelial tissues and epithelial tumors have been noted [2, 6, 8, 12]. In addition to broadly cross-reacting antibodies which stain virtually all types of epithelial tissues, monoclonal keratin antibodies have been developed which cross-react with only one of the 19 keratin polypeptides and which show a more tissue-specific staining pattern. For example, monoclonal antibodies to keratin 18 [8] recognize glandular epithelia but not stratified squamous epithelia, and can therefore distinguish adenocarcinomas from squamous cell carcinomas. An antibody to keratin 10, on the other hand, stains only keratinizing cells, and can also recognize such cells in squamous cell carcinomas [9]. Antibodies to keratin 7 have been found to react with only a subfraction of the glandular cells present in the human body [12]. As a result, such antibodies react with a restricted number of adenocarcinomas. For example, adenocarcinomas from prostate, stomach, colon, and small intestine as well as renal cell carcinomas are negative when tested with keratin 7 antibodies in immunohistochemical assays. Also, most squamous cell carcinomas are negative, with the exception of lung squamous cell carcinomas which may be focally positive. By contrast, ovarian carcinomas, transitional cell carcinomas of the urinary bladder, adenocarcinomas of the lung, and breast carcinomas are positive. These antibodies can be of great help for tumor diagnosis, especially in cases of anaplastic carcinoma metastases. Figure 1 illustrates the usefulness of keratin monoclonal antibodies in histopathology.

Serous Effusions

In the application of antibodies to IFP in diagnostic cytopathology of body cavity effusions, broadly cross-reacting keratin antibodies and vimentin antisera can be of value in those cases where the malignancy of the underlying process has been determined, but where routine cytopathologic methods cannot give conclusive results about the nature of the neoplastic or metastatic cells. In problematic cases where, for example, a differential diagnosis has to be made between anaplastic small cell carcinoma and malignant lymphoma, a positive reaction of the tumor cells with keratin antibodies is a very strong indication of a carcinomatous process. A negative reaction of the tumor cells for keratin together with a positive reaction for vimentin, however, is indicative of a lymphoma or a sarcoma. It should be kept in mind that both epithelial tumor cells and malignant and reactive mesothelial cells can coexpress both types of IFP when growing under such circumstances. We have noted such coexpression in several types of adenocarcinomas [7, 10] but, interestingly, so far not in squamous cell carcinomas and transitional cell carcinomas. Some authors have suggested that in serous effusions only mesothelial cells coexpress keratins and vimentin, in contrast to, for example, ovarian carcinoma cells (see Fig. 2). In 10 out of 19 cases of epithelial ovarian cancers in ascites, we found coexpression of vimentin and keratin in variable numbers of tumor cells. These results clearly showed that since tumor cells may coexpress keratins and vimentin, this phenomenon cannot be used to distinguish benign or malignant mesothelial cells from adenocarcinoma cells. In the case of metastatic carcinoma, use of monoclonal antibodies directed against specific keratin polypeptides may give important information about the type of epithelial differentiation or the nature of the tumor cells. For example, monoclonal antibodies directed against keratin 18 can in general distinguish adenocarcinoma cells (Fig. 2) from cells of squamous cell carcinomas [8]. It should be kept in mind, however, that some squamous cell tumors such as moderately and poorly differentiated squamous cell carcinomas of the lung may contain a significant fraction of cells positive for keratin 18 [2].

In the differential diagnosis of cells present in an effusion, antibodies to keratin 7 have already proven useful in problem cases where it was not possible to distinguish between renal cell carcinoma and transitional cell carcinoma. In differential diagnosis of a carcinoma of the stomach or colon on one hand and an ovarian carcinoma on the other hand, keratin 7 antibodies are very helpful in that they react positively with the ovarian tumor (see Fig. 2) but do not stain several of the gastrointestinal tract tumors. Mesothelial cells also contain keratin 7.

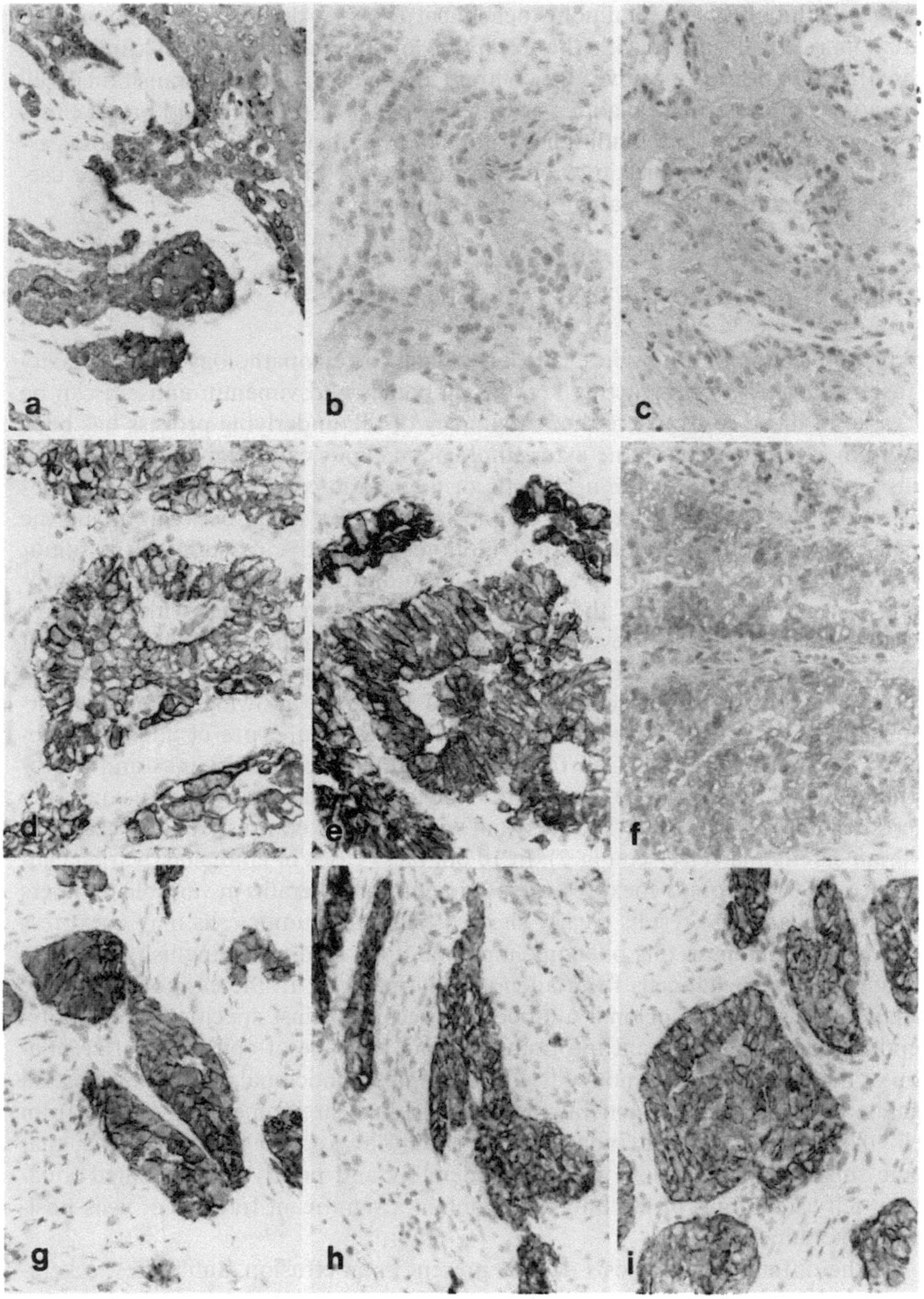

Fig. 1 a–i. Immunohistochemical staining pattern (peroxidase reaction) of **a–c** a squamous cell carcinoma, **d–f** a carcinoma of the large intestine, and **g–i** an ovarian carcinoma with monoclonal keratin antibodies. **a, d, g** All three epithelial tumor types react positively with a broadly cross-reacting keratin antibody. **b, e, h** A monoclonal antibody to keratin 18 stains only the adenocarcinomas. **c, f, i** A monoclonal antibody to keratin 7 recognizes only a subpopulation of the adenocarcinomas

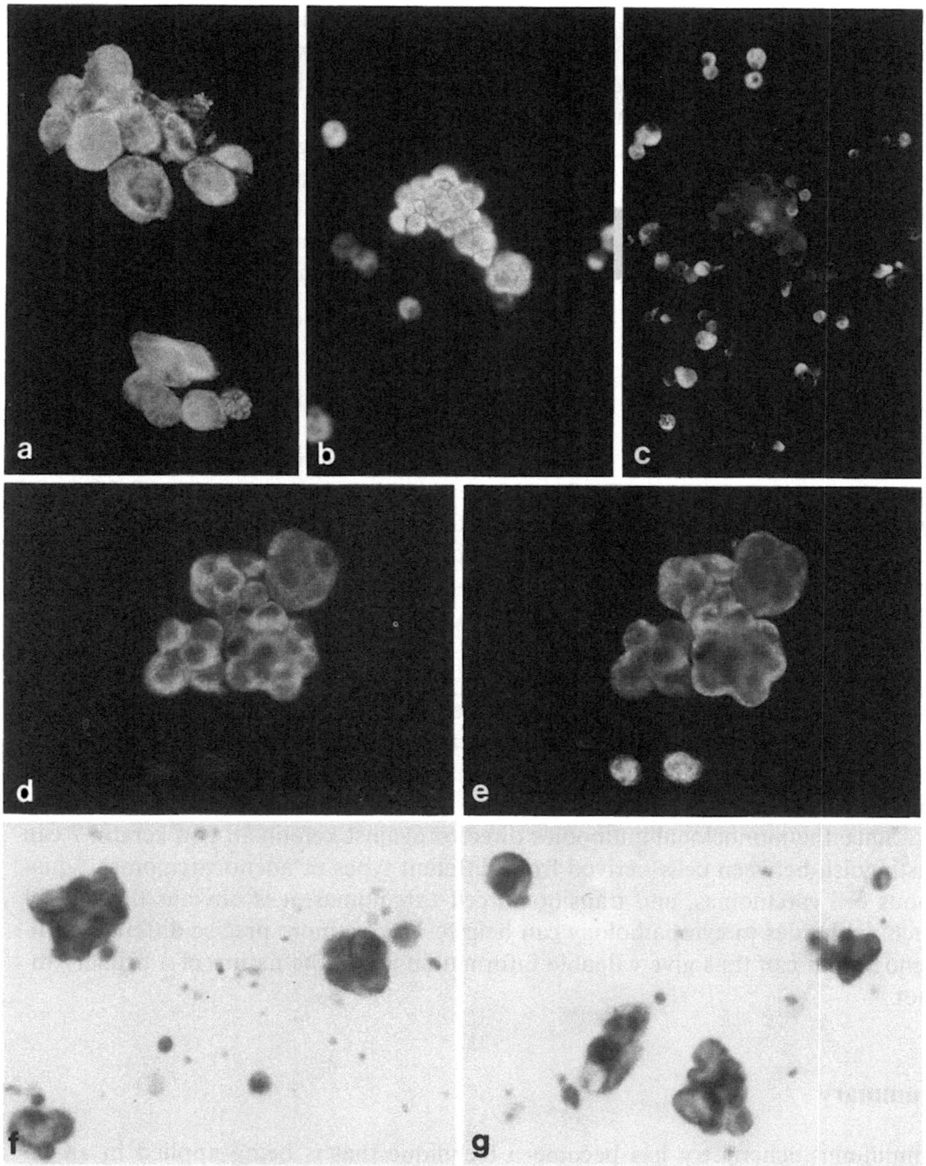

Fig. 2. **a–e** Immunofluorescence and **f, g** immunoperoxidase reaction patterns of: **a** stomach carcinoma, **b, c** pancreas carcinoma, **d, e** mesothelioma, and **f, g** ovarian carcinoma, with antibodies to keratin 18 (**a, b, d, f**), keratin 7 (**g**), and vimentin (**c, e**). **a, f, g** ×240; **b, c** ×145; **d, e** ×320

One of the major pitfalls in cytopathologic examination of effusions is the differential diagnosis between reactive atypical mesothelial cells and single malignant cells. Our experience with the HMFG-2 antibody [14] is most promising in this respect. We tested the usefulness of HMFG-2 for the distinction between carcinoma cells and reactive or atypical mesothelial cells in more than 100 cases of pleural and peritoneal effusions. In 86% of the cases examined using HMFG-2 the expected reaction pattern was found. That means there was no reaction in those cases only containing atypical mesothelial cells and a positive reaction with the tumor cells in cases of malignant effusions.

In seven out of 11 cases that initially appeared to give false-positive results with HMFG-2 it was shown, after revision of the slides, after autopsy, or after examination of effusions obtained at later stages, that the patient had a malignant epithelial process. The three false-negative cases concerned one breast carcinoma and two small cell anaplastic lung carcinomas that had metastasized to body cavities.

Lymph node Aspirates

One of the areas in cytology where the use of IFP antibodies seems most promising is in lymph node aspirates. In all cases lymphocytes are negative for keratin antibodies, but (weakly) positive for vimentin. As a result, the detection of keratin-positive cells in aspirates from lymph nodes is a very strong indication for the presence of metastatic carcinoma cells. Normally, no keratin-positive cells can be observed in cases of chronic lymphadenitis or malignant lymphomas. Instead, the cells are vimentin positive.

When using monoclonal antibodies to keratin 18 no reaction with cells from metastatic squamous cell carcinomas is seen in most cases. A rather strong reaction, however, is seen in cells from metastases of adenocarcinomas. Keratinizing cells from well-differentiated squamous cell carcinoma metastases react with an antibody to keratin 10.

Since the monoclonal antibodies directed against keratin 18 and keratin 7 can distinguish between cells derived from different types of adenocarcinomas, squamous cell carcinomas, and transitional cell carcinomas, it is obvious that use of such antibodies in cytopathology can help to make a more precise differential diagnosis and can thus give valuable information about the nature of a primary tumor.

Summary

Immunohistochemistry has become a technique that is being applied in an increasing number of cases in cytopathologic diagnosis. Because fine-needle aspiration biopsy smears, sputum smears, or cytologic preparations from body cavity effusions are often partially or entirely devoid of any morphologic information, and since the diagnosis of pathologic changes by cytologic examination of exfoliated or aspirated cellular material is becoming an increasingly important part of diagnostic pathology, considerable effort has been put into the development of immunohistochemical methods for use in cytology during the past few years. Using monoclonal antibodies to tissue-specific antigens, several tumor cell types can be distinguished on the basis of their immunohistochemical reaction patterns. For example, application of antibodies to the intermediate filament proteins keratin and vimentin can separate epithelial from nonepithelial (tumor) cell types. Monoclonal antibodies to subtypes of keratin can identify certain types of carcinoma cells. An antibody to keratin 18 can separate adenocarcinomas from squamous cell carcinomas, while an antibody to keratin 7 can distinguish ovarian carcinoma from stomach carcinoma or transitional cell carcinoma of the bladder from renal cell carcinoma. For the distinction between benign mesothelial cells and adenocarcinoma cells the HMFG-2 antibody has given promising results.

Acknowledgements. Olof Moesker, Gert Schaart, and Kiek Verrijp are acknowledged for their technical assistance in the preparation and specificity testing of intermediate filament antisera. We are grateful to Anne Wiersma-van Tilburg, Ineke van Zuylen, and Tiny Heynen-Wijnen for preparing the cell smears, and to Anneke van Hoffen for help with immunohistochemical tests on fine-needle aspirates of lymph nodes. We thank Renate Möhrling for excellent secreterial support in the preparation of the manuscript.

References

1. Altmannsberger M, Osborn M, Treuner J, Hölscher A, Weber K, Schauer A (1982) Diagnosis of human childhood rhabdomyosarcoma by antibodies to desmin, the structural protein of muscle specific intermediate filaments. Virchows Arch [B] 39: 203-215
2. Broers JLV, Klein Rot M, Oostendorp T, Huysmans A, Wagenaar SS, Wiersma-van Tilburg AJM, Vooijs GP, Ramaekers FCS (1987) Immunocytochemical detection of lung cancer heterogeneity using antibodies to epithelial, neuronal, and neuroendocrine antigens. Cancer Res 47: 3225-3234
3. Chess Q, Hajdu SI (1986) The role of immunoperoxidase staining in diagnostic cytology. Acta Cytol (Baltimore) 30: 1-7
4. Churg A (1985) Immunohistochemical staining for vimentin and keratin in malignant mesothelioma. Am J Surg Pathol 9: 360-365
5. Moll R, Franke WW, Schiller DL, Geiger B, Krepler R (1982) The catalog of human keratins: pattern of expression in normal epithelia, tumors and cultured cells. Cell 31: 11-24
6. Osborn M, Weber K (1983) Tumor diagnosis by intermediate filament typing: a novel tool for surgical pathology. Lab Invest 48: 372-394
7. Ramaekers FCS, Haag D, Kant A, Moesker O, Jap PHK, Vooijs GP (1983) Coexpression of keratin- and vimentin-type intermediate filaments in human metastatic carcinoma cells. Proc Natl Acad Sci USA 80: 2618-2622
8. Ramaekers FCS, Huijsmans A, Moesker O, Kant A, Jap PHK, Herman C, Vooijs GP (1983) Monoclonal antibodies to keratin filaments, specific for glandular epithelia and their tumors. Use in surgical pathology. Lab Invest 49: 353-361
9. Ramaekers FCS, Puts JJG, Moesker O, Kant A, Huijsmans A, Haag D, Jap PHK, Herman CJ, Vooijs GP (1983) Antibodies to intermediate filament proteins in the immunohistochemical identification of human tumours. An overview. Histochem J 15: 691-713
10. Ramaekers FCS, Haag D, Jap PHK, Vooijs GP (1984) Immunochemical demonstration of keratin and vimentin in cytological aspirates. Acta Cytol (Baltimore) 28: 385-392
11. Ramaekers FCS, Vroom TM, Moesker O, Kant A, Scholte G, Vooijs GP (1985) The use of antibodies to intermediate filament proteins in the differential diagnosis of lymphoma versus metastatic carcinoma. Histochem J 17: 57-70
12. Ramaekers FCS, Huijsmans A, Schaart G, Moesker O, Vooijs GP (1987) Tissue distribution of keratin 7 as monitored by a monoclonal antibody. Exp Cell Res 170: 235-249
13. Ramaekers FCS, Pruszczynski M, Smedts F (1988) Cytokeratins in smooth muscle cells and smooth muscle tumours. Histopathology 12: 558-561
14. Taylor-Papadimitriou J, Peterson JA, Arklie J, Burchell J, Ceriani RL, Bodmer WF (1981) Monoclonal antibodies to epithelium-specific components of the human fat globule membrane: production and reaction with cells in culture. Int J Cancer 28: 17-21

The Cytoskeleton: Diagnostic Possibilities and Limitations

K. Schwechheimer

Introduction

The cytoskeleton comprises the three-dimensional network of cellular structures including nuclei, organelles, fibrils, and membranes. Among these, the intracytoplasmic filaments form the framework of the cytoskeleton in a narrower sense. The ubiquitous intermediate filaments (7–11 nm) whose diameters lie approximately between those of microfilaments (6 nm) and microtubules (25 nm) can be separated biochemically and immunologically into five classes. Each kind of intermediate filament polypeptide is associated with a well defined cell and tissue type. Cytokeratin polypeptides can be identified in epithelial cells, vimentin predominantly in mesenchymal cells, desmin in muscle cells, glial fibrillary acidic protein (GFAP) typically in astrocytes, and neurofilament polypeptides (NF) mainly in neurons (Table 1; Osborn and Weber 1983; Traub 1985).

In addition to the epithelial cytokeratin intermediate filaments, there are desmosomes which form a specialized intercellular junctional complex associated

Table 1. Classification of intermediate filaments

Type	Proteins	Examples in situ
1 Epithelial	Cytokeratins, multiple polypeptides 40000 to 68000 daltons	Keratinizing and nonkeratinizing epithelia (can be subdivided further)
2 Neuronal	NF, 68000, 160000, 200000 daltons	Most but probably not all neurons in central and peripheral nervous system
3 Glial	GFA, 55000 daltons	Astrocytes, Bergmann glia (some GFA+V+)
4 Muscle	Desmin, 53000 daltons	Striated muscle, smooth muscle; vascular smooth muscle cells can be D+V+, D+V−, or D−V+
5 Mesenchymal	Vimentin, 57000 daltons	Fibroblasts, chondrocytes, macrophages, endothelial cells, etc.
6 No IF		Cells of the early embryo, certain neurons

NF, neurofilaments; GFA, glial filament; V, vimentin; D, desmin. From Osborn and Weber (1983)

with epithelial differentiation. Several polypeptides of the desmosomal plaque have been characterized biochemically; their major constituents have been designated "desmoplakins" (Franke et al. 1982, 1983; Mueller and Franke 1983). Desmoplakins which can be identified at the light microscopic level by monoclonal antibodies may be regarded as general marker polypeptides of epithelial differentiation.

The stability and excellent conservation of intermediate filaments and desmosomes after neoplastic transformation of cells and tissues form the basis of the use of antisera and monoclonal antibodies against these structural polypeptides for the identification, classification, and differential diagnosis of tumors in surgical pathology. These cellular marker polypeptides can be used both in histopathology and cytology.

Cytokeratins and Desmoplakins

Cytokeratin intermediate filaments and desmoplakins are two independent structural marker proteins indicating epithelial differentiation. They can easily be recognized by monoclonal antibodies and antisera in benign and malignant epithelial tumors (Fig. 1 a-d). Cytokeratin polypeptides and desmoplakins have been demonstrated in numerous squamous cell carcinomas and adenocarcinomas, epithelial neuroendocrine tumors such as Merkel cell carcinomas and carcinoid tumors, and germ cell tumors (for reviews see Moll 1986, 1987).

The remarkable stability of both epithelial markers in undifferentiated and anaplastic carcinomas is very important for differential diagnosis. In most cases, cytokeratins are the only intermediate filament type of epithelial tumors. Coexpression of cytokeratins and vimentin, however, is being observed in a small but increasing number of carcinomas. These include adenocarcinomas of salivary and thyroid glands, lung, kidney, endometrium, ovary, prostate, malignant mesotheliomas, and plexus papillomas of the brain (Table 2, Fig. 1 e, f; Moll 1986, 1987; Schwechheimer 1987).

In addition, cytokeratin polypeptides can be expressed with neurofilament polypeptides in epithelial neuroendocrine tumors of skin (Merkel cell carcinoma) and lung (carcinoid tumor, small cell carcinoma), in parathyroid adenomas, in islet cell tumors of pancreas, and in medullary carcinomas of thyroid gland.

In exceptional cases, cytokeratins are colocalized with GFAP and vimentin in some myoepithelial cells of human parotid gland and in pleomorphic adenomas as well as in single tumor cells of brain choroid plexus papillomas or carcinomas (Achtstätter et al. 1986; Miettinen et al. 1986; Born et al. 1987; Schwechheimer 1987).

In contrast to all other classes of intermediate filaments, cytokeratins represent a multigene family of polypeptides. Using two-dimensional gel electrophoresis, 19 different polypeptides can be identified in epithelial cells (cytokeratin polypeptides 1-19). These can be subdivided into a basic (polypeptides 1-8) and an acidic family (polypeptides 9-19). At least one basic and one acidic cytokeratin filament takes part in the formation of a heterotypic tetramer which constitutes the cytokeratin intermediate filament subunit. In nonneoplastic epithelial cells and tissues

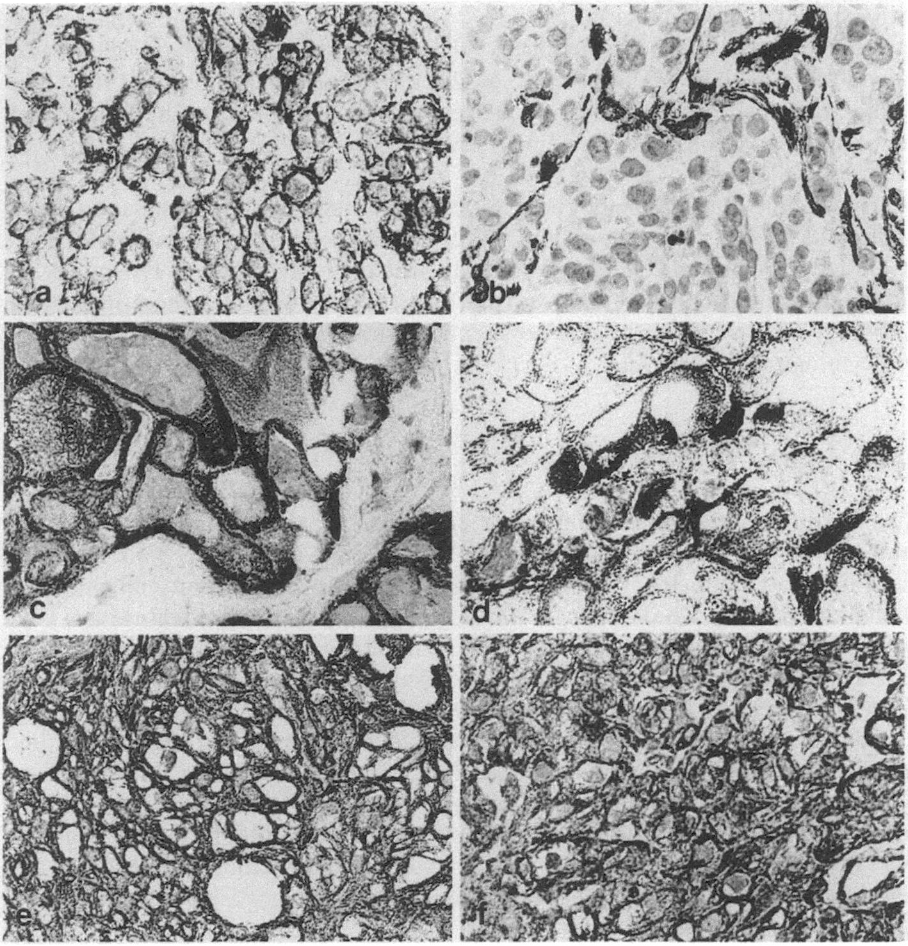

Fig. 1. **a, b** Ductal carcinoma of the breast: **a** strong intracytoplasmic cytokeratin immunoreactivity in the tumor cells; **b** the stroma is vimentin-positive. **c, d** Squamous cell carcinoma of the lung: **c** heterogeneous expression of cytokeratin filaments and **d** punctate pattern of desmoplakin immunoreactivity indicating the presence of abundant desmosomes. **e, f** Coexpression of **e** cytokeratin and **f** vimentin in a renal cell carcinoma (clear cell type). Avidin-biotin complex (ABC), 3-amino-9-etylcarbazole (AEC), hematoxylin counterstain; original magnification, ×25

cytokeratins are expressed in a differentiation-dependent manner which allows subdivision of cytokeratin polypeptides into the stratified and simple epithelia types (Franke et al. 1981; Moll et al. 1982; for reviews see Sun et al. 1984; Cooper et al. 1985; Quinlan et al. 1985; Moll 1986, 1987). According to preliminary biochemical and immunocytochemical studies the cytokeratin polypeptide pattern in tumors is more complex and heterogeneous than in nonneoplastic epithelial tissues. The principal differences in cytokeratin polypeptide distribution, however, are conserved in the neoplastic state thus allowing discrimination between squamous cell carcinomas and adenocarcinomas. At present it is not completely clear whether a subtle immunocytochemical differentiation of adenocarcinomas is pos-

Table 2. Coexpression of cytokeratin and vimentin in epithelial tumors

Organ	Tumor type
Thyroid gland	Carcinomas
Lung	Adenocarcinomas[a] Large cell carcinomas[a]
Kidney	Renal cell carcinomas Nephroblastomas Malignant rhabdoid tumors
Salivary gland	Adenoid-cystic carcinomas
Urogenital tract	Adenocarcinomas of endometrium Carcinomas of ovary[a] Granulosa cell tumors of ovary Adenocarcinomas of prostate
Mesothelium	Malignant mesotheliomas
Brain[b]	Choroid plexus papillomas/carcinomas Craniopharyngiomas[a]

For review and further references see Moll (1987)
[a] Coexpression of cytokeratin and vimentin only in some tumor cells
[b] for review see Schwechheimer (1987)

sible on the basis of reliable differences in cytokeratin polypeptide patterns of diverse carcinoma types (Moll et al. 1983; Moll and Franke 1986; Osborn et al. 1986; for reviews see Moll 1986, 1987). To demonstrate the true importance of cytokeratin polypeptide typing in the differential diagnosis of carcinomas further studies on large tumor series will be necessary.

Vimentin

Vimentin is a type of intermediate filament which is characteristic for non-muscular mesenchymal cells (Brown et al. 1976; Osborn and Weber 1977). Most benign and malignant soft tissue tumors and bone neoplasms express vimentin (Fig. 2b, e; Altmannsberger et al. 1986 a, b). This is also true for the majority of Hodgkin's and non-Hodgkin lymphomas. Negative results (Giorno 1985; Giorno and Sciotto 1985) could not be confirmed by Altmannsberger et al. (1986a, b). Finally, vimentin is the only intermediate filament type in malignant melanomas (Fig. 2a; Caselitz et al. 1983). Therefore, cytokeratin- and desmoplakin-positive carcinomas can be distinguished from vimentin-positive malignant melanomas.

In addition to vimentin, cytokeratin and possibly desmoplakins can be detected in some mesenchymal neoplasms (Table 3). These include synovial sarcomas (Fig. 2c, d; Miettinen et al. 1982a), epithelioid sarcomas (Fig. 2f; Chase et al. 1984), adamantinomas of the tibia (Rosai and Pinkus 1982), and chordomas (Miettinen et al. 1983). In these tumors cytokeratin is coexpressed with vimentin (Fig. 2 b, e). Using monoclonal antibodies against desmoplakins and an indirect immunofluorescence technique we also observed small fluorescent spots on the apicolateral membrane of cytokeratin-positive epithelial cells in the epithelial

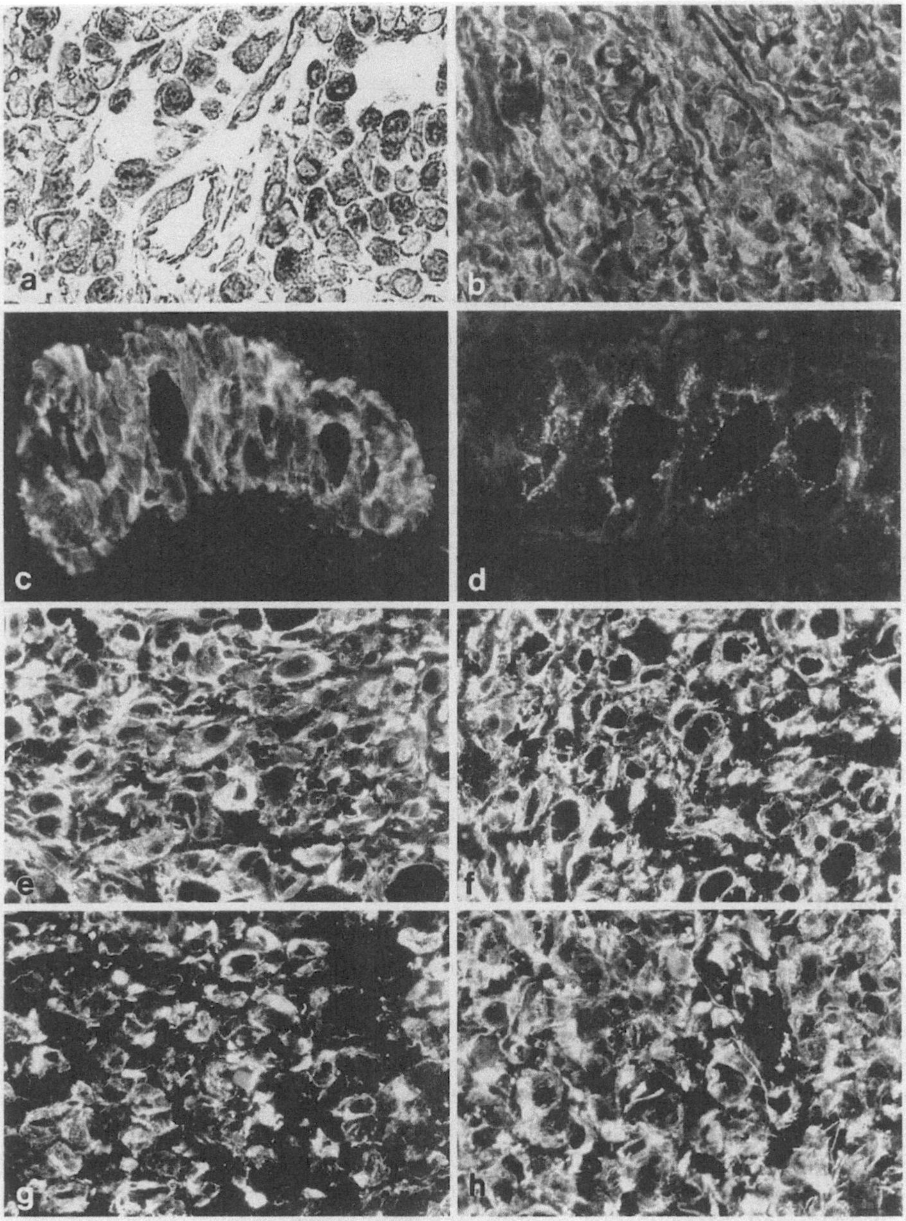

Fig. 2. **a** Vimentin immunoreactivity in a malignant melanoma. **b–d** Biphasic synovial sarcoma: **b** consistent vimentin expression in the spindle cell compartment; **c** cytokeratin and **d** desmoplakin immunoreactivity in the tubular epithelial structures. **e, f** Vimentin (**e**) and cytokeratin (**f**) coexpression in an epithelioid sarcoma. **g, h** Rhabdomyosarcoma with coexpression of **g** desmin and **h** vimentin intermediate filaments. **a** ABC, AEC, hematoxylin counterstain, ×25; **b–h** indirect immunofluorescence, tetramethylrhodamine isothiocyanate coupled secondary antibodies (TRITC); original magnification, ×40

Table 3. Cytokeratin expression in soft tissue tumors

Synovial sarcoma
Epithelioid sarcoma
Adamantinoma of the tibia
Chordoma
Ewing's sarcoma

compartment of biphasic synovial sarcoma, indicating the presence of true desmosomes (Fig. 2d). Only recently Moll et al. (1987) reported on a variable number of cytokeratin- and desmoplakin-positive cells in some cases of vimentin-positive Ewing's sarcomas. Immunocytochemical studies on alveolar soft part sarcomas have yielded controversial results (for review see Altmannsberger and Osborn 1987).

Desmin

Desmin is the muscle-specific type of intermediate filaments. It is the exclusive intermediate filament class in skeletal, cardiac, and visceral smooth muscle cells, whereas vascular smooth muscle cells can be vimentin-, vimentin- and desmin-, or desmin-positive (for review see Altmannsberger and Osborn 1987).

Desmin is the most reliable marker for benign and malignant myogenic tumors such as rhabdomyomas, rhabdomyosarcomas, leiomyomas, and leiomyosarcomas (Altmannsberger et al. 1982, 1985; Miettinen et al. 1982 b, c; Evans et al. 1983). In a variable number of tumor cells, vimentin can be found coexpressed with desmin (Fig. 2g, h; Altmannsberger et al. 1985). The occurrence of a vimentin-only variant of rhabdomyosarcoma is still a matter of debate (Gabbiani et al. 1981; Denk et al. 1983; Molenaar et al. 1985; for review see Altmannsberger and Osborn 1987).

Glial Fibrillary Acidic Protein (GFAP)

GFAP was first isolated by Eng and co-workers (1971) from multiple sclerosis plaques. In the central nervous system, GFAP is specific for astroglia, while oligodendroglia lack GFAP intermediate filaments (Bignami and Dahl 1977; Schachner et al. 1978; de Armond et al. 1980). In protoplasmic and fibrillary astrocytes GFAP and vimentin are coexpressed (Schnitzer et al. 1981; Shaw et al. 1981; Yen and Fields 1981; Osborn et al. 1982b).

At present, GFAP represents the most reliable marker for astrocytic tumors including fibrillary, protoplasmic, gemistocytic, and pilocytic astrocytomas as well as the astrocytic compartment of oligoastrocytomas (Fig. 3a). It can also be demonstrated in anaplastic astrocytomas and glioblastomas (Fig. 3b). The glial filament, however, is not specific for astrocytomas. Focal GFAP expression has been observed in oligodendrogliomas, ependymomas, subependymomas, plexus papillo-

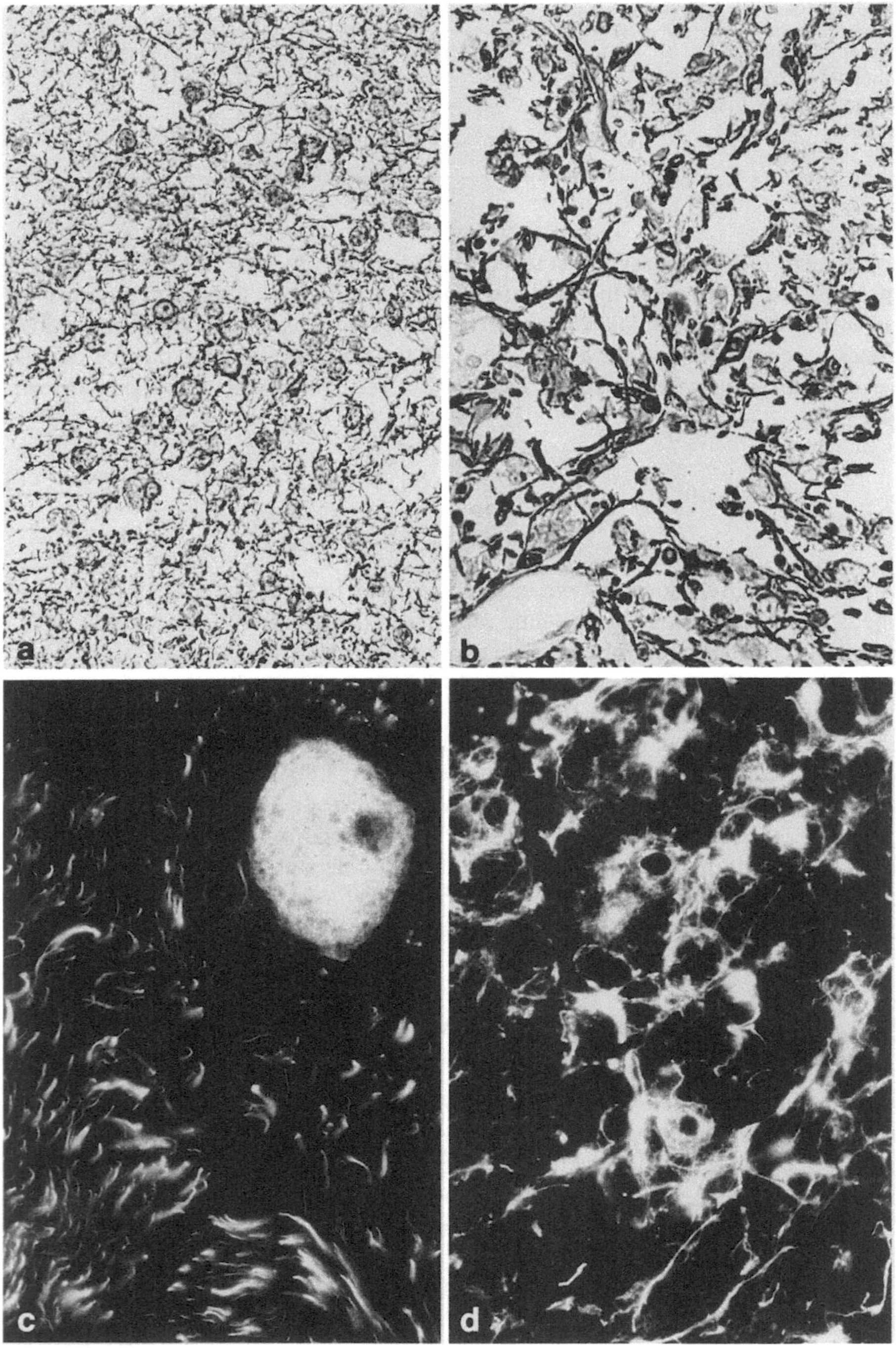

Fig. 3. a, b Strong filamentous GFAP immunoreaction in **a** a fibrillary astrocytoma and **b** a glioblastoma multiforme. **c, d** Neurofilament expression in axons and the cytoplasm of **c** a ganglion cell *(upper right)* of a ganglioneuroma and **d** in a neuroblastoma. **a, b** ABC, AEC, hematoxylin counterstain; original magnification, ×25. **c, d** Indirect immunofluorescence, TRITC; original magnification, ×40

mas, and other histological variants of gliomas (for recent reviews see Schwechheimer 1986, 1987; Kleihues et al. 1987).

GFAP is very useful as an indicator of astrocytic differentiation in medulloblastomas and other primitive neuroectodermal tumors (for reviews see Burger et al. 1987; Kleihues et al. 1987). As in nonneoplastic glia, coexpression of GFAP and vimentin is common in gliomas (Roessmann et al. 1983; Yung et al. 1985; Herpers et al. 1986; Schiffer et al. 1986; Reifenberger et al. 1987).

GFAP is not restricted to the central nervous system. It has also been found in some extraglial cell types such as Schwann and satellite cells as well as in folliculostellate cells of the adenohypophysis, myoepithelial cells of the human parotid gland, and others (for recent reviews see Achtstätter et al. 1986; Budka 1986; Born et al. 1987).

In conclusion, GFAP cannot be regarded as a specific marker protein for glial differentiation, but can be regarded as typical. It is therefore of practical value in the differential diagnosis of gliomas against cytokeratin- and desmoplakin-positive carcinomas and vimentin-positive sarcomas as well as their metastases.

Neurofilament Polypeptides

Neurofilament polypeptides form the neuron-characteristic type of intermediate filaments (Table 1; for review see Osborn et al. 1982b). On biochemical and immunological grounds, three different polypeptides (neurofilament triplet) can be distinguished (Lee et al. 1982). Neurofilament polypeptides are consistently localized in well-differentiated neuronal tumors such as gangliocytomas, ganglioneuromas (Fig. 3c), and gangliogliomas, and in pheochromocytomas, paragangliomas, and neuroblastomas (Fig. 3d; Osborn et al. 1982a; Trojanowski et al. 1984; Gould et al. 1987; Reifenberger et al. 1987; Schwechheimer 1987; Schwechheimer et al. 1987).

In addition, neurofilament expression was observed in some cases of primitive neuroectodermal neoplasms such as medulloblastomas, olfactory neuroblastomas, pinealoblastomas, and retinoblastomas as well as in cytokeratin-positive neuroendocrine tumors especially in carcinoids and small cell carcinomas of the lung (Table 4; for reviews see Schwechheimer 1986, 1987; Kleihues et al. 1987).

Table 4. Neurofilament-positive tumors

Consistently positive	Sometimes positive
Well-differentiated neuronal tumors:	Medulloblastoma
	Olfactory neuroblastoma
Gangliocytoma	Pineocytoma - pinealoblastoma
Ganglioneuroma	Retinoblastoma
Ganglioglioma	Cytokeratin-positive neuroendocrine and small cell carcinoma of the lung, Merkel cell carcinoma of the skin, parathyroid adenoma, islet cell tumor of pancreas, medullary carcinoma of thyroid gland
Pheochromocytoma	
Paraganglioma	
Neuroblastoma	

Table 5. Differential expression of intermediate filaments in tumors

Tumor type	CK	DPL	VIM	DES	GFAP	NF
Carcinoma	+	+	–[a]	–	–	–
Malignant melanoma	–	–	+	–	–	–
Nonmyogenic soft tissue tumor	–[a]	–[a]	+	–	–	–
Myogenic soft tissue tumor	–	–	(+)	+	–	–
Glioma	–	–	+	–	+	–
Neuronal tumor[a]	–	–	–	–	–	+

CK, cytokeratins; DPL, desmoplakins; VIM, vimentin; DES, desmin; GFAP, glial fibrillary acidic protein; NF, neurofilament polypeptides
[a] For exceptions and modifications see text

Diagnostic Possibilities and Limitations

Among the components of the cytoskeleton, intermediate filaments and desmosomal proteins such as desmoplakins can be used as structural markers in diagnosis and differential diagnosis of tumors. As these polypeptides are highly conserved in anaplastic neoplasms, their immunocytochemical demonstration is of inestimable value in surgical pathology and cytology. By the use of monoclonal antibodies and antisera against the different classes of intermediate filaments, carcinomas (cytokeratin- and desmoplakin-positive), malignant melanomas (vimentin-positive, cytokeratin- and desmoplakin-negative), nonmyogenic soft tissue tumors (vimentin-positive), myogenic soft tissue tumors (desmin-positive, vimentin-positive?), gliomas (GFAP- and vimentin-positive), and neuronal tumors (neurofilament polypeptide-positive) can consistently be differentiated (Table 5).

The limited value of vimentin as a tumor marker is evident as this intermediate filament type can also be coexpressed with cytokeratin, desmin, and GFAP.

Unfortunately, a further subdivision of mesenchymal, myogenic, glial, and neuronal tumors is only possible in exceptional cases (i.e., cytokeratin-positive soft tissue sarcomas). The immunocytochemical demonstration of selective cytokeratin polypeptides leading to typical cytokeratin patterns could possibly be used for more detailed analysis and further differentiation of epithelial tumors.

It has to be kept in mind that a loss of intermediate filaments in highly anaplastic tumors was sometimes observed. Finally, in spindle cell carcinomas of the aerodigestive tract a neo-expression of vimentin has been documented which can be interpreted as a change of the intermediate filament type related to altered morphology (Ellis et al. 1987).

Acknowledgements. The excellent technical assistance of C. Wahl and A. Sommer is gratefully acknowledged. We are indebted to U. Wiehle for photographic work and I. Alffermann for carefully typing the manuscript. This paper was supported by a grant of the Tumorzentrum Heidelberg/Mannheim, FRG.

References

Achtstätter T, Moll R, Anderson A, Kuhn C, Pitz S, Schwechheimer K, Franke WW (1986) Expression of glial filament protein (GFP) in nerve sheaths and non-neural cells re-examined using monoclonal antibodies, with special emphasis on the co-expression of GFP and cytokeratins in epithelial cells of human salivary gland and pleomorphic adenomas. Differentiation 31: 206-227

Altmannsberger M, Osborn M (1987) Mesenchymal tumor markers. In: Seifert G (ed) Morphological tumor markers. General aspects and diagnostic relevance. Springer, Berlin Heidelberg New York, pp 155-178

Altmannsberger M, Osborn M, Treuner J, Hölscher A, Weber K, Schauer A (1982) Diagnosis of human childhood rhabdomyosarcoma by antibodies to desmin, the structural protein of muscle specific intermediate filaments. Virchows Arch [B] 39: 203-215

Altmannsberger M, Weber K, Droste R, Osborn M (1985) Desmin is a specific marker for rhabdomyosarcomas of human and rat origin. Am J Pathol 118: 85-95

Altmannsberger M, Dirk M, Osborn M, Weber K (1986a) Immunohistochemistry of cytoskeletal filaments in the diagnosis of soft tissue tumors. Semin Diagn Pathol 3: 306-316

Altmannsberger M, Alles JU, Fitz H, Jundt G, Osborn M (1986b) Mesenchymale Tumormarker. Verh Dtsch Ges Pathol 70: 51-63

Bignami A, Dahl D (1977) Specifity of the glial fibrillary acidic protein for astroglia. J Histochem Cytochem 25: 466-469

Born IA, Schwechheimer K, Maier H, Otto HF (1987) Cytokeratin expression in normal salivary glands and in cystadenolymphomas demonstrated by monoclonal antibodies against selective cytokeratin polypeptides. Virchows Arch [A] 411: 583-589

Brown S, Levinson W, Spudich JA (1976) Cytoskeletal elements of chick embryo fibroblasts revealed by detergent extraction. J Supramol Struct 5: 110-130

Budka H (1986) Non-glial specificies of immunocytochemistry for the glial fibrillary acidic protein (GFAP). Triple expression of GFAP, vimentin and cytokeratins in papillary meningioma and metastasizing renal carcinoma. Acta Neuropathol (Berl) 72: 43-54

Burger PC, Grahmann FC, Bliestle A, Kleihues P (1987) Differentiation in the medulloblastoma. A histological and immunohistochemical study. Acta Neuropathol (Berl) 73: 115-123

Caselitz J, Jänner M, Breitbart E, Weber K, Osborn M (1983) Malignant melanomas contain only the vimentin type of intermediate filaments: implications for histogenesis and diagnosis. Virchows Arch [A] 400: 43-51

Chase DR, Weiss SW, Enzinger FM, Langloss IM (1984) Keratin in epithelioid sarcoma. Am J Surg Pathol 8: 435-441

Cooper D, Schermer A, Sun T-T (1985) Biology of disease. Classification of human epithelia and their neoplasms using monoclonal antibodies to keratins: strategies, applications, and limitations. Lab Invest 52: 243-256

De Armond SJ, Eng LF, Rubinstein LJ (1980) The application of glial fibrillary acidic (GFAP) protein immunohistochemistry in neurooncology. A progress report. Pathol Res Pract 168: 374-394

Denk H, Krepler R, Artlieb U, Gabbiani G, Rungger-Brändle E, Leoncini P, Franke WW (1983) Proteins of intermediate filaments: an immunohistochemical and biochemical approach to the classification of soft tissue tumors. Am J Pathol 110: 193-208

Ellis GL, Langloss JM, Heffner DK, Hyams VJ (1987) Spindle-cell carcinoma of the aerodigestive tract. An immunohistochemical analysis of 21 cases. Am J Surg Pathol 11: 335-342

Eng LF, Vanderhaeghen JJ, Bignami A, Gerstl B (1971) An acidic protein isolated from fibrous astrocytes. Brain Res 28: 351-354

Evans DJ, Lambert IA, Jacob M (1983) Intermediate filaments in smooth muscle tumors. J Clin Pathol 36: 57-61

Franke WW, Denk H, Kalt R, Schmid E (1981) Biochemical and immunological identification of cytokeratin proteins present in hepatocytes of mammalian liver tissue. Exp Cell Res 131: 299-319

Franke WW, Moll R, Schiller DL, Schmid E, Kartenbeck J, Mueller H (1982) Desmoplakins of epithelial and myocardial desmosomes are immunologically and biochemically related. Differentiation 23: 115-127

Franke WW, Moll R, Mueller H, Schmid E, Kuhn C, Krepler R, Artlieb U, Denk H (1983) Immunocytochemical identification of epithelium derived human tumors with antibodies to desmosomal plaque proteins. Proc Natl Acad Sci USA 80: 543-547
Gabbiani G, Kapanci Y, Barrazone P, Franke WW (1981) Immunochemical identification of intermediate sized filaments in human neoplastic cells. A diagnostic aid for the surgical pathologist. Am J Pathol 104: 206-216
Giorno R (1985) Immunohistochemical analysis of the distribution of vimentin in human peripheral lymphoid tissues. Anat Rec 211: 43-47
Giorno R, Sciotto CG (1985) Use of monoclonal antibodies for analyzing the distribution of the intermediate filaments protein vimentin in human non-Hodgkin lymphomas. Am J Pathol 120: 351-355
Gould VE, Wiedenmann B, Lee I, Schwechheimer K, Dockhorn-Dworniczak B, Radosevich JA, Moll R, Franke WW (1987) Synaptophysin expression in neuroendocrine neoplasms as determined by immunocytochemistry. Am J Pathol 126: 243-257
Herpers MJHM, Ramaekers FCS, Aldeweireldt J, Moesker O, Slooff J (1986) Co-expression of glial fibrillary acidic protein- and vimentin-type intermediate filaments in human astrocytomas. Acta Neuropathol (Berl) 70: 333-339
Kleihues P, Kiessling M, Janzer RC (1987) Morphological markers in neuro-oncology. In: Seifert G (ed) Morphological tumor markers. General aspects and diagnostic relevance. Springer, Berlin Heidelberg New York, pp 307-338
Lee V, Wu HL, Schlaepfer WW (1982) Monoclonal antibodies recognize individual neurofilament triplet proteins. Proc Natl Acad Sci USA 79: 6089-6092
Miettinen M, Lehto VP, Virtanen I (1982a) Keratin positivity in the epithelial-like cells of classical biphasic synovial sarcoma. Virchows Arch [B] 40: 157-161
Miettinen M, Lehto VP, Badley RA, Virtanen I (1982b) Expression of intermediate filaments in soft tissue sarcomas. Int J Cancer 30: 541-546
Miettinen M, Lehto VP, Badley RA, Virtanen I (1983c) Alveolar rhabdomyosarcoma. Demonstration of the muscle type of intermediate filament protein desmin as a diagnostic aid. Am J Pathol 108: 246-251
Miettinen M, Lehto VP, Virtanen I (1983) Monophasic synovial sarcoma of spindle-cell type. Virchows Arch [B] 44: 187-199
Miettinen M, Clar R, Virtanen I (1986) Intermediate filament proteins in choroid plexus, ependyma and their tumors. Am J Pathol 123: 231-240
Molenaar WM, Oosterhuis JW, Oosterhuis AM, Ramaekers FCS (1985) Mesenchymal and muscle-specific intermediate filaments (vimentin and desmin) in relation to differentiation in childhood rhabdomyosarcomas. Hum Pathol 16: 838-843
Moll R (1986) Epitheliale Tumormarker. Verh Dtsch Ges Pathol 70: 28-50
Moll R (1987) Epithelial tumor markers: cytokeratins and tissue polypeptide antigen (TPA). In: Seifert G (ed) Morphological tumor markers. General aspects and diagnostic relevance. Springer, Berlin Heidelberg New York, pp 71-101
Moll R, Franke WW (1986) Cytochemical cell typing of metastatic tumors according to their cytoskeletal proteins. In: Lapis K, Liotta LA, Rabson AS (eds) Biochemistry and molecular genetics of cancer metastasis. Nijhoff, The Hague, pp 101-114
Moll R, Franke WW, Schiller DL, Geiger B, Krepler R (1982) The catalog of human cytokeratins: patterns of expression in normal epithelia, tumors and cultured cells. Cell 31: 11-24
Moll R, Krepler R, Franke WW (1983) Complex cytokeratin polypeptide patterns observed in certain human carcinomas. Differentiation 23: 256-269
Moll R, Lee I, Gould VE, Berndt R, Roessner A, Franke WW (1987) Immunocytochemical analysis of Ewing's tumor. Pattern of expression of intermediate filaments and desmosomal proteins indicate cell type heterogeneity and pluripotential differentiation. Am J Pathol 127: 288-304
Mueller H, Franke WW (1983) Biochemical and immunological characterization of desmoplakins I and II, the major polypeptides of the desmosomal plaque. J Mol Biol 163: 647-671
Osborn M, Weber K (1977) The detergent-resistant cytoskeleton of tissue culture cells includes the nucleus and the microfilament bundles. Exp Cell Res 106: 339-349
Osborn M, Weber K (1983) Tumor diagnosis by intermediate filament typing: a novel tool for surgical pathology. Lab Invest 48: 372-394
Osborn M, Altmannsberger M, Shaw G, Schauer A, Weber K (1982a) Various sympathetic de-

rived human tumors differ in neurofilament expression. Use in diagnosis of neuroblastoma, ganglioneuroblastoma and pheochromocytoma. Virchows Arch [B] 40: 141-156
Osborn M, Geisler N, Shaw G, Sharp G, Weber K (1982b) Intermediate filaments. Cold Spring Harbor Symp Quant Biol 46: 413-429
Osborn M, van Lessen G, Weber K, Klöppel G, Altmannsberger M (1986) Differential diagnosis of gastrointestinal carcinomas by using monoclonal antibodies specific for individual keratin polypeptides. Lab Invest 55: 497-504
Quinlan RA, Schiller DL, Hatzfeld M, Achtstätter T, Moll R, Jorcano JL, Magin TM, Franke WW (1985) Patterns of expression and organization of cytokeratin intermediate filaments. Ann NY Acad Sci 455: 282-306
Reifenberger G, Szymàs J, Wechsler W (1987) Differential expression of glial- and neuronal-associated antigens in human brain tumors of the central and peripheral nervous system. Acta Neuropathol (Berl) 74: 105-123
Roessmann U, Velasco ME, Gambetti P, Autilio-Gambetti L (1983) Vimentin intermediate filaments are increased in human neoplastic astrocytes. J Neuropathol Exp Neurol 42: 309
Rosai J, Pinkus CS (1982) Immunohistochemical demonstration of epithelial differentiation in adamantinoma of the tibia. Am J Pathol 6: 427-434
Schachner M, Smith C, Schoonmaker G (1978) Immunological distinction between neurofilament and glial fibrillary acidic proteins by mouse antisera and their immunohistological characterizatiion. Dev Neurosci 1: 1-14
Schiffer D, Giordana MT, Mauro A, Migheli A, Germano I, Giaccone G (1986) Immunohistochemical demonstration of vimentin in human cerebral tumors. Acta Neuropathol (Berl) 70: 209-219
Schnitzer J, Franke WW, Schachner M (1981) Immunocytochemical demonstration of vimentin in astrocytes and ependymal cells of developing and adult mouse nervous system. J Cell Biol 90: 435-447
Schwechheimer K (1986) Nervale Tumormarker. Verh Dtsch Ges Pathol 70: 82-103
Schwechheimer K (1987) Immuncytochemische Untersuchungen an Tumoren des zentralen, peripheren und autonomen Nervensysteme. Habilitation-Thesis, Fakultät für Theoretische Medizin der Universität Heidelberg
Schwechheimer K, Wiedenmann B, Franke WW (1987) Synaptophysin: a reliable marker for medulloblastomas. Virchows Arch [A] 411: 53-59
Shaw G, Osborn M, Weber K (1981) An immunofluorescence microscopical study of the neurofilament triplet proteins, vimentin and glial fibrillary acidic protein within the adult rat brain. Eur J Cell Biol 26: 68-82
Sun T-T, Eichner R, Schermer A, Cooper D, Nelson WG, Weiss RA (1984) Classification, expression and possible mechanisms of evolution of mammalian epithelial keratins: a unifying model. In: The transformed phenotype, cancer cells, vol 1. Cold Spring Harbor Laboratory, Cold Spring Harbor, pp 169-176
Traub P (1985) Intermediate filaments. A review. Springer Berlin Heidelberg New York
Trojanowski JQ, Lee VM-Y, Schlaepfer WW (1984) An immunohistochemical study of human central and peripheral nervous system tumors, using monoclonal antibodies against neurofilaments and glial filaments. Hum Pathol 15: 248-257
Yen S-H, Fields KL (1981) Antibodies to neurofilament, glial filament, and fibroblast intermediate filament proteins bind to different cell types of the nervous system. J Cell Biol 88: 115-126
Yung W-KA, Luna M, Borit A (1985) Vimentin and glial fibrillary acidic protein in human brain tumors. J Neurooncol 3: 35-38

Intermediate Filament Typing vs. Electron Microscopy in the Diagnosis of Major Tumor Types in Fine Needle Aspirates

W. Domagala, J. Lubinski, K. Weber, and M. Osborn

Introduction

In aspiration cytology, the accuracy of diagnosis of tumor type depends on the expression of morphologic signs of differentiation of tumor cells at the light microscopic level and on the availability of reliable clinical, radiological, and other relevant information. If differentiation of tumor cells in fine needle aspirate (FNA) is clearly expressed, the specific diagnosis of major tumor type such as carcinoma, sarcoma, malignant lymphoma, or malignant melanoma can easily be made by an experienced cytopathologist [21]. However, the diagnosis of tumor type in FNA has its limitations. If perceptible morphologic features of differentiation of tumor cells are absent, the "descriptive tumor typing" is of little use. This is supported by poor interobserver reproducibility among several cytopathologists in the diagnosis of tumor type in difficult aspirates with no detectable features indicating differentiation of tumor cells [9] and also by poor interobserver agreement as to malignant cell types [36]. In such cases, "histogenetic tumor typing" based on information provided by differentiation or tissue-specific markers, when combined with analysis of morphology of tumor cells and clinical data can markedly improve cytologic diagnosis of tumor type in FNA [9, 12, 14, 19]. In aspiration cytology such information can be obtained by exploiting monoclonal antibodies (MAbs) to intermediate filaments (IF) [2, 3, 6-14, 33] and to other cell antigens [18, 19, 28] as well as by transmission electron microscopy (EM) [1, 4, 5, 16, 20, 24-26, 30, 34, 37].

Here, on the basis of our own material and of the literature (Table 1) we com-

Table 1. Number of FNA tested by IF and EM

Tumor type	No. of FNA studied by			
	Authors		Others[a]	
	IF	EM	IF	EM
Carcinoma	218	49	42	105
Malignant lymphoma	72	21	2	44
Malignant melanoma	26	8	1	9
Sarcoma	57	23	7	91
Total	373	101	52	249

[a] IF [2, 3, 14, 33]; EM [1, 4-5, 16, 20, 30, 34, 37]

pare how IF and EM can be used for tumor typing in FNA and discuss advantages and limitations of both methods.

Intermediate Filaments and Electron Microscopy in the Diagnosis of Major Tumor Types in FNA

Carcinomas

Intermediate Filaments

Tumor cells in FNA from poorly or well-differentiated carcinomas are positive when tested with broad specificity keratin MAbs [2, 3, 6-12, 14, 33]. In our experience the proportion of tumor cells in the smears that stained with the immunofluorescence, immunoperoxidase or immunoalkaline phosphatase has been very high (well over 80% of cells in the vast majority of cases). Usually, the positive staining is strong (Figs. 1 A, 2 A, 3 A), the background is negligible, and at high magnification filaments can be distinguished within the cytoplasm using immunofluorescence. Lymphocytes and macrophages which are present in almost every smear serve as a negative "built-in" control since they are keratin-negative and vimentin-positive (Fig. 3 A, B). Undifferentiated carcinomas, including neuroendocrine (Merkel cell) skin carcinoma [8], small cell carcinoma of the lung and other organs, or nasopharyngeal carcinoma, may have only a few keratin-positive filaments in the narrow rim of their cytoplasm; nevertheless, in most cases they can be easily and reliably distinguished.

The biochemically documented heterogeneity of human keratins [29], as well as the characteristic morphologic arrangement of keratin filaments in most carcinomas and coexpression of keratin with other types of IF in some carcinomas, offer additional possibilities for subdividing carcinoma. For example, carcinomas from simple epithelia (mostly adenocarcinomas) and from transitional epithelium are decorated by cytokeratin 2 (CK2) MAbs (Fig. 2 A), whereas CKS1 MAb recognizes squamous epithelia (and hence some squamous carcinomas) [7, 9, 14]. Other MAbs can also permit subdivision of carcinomas [19, 35]. Tumor cells of neuroendocrine (Merkel cell) carcinoma of the skin are characterized by "IF buttons", i.e., distinct round IF aggregates located preferentially in the perinuclear area in a cap-like fashion or detached from the nucleus and isolated in the smear [8]. The IF buttons in this tumor coexpress keratin and neurofilaments (for details on differential diagnosis of Merkel cell carcinoma in FNA see [8]). Finding coexpression of IF in tumor cells in an aspirate can be indicative of tumor type or site of origin of carcinoma [3, 13, 14] (for diagnostic significance of true, pseudo, or false coexpression in FNA see [11]). Tumor cells in FNA of medullary thyroid carcinoma coexpress keratin, vimentin, and neurofilaments [3, 13]. True coexpression of keratin and vimentin was documented in FNA of the majority of cases of renal cell (Fig. 3 A, B), endometrial, thyroid, and ovarian serous carcinoma, whether primary or metastatic [3, 7, 9, 12-14]. It can also be found in a small percentage of carcinoma from other sites, e.g., lung, breast, or adrenal gland [12]. Hence, if true coexpression of keratin and vimentin is found in a metastatic carcinoma of unknown origin, the search for the primary carcinoma should be first concentrated on those

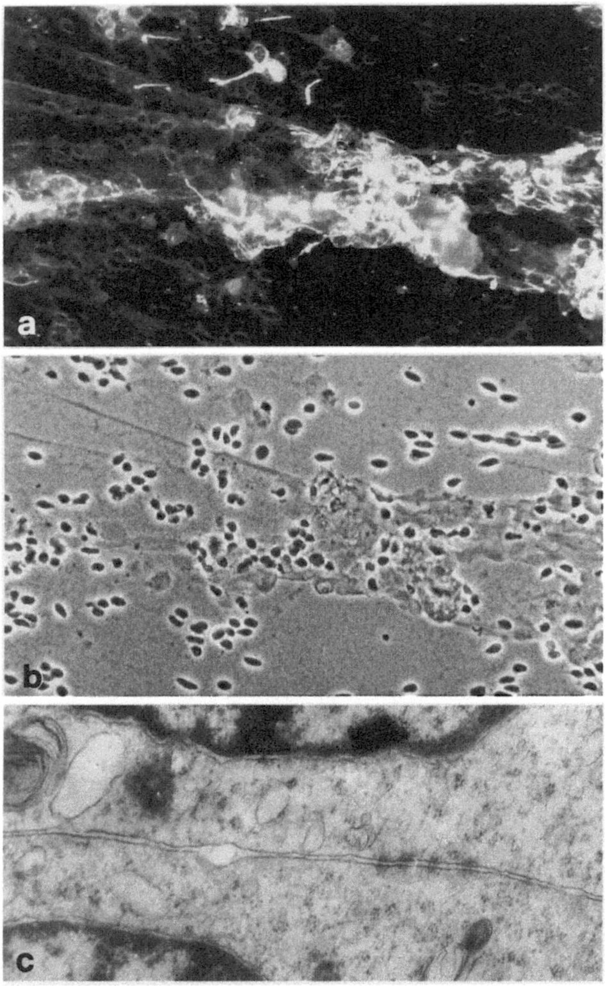

Fig. 1 A–C. Small cell undifferentiated carcinoma (FNA, lymph node). **A** A clump of tumor cells and few isolated cells strongly labeled by KL1 keratin MAb (immunofluorescence, ×250). **B** Phase-contrast micrograph of cells shown in **A** (×250). **C** Poorly developed cell junctions joining two cancer cells. They were the only EM evidence of epithelial differentiation in this case (EM; ×23400). For details of methods used for preparation of FNA see [7, 11] for IF, and [23] for EM

sites where the coexpression is frequent, i.e., thyroid, kidney, endometrium, and ovary. It has been shown on tissue sections that MAbs recognizing keratin 7 and keratin 19 can distinguish primary hepatocarcinoma from metastatic colorectal and pancreatic duct adenocarcinoma [15]. We have obtained similar results on FNA from primary and metastatic carcinomas of the liver (unpublished results). The differential diagnosis of primary vs. metastatic liver carcinoma can further be aided by using vimentin and endothelial cell-specific MAbs, since numerous endothelial cells have been found in FNA from primary hepatocarcinomas while they were rare or absent in metastatic cancers [10].

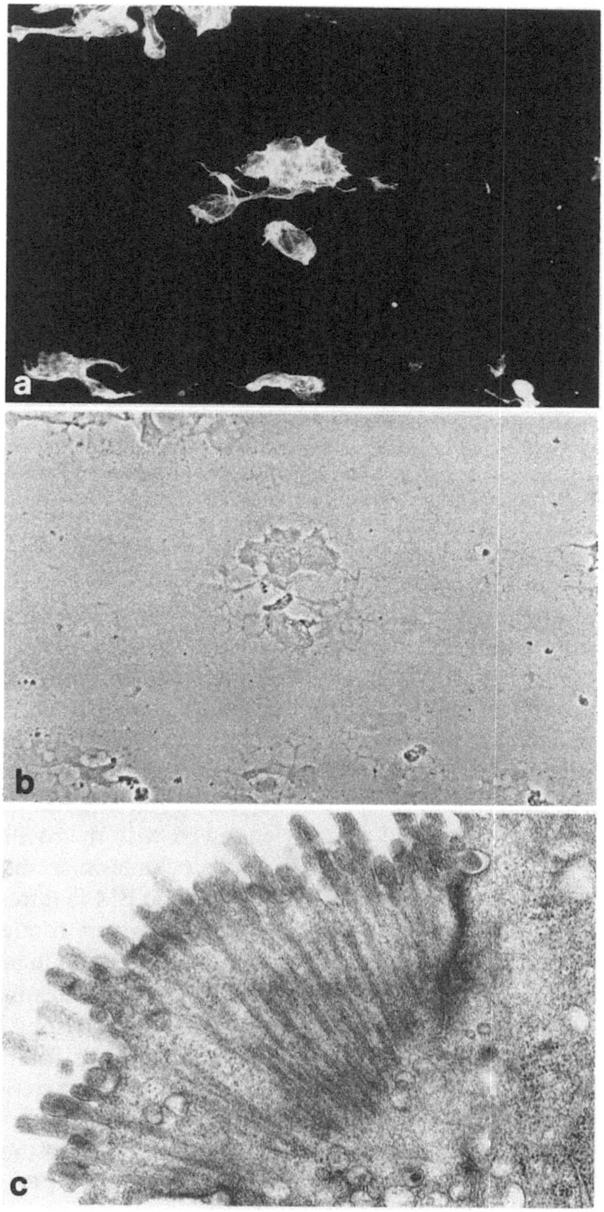

Fig. 2 A-C. Adenocarcinoma of the colon (FNA, lymph node). **A** Tumor cells labeled by CK2 keratin MAb (immunofluorescence, ×250). **B** Phase-contrast micrograph of cells shown in **A** (×250). **C** Filamentous core rootlets in a tumor cell (EM; ×25000)

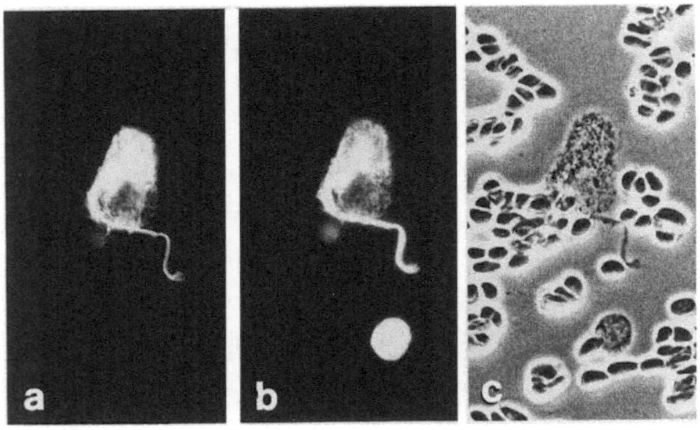

Fig. 3A-C. Renal cell carcinoma (FNA, lymph node). **A, B** Tumor cell labeled by CK2 keratin MAb (**A**) and by guinea pig vimentin antibody (**B**) coexpresses keratin and vimentin while a lymphocyte *(right lower corner)* is stained only with antibody to vimentin (double immunofluorescence staining on the same smear; ×400). **C** Phase-contrast micrograph of cells shown in **A** and **B**. Note prominent nucleolus in tumor cell nucleus (×400)

Electron Microscopy

Typically, carcinoma cells in aspirates form clusters in which carcinoma cells are usually connected by a number of variously developed cell junctions [1, 4, 20, 24-26, 30, 34, 37]. In well-differentiated carcinoma one finds well-developed desmosomes; in undifferentiated carcinoma attenuated desmosomes or poorly developed cell junctions are commonly seen (Fig. 1C). Well-differentiated squamous carcinomas are characterized by numerous well-developed desmosomes and bands of tonofilaments [1, 4, 25]. Adenocarcinomas display inter- and intracytoplasmic tubules lined by cells endowed with microvilli and joined by junctional complexes [1, 4, 16, 24, 34, 37]. In the cytoplasm secretory granules can be found. Some adenocarcinomas have quite specific EM features indicating the primary tumor. Microvilli with prominent filamentous core rootlets (Fig. 2C), surface glycocalyx, and glycocalyceal bodies are typically found in adenocarcinoma of the large bowel, gallbladder, pancreatic ducts, and in the intestinal type of adenocarcinoma of the stomach [4, 24, 37]. Numerous intracytoplasmic lumina frequently occur in breast carcinoma [24, 37], but they can also be found in other carcinomas, although with much lower frequency [4, 37]. Characteristic large, dense, secretory granules are found in acinic carcinoma of the salivary gland [16]; neurosecretory granules are found in neuroendocrine carcinoma of the skin, lung, and thyroid [4, 8, 16, 23, 26, 37].

Malignant Lymphomas

Intermediate Filaments

Malignant lymphoma cells in FNA are vimentin-positive and keratin-negative [2, 7, 9, 12, 14]. All malignant lymphomas we have tested (over 70 cases) were vimentin-positive (50%-100% positive tumor cells; Fig. 4A) and all tested with keratin MAb were keratin-positive. Malignant lymphomas in FNA, as in tissue sections [22], express leukocyte common antigen (LCA). Tumor cells in FNA from all ma-

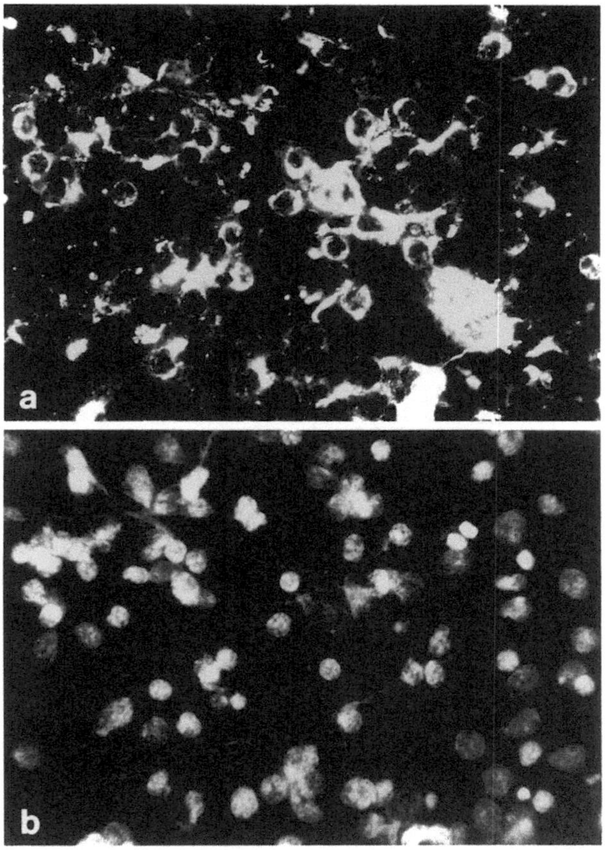

Fig. 4 A, B. Malignant lymphoma (FNA, lymph node). **A** The cytoplasm of tumor cells is labeled by vimentin MAb (immunofluorescence; ×250). **B** Hoechst stain of nuclei of cells shown in **A** (×250)

lignant lymphomas in our material were LCA-positive (20%-100% positive tumor cells in various cases). In immunofluorescence, the granular staining pattern of LCA which decorates plasma membrane (Fig. 5 A) differs from filamentous staining of vimentin which is found in the cytoplasm (Fig. 4 A). Additional immunophenotyping of malignant lymphomas can be done in FNA using markers for B and T cells [28].

Electron Microscopy

Tumor cells are usually isolated (Fig. 5 C), but small clusters of cells can sometimes be found [37]. There are no desmosomes [1, 4, 16, 37]. Numerous nuclear pockets, if found, favor the diagnosis of malignant lymphoma [17, 37]. However, if they are not evident, and if real or artificial (as a result of cell concentration in the initial stages of specimen preparation) clumps of cells are seen, one can never be sure - even after a meticulous search - whether a few desmosomes might not be found if more cells were examined. In general, the diagnosis of malignant lymphoma is

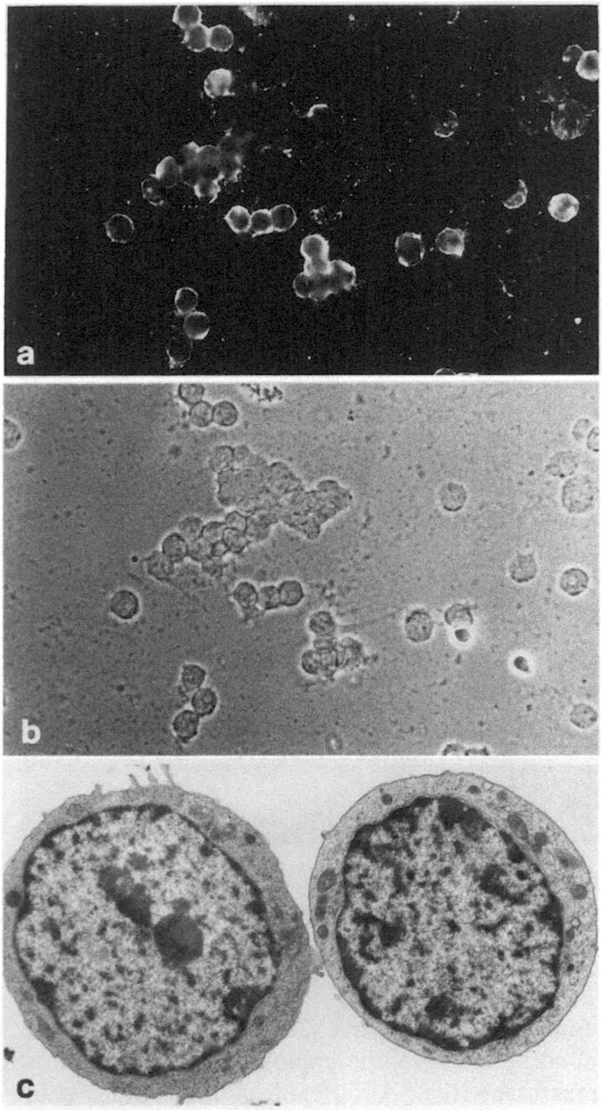

Fig. 5 A–C. Malignant lymphoma (FNA, lymph node). **A** Surface staining of tumor cells labeled with LCA MAb (immunofluorescence; ×250). **B** Phase-contrast micrograph of cells in **A** (×250). **C** Two tumor cells without nuclear pockets (EM; ×5600)

suggested by the lack of EM features characteristic for other tumor types in a tumor composed of dispersed lymphoid-like or undifferentiated cells [4, 16, 37]. Because of these limitations, despite some claims to the contrary [1], we share the opinion of Mackay [27] that malignant lymphoma cannot be diagnosed by EM, although it can be suggested with a certain degree of probability in some cases.

Sarcomas

Intermediate Filaments

Tumor cells of the majority of nonmuscle sarcomas, e.g., fibrosarcoma, malignant fibrous histiocytoma, liposarcoma, chondrosarcoma, osteosarcoma, and angiosarcoma, contain only vimentin, whether in FNA [6, 12] or in tissue sections [31, 32]. Rhabdomyosarcoma (Fig. 6A) and leiomyosarcoma are desmin-positive [2, 9, 12,

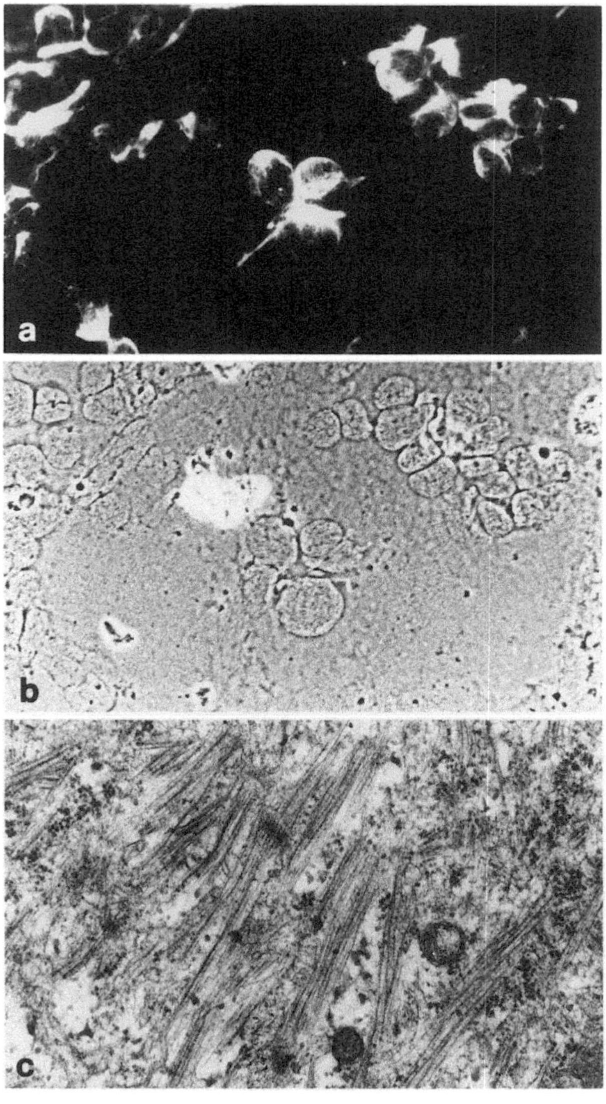

Fig. 6A-C. Embryonal rhabdomyosarcoma (FNA, lung). **A** Clusters of tumor cells positively stained with MAb to desmin (immunofluorescence; × 400). **B** Phase-contrast micrograph of cells shown in **A** (× 400). **C** A fragment of tumor cell cytoplasm. Z lines and thin and thick filaments (EM; × 30000)

14, 31, 32]. A proportion of these tumors also coexpress vimentin [2, 9, 12]. Tumor cells of synovial and epithelioid sarcoma, as well as malignant rhabdoid tumors of soft tissues, can be easily distinguished from other sarcomas since they coexpress vimentin and keratin [11]. IF typing cannot help to distinguish these sarcomas from soft tissue metastases of carcinoma coexpressing keratin and vimentin. Such dilemmas, however, are very rare. It is useful to remember that among carcinomas which coexpress keratin and vimentin, large cell anaplastic carcinoma of the thyroid and renal carcinoma are most likely to be misinterpreted as sarcoma by light microscopy.

Electron Microscopy
In FNA, as in tissue sections, individual sarcoma cells are usually separated by stroma. In cell clusters, intermediate type cell junctions can be found [1, 20, 30]. Some sarcomas display quite characteristic ultrastructure. Myofibrils composed of thick and thin filaments (Fig. 6C), Z lines, and glycogen highlight rhabdomyosarcoma [5, 30]. Leiomyosarcoma contains longitudinally oriented bundles of thin myofilaments with focal densities, micropinocytotic vesicles, and often discontinuous basal lamina around tumor cells [20, 30]. Malignant fibrous histiocytoma contains histiocyte-like cells in addition to fibroblast-like and myofibroblast-like cells usually found in fibrosarcoma [20, 30]. Granulofibrillary matrix, scalloped cellular margins, abundant glycogen, and ergastoplasm are found in FNA from chondrosarcoma [20]. In general, EM allows for subclassification of sarcoma, but unfortunately in poorly differentiated sarcoma many of the EM characteristics listed above are often poorly expressed or lost.

Malignant Melanomas

Intermediate Filaments
Tumor cells of amelanotic or melanin-containing malignant melanomas in FNA are strongly positive when tested with vimentin MAb and are negative for keratin [6, 7, 9, 12, 14]. Over 90% of malignant melanoma cells in aspirates were positive for vimentin in each case we have tested (Fig. 7A). They are also LCA-negative. Since the majority of nonmuscle sarcomas and all malignant lymphomas are also vimentin-positive and keratin-negative, vimentin positivity in tumor cells supports the diagnosis of malignant melanoma only if the clinical context and morphology of tumor cells are suggestive of this tumor type. Specific anti-malignant melanoma MAbs can be used to diagnose malignant melanoma directly [18] whereas LCA used to exclude malignant melanoma and confirm malignant lymphoma [12].

Electron Microscopy
The cytoplasm of tumor cells contains variable numbers of stage II melanosomes, i.e., oval membranous structures with series of parallel or concentric, dotted, or striated lines (Fig. 7B) [1, 16, 20, 30, 37]. They can be better visualized in thin sections by Warthin-Starry stain (Fig. 7C). Stage III or IV melanosomes can be found in pigmented malignant melanoma. Although occasionally melanosomes can be sparse and occur only in a small proportion of tumor cells, we agree with the re-

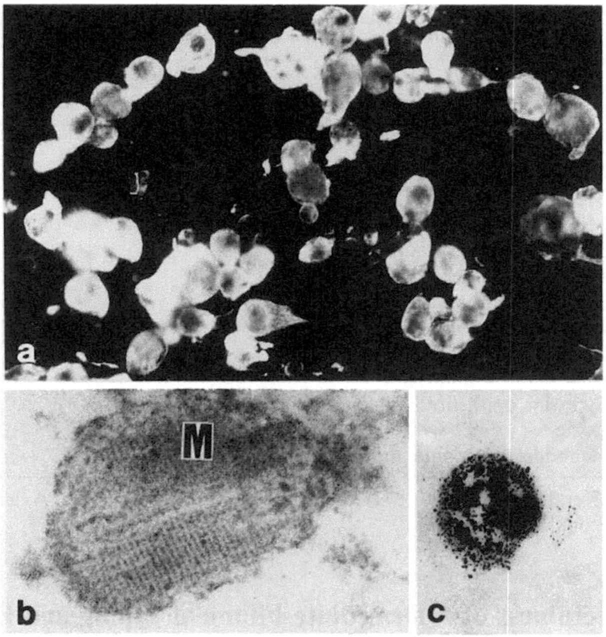

Fig. 7 A–C. Malignant melanoma (FNA, lymph node). **A** All tumor cells strongly stained with vimentin MAb (immunofluorescence; ×250). **B** Stage II melanosome *(M)* (EM; ×105000). **C** Stage II melanosome (EM; Warthin-Starry stain; ×42000)

ports indicating that EM can effectively be used to diagnose malignant melanoma in FNA [20, 37], especially if the distinction between malignant melanoma and carcinoma cannot be made by light microscopy and the cells are vimentin-positive.

Summary of Results

The results of the IF and LCA typing in FNA of major tumor types are summarized in Table 2. It is obvious that tumor typing by IF and LCA can be of great help in the differential diagnosis among carcinoma, malignant lymphoma, malignant melanoma, and sarcoma. For reasons specified above and in the next section, we believe that currently there is no role for EM in the diagnosis of malignant lymphoma, carcinoma, and sarcoma in FNA. However, EM may be useful in the diagnosis of malignant melanoma and in subdividing nonmuscle sarcoma and some adenocarcinomas by criteria described on pp. 179–182 and 185–186. The successful use uf immunocytochemistry in the diagnosis of major tumor types in FNA depends primarily on how it is introduced into the diagnostic process. First, the analysis of the morphologic features of tumor cells and the available clinical data should lead to the precise question specifying the possible tumor types. Then the panel of MAbs to IF and LCA is used to answer the question. Finally, the combined results (light microscopy + clinical history + IF and LCA typing) lead to the diagnosis of tumor type.

Table 2. IF typing in the diagnosis of tumor type in FNA[a]

Tumor type	Results of typing with				
	K	V	NF	D	LCA
Carcinoma					
Merkel cell	+	−	+		−
Medullary thyroid	+	+	+		
Majority of renal, ovarian, endometrial, and thyroid	+	+	−		
Some lung, breast, adrenal	+	+	−		
Majority of other	+	−	−		
Malignant lymphoma	−	+	−		+
Malignant melanoma	−	+	−		−
Sarcoma					
Majority of nonmuscle	−	+	−	−	
Synovial, epithelioid, malignant rhabdoid tumor of soft tissue	+	+	−	−	
Rhabdomyosarcoma and leiomyosarcoma	−	−/+	−	+	

K, Keratin; V, vimentin; NF, neurofilaments; D, desmin; LCA, leukocyte common antigen
[a] From [2, 3, 6–14, 33]

Usefulness of Intermediate Filament Typing and Electron Microscopy in Diagnosis

Our experience, and that of others, proves that the results of IF typing and EM in FNA closely parallel those obtained on tissue sections [17, 31, 32]. In general, the specific information necessary for the diagnosis of major tumor types in FNA, which is inherent in the ultrastructure, is almost the same as that which can be obtained by IF typing of tumor cells. However, in the case of EM, the retrieval of this information and its meaningful interpretation for diagnostic purposes in FNA is greatly hampered by the following.

1. Many EM features characteristic for the particular tumor type, e.g., desmosomes and myofilaments, are well expressed in well-differentiated tumors (which can be reliably diagnosed by light microscopy), while they are often lost or poorly expressed in undifferentiated tumors (where diagnosis of tumor type by light microscopy is uncertain or impossible). By contrast, IF are clearly expressed in both well and poorly differentiated tumors.
2. EM diagnosis of certain tumor types, e.g., malignant lymphoma, is often based not on a positive identification but on lack of the ultrastructural features specific for other tumor types. This makes the differential diagnosis with poorly differentiated tumors all the more difficult.

   For the above reasons we could not unequivocally resolve a differential diagnosis of carcinoma vs. malignant lymphoma by EM in 11 FNA positively identified as malignant lymphomas by IF typing.

3. Lack of specificity of several EM features, e.g., glycogen.
4. EM assessment of tumor cells is a ten-step procedure, beginning with an analy-

sis of topographic cellular relationships and ending with the interpretation of nuclear and nucleolar morphology [17]. In this analysis a large number of ultrastructural features have to be taken into account; for example, at least 16 types of cytoplasmic granules are known to exist. Correct interpretation of them, which is already difficult in itself, is further compounded by the occurrence of abortive variants found in tumors. Thus the interpretation of the ultrastructure requires special training and experience and in practice is very difficult for those not routinely familiar with EM. This is in sharp contrast to the interpretation of the immunocytochemistry, where one has to learn basically how to distinguish positive from negative staining, a task greatly facilitated by simultaneously run controls, some of them "built into" the smear.
5. Although quite an impressive number of tumor cells can be examined in thin sections from well prepared aspirates, (e.g. [37]) many more cells are present in smears treated for immunocytochemistry.

The routine use of EM in the cytology laboratory is further complicated by the need for special *fixation*, methods for preliminary concentration of cells, removal of erythrocytes, examination by light microscopy of the plastic semithin sections, etc. In comparison, simple alcohol fixation, exactly the same as the one used for routine hematoxylin-eosin and Pap stained smears, is used for IF typing, LCA typing, and some other immunocytochemical tests. One or two smears can be left in alcohol until a decision is reached whether IF typing is necessary. Although methods of preparation of aspirates for EM have been much simplified recently [23, 37], they are still more elaborate than immunocytochemistry as applied to FNA.

The *results* of IF typing can be known within 2-3 h; those of EM take 4-24 h. This fact additionally favors immunocytochemistry.

Costs involved in applying immunocytochemistry or EM to FNA comprise not only the cost of reagents and equipment but also training of technicians and cytopathologists as well as the *time* required to obtain and to interpret the results.

Finally, immunocytochemistry offers the possibility of *development of new MAbs* in the future, which may further improve and expand the possibilities of diagnosis of tumor type in FNA. In contrast, new opportunities in this respect can hardly be expected from morphologic examination of cell ultrastructure. Thus, the above factors clearly favor using immunocytochemistry instead of EM for tumor typing in FNA.

Conclusions

Immunocytochemistry in FNA, in particular the use of MAbs to IF and LCA, is a quick, reliable, unambiguous, relatively simple, and inexpensive method which can help to confirm, revise, or refine typing of major tumor categories [9, 12] and as such improves the accuracy of cytologic diagnosis of tumor type. It is clearly superior to tumor typing by EM.

References

1. Akhtar M, Ali MA, Owen E (1981) Application of electron microscopy in interpretation of fine-needle aspiration biopsies. Cancer 48: 2458-2463
2. Altmannsberger M, Osborn M, Droese M, Weber K, Schauer A (1984) Diagnostic value of intermediate filament antibodies in clinical cytology. Klin Wochenschr 62: 114-123
3. Altmannsberger M, Dralle H, Weber K, Osborn M, Droese M (1987) Intermediate filaments in cytological specimens of thyroid tumors. Diagn Cytopathol 3: 210-219
4. Berkman WA, Chowdhury L, Brown NL, Padleckas R (1983) Value of electron microscopy in cytologic diagnosis of fine needle biopsy. Am J Roentgenol 240: 1253-1258
5. Collins VP, Ivarsson B (1981) Tumor classification by electron microscopy of fine needle aspiration biopsy material. Acta Pathol Microbiol Scand [A] 89: 103-105
6. Domagala W, Lubinski J, Weber K, Osborn M (1986) Intermediate filament typing of tumor cells in fine needle aspirates by means of monoclonal antibodies. Acta Cytol (Baltimore) 30: 214-224
7. Domagala W, Weber K, Osborn M (1986) Differential diagnosis of lymph node aspirates by intermediate filament typing of tumor cells. Acta Cytol (Baltimore) 30: 225-234
8. Domagala W, Lubinski J, Lasota J, Giryn I, Weber K, Osborn M (1987) Neuroendocrine (Merkel cell) skin carcinoma: cytology, intermediate filament typing and ultrastructure of tumor cells in fine needle aspirates. Acta Cytol (Baltimore) 31: 267-275
9. Domagala W, Lubinski J, Lasota J, Woyke S, Wozniak L, Szadowska A, Weber K, Osborn M (1987) Decisive role of intermediate filament typing of tumor cells in the differential diagnosis of difficult fine needle aspirates. Acta Cytol (Baltimore) 31: 253-266
10. Domagala W, Lasota J, Weber K, Osborn M (1987) Endothelial cells are an often unrecognized part of the cytologic pattern of primary hepatocellular carcinoma in fine needle aspirates: an immunofluorescence study with vimentin and endothelial cell specific antibodies. (submitted for publication)
11. Domagala W, Weber K, Osborn M (1988) Diagnostic significance of coexpression of intermediate filaments in fine needle aspirates of human tumors. Acta Cytol (Baltimore) 32: 49-59
12. Domagala W, Lasota J, Chosia M, Szadowska A, Weber K, Osborn M (1988) Diagnosis of major tumor categories in fine needle aspirates is more accurate when light microscopy is combined with intermediate filament typing: a study of 403 cases. Cancer (submitted for publication)
13. Domagala W, Lasota J, Wolska H, Lubinski J, Weber K, Osborn M (1988) Diagnosis of metastatic renal cell and thyroid carcinomas by intermediate filament typing and cytology of tumor cells in fine needle aspirates. Acta Cytol (Baltimore) (in press)
14. Droese M, Altmannsberger M, Kehl A, Lankisch PG, Weiss R, Weber K, Osborn M (1984) Ultrasound-guided percutaneous fine needle aspiration biopsy of abdominal and retroperitoneal masses. Acta Cytol (Baltimore) 28: 368-384
15. Fisher HP, Altmannsberger M, Weber K, Osborn M (1987) Keratin polypeptides in malignant epithelial liver tumors. Differential diagnosis and histogenetic aspects. Am J Pathol 127: 530-538
16. Hagelqvist E (1978) Light and electron microscopic studies on material obtained by fine needle biopsy. Acta Otolaryngol [Suppl] (Stockh) 354: 1-75
17. Henderson DW, Papadimitriou JM, Coleman M (1986) Ultrastructural appearances of tumours. Churchill Livingstone, Edinburgh
18. Johnston WW, Borowitz MJ, Stuhmiller GM, Seigler HF (1985) Expression of a melanoma-tumor-associated antigen as demonstrated by a monoclonal antibody (D6.1) in cytopathologic preparations of human tumor cells from effusions and needle aspirates. Anal Quant Cytol 7: 72-80
19. Johnston WW, Szpak CA, Lottich SC, Thor A, Schlom J (1986) Use of a monoclonal antibody (B72.3) as a novel immunohistochemical adjunct for the diagnosis of carcinomas in fine needle aspiration biopsy smears. Hum Pathol 17: 501-513
20. Kindblom LG (1983) Light and electron microscopic examination of embedded fine-needle aspiration biopsy specimens in the preoperative diagnosis of soft tissue and bone tumors. Cancer 51: 2264-2277

21. Koss LG, Woyke S, Olszewski W (1984) Aspiration biopsy: cytologic interpretation and histologic bases. Igaku-Shoin, New York
22. Kurtin PJ, Pinkus GG (1985) Leukocyte common antigen - a diagnostic discriminant between hematopoietic neoplasms in paraffin sections using monoclonal antibodies. Correlation with immunologic studies and ultrastructural localization. Hum Pathol 16: 353-365
23. Lubinski J, Domagala W (1985) Rapid method of preparation of fine needle aspirates for transmission electron microscopy (in Polish). Patol Pol 36: 195-199
24. Lubinski J, Domagala W (1986) Differential diagnosis of carcinomas in fine needle aspirates by transmission electron microscopy. I. Adenocarcinomas (in Polish). Patol Pol 37: 197-204
25. Lubinski J, Domagala W (1986) Differential diagnosis of carcinomas in fine needle aspirates by transmission electron microscopy. II. Squamous cell carcinomas (in Polish). Patol Pol 37: 324-334
26. Lubinski J, Domagala W (1987) Differential diagnosis of carcinomas in fine needle aspirates by transmission electron microscopy. III. Anaplastic carcinomas (in Polish). Patol Pol 38: 128-137
27. Mackay B (1985) Ultrastructural diagnosis of lymphomas and leukemias. Ultrastruct Pathol 9: 209-214
28. Martin SE, Zhang HZ, Magyarosy E, Jaffe ES, Hsu SM, Chu EW (1984) Immunologic methods in cytology: definitive diagnosis of non-Hodgkin's lymphomas using immunologic markers for T and B cells. Am J Clin Pathol 82: 666-673
29. Moll R, Franke WW, Schiller DL, Geiger B, Krepler R (1982) The catalogue of human cytokeratins, patterns of expression in normal epithelia, tumors and cultured cells. Cell 31: 11-24
30. Nordgren H, Åkerman M (1982) Electron microscopy of fine needle aspiration biopsy from soft tissue tumors. Acta Cytol (Baltimore) 26: 179-188
31. Osborn M, Weber K (1983) Biology of disease. Tumor diagnosis by intermediate filament typing: a novel tool for surgical pathology. Lab Invest 48: 372-394
32. Osborn M, Altmannsberger M, Debus E, Weber K (1985) Differentiation of the major human tumor groups using conventional and monoclonal antibodies specific for individual intermediate filament proteins. Ann NY Acad Sci 455: 649-668
33. Ramaekers F, Haag D, Jap P, Vooijs PG (1984) Immunochemical demonstration of keratin and vimentin in cytologic aspirates. Acta Cytol (Baltimore) 28: 385-392
34. Sehested M, Francis D, Hainau B (1983) Electron microscopy of trans-thoracic fine needle aspiration biopsies. Acta Pathol Microbiol Immunol Scand [A] 91: 457-461
35. Shah KD, Tabibzadeh SS, Gerber MA (1987) Comparison of cytokeratin expression in primary and metastatic carcinomas. Diagnostic application in surgical pathology. Am J Clin Pathol 87: 708-715
36. Taft PD, Szyfelbein WM, Greene R (1980) A study of variability in cytologic diagnoses based on pulmonary aspiration specimens. Am J Clin Pathol 73: 36-40
37. Wills EJ, Carr S, Philips J (1987) Electron microscopy in the diagnosis of percutaneous fine needle aspiration specimens. Ultrastruct Pathol 11: 361-387

Application of Immunocytochemical Methods to Cytological Specimens for Differential Diagnosis of Malignant Tumors

P. Dalquen and D. Bittel

Introduction

Immunocytochemistry has turned out to be of great help in the differential diagnosis of malignant tumors. In this paper the results of 100 consecutive examinations performed on cytological material between 1984 and 1988 are reported. All examinations were made for diagnostic purposes.

Material and Methods

The origins of the 100 cytological specimens are listed in Table 1. Sixty-seven specimens were fine-needle aspirates. Forty-eight specimens were sent from outside our hospital and were spray fixed. Most of the other specimens were wet fixed by immersion in Delaunay's solution (equal parts acetone and ethanol with an admixture of 0.5 ml 1 *M* trichloracetic acid). The fixed specimens were first stained with Papanicolaou's stain and then examined by light microscopy as usual. When immunocytochemical examination was indicated the slides were uncovered and

Table 1. Origin of cytological material (100 specimens) examined using immunocytochemical methods

Fine-needle aspirates	67
Lymph nodes	17
Retroperitoneum	
Mediastinum	16
Liver	14
Skin	10
Thyroid gland	5
Lung	4
Breast	1
Exfoliated material	28
Serous cavities	16
Bronchial system	5
Gastrointestinal tract	4
Cerebrospinal fluid	3
Imprints from fresh tissue	5

incubated, usually using the avidin-biotin complex (ABC) method. Subsequently they were restained and covered again [1, 2].

Immunocytochemical methods were used with cytological specimens mainly in cases of poorly differentiated small or large cell tumors which were difficult to classify by light microscopy alone. The aim of the immunocytochemical examinations was to differentiate between carcinoma and sarcoma, between small cell carcinoma and lymphoma, and between malignant melanoma and carcinoma. In some cases neuroendocrine or other special cell differentiations were looked for. Whenever possible, two samples were incubated with different antibodies, the choice of antibody depending on the tumors to be differentiated. At least one antibody was chosen which was expected to react positively with the tumor cells or other cells on the smear. Only one antibody was used in ten specimens, two antibodies in 31, and more than two in 59.

Only a small panel of antibodies was used (Table 2). The decision about which antibodies to use in a given case was made according to the scheme shown in Fig. 1 and always depended on the tentative diagnosis made by conventional light microscopy.

Most frequently, either one slide was incubated with the panepithelial antibody lu-5 and the other with anti-vimentin, or one slide was incubated with lu-5 and the other with anti-LCA (leukocyte common antigen).

A reaction was judged as positive if the cytoplasm of the tumor cells showed a granular reaction product. The assessment of a positive reaction was sometimes difficult, especially in small cell tumors when the atypical cells had only a small rim of delicate cytoplasm. It could also be difficult if the background of the smear contained much cytoplasmic debris or blood. The reliability of the immunocytochemical results was checked by reviewing all available clinical and histological data.

Table 2. Panel of monoclonal (MAb) and polyclonal antibodies (PAb) used for testing cytological preparations

Antibody	Concentration	Recognizes
MAb lu-5[a]	undiluted	Epithelium, mesothelium
MAb b-12[a]	1:5000	Secretory epithelium
MAb c-24[a]	100 mg/ml	Carcinoembryonic antigen
MAb VIM[b]	1:2	Vimentin
MAb LCA[c]	1:5	Leukocytic common antigen
PAb CT[c]	1:500	Calcitonin
PAb AAT[c]	1:250	Al antitrypsin
PAb SPP[d]	1:500	Acid prostate phosphatase
PAb NSE[e]	1:10	Neuron-specific enolase
PAb Ker[c]	1:500	Cytokeratin
PAb S100[c]		S100 protein
PAb TG[c]	1:500	Thyroglobulin

Sources: [a] Hoffmann-La Roche, Basel, Switzerland; [b] Boehringer, Mannheim, FRG; [c] DAKO, Zürich, Switzerland; [d] Ortho Diagnostic Systems, Neckargemünd, FRG; [e] Immuno Nuclear Corporation, Stillwater, USA.

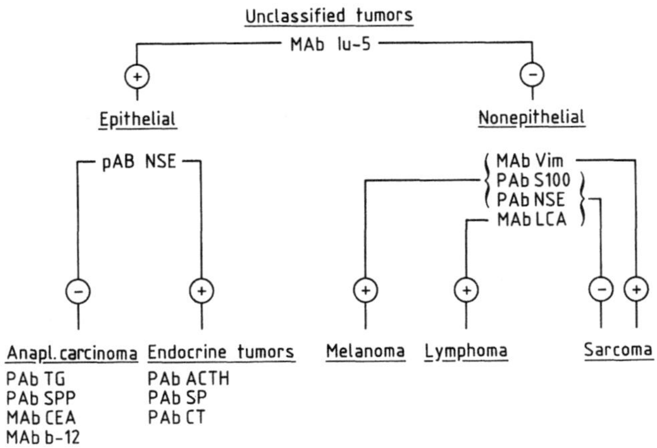

Fig. 1. Plan of application of immunocytochemical examinations on cytological specimens [2]

Results

In 71 of the 100 cases the results of the immunocytochemical examinations enabled a diagnosis to be made. There was no difference in the diagnostic efficacy between the specimens sent from outside and those which came from the Kantonsspital Basel.

In the 29 cases where the results did not enable a diagnosis to be made, the failure was probably caused by drying artifacts in 14 cases, application of an inadequate combination of antibodies in seven, nonreactivity of the antibody in seven, and an "aberrant" reaction in one.

Discussion

In contrast with histological examinations, where a nearly unlimited number of sections can be made for additional examinations, only a limited number of specimens from fine-needle aspirates or exfoliated material are available in diagnostic cytology. Therefore, the combination of immunocytochemical examinations and Papanicolaou staining of the same preparation is necessary [2]. This procedure allows differentiation of benign and malignant cells by the usual cytological criteria, even after the specimens have been incubated with mono- or polyclonal antibody.

There are, however, many sources of error, so in our material the immunocytochemical examinations of cytological specimens enabled a diagnosis to be made in only 71 of the 100 cases. This is similar to the experience of others [3]. Aberrant results may be false-positive or false-negative. The rate of false-positive results can be kept down by cautious interpretation of unequivocal reactions. False-negative reactions are easily recognized if the background cells do not react adequately with the antibodies or in specimens without such internal controls, if none of two "complementary" antibodies gives a positive reaction with the tumor cells. In the second case the cause of the aberrant reaction is often difficult to determine.

Negative reactions may be caused by technical factors, by a nonreactive antibody sample or by uneven expression of the markers within a tumor cell population. The most important cause of false-negative reactions in our material was inadequate spray fixation and/or air-drying of the fixed smears over more than 24 h. In our laboratory we have been able continuously to improve the diagnostic efficacy of immunocytochemical examinations [2]. The results presented here confirm our earlier findings [1, 2] that satisfactory results of immunocytochemical examinations in cytology can only be expected if the cells are well preserved and have not dried before or after fixation, the antibodies have been previously tested, and at least two "complementary" antibodies are used.

Summary

This study reports on the results of combining immunocytochemical examinations with Papanicolaou's staining in 100 tumor cases. The examinations enabled a diagnosis to be made in 71 cases. Sources of error were drying artifacts, use of the wrong antibodies, decay of antibody reactivity, and, rarely, an "aberrant" reaction.

References

1. Dalquen P, Bittel D, Gudat F, von Overbeck J, Heitz PU (1986) Combined immunoreaction and Papanicolaou's stain on cytological smears. Pathol Res Pract 181: 50-54
2. Dalquen P (1987) Immunocytochemical cytodiagnosis of tumors. In: Seifert G, Huebner K (eds) Pathology of cell receptors and tumor markers; application of immunocytochemistry and hybridization in tumor diagnosis. Fischer, Stuttgart, pp 196-201
3. Chess Q, Hajdu SI (1986) The role of immunoperoxydase staining in diagnostic cytology. Acta Cytol 30: 1-7

VI. Immunocytochemistry – Applications
1. Fine-Needle Aspiration Biopsy

Tumour Typing in Fine-Needle Aspiration Biopsies by Immunocytochemistry

T. Dirk and M. Droese

Introduction

Rapidly increasing use of fine-needle aspiration techniques has also expanded the problems in differential diagnosis of lesions. The cytopathologist has to decide whether a lesion is benign or malignant. Furthermore the detected tumour has to be classified by cytology. As has been shown recently, intermediate filaments (IF) typing of tumour cells may be of great value in classifying malignant primary tumours or metastases with respect to histogenetic origin [1, 3-5, 14].

In general, tumours express the IF profile typical of the cell of origin and do not express new types of IF proteins as a result of malignant transformation. Thus, IF typing provides information relevant to the histogenetic origin of tumours. Well-known studies performed on sections of tumour tissue provided the basis for the diagnostic use of antibodies directed against IF proteins in both histopathological and cytopathological examination [11, 12]. The results are summarized in Table 1.

This study presents some of the major problems of differential diagnosis in cytopathology in which IF typing has proved of great help.

Methods

In our laboratory we usually save up to half of the air-dried smears for immunostaining, depending on the diagnostic problem in question. The best results in immunostaining are obtained using fresh smears processed either immediately

Table 1. Cell and tumour specifity of IF proteins

Type of IF protein	Cell type	Tumour type
Cytokeratins	Epithelial cells	Carcinomas
Vimentin	Mesenchymal cells	Sarcomas Lymphomas Melanomas
Desmin	Muscle cells	Myosarcomas
Neurofilaments	Neuronal cells	Neuronal and neuroendocrine tumours
Glial fibrillary acidic protein	Astroglial cells	Astrocytomas

Table 2. Monoclonal antibodies used in this study

Monoclonal antibody	Clone	Source	Specifity
Pan-cytokeratin	LP 34	DAKO	Most cytokeratin polypeptides, including keratin 6 and 18
Vimentin	V 9	DAKO	57 kD IFP of mesenchymal cells
Desmin	DE-R-11	DAKO	53 kD IFP of muscle cells
Neurofilament protein	NR 4	DAKO	68 kD, 160 kD, 200 kD polypeptide subunit of neurofilaments
Cytokeratin (glandular epithelia)	a	Amersham	Cytokeratin No. 7
Cytokeratin 8	a	Amersham	Cytokeratin No. 8
Cytokeratin (simple epithelia)	a	Amersham	Cytokeratin No. 18
Cytokeratin 19	a	Amersham	Cytokeratin No. 19
Synaptophysin	b	Boehringer	Pan-neuroendocrine

[a] See [8]
[b] See [7]

or within a few hours after preparation. If immediate processing is impossible, smears are stored at −20 °C or below.

For first years in which we used visualization of immunoreactions we used to apply both fluorescein isothiocyanate and immunoperoxidase staining (for references see [10, 11]). Now we prefer the alkaline phosphatase-anti-alkaline phosphatase (APAAP) method as well as the avidin-biotin complex (ABC) method, principally in order to avoid difficulties in assessing cell morphology or to reduce background reactions. For diagnostic purposes we use monoclonal antibodies which are available from commercial sources (Table 2).

APAAP procedure: Smears were air dried and fixed in acetone at −10 °C for 10 min and then washed with Tris-buffered saline (TBS, 0.05 *M* TRIS, 0.10 *M* NaCl, pH 7.6). The slides were incubated with the first antibody for 30 min at 20 °C. After washing with TBS a rabbit anti-mouse IgG followed by the APAAP complex (DAKO, Denmark) was used. The APAAP complex was visualized with new fuchsin. Smears were counterstained with Mayer's acid hemalum.

Results

Fifty-three selected aspiration biopsies from different tumours with cytological features of malignancy in conventional staining were investigated by IF typing. In these cases it had been impossible to classify the tumour or to distinguish between two possible types. Furthermore, in cases of confirmed metastasis, the localization of the primary tumour had to be determined.

The results of IF typing are summarized in Tables 3-6. Immunostaining was considered positive when 20%-90% of the tumour cells were labelled by the antibody used. All diagnoses were confirmed by histology or clinical follow up.

Histogenetic Origin of Tumours Confirmed by Detection of One IF Protein

Cytokeratin is the most common and specific epithelial marker. Broad-specificity cytokeratin antibodies recognize cytokeratin in all epithelial cells. Corresponding-

Table 3. Histogenetic origin of tumours defined by detection of one IF protein

Case	Definitive diagnosis	Diagnostic problem	IF typing
1	Anaplastic carcinoma	Anaplastic carcinoma vs. large cell lymphoma	CK+, Vi−
2	Anaplastic carcinoma	Anaplastic carcinoma vs. large cell lymphoma	CK+, Vi−
3	Anaplastic carcinoma	Anaplastic carcinoma vs. paraganglioma	CK+, Vi−
4	Malignant melanoma	Anaplastic carcinoma vs. malignant melanoma	CK−, Vi+
5	Malignant melanoma	Anaplastic carcinoma vs. malignant melanoma	CK−, Vi+
6	Malignant fibrous histiocytoma	Anaplastic carcinoma vs. malignant fibrous histiocytoma	CK−, Vi+
7	Malignant schwannoma	Metastasis of malignant schwannoma?	CK−, Vi+
8	Liposarcoma	Metastasis of carcinoma of collum uteri?	CK−, Vi+
9	Leiomyosarcoma	Metastasis of leiomyosarcoma?	Des+
10	Leiomyosarcoma	Metastasis of leiomyosarcoma?	Des+
11	Leiomyosarcoma	Leiomyosarcoma vs. malignant schwannoma	Des+, Vi+
12	Rhabdomyosarcoma	Small cell tumour (child), Ewing's sarcoma?	Des+
13	Rhabdomyosarcoma	Small cell tumour (child), neuroblastoma?	Des+, Vi+, NF−
14	Extraskeletal Ewing's sarcoma	Small cell tumour (child)	Des−, Vi+, NF−, CK−
15	Neuroblastoma	Small cell tumour (child)	Des−, Vi−, NF+, CK−
16	Pheochromocytoma	Pheochromocytoma?	NF+, CK−

CK, pan-cytokeratin; Vi, vimentin; Des, desmin; NF, neurofilaments

Table 4. Tumours characterized by coexpression of cytokeratin and vimentin

Renal cell carcinomas (7)
Carcinomas of the adrenal cortex (3)[a]
Granulosa cell tumours (3)

( ) Number of cases
[a] One highly differentiated tumour expressed cytokeratin only

Table 5. Tumours characterized by coexpression of cytokeratin and neurofilaments

Carcinoids (3)
Gastrinomas (2)[a]
Neuroendocrine carcinomas (2)
Neuroendocrine (Merkel cell) carcinoma of the skin (1)[a]
Small cell carcinomas of the lung (3)[a, b]

( ) Number of cases
[a] Monoclonal antibodies to synaptophysin were used and showed positive immunostaining
[b] Tumour cells of one anaplastic case also expressed vimentin

ly, antibodies of this type react with cells of all epithelial tumours. Even cells of undifferentiated or anaplastic carcinomas are identified. Thus, antibodies to pan-cytokeratin and vimentin are very helpful in the differential diagnosis of anaplastic carcinoma vs. lymphoma, melanoma and sarcoma, as shown in Table 3 (cases 1-6).

IF typing is also very useful when discrimination between spindle cell tumours is required (see cases 7-11). For example, tumour cells of leio- and rhabdomyosarcomas contain desmin, sometimes coexpressed with vimentin, whereas malignant melanomas, fibrosarcomas, liposarcomas and malignant schwannomas display only vimentin.

Correct classification of round cell tumours in childhood is of the utmost clinical importance. To achieve a definitive diagnosis the complete battery of antibodies to IFs is necessary, as shown by cases 12-15.

Tumours Characterized by Coexpression of Cytokeratin and Vimentin

It is well known that a distinct subgroup of epithelial tumours (including pleomorphic adenomas, adenoid cystic carcinomas, some carcinomas of the thyroid, malignant mesotheliomas, synovial sarcomas, epitheloid sarcomas, nephroblastomas and the entities listed in Table 4) is characterized by the frequent coexpression of cytokeratin and vimentin. This is of great value in cases of metastases and unknown primary tumours.

Tumours Characterized by Coexpression of Cytokeratin and Neurofilaments

Certain epithelial neuroendocrine tumours (summarized in Table 5) show coexpression of cytokeratin and neurofilaments. This is a valuable aid to distinguishing neuroendocrine tumours from carcinomas and malignant lymphomas. In some of these cases in which smears were still available after IF typing, tumour cells were also clearly labelled by a monoclonal antibody to synaptophysin, a new pan-neuroendocrine marker [see 7].

It has been shown previously that monoclonal antibodies to neurofilaments are very useful for the diagnosis of medullary carcinomas of the thyroid [2].

Cytokeratin Subtyping

Epithelial tumours may be further subtyped by the analysis of cytokeratin polypeptide pattern. So far 19 different cytokeratin polypeptides have been found in human tissues. Moll and coworkers have studied the cytokeratin polypeptide pattern in normal tissues as well as in carcinomas using two-dimensional gel electrophoresis [7].

Table 6. Differential diagnosis of malignant epithelial liver tumours by cytokeratin subtyping

	Cytokeratin patterns
Primary tumours	
Hepatocellular carcinoma (1)	8, 18
Cholangiocarcinoma (1)	7, 8, 18, 19
Metastasis	
Gastric carcinoma (1)	8, 18, 19
Colorectal carcinoma (3)	8, 18, 19
Ductal carcinoma of the pancreas (4)	7, 8, 18, 19
Carcinoma of the ovary (2)	7, 8, 18, 19
Ductal breast carcinoma (1)	7, 8, 18, 19

( ) Number of cases

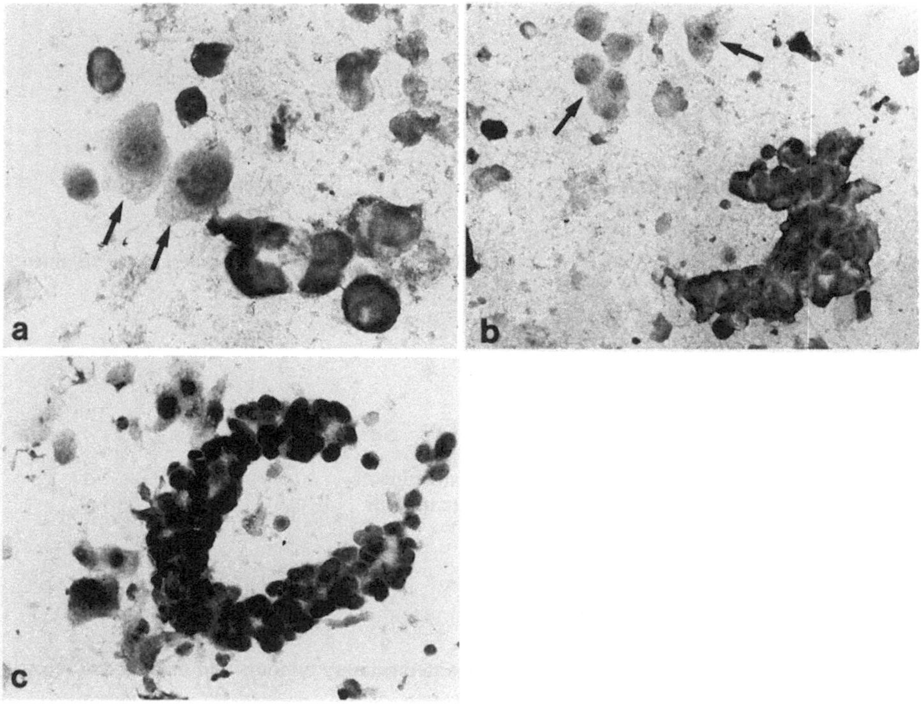

Fig. 1 a–c. Aspiration biopsy of a cholangiocellular carcinoma stained with monoclonal antibodies specific for CK 7 (**a, c**) and CK 19 (**b**). Tumour cells were clearly labelled by antibodies to CK 7 (**a**) and CK 19 (**b**), whereas immunostaining of hepatocytes *(arrows)* was negative. In this case an epithelial liver tumour was suspected clinically. Hepatocellular carcinoma and a metastasis of colorectal carcinoma could be excluded by immunostaining. The CK subtyping results combined with a bile duct-like formation of tumour cells (**c**) suggesting a cholangiocellular carcinoma. (APAAP; **a** ×350; **b, c** ×220)

To the extent that monoclonal antibodies to cytokeratin polypeptides are commercially available, cytokeratin typing can be applied routinely. This has been achieved for cytokeratins 7, 8, 18 and 19 and permits particularly a differential diagnosis of malignant epithelial liver tumours, a problem the cytopathologist is frequently confronted with. This is illustrated in Table 6: cells of hepatocellular carcinoma are decorated by antibodies to cytokeratins 8 and 18 only. However, cholangiocarcinoma and metastatic ductal carcinoma of the pancreas display cytokeratins 7, 8, 18 and 19. In contrast to these, gastric and colorectal carcinomas lack cytokeratin 7 [6, 13].

Conclusions

IF typing using well-characterized monoclonal antibodies can be very useful in cytopathology, especially in determination of the histogenetic origin of tumour cells. Cytokeratin subtyping may help in recognition of the primary tumour when metastases are analysed. Figure 2 summarizes the decision-making process in the

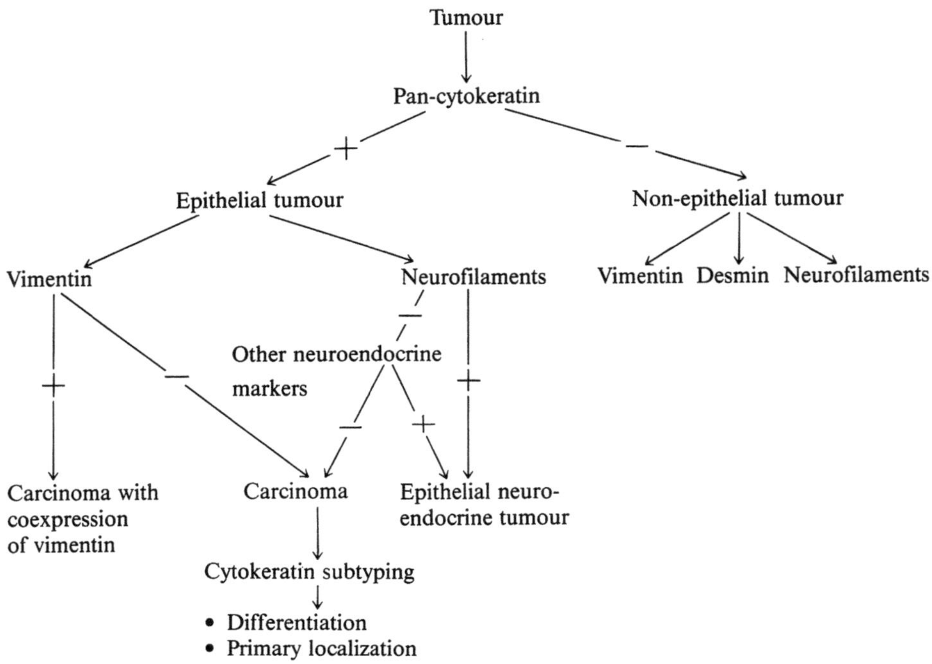

Fig. 2. Intermediate filament typing of undifferentiated primary tumours and metastases (according to [8])

differential diagnosis of undifferentiated primary tumours and metastases. The classification of carcinomas is particularly emphasized. The step-by-step analysis, however, in practice is often time consuming and uneconomical. Therefore, in the individual case the analytical process should always be guided by the cytological features in conventional staining.

Summary

Tumour typing by means of well characterized monoclonal antibodies to intermediate filaments (IF) is a most useful aid in diagnostic cytopathology, especially in cases in which the histogenetic origin of tumour cells cannot be defined by conventional stains. The results of IF typing should be evaluated only in context with cytological features obtained in conventionally stained smears. The alkaline phosphatase-anti-alkaline phosphatase (APAAP) method combines high sensitivity with high specifity.

References

1. Altmannsberger M, Osborn M, Droese M, Weber K, Schauer A (1984) Diagnostic value of intermediate filament antibodies in clinical cytology. Klin Wochenschr 62: 114-123

2. Altmannsberger M, Dralle H, Weber K, Osborn M, Droese M (1987) Intermediate filaments in cytological specimens of thyroid tumors. Diagn Cytopathol 3: 210-214
3. Domagala W, Lubinsky J, Weber K, Osborn M (1986) Intermediate filament typing of tumor cells in fine needle aspirates using monoclonal antibodies. Acta Cytol 30: 214-224
4. Domagala W, Lubinski J, Lasota J, Woyke S, Wozniak L, Szadowska A, Weber K, Osborn M (1987) Decisive role of intermediate filament typing of tumor cells in the differential diagnosis of difficult fine needle aspirates. Acta Cytol 31: 253-266
5. Droese M, Altmannsberger M, Kehl A, Lankisch PG, Weiss R, Weber K, Osborn M (1984) Ultrasound-guided percutaneous fine needle aspiration biopsy of abdominal and retroperitoneal masses. Acta Cytol 28: 368-384
6. Fischer HP, Altmannsberger M, Weber K, Osborn M (1987) Keratin polypeptides in malignant epithelial liver tumors. Differential diagnostic and histogenetic aspects. Am J Pathol 127: 530-537
7. Gould VE (1987) Synaptophysin: a new and promising pan-neuroendocrine marker. Arch Pathol Lab Med 111: 791-794
8. Moll R, Franke WW, Schiller DL, Geiger B, Krempler R (1982) The catalogue of human cytokeratins, patterns of expression in normal epithelia, tumors and cultured cells. Cell 31: 11-24
9. Moll R (1986) Epithelial tumor markers. Verh Dtsch Ges Path 70: 28-50
10. Osborn M, Weber K (1982) Immunofluorescence and immunocytochemical procedures with affinity-purified antibodies: tubulin-containing structures. In: Wilson L (ed) The cytoskeleton. Academic, New York, pp 97-132 (Methods in cell biology, vol 24)
11. Osborn M, Weber K (1983) Biology of disease. Tumor diagnosis by intermediate filament typing: a novel tool for surgical pathology. Lab Invest 48: 372-394
12. Osborn M, Altmannsberger M, Debus E, Weber K (1985) Differentiation of the major human groups using conventional and monoclonal antibodies specific for individual intermediate filament proteins. Ann NY Acad Sci 455: 649-668
13. Osborn M, van Lessen G, Weber K, Klöppel G, Altmannsberger M (1986) Methods in laboratory investigation. Differential diagnosis of gastrointestinal carcinomas by using monoclonal antibodies specific for individual keratin polypeptides. Lab Invest 55: 497-504
14. Ramaekers F, Haag D, Jap P, Vooijs PG (1984) Immunochemical demonstration of keratin and vimentin in cytologic aspirates. Acta Cytol 28: 385-392

Immunocytochemical Determination of the Proliferation Fraction in Fine-Needle Aspiration Smears of Breast Tumors and Their Metastases

V. Kuenen-Boumeester, D. I. Blonk, and T. H. Van Der Kwast

Introduction

Since breast carcinomas have a variable clinical course, the search for independent prognostic factors in histological and cytological material is considered highly relevant to prognosis. The lymph node status is the most important, but other factors like the hormonal receptor status and factors concerning proliferation, like the mitotic activity index and the thymidine labelling index, have been shown to be independent prognostic factors as well [1, 5, 7].

A mouse monoclonal antibody (Ki-67) has become available defining a nuclear antigen present in proliferating cells throughout the cell cycle and absent in G0 and early G1. Thus, Ki-67 enables the immunocytochemical detection of cycling cells without the need of external administration of radioactively labelled nucleotides or mutagenic substances such as bromodeoxyuridine or iododeoxyuridine. It might also give a better indication of proliferation than the counting of mitoses, since only a small fraction of cycling cells is in mitosis.

Most studies have used this antibody on histological material [2, 3] therefore we investigated the feasibility of the use of this antibody on fine-needle aspiration (FNA) smears obtained from benign and malignant tumours as well as from distant metastases of breast carcinomas. In addition, we wanted to do a preliminary analysis of a possible relationship between the Ki-67 determined growth fraction and other clinical, morphometrical and hormonal features.

Material and Methods

Tumour Specimens and Immunocytochemistry

The material consisted of FNA smears of 38 histologically proven breast carcinomas, 20 benign breast tumours and 26 metastases of breast carcinomas. The air-dried smears were fixed in acetone for 5 min and immunostained with the mouse monoclonal antibody Ki-67 (Dako, Denmark) diluted 1:10 in phosphate-buffered saline (PBS), pH 7.4, containing 0.05% gelatin and 0.1% NaN_3, using the indirect conjugated immunoperoxidase method. After addition of diaminobenzidine as substrate, brown speckled staining of nuclei or nucleoli was considered as a positive reaction with Ki-67. Mitotic figures were also stained. Nuclear counterstaining was achieved by 1-min incubation in Mayer's haematoxylin. The percentage of Ki-67-positive nuclei was determined by counting 500 cells at a magnification of 1000 ×.

Primary Carcinomas

From each Giemsa-stained smear 100 nuclei were measured, using a Videoplan image analysis system. Area, perimeter and maximum diameter were measured. In 32 patients the oestrogen receptor and in 26 patients the progesterone receptor content was measured by the dextran-coated charcoal assay according to the European Organisation for Research and Treatment of Cancer (EORTC) guidelines after surgical excision. Menopausal status, tumour size, lymph node status and tumour type were also recorded.

Results

General

The FNA smears of the benign tumours contained a low percentage of immunostained nuclei, ranging from 0.0%-6.7% per smear, with a mean of 1.1% and a median of 0.4%. By contrast, in only 4 of the malignant tumours did less than 2% of nuclei stain; in 21 tumours, staining occurred in 2%-10% of nuclei, and in 13 the percentage was even higher than this (Fig. 1), with a mean of 10.5% and a median of 7.9%. In comparison with the malignant primary tumours, the metastases stained even more highly, with a mean of 14.3% and a median of 12.3%. The dif-

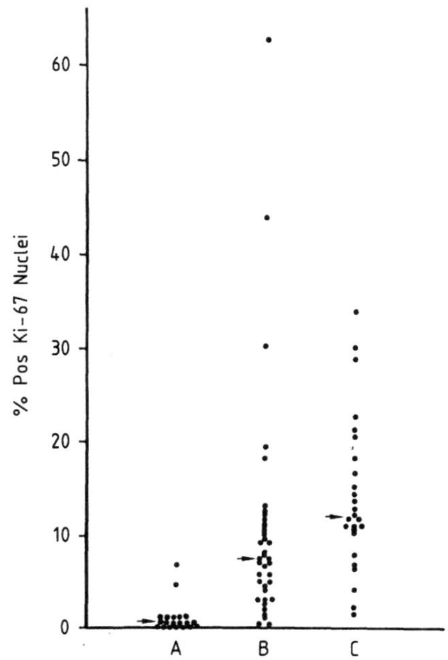

Fig. 1. The percentages of Ki-67-stained nuclei in benign, malignant and metastatic breast tumours. *A*, benign ($n=20$); *B*, malignant ($n=38$); *C*, metastases ($n=26$). *Arrows* signify medians. (From [4])

ference between the last two median values was statistically significant (Mann-Whitney test; $p < 0.05$)

Primary Carcinomas

We found no correlation between tumour size, menopausal status, or lymph node status and percentage of Ki-67-labelled nuclei. Preliminary data indicated a negative correlation between receptor status and Ki-67 score (Fig. 2) and a positive correlation between the standard deviation of the nuclear area and the Ki-67 score (Fig. 3).

Discussion

Determination of the fraction of proliferative tumour cell fraction in FNA smears of malignant breast lesions by Ki-67 immunostaining led to essentially similar results to those recently reported [2, 3]. The extent of variation in nuclear area has been shown to correlate strongly with the frequency of mitosis and therefore with

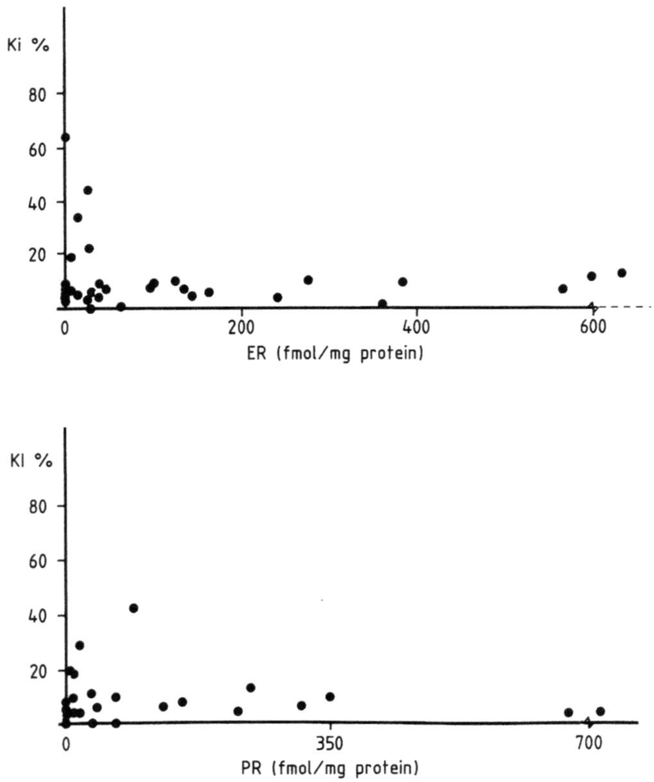

Fig. 2. Correlations of percentages of Ki-67-stained nuclei of primary carcinomas with *(above)* oestrogen receptor status and *(below)* progesterone receptor status

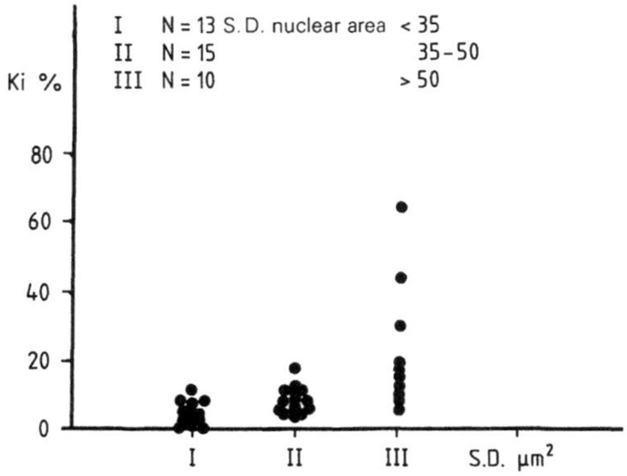

Fig. 3. Percentages of Ki-67-stained nuclei of primary carcinomas vs. standard deviations of nuclear area

proliferation [6]. This accords with our preliminary finding that the group of patients with the highest standard deviation of nuclear area had the highest proportion of Ki-67 staining (Fig. 3).

Baak et al. [1] have reported that the mitotic activity index in histological material is the strongest independent prognostic factor in their group of patients. Although the prognostic significance of the proliferative tumour cell fraction as determined by Ki-67 remains to be proved, we think that determination of the total cycling fraction of a tumour gives more information about prognosis than assessment of numbers of mitotic figures alone. This holds especially true for FNA smears, since they rarely contain mitotic tumour cells. Further support for the hypothesis that the Ki-67-determined proliferative fraction may be of prognostic relevance can be drawn from the in vitro [^{3}H]thymidine labelling studies of Meyer et al. [5] and Tubiana et al. [7]. These authors incubated small specimens of freshly obtained breast tumour tissue with [^{3}H]thymidine and counted the number of labelled nuclei from autoradiographed microscopic sections. In the study of Meyer et al. [5], with a follow-up period of 4 years, both lymph node status and the [^{3}H]thymidine labelling index appeared to be the strongest independent indicators of early relapse. In the long-term prospective study of Tubiana et al. [7] the [^{3}H]thymidine labelling index appeared to be the most predictive independent indicator with respect to relapse-free survival and total survival. It should be noted that Ki-67 immunostaining is only indirect evidence of proliferation, while mitotic index and [^{3}H]thymidine incorporation relate directly to proliferative activity. Furthermore, Ki-67 stains cells in G1 phase; cells arrested in this phase of the cell cycle may also be stained by Ki-67, leading to an overestimation of the fraction of actually proliferating cells.

In agreement with the study of Barnard et al. [2] we found no relation between lymph node status and percentage of Ki-67-labelled tumour cells. The thymidine labelling studies [5, 7] did not show any relation to lymph node status either.

Oestrogen receptor status has been shown to have a negative correlation with the thymidine labelling index [5]. This is in agreement with our preliminary results (see Fig. 2). The same is true for progesterone receptor status and Ki-67 immunostaining.

We found no relation between tumour size or menopausal status and percentage of Ki-67-immunostained nuclei. Probably the number of investigated tumours was too low to establish such a relationship. It was interesting to see, however, that the two really high values (above 40%) were only seen in premenopausal women. In addition, the two highest values (30.5% and 19%) in the group of postmenopausal women were medullary carcinomas. It is known that the latter tumours have a high mitotic activity and display a different biological behaviour than infiltrating duct carcinomas.

Our observation of significantly higher percentages of Ki-67-labelled tumour cells in metastases than in primary carcinomas may point to two different possibilities: (a) tumours with a higher proliferative activity have more aggressive biological behaviour, or (b) the higher proliferative activity of metastasized tumour cells is enhanced by a more favourable environment. In support of the first possibility is the prospective study of Tubiana et al. [7], which showed that the [^{3}H]thymidine labelling index is related to the probability of metastatic dissemination. Nevertheless, it would be interesting to compare the staining patterns of the aspirates of the primary tumour and the metastases from the same patient.

Further studies seem to be indicated to investigate the prognostic relevance of the monoclonal antibody Ki-67 described here, particularly since this antibody is not strongly related to other well-known factors used in predicting the biological behaviour of breast carcinomas.

Summary

To evaluate the difference in proliferation of benign breast tumours, breast carcinomas and metastases of breast carcinomas, fine-needle aspiration smears were immunostained with monoclonal antibody Ki-67, defining a proliferation-associated nuclear antigen. In most benign tumours, less than 2% of nuclei stained positively (median 0.4%); 2%-10% stained in the majority of carcinomas (median 7.9%), whereas in most metastases above 10% stained positively (median 12.3%). We found no correlation between lymph node status, menopausal status or tumour size and percentage of Ki-67-labelled nuclei. We found a negative correlation with hormonal receptor status and a positive correlation with the standard deviation of the nuclear area. These results are similar to those found in thymidine labelling studies; staining of breast carcinomas with Ki-67 may therefore provide prognostic information.

Acknowledgements. We wish to express our gratitude to Miss C. Trappenburg and A. C. M. van Nispen for their excellent technical assistance.

References

1. Baak JPA, van Dop H, Kurver PHJ, Hermans J (1985) The value of morphometry to classic prognosticators in breast cancer. Cancer 56: 374-382
2. Barnard NJ, Hall PA, Lemoine NR, Kadar N (1987) Proliferative index in breast carcinoma de-

termined in situ by Ki-67 immunostaining and its relationship to clinical and pathological variables. J Pathol 152: 287-295
3. Gerdes J, Lelle RJ, Pickartz H, Heidenreich W, Schwarting R, Kurtsiefer I, Stauch G, Stein H (1986) Growth fractions in breast cancers determined in situ with monoclonal antibody Ki-67. J Clin Pathol 39: 977-980
4. Kuenen-Boumester V, Blonk DI, Van Der Kwast TH (1988) Immunocytochemical staining of proliferating cells in fine-needle aspiration smears of primary and metastatic breast tumours. Br J Cancer (in press)
5. Meyer JS, Friedman E, McCrate MM, Bauer WC (1983) Prediction of early course of breast carcinoma by thymidine labelling. Cancer 51: 1879-1886
6. Stenkvist B, Bengtsson E, Eriksson O, Jarkrans T, Nordin B, Westman-Naeser S (1981) Correlation between cytometric features and mitotic frequency in human breast carcinoma. Cytometry 1: 287-291
7. Tubiana M, Pejovic MH, Chavaudra N, Contesso G, Malaise EP (1984) The long term prognostic significance of the thymidine labelling index in breast cancer. Int J Cancer 33: 441-445

Growth Fraction as Determined in Cytologic Specimens of Breast Carcinomas: A Study with the Monoclonal Antibody Ki-67

R. J. Lellé, W. Heidenreich, G. Stauch, and J. Gerdes

Introduction

There is a growing awareness today of the importance of cell kinetic measurements especially in breast cancer. The European Congress of Cytology produced a number of interesting papers, some of them dealing with DNA measurements. However, such techniques are expensive and require extensive technical equipment. This has been an obstacle for widespread application in the past.

Materials and Method

We routinely apply the proliferation marker Ki-67 to both histologic and cytologic material from breast cancer. Ki-67 is a murine monoclonal antibody, developed by Gerdes and Stein from the Institute of Pathology in West Berlin (Gerdes et al. 1983). This antibody binds to a nuclear antigen present in all active cells of the cell cycle, but not in resting cells: Ki-67 is expressed only in the G1, S, G2 and M phases of continuously cycling cells but is absent in G0 cells (Gerdes et al. 1984). Thus, the "Ki-67 growth fraction" or "Ki-67 GF" of a tumor can be defined as the number of Ki-67-positive cells divided by the total number of cells.

The Ki-67 GF can be determined by immunostaining of fresh tissue material. First results obtained on breast tissue have been encouraging (Lellé 1987; Lellé et al. 1987 a, b). Since then, 261 primary invasive breast carcinomas and 126 benign breast lesions have been studied using the indirect three step immunoperoxidase method described in detail by Stein and coworkers (1982) or the so-called APAAP method (Cordell et al. 1984). We prefer APAAP staining as it yields a red or pink color of positive cells, which can be clearly discerned from negative cells after counterstaining with haematoxylin.

Figure 1 illustrates the principle of APAAP staining: Ki-67 as the primary antibody binds to the antigen located in the cell nucleus. The alkaline phosphatase-anti-alkaline phosphatase (APAAP) complex is attached to Ki-67 by a bridging antibody. The advantage of this method ist the possibility of enhancing sensitivity by attaching two or even more APAAP complexes to the primary antibody. Two APAAP complexes are used for histologic examinations and three complexes for cytologic examinations.

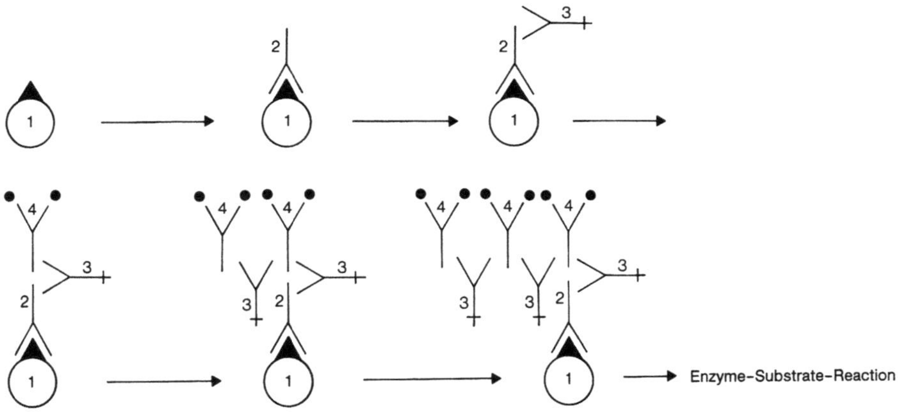

Fig. 1. Alkaline phosphatase-anti-alkaline phosphatase (APAAP) staining. The Ki-67 antibody *(2)* binds to an antigen located in the cell nucleus *(1)*. The APAAP complexes *(4)* are attached by bridging antibodies *(3)*. Two APAAP complexes are used for histologic studies and three complexes for cytologic studies

Results

In 308 cases of benign and malignant breast disease tissue material was studied both histologically and cytologically and the results evaluated separately (Table 1). As was to be expected, both methods yielded similar results; the Pearson correlation coefficient is 0.83. The slope of the regression line is 0.82, pointing to a lower Ki-67 GF in imprint cytology.

As Ki-67 GFs are not normally distributed, median and range are given. Ki-67 GF is calculated as the average number of Ki-67 positive cells in both histologic and cytologic examinations. The Ki-67 GF in malignant breast tumors is 4-5 times higher than in benign tumors: whereas the growth fraction of benign tumors never exceeds about 10%, the Ki-67 GF of malignant tumors can reach up to 65%. Growth fractions are slightly lower in cytologic specimens than in frozen sections,

Table 1. Ki-67 growth fraction (Ki-67 GF) of benign and malignant breast lesions

	Benign (n=126) %	Malignant (n=261) %	Regression statistics
Ki-67 GF/C	2.0 (0-9.5)	10.25 (0-70.0)	r=0.83 Slope=0.82
Ki-67 GF/H	3.0 (0-13.0)	14.0 (1.0-70.0)	
Ki-67 GF[a]	2.5 (0.25-10.75)	12.5 (0.75-65.0)	

Figures given are the median (minimun - maximum)
[a] Ki-67 GF is calculated as the mean value of cytologic (Ki-67 GF/C) and histologic (Ki-67 GF/H) results.

both in benign and malignant tumors. The reason for this is unknown. One might rather expect increased exfoliation of proliferating cells.

The following results are based on histologic as well as cytologic studies but can be reproduced by cytology alone.

Tumors were subdivided by the median Ki-67 GF (12.5%) into groups with "low" and "high" growth fractions. Looking at histologic grading, there is a higher proportion of tumors with high proliferation in the groups with less differentiation (Fig. 2).

A similar correlation exists with hormone receptor content (Fig. 3). Tumors positive for both estrogen and progesterone receptors more often have low Ki-67 GFs in comparison to hormone receptor-negative tumors. This is true for postmenopausal patients (age > 50 years) but not for younger patients. This may be due to endogenous hormone production in premenopausal women.

Some prognostic data are now available, although with a very short observation period: 141 patients have been followed up for 6–42 months (median 22 months). Nineteen patients have died from breast cancer. Figure 4 illustrates that a high growth fraction in the primary tumor may indicate an increased risk of early metastasis and death, as 79% of nonsurvivors had Ki-67 GFs above the median, compared to 57% of survivors. The difference is statistically significant.

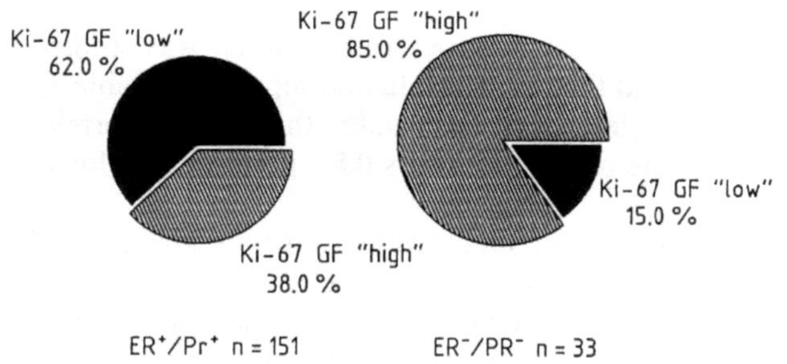

Fig. 2. Correlation of Ki-67 growth fraction (Ki-67 GF) and histological grading. A significantly higher proportion of tumors with high proliferation is found in the groups with less differentiation

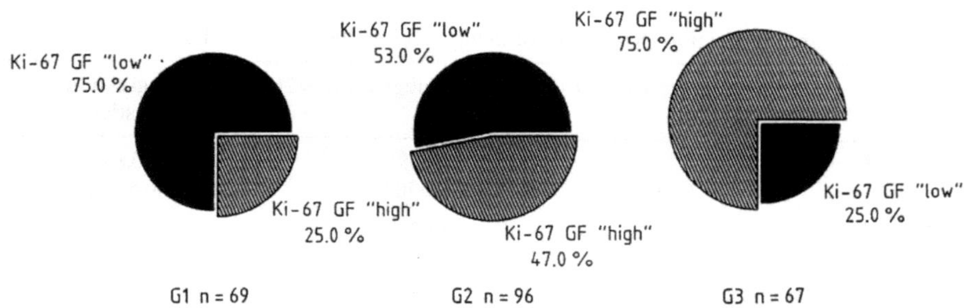

Fig. 3. Correlation of Ki-67 GF with hormone receptor content in postmenopausal patients (age > 50 years). Carcinomas negative for both estrogen and progesterone receptors *(ER–/PR–)* are more often associated with high Ki-67 GFs than hormone receptor-positive *(ER+/PR+)* carcinomas ($p = 0.0008$)

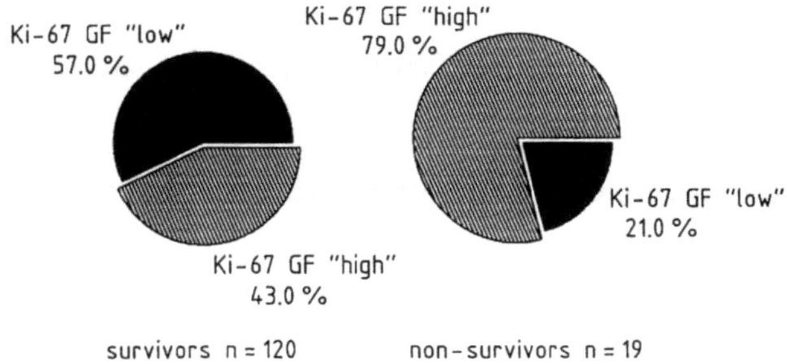

Fig. 4. Correlation of Ki-67 GF of breast cancer with survival. 141 patients were followed up for 6-42 months (median: 22 months). Patients who died from breast cancer were more likely to have primary tumors with high Ki-67 GFs ($p = 0.0155$)

Conclusions

The Ki-67 growth fraction gives important information on the biological nature of a malignant tumor. Immunostaining with the monoclonal antibody Ki-67 is an easy and rapid technique which is ideal for routine use. Information on tumor growth can be obtained by determination of Ki-67 GF by aspiration cytology even before surgery. Furthermore, in combination with other parameters, Ki-67 immunostaining may be useful in the selection of patients for various treatment modalities. Measurement of the growth fraction can be recommended and should be included in future studies of breast cancer.

References

Cordell JL, Falini B, Erber WN, Ghosh AK, Abdulaziz Z, McDonald S, Fulford KAF, Stein H, Mason DY (1984) Immunoenzymatic labeling of monoclonal antibodies using immune complexes of alkaline phosphatase and monoclonal anti-alkaline phosphatase (APAAP complexes). J Histochem Cytochem 32: 219-229

Gerdes J, Schwab U, Lemke H, Stein H (1983) Production of a mouse monoclonal antibody reactive with a human nuclear antigen associated with cell proliferation. Int J Cancer 31: 13-20

Gerdes J, Lemke H, Baisch H, Wacker H-H, Schwab U, Stein H (1984) Cell cycle analysis of a cell proliferation-associated human nuclear antigen defined by the monoclonal antibody Ki-67. J Immunol 133: 1710-1715

Lellé RJ (1987) Zellkinetische Parameter als Prognosefaktoren beim Mammakarzinom. Ärztl Lab 33: 157-164

Lellé RJ, Heidenreich W, Stauch G, Gerdes J (1987a) The correlation of growth fractions with histologic grading and lymph node status in human mammary carcinoma. Cancer 59: 83-88

Lellé RJ, Heidenreich W, Stauch G, Wecke I, Gerdes J (1987b) Determination of growth fractions in benign breast disease (BBD) with monoclonal antibody Ki-67. J Cancer Res Clin Oncol 113: 73-77

Stein H, Gerdes J, Schwab U, Lemke H, Mason DY, Ziegler A, Schienle W, Diehl V (1982) Identification of Hodgkin and Sternberg-Reed cells as a unique cell type derived form a newly detected small-cell population. Int J Cancer 30: 445-459

Immunocytochemical Detection of Estrogen Receptors in Fine-Needle Aspirates of Breast Cancer

H. C. Kübler, W. Kühn, H. H. Rummel, M. Kaufmann, E. Krapfl, and K. Klinga

Introduction

The biochemical detection of estrogen receptors (ER) in breast cancer tissues (Jensen 1962; King and Green 1984) is a new method for determining the prognosis of breast cancer, presence of receptors having a clear influence on survival and tumor recurrence (Kaufmann and Klinga 1982), that can be used in addition to conventional prognostic factors for breast cancer such as tumor size, lymph node involvement, and histological grading. The clinical application of hormone receptor measurements in systemic endocrine treatment of metastasizing breast cancer showed that 60% of ER-positive breast cancers responded to therapy. Adjuvant hormone treatment of postmenopausal women with affected lymph nodes resulted in significantly longer overall survival of patients with ER-positive tumors.

Conventional biochemical ER assays have various methodological disadvantages: a relatively large tissue mass is necessary, the results are influenced by assessment of nonrepresentative tissue, and results are affected in premenopausal women by endogenous blocking of receptor binding sites. The development of an immuncytochemical assay for detection of ER was therefore necessary.

Production of monoclonal antibodies to the ER protein (Green et al. 1980) has allowed the development of ER assays based on direct antigenic recognition. This procedure permits the detection of ER in very small breast cancers, punch biopsy specimens, and even fine-needle aspirates. Although several studies have already been conducted comparing biochemical and immunohistochemical assays for ER analysis in frozen sections of breast cancers, there have been only few investigations on the relation of immunocytochemical ER assays and biochemical ER assays in fine-needle aspirates. The aim of our study was therefore to increase our knowledge of ER assays in fine-needle aspirates.

Material and Methods

Preoperative fine-needle aspirates from clinically and radiologically suspicious breast lesions were used. Each cell specimen was divided onto four slides. Two of them were Papanicolaou-stained after alcohol fixation and the others were deep frozen immediately after preparation. The cryopreserved slides with cytologically positive reference slides were assayed using the Abbott ER-ICA (estrogen receptor immunocytochemical assay) method. This assay uses rat monoclonal antibodies against the ER protein of breast cancer cells. Evaluation of immunocytochemical slides was on the basis of the immunoreactive score (IRS) established by Remmele and Stegner (1987) which has a hypothetical maximum of 12. The results of immunocytochemical staining were compared with biochemical ER measurements. We defined an IRS of 0-1 as ER-negative, corresponding to a biochemical value of < 20 fmol bound estradiol/mg protein (Kaufmann and Klinga 1982). Evaluation of the slides was performed preoperatively by the same investigator.

We used the dextran-coated charcoal assay (DCCA) for biochemical hormone receptor measurements. For small amounts of cytosol a single measurement was made but if there was sufficient material a five-point Scatchard plot was established. The receptor concentration was measured in fmol bound estradiol/mg cytosol protein. Steroid binding of > 20 fmol/mg cytosol protein led to classification of the tumor as ER-positive (Kaufmann and Klinga 1982).

Results

In 41 of the 51 cryoconserved slides with cytologically positive reference slides enough tissue was available for immunocytochemical analysis. In 32 of these 41 cases the ER-ICA staining corresponded in semiquantitative assessment to the biochemical ER measurements (Fig. 1). The nine discrepant cases had IRSs in the range 0-6 (ER--ICA) and 3-82 fmol/mg protein (DCCA). The four cases with positive results by ER-ICA and negative results by biochemical assay were:

1. ER (DCCA) 18 fmol/mg and IRS 4 (ER-ICA) with an intensity of staining grade 2 (moderate), 20% nuclear staining and 5% nuclear staining of high intensity.
2. ER (DCCA) 11 fmol/mg and IRS 2 (ER-ICA) with a staining intensity grade 1 (low) and 20% nuclear staining.
3. ER (DCCA) 3 fmol/mg and IRS 2 (ER-ICA) with a staining intensity grade 1 (low) and 20% nuclear staining. The staining in this case was markedly heterogeneous.
4. ER (DCCA) 15 fmol/mg and IRS 6 (ER-ICA) with a staining intensity grade 2 (moderate) and 70% nuclear staining.

The fife cases with negative results by ER-ICA and slightly positive results of biochemical assay were:

1. ER (DCCA) 20 fmol/mg and IRS 0 (ER-ICA) with no nuclear staining although enough tumor cells were available.
2. ER (DCCA) 25 fmol/mg and IRS 1 (ER-ICA) with < 10% nuclear staining of low intensity.
3. ER (DCCA) 34 fmol/mg and IRS 1 (ER-ICA) with < 10% nuclear staining of low intensity.
4. ER (DCCA) 34 fmol/mg and IRS 0 (ER-ICA) with no nuclear staining although there were enough, but slightly damaged tumor cells.

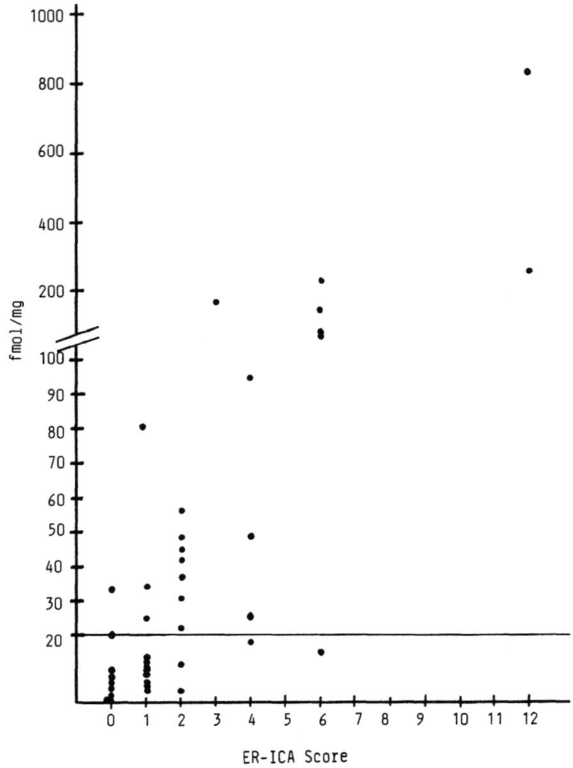

Fig. 1. Comparison of ER-ICA score and biochemical ER estimations (DCCA). The borderline between ER-positive and ER-negative cases was set at 20 fmol/mg

Table 1. Cytologically positive fine-needle aspirates of the breast (n=41); $\chi^2=12.897$, $p<0.001$

	DCC+		DCC−	
	(n)	(%)	(n)	(%)
ER-ICA +	17	41.5	4	9.7
ER-ICA −	5	12.2	15	36.6

5. ER (DCCA) 82 fmol/mg and IRS 1 (ER-ICA) with a staining intensity grade 1 (low) and 5% nuclear staining.

Discussion

The results of the ER-ICA of fine-needle aspirates of breast cancer were largely in agreement with those of the biochemical receptor assay in this study. This is in concordance with the results of Beck et al. (1986) and Remmele (1986) on frozen sections of breast cancer tissue. The main advantage of the immunocytochemical method is the small amount of tissue necessary. A good evaluation is thus possible even with fine-needle aspirates. In contrast to DCCA, the receptor protein is made visible. The results of the Abbott ER-ICA show the heterogeneity of the tumor cell population with respect to ER content. The results of the two methods were similar in 32 of 41 cases (78%), being in the range observed by Pollow (1987) and Remmele (1986). Discrepancies were found in four cases where results of ER-ICA were positive and of DCCA were negative and in five cases where results of ER-ICA were negative and of DCCA were positive. The lack of coincidence is, however, not the result of error of one of these procedures. At present the two methods are complementary to each other.

One of the four ER-ICA-positive patients with < 20 fmol/mg protein according to DCCA was receiving tamoxifen treatment. In this case, as expected, ER were not detected biochemically although results of immunocytochemical ER assay were clearly positive. This confirms that the monoclonal antibody method is not dependent on the influence of antiestrogens, as already stated by Remmele (1986) and by Taylor et al. (1980). In a second case, the immunocytochemical method resulted in a highly heterogeneous pattern of nuclear staining. The tissue sample probably originated from a section with a low ER content.

In the other five discrepant cases, where results of biochemical ER assay were slightly positive and results of ER-ICA were negative, it must be assumed that the ER content was too low for appropriate immunocytochemical staining. Furthermore, it is possible that cell specimens were sufficiently intact for cytological examination but not for the monoclonal antibody test.

The discrepancies between the results of the biochemical and immuncytochemical assays were in the so-called threshold area, i.e., there was good agreement in ER-negative and highly ER-positive cases. Possible explanations of the discrepancies are the existence of sensitivity limits for visualizing low nuclear ER contents with immunocytochemical methods and the unrepresentative or unknown tumor cell content of the tissues in the biochemical procedures. A reference score for a clear distinction between ER-positive and ER-negative results ER-ICA has still to be defined.

Summary

The biochemical detection of estrogen receptors in breast cancer is of great value for prognosis and in planning treatment. The conventional biochemical assays of ER are expensive, time consuming, and give little information on the topography and heterogeneity of the receptor protein in the tissue. Other groups have shown good agreement between the results of the biochemical assay and the monoclonal antibody method in frozen sections. We used these detection methods on 51 fine-needle aspirates from patients with breast cancer. Forty-one samples (80.4%) contained sufficient cells for analysis. In 32 of these 41 fine-needle aspirates (78%) the results of the biochemical assay and the monoclonal method were in agreement. Cytological examinations including ER detection allow determination not only of the receptor status but also of the malignancy of the tumor. Preoperative knowledge of ER status allows more accurate decisions about therapeutic procedure.

References

Beck T, Pollow K, Heubner A (1986) Monoklonale Antikörper zum immunhistochemischen Nachweis des Östrogenrezeptorstatus am Gewebeschnitt primärer Mammacarcinome. Geburtsh Frauenheilk 46: 490-494

Greene CL, Nolan C, Engler JP, Jensen EV (1980) Monoclonal antibodies to human estrogen receptor. Proc Natl Acad Sci USA 77: 5115-5119

Jensen EV, Jacobson HI (1962) Basic guides to the mechanism of estrogen receptor action. Recent Prog Horm Res 18: 387-414

Kaufmann M, Klinga K (1982) Mögliche prognostische Kriterien zur Therapieverbesserung beim primären Mammacarcinom. Geburtsh Frauenheilk 42: 501-509

King WJ, Greene CL (1984) Monoclonal antibodies localize estrogen receptor in the nuclei of target cells. Nature 307: 745-747

Pollow K (1987) Stellungnahme zu "Mammacarcinom: Prognostisch wichtige Rezeptorbestimmungen." Dtsch Aerztebl 84: 1941-1942

Remmele W (1986) Mammacarcinom: Prognostisch wichtige Rezeptorbestimmungen. Dtsch Aerztebl 83: 3359-3362

Remmele W, Stegner HE (1987) Immunhistochemischer Nachweis von Östrogenrezeptoren (ERICA) in Mammacarcinomgewebe: Vorschlag zur einheitlichen Bewertung des Untersuchungsbefundes. Frauenarzt 2: 41-43

Taylor RE, Powles TJ, Humphreys J, Bettelheim R, Dowsett M, Casey AJ, Neville AM, Coombes RC (1980) Effects of endocrine therapy on steroid-receptor content of breast cancer. Br J Cancer 45: 80-85

VII. Immunocytochemistry – Applications
2. Effusions

Immunocytochemical Differentiation of Lymphocytic Effusions

P. Spieler and U. Schmid

Introduction

The cytological diagnosis of malignant lymphoma in serous effusions can be difficult because reactive lymphocytes may be morphologically indistinguishable from malignant cells in lymphocytic and other low-grade non-Hodgkin's lymphomas (NHL). Possible causes of lymphocyte-rich effusions are: viral pleuropneumonia, tuberculous pleurisy, infarct (lung), trauma (repeated paracentesis), irradiation, malignant lymphoma, carcinosis of serous membranes. A few recent studies have shown that some of the diagnostic problems can be clarified with B and T cell enumeration using immunocytochemical techniques (Ghosh et al. 1985; Martin et. al. 1984). The immunocytochemical labelling technique can be applied to routinely performed cytological smears. The laboratory procedure is simple, the material may be stored for long periods, and the labelled cells exhibit well-preserved morphological details.

Material and Methods

During 1985 and 1986, 44 consecutive cytological specimens, including 39 pleural, 2 pericardial, and 3 ascitic fluids, were studied with a panel of monoclonal anti-B- and anti-T-cell antibodies (PAN B/CD22, κ, λ, T1/CD5, T2/CD7, OKT4/CD4, T8/CD8). The samples had to be rich in well-preserved cells with 80% or more lymphoid elements. Based upon the final diagnosis (cytomorphology, clinical information, microbiological investigation) the specimens were divided into seven diagnostic groups:

Nonspecific reactive lymphocytic fluid	$n = 14$
Tuberculous pleurisy	$n = 4$
Known carcinoma, no tumor cells in effusion	$n = 7$
Lymphocytic effusion with carcinoma cells	$n = 6$
NHL, B cell	
definitive diagnosis cytologically	$n = 5$
suspicious cytologically	$n = 5$
NHL, T cell (suspicious)	$n = 1$
Hodgkin's disease	$n = 2$

There were 11 patients with effusions immunocytochemically and/or cytologically positive for NHL, in 10 of whom there was histological confirmation. Histological typing according to the Kiel classification was centroblastic-centrocytic lymphoma in six cases, centrocytic lymphoma in one, immunocytoma in two, and T-cell lymphoma in one.

Routine Laboratory Procedure. After centrifugation at 2800 rpm for 10 min, the supernatant was decanted and three direct smears were prepared from the button for conventional cytological examination. The remaining cell material was washed at least three times with physiological saline and then evenly smeared. A minimum of 12 slides were prepared. Half of the slides were fixed and stained according to the Papanicolaou method for subsequent intracytoplasmic immunoglobulin staining. The other half were air dried for May-Grünwald-Giemsa staining (one or two smears) and for surface marker studies. When the immunocytochemical procedure was delayed, air-dried smears were wrapped in aluminium foil and stored at −20 °C. From blood-rich specimens the mononuclear cells were isolated by density centrifugation using Lymphoprep. The immunocytochemical staining procedure and data analyses are described in detail elsewhere (Spieler et. al. 1986).

Results

Reactive lymphocytic effusions were characterized by a predominance of T cells (80% of all lymphocytes) with an excess of helper/inducer cells (mean helper to suppressor ratio 3.0) and by a surface κ to surface λ ratio of 1.6 on B cells. Tuberculous effusions showed a similar distribution of lymphocyte subpopulations. Hence, immunocytochemical investigations offer no advantages in distinguishing between specific and nonspecific reactive lymphocytosis.

Most of the lymphocytic fluids with carcinomatous lesions had a lower percentage of T cells (lowest value 67%), a lower helper to suppressor ratio (mean 2.0) with sporadic extreme values, and occasional κ to λ ratios that tended to monoclonality. Interpretation of these findings is not possible because of the small number of cases and the inhomogeneity of the material. Further investigations with more cases may show that T-cell enumeration could be a prognostic method for serous membrane-infiltrating carcinoma.

Lymphoid cells in samples of ten B-cell lymphomas were characterized by T-cell depression ($\leq$70%). B cells in six cases expressed clear-cut light-chain monoclonality (surface κ and λ immunoglobulin), which was at least suggested in the other four cases. That clear monoclonality was not found may be because of technical reasons (cell evaluation regardless of atypias), combined with there being a mixed population of malignant and reactive lymphoid cells.

In one case of peripheral T-cell lymphoma, the immunological results were indistinguishable from those of reactive effusions. The two fluids accompanying Hodgkin's disease that were investigated had inconclusive immunocytochemical results.

Our investigations show that diagnostic accuracy can be improved by means of B- and T-cell enumeration using an immunoalkaline-phosphatase method (IAP).

Our results are comparable with those reported in previous studies using either immunofluorescence or immunocytochemical labelling techniques. For B-cell lymphomas a marked reduction in the percentage of T cells (Domagala et. al. 1981; Krajewski et. al. 1982) and a preponderance of B cells (Ghosh et. al. 1985) have been reported. Futhermore, monoclonal light-chain immunoglobulin expression was observed in six of nine lymphomas by Krajewski et. al. (1982), in 16 of 20 cases by Martin et. al. (1984), and in two of four multiple myelomas by Walts and Said (1983).

Conclusions

It is concluded that: (1) The chosen panel of antibodies in combination with the IAP labelling method is of great value in identifying B-cell lymphomas. The immunological pattern consistent with B-cell lymphoma shows T-cell depression to below 70% and monoclonality of κ λ light chains. (2) The immunological pattern of peripheral T-cell lymphoma may be identical to that of reactive lymphocytic fluids. (3) The immunological patterns in lymphocytic effusions of nonspecific and of tuberculous origin are identical. (4) Occasional equivocal immunological results are obtained in lymphocyte-rich fluids due to carcinoma.

References

Domagala W, Emeson EE, Koss LG (1981) T and B lymphocyte enumeration in the diagnosis of lymphocyte-rich pleural fluids. Acta Cytol (Baltimore) 25: 108-110

Ghosh AK, Spriggs AI, Mason DY (1985) Immunocytochemical staining of T and B lymphocytes in serous effusions. J Clin Pathol 38: 608-612

Krajewsky AS, Dewar AE, Ramage EF (1982) T and B lymphocyte markers in effusions of patients with non-Hodgkin's lymphoma. J Clin Pathol 35: 1216-1219

Martin SE, Zhang HZ, Magyarosy E, Jaffe ES, Hsu SM, Chu EW (1984) Immunologic methods in cytology: definitive diagnosis of non-Hodgkin's lymphomas using immunologic markers for T- and B-cells. Am J Clin Pathol 82: 666-673

Spieler P, Kradolfer D, Schmid U (1986) Immunocytochemical characterization of lymphocytes in benign and malignant lymphocyte-rich serous effusions. Virchows Arch [A] 409: 211-221

Walts AE, Said JW (1983) Specific tumor markers in diagnostic cytology. Acta Cytol (Baltimore) 27: 408-416

Immunocytochemistry: Methods and Applications

G. Mechtersheimer and P. Möller

Introduction

There is accumulating evidence that tumor cells have defects in gene regulation leading to changes in the expression, composition, and structure of cell surface molecules. The cell membrane itself is the mediator of intercellular cooperation, which for epithelial cells mainly means cohesion. The surface membrane exposes molecules defining "self" and "non-self" that are recognized as such by the immune system. These include receptors and many enzymes. Thus, the cell membrane, metaphorically speaking, is a pelt of functional structures. These considerations alone make it evident that theoretical oncology is mainly concerned with the cell membrane. The cell membrane also inspires clinical scientists since substances reacting with structures of the cell membranes can be used to detect and modify these structures. Monoclonal antibodies have rendered the immunoglobulin molecule a powerful tool for generating scientific and medical knowledge. The scientific community presently sees the world by the afferent pathway of the murine immune system: as an arrangement of antigens. Antigens on cell surfaces are ideal tools for immunodiagnosis of cells in suspension.

We describe here in detail the immunostaining procedure adopted for mouse monoclonal antibodies in our laboratory. The major task for its diagnostic use is to set up a reasonable panel of well-characterized, readily available antibodies to well-known antigens (Table 1). Although there are cell type specific antigens, mostly specific secretion products such as hormones, these are not an integral part of the surface membrane of the producer cell. In contrast, cell line specific surface antigens that are conserved during neoplastic transformation and do not show lineage promiscuity in neoplasia are scare.

Leucocyte Differentiation Antigens

Some of the leucocyte differentiation antigens are in fact cell line specific, and some of them (e.g. CD45, CD19, CD20, CD3) - CD nomenclature for leucocyte antigens used according to the recommendations of the nomenclature committee of the Third Workshop on Leucocyte Typing; McMichael et. al. 1987 - are expressed from early stages of lymphocyte maturation up to very late steps of differentiation. These antigens have therefore been designated as "pan-B" or "pan-T"

Table 1. Cell surface antigens and corresponding commercialized monoclonal antibodies suitable for the immunophenotypic characterization of cells in suspension

Antigen	Clone	Source
Leucocyte antigens		
Leucocyte common antigen		
CD45	2H4	Dakopatts
B-lymphocyte antigens		
CD19	HD37	Behring
CD20	B1	Coulter
CD22	HD39	Behring
T-lymphocyte antigens		
CD2	OKT11	Ortho
CD3	OKT3	Ortho
CD4	OKT4	Ortho
CD8	OKT8	Ortho
Myelomonocytic antigens		
CD13	My7	Coulter
CD15	LeuM1	Becton Dickinson
CD33	My9	Coulter
Major histocompatibility complex antigens		
HLA-A, B, C	W6/32	Dakopatts
HLA-DR	L243	Becton Dickinson
HLA-DP	B7/21	Becton Dickinson
HLA-DQ	TÜ22	Biotest
Epithelial surface antigens		
Egp34	HEA125	Progen
EMA	HMFG-2	Dakopatts
CEA	FO23C5	Immunotech
Sialyl-Lewis[a]	Ca19-9	Centocor
>200kd	Ca12-5	Centocor

These antigens are not cell lineage specific. Some EMA antibodies have been reported to react with Hodgkin cells and plasma cells. LeuM1 recognizing the X-hapten binds to various kinds of epithelial cells.

molecules. Others are only expressed during a short phase of development and therefore characterize a particular step in differentiation. Although there are some monocyte- or macrophage-specific antigens (e.g. CD19 and CD33), "pan-mφ" antigens have not yet been found. Furthermore, there is neither a broadly expressed nor a cell type specific surface marker for mesenchymal cells.

Epithelial Surface Antigens

For epithelial cells we are somewhat better off. Of the several hundered published monoclonal antibodies for which specificity for epithelial cells has been claimed by the authors (recently reviewed by Möller et. al. 1987), all those whose specificity has been tested only on cell lines will have to be used with caution. We consider that for immunodiagnostic purposes extensive immunohistochemical characterization of the binding pattern in normal tissues is necessary. One of the antigens meeting the requirements of a reliable epithelial marker is the epithelial glycoprotein Egp34 recognized by the antibody HEA125 (Momburg et al. 1987; Moldenhauer et al. 1987). HEA125 reacts with all kinds of normal epithelial cells, hepatocytes, keratinocytes and cortical thymic epithelial cells being excluded. Egp34 is highly conserved in epithelial neoplasias and is therefore an excellent marker positively defining the epithelial nature of a cell.

Antibodies reactive with CEA (carcinoembryonic antigen) have to be examined for their cross-reactivity with NCA (non-specific cross-reactive activity) before they can be used for cytodiagnosis. Antibodies to EMA (epithelial membrane antigen) may be used for the diagnosis of carcinoma cells in suspension, keeping in mind that EMA has been found to be expressed on plasma cells, Hodgkin cells (Delson et al. 1984) and so-called Ki-1-positive large cell lymphomas (i.e. lymphomas expressing CD30 antigen; Fujimoto et al. 1988). Ca12-5 and Ca19-9 are epithelium specific antigens but exhibit a restricted binding pattern. Antibody to Ca12-5 frequently binds to ovarian and pancreatic carcinoma (Bast et al. 1981). Antibody to Ca19-9 binds to colon carcinoma cells although the antigen is not so frequently expressed as CEA (Quentmeier et al. 1987).

Carcinoma, Lymphoma, Sarcoma

Economic strategies have been developed for the differential diagnosis between carcinoma and lymphoma (e.g. Gatter et al. 1985; Mason and Gatter 1987). In most studies antibodies to cytokeratin have been used to prove or disprove the epithelial nature of a proliferation (Dean et al. 1987). For several reasons, one of them being systematic since we want to restrict the line of argument to cell membrane structures, we used and recommend the pair of antibodies to CD45 (which is a pan-leucocyte antigen; Borowitz et al. 1984; Battifiora and Trowbridge 1983; Pizzolo et al. 1980) and to Egp34. If neither antigen is detectable, possible diagnoses still include squamous cell carcinoma of the skin (being mostly Egp34-negative), Hodgkin's disease (Hodgkin cells often lack CD45; Michels et al. 1987), mesothelioma and sarcoma. In such a case the additional presence of EMA would rule out the first and the fourth kinds of tumor, and the presence of CD30 or CD15 would then exclude the mesothelioma and lead to the final diagnosis of Hodgkin's disease. The positive diagnosis of a sarcoma, however, is not possible via surface antigens for the moment.

Mesothelial Cells, Mesothelioma Cells, Carcinoma Cells

From the practical point of view , the crucial pitfall in working with cells in effusions is the distinction between mesothelial cells, mesothelioma cells and carcinoma cells. Normal mesothelial cells are Egp34-negative whereas every carcinoma causing malignant effusions will be Egp34-positive. However, the possibility that some epithelioid mesotheliomas express this antigen cannot be excluded. EMA and CEA are of no help in the differential diagnosis of adenocarcinoma and epithelioid mesothelioma since these two are inconstantly expressed on both types of malignancy (Strickler et al. 1987). In this rather critical situation the additional detection of intermediate filaments like keratin and vimentin may be misleading for various reasons, whereas the additional expression of the Ca12-5 antigen would favour the diagnosis of ovarian cancer. The problem arises from the potential of mesothelial cells to change their antigenic profile dramatically in the context of inflammation; they may loose vimentin and express both CEA and EMA (but not Egp34!).

Leukemia and Lymphoma Typing

Against the above-mentioned background, leukemia and lymphoma typing from cells in suspension is an easy task. Together with CD45 the antigen CD13 (My7) is a very good marker for early myelomonocytic proliferation (Davey et al. 1987), and CD15 (LeuM1) or CD33 (My9) for the more mature forms. CD19, CD20, and CD22 positively define a B-cell lymphoma and CD2 and CD3 are reliable T-cell markers that are, however, not yet expressed in immature T-cell tumors. In this case antibodies to CD1 and CD38 may lead to the final diagnosis of a leukemia or lymphoma of the immature thymus cortex type. CD2 and/or CD3 may furthermore be lost during the course of malignant transformation although the disappearance of both seems to be a very rare event (Warnke and Rouse 1985).

In summary, despite some differential diagnostic problems that arise from physiological changes of antigenic profiles during maturation, activation and inflammation, and from pathological changes of the immunophenotype during malignant transformation, the present state of the art allows a reliable answer to many a question in diagnostic cytology. But further progress in this field is ready to come.

Materials and Methods

Materials

Cell suspensions are obtained from heparinized peripheral blood or from pleural or peritoneal effusions.

Cell Separation

Effusions. Pleural and peritoneal effusions are cytocentrifuged at 120 *g* for 10 min. The pellet is then resuspended in phosphate-buffered saline (PBS, pH 7.4) at a final concentration of 1×10^6/ml.

Peripheral blood. For isolation of mononuclear cells from heparinized (40 IU/ml) venous blood we use Ficoll-Paque density gradient centrifugation: 3 ml Ficoll-Paque (density = 1.007 g/ml; Pharmacia Fine Chemicals AB, Uppsala, Sweden) is added to a centrifuge tube. Then 4 ml whole blood diluted 1:1 (v/v) with PBS is carefully layered on the Ficoll-Paque and centrifuged at 400 *g* for 30 min at room temperature. Subsequently, the mononuclear cells at the interphase are collected using a Pasteur pipette. After isolation, the cells are washed three times in PBS during 10 min centrifugation at 120 *g*. Subsequently, they are resuspended in PBS at a final concentration of 1×10^6/ml.

Cytospin Preparation

Cytospins are done with a Shandon-Elliott centrifuge (Shandon, London, UK). Cells are centrifuged for 7 min at 55 *g* using a constant cell number (10^5) in a constant volume (100 µl) to obtain slides with a standard cell density. Cytocentrifuged cells are air-dried for 1 h, subsequently fixed in acetone at room temperature for 10 min and immunostained immediately or stored at −20 °C for a maximun of 2 weaks. For conventional histological examination the slides are stained with May-Grünwald and Giemsa stains according to Pappenheim.

Immunostaining Procedure

After rehydration with PBS, the cytospin preparations are incubated for 1 h with purified monoclonal antibodies (see Table 1) diluted to appropriate concentrations in PBS.

Single-Staining Procedures. In cases poor in granulocytes, the slides are then incubated with biotinylated anti-mouse immunoglobulin (1:50) and streptavidin-biotinylated peroxidase complex (1:100; both from Amersham Buchler, Braunschweig, FRG) for 30 min.

In cases rich in granulocytes, immunolabelling is performed with the alkaline phosphatase/anti-alkaline phosphatase method using rabbit anti-mouse IgG & IgM (1:50) as bridging antibody, followed by incubation with monoclonal alkaline phosphatase/anti-alkaline phosphatase complexes (1:30; both from Jackson Immuno Research Laboratories Inc., Avondale, USA) to avoid irritating non-specific staining due to excessive endogenous peroxidase reactivity. The incubation times are 30 min in each case. All incubation steps are carried out in a humid chamber at room temperature and are followed by rinsing twice in PBS.

The peroxidase reaction product is reddish-brown in color when 3-amino-9-ethylcarbazole (AEC; Sigma, St. Louis, USA) and H_2O_2 are used as substrates (Koretz et al. 1987): 4 mg AEC is dissolved in 500 µl dimethylformaide (DMF; Sigma, St. Louis, USA). Subsequently, 9.5 ml sodium acetate buffer (0.05 *M*; pH 5.0) is added. Just before use, 5 µl H_2O_2 (30%) is added. The alkaline phosphatase is developed with naphthol-AS-BI-phosphate (Fluka AG, Buchs, Switzerland) as substrate and New Fuchsin (Merck, Darmstadt, FRG) as coupler, resulting in intense red labelling of the tissue binding site. Immediately before staining, 150 µl 5% New Fuchsin and 250 µl 4% sodium nitrite are mixed and added to 50 ml propandiol buffer (pH 9.75). The final pH is adjusted to 8.7. Finally, 25 mg naphthol-AS-BI-phosphate dissolved in 300 µl DMF is added. 1 m*M* Levamisole (Sigma, St. Louis, USA) is added to inhibit endogenous alkaline phosphatase activity. The incubation time is 10-15 min. The slides are then rinsed in distilled water, counterstained with Harris' haematoxylin and mounted in glycerol gelatin.

Double-Staining Procedures. The sequential application of the two above-mentioned staining methods allows the simultaneous demonstration of two different binding sites in a single cytospin preparation (Mechtersheimer et al. 1986). For the demonstration of the first binding site we use the streptavidin-biotin-peroxidase method. AEC and H_2O_2 yield a reddish-brown reaction product as described above. Subsequently, for the second binding site, the alkaline phosphatase/anti-alkaline phosphatase method is used. To obtain a colour precipitate contrasting well with the reddish-brown reaction product of the first staining sequence, in the double-staining procedures the alkaline phosphatase is developed with naphthol-AS-phosphate as substrate and fast blue BB salt as coupler resulting in blue labelling of the binding site: 25 mg naphthol-AS-phosphate (Fluka, Buchs, Switzerland) and 25 mg fast blue BB salt (Sigma, St. Louis, USA) are each dissolved in 1250 μl DMF and diluted with 50 ml propandiol buffer (pH 9.75, finally adjusted to pH 8.7). Again, 1 m*M* levamisole is added to inhibit endogenous alkaline phosphatase activity. Double-stained slides are not counterstained.

Controls

Negative controls are performed by omitting the primary antibody.

References

Bast RC JR, Feeney M, Lazarus H, Nadler LM, Colvin RB, Knapp RC (1981) Reactivity of a monoclonal antibody with human ovarian carcinoma. J Clin Invest 68: 1331-1337

Battifora H, Trowbridge IS (1983) A monoclonal antibody useful for the differential diagnosis between malignant lymphoma and nonhematopoietic neoplasms. Cancer 51: 816-821

Borowitz MJ, Stevanovic G, Gottfried M (1984) Differential diagnosis of undifferentiated malignant tumors with monoclonal antibody T29/33. Hum Pathol 15: 928-934

Davey FR, Erber WN, Gatter KC, Mason DY (1987) Immunophenotyping of acute myeloid leukemia by immuno-alkaline phosphatase (APAAP) labeling with a panel of antibodies. Am J Hematol 26: 157-166

Dean PJ, Moinuddin SM, Emerson LD (1987) Application of anti-leukocyte common antigen and anti-cytokeratin antibodies to the biopsy diagnosis of gastric large cell lymphoma. Hum Pathol 18: 918-923

Delsol G, Gatter KC, Stein H, Erber WN, Pulford KAF, Zinne K, Mason DY (1984) Human lymphoid cells express epithelial membrane antigen. Implications for diagnosis of human neoplasms. Lancet ii: 1124-1128

Fujimoto J, Hatu I, Ishii E, Kiyokawa N, Tanaka S, Morikawa Y, Shimizu K, Hajikano H (1988) Ki-1 lymphomas in childhood: immunohistochemical analysis and the signification of epithelial membrane antigen (EMA) as a new marker. Virchows Arch [A] 412: 307-314

Gatter KC, Heryet A, Alcock C, Mason DY (1985) Clinical importance of analysing malignant tumours of uncertain origin with immunohistological techniques. Lancet i: 1302-1305

Koretz K, Leman J, Brandt I, Möller P (1987) Metachromasia of 3-amino-9-ethylcarbazole (AEC) and its prevention in immunoperoxidase techniques. Histochemistry 86: 471-478

Mason DY, Gatter KC (1987) The role of immunocytochemistry in diagnostic pathology. J Clin Pathol 40: 1042-1054

McMichael AJ et al. (eds) (1987) Leucocyte typing III. White cell differentiation antigens. Oxford University Press, Oxford

Mechtersheimer G, Möller P, Momburg F, Moldenhauer G, Schwechheimer K (1986) Simultaneous demonstration of lectin-binding sites and antigens detected by monoclonal antibodies in a parallelized double-staining technique: a highly discriminative and quickly developing technique for frozen sections. J Histochem Cytochem 34: 399-402
Michels S, Swanson PE, Frizzera G, Wick MR (1987) Immunostaining for leukocyte common antigen using an amplified avidin-biotin-peroxidase complex method and paraffin setions. A study of 735 hematopoietic human neoplasms. Arch Pathol Lab Med 111: 1035-1039
Moldenhauer G, Momburg F, Möller P, Schwartz R, Hämmerling GJ (1987) Epithelium-specific surface glycoprotein of Mr 34,000 is a widely distributed human carcinoma marker. Br J Cancer 56: 714-721
Möller P, Momburg F, Moldenhauer G (1987) Epithelial membrane markers: Review, presentation of own monoclonal antibodies, and perspectives of application in histopathology. In: Seifert G, Hübner K (eds) Pathology of cell receptors and tumor markers. Application of immunocytochemistry and hybridization in tumor diagnosis. Fischer, Stuttgart, p 110
Momburg F, Moldenhauer G, Hämmerling GJ, Möller P (1987) Immunohistochemical study of the expression of a Mr 34,000 human epithelium-specific surface glycoprotein in normal and malignant tissues. Cancer Res 47: 2883-2891
Pizzolo G, Sloane J, Beverley P, Thomas JA, Bradstock KF, Mattingly S, Janossy G (1980) Differential diagnosis of malignant lymphoma and nonlymphoid tumors using monoclonal anti-leucocyte antibody. Cancer 46: 2640-2647
Quentmeier A, Möller P, Schwarz V, Abel U, Schlag P (1987) Carcinoembryonic antigen, CA 19-9, and CA 125 in normal and carcinomatous human colorectal tissue. Cancer 60: 2261-2266
Strickler JG, Herndier BG, Rouse RV (1987) Immunohistochemical staining in malignant mesotheliomas. Am J Clin Pathol 88: 610-614
Warnke RA, Rouse RV (1985) Limitations encoutered in the application of tissue section immunodiagnosis to the study of lymphomas and related disorders. Hum Pathol 16: 326-331

Improvement of Immunocytochemical Diagnosis of Neoplastic Effusions by Short-Term Culture of the Fluid: A Case Report*

A. M. Cianciulli, M. Mottolese, I. Venturo, R. Perrone Donnorso, C. Gallo Curcio, M. Rinaldi, and P. G. Natali

Introduction

The metastatic nature of effusions from solid tumors may remain undetected in about 20% of instances when they are analyzed by conventional cytological techniques [5]. Such failures are mostly due to the scarcity of tumor cells, which may be masked by mesothelial and lymphoid elements. We have recently shown that immunocytochemical methods employing monoclonal antibodies (MAb) to tumor-associated antigens (TAAs) can greatly improve the accuracy of cytodiagnosis of metastatic effusions [7].

In contrast to the standard method, we have made short-term cultures from effusions in which the results of morphological and immunochemical methods were inconclusive because of the low cellularity, with the aim of increasing the tumor cell population.

We report here a clinical case studied with this technique which exemplifies the diagnostic value of this novel approach.

Materials and Methods

Sampling of Effusion and Short-Term Culture. Samples of pleural effusions were collected with heparin under sterile conditions. Cytospins were prepared by centrifugation at 160 *g* for 10 min with a Shandon cytocentrifuge (Shandon, Runcorn, UK). They were used for conventional morphological diagnosis after staining using the Papanicolaou and May-Grünwald methods or were fixed for 10 min in cold absolute acetone for immunocytochemical evaluation. The remaining cells were cultivated in 25-cm^2 plastic flasks (Falcon, Oxnard, CA, USA) in RPMI 1640 medium (Gibco, Gran Island, NY; USA) supplemented with 10% inactivated fetal calf serum (Gibco), 1 m*M* L-glutamine (Gibco), 200 μg streptomycin/ml (Gibco), and 20 IU penicillin/ml (Sigma, St. Louis, USA) at 37 °C and with a 5% CO_2 atmosphere. After two passages, cultured cells were grown further on microchamber culture slides (Labtek Division, Miles Laboratories Napierville, Ill., USA). When confluent, the cells were used for immunocytochemical evaluation after fixation with absolute acetone for 10 min.

---

* Supported by CNR PF "Oncologia" no. 870155944, by AIRC, and by Tecnobiomedica

Table 1. Monoclonal antibodies used for the immunocytochemical analysis of serous effusion

Antibody	Isotype	Molecular weight (daltons)	Major reactive tissue	References
B72.3	IgG1	$> 10^6$	Adenocarcinomas	[3, 9]
B6.2	IgG1	9×10^4	Breast carcinoma	[3]
MBr1	IgM		Breast carcinoma	[2]
OC-125	IgG1	$> 0.5 \times 10^6$	Ovarian carcinoma	[1]
MOv19	IgG2a	3.8×10^4	Ovarian carcinoma	[6]
anti-keratin A	IgM	5.4×10^4	Keratin of non-squamous epithelium	[4]
anti-keratin B	IgG1	6.6×10^4	Keratin of squamous and ductal epithelium	[4]

Monoclonal Antibodies. MAb MBr1, MOv19, OC-125, B72.3, and B6.2 [1, 2, 3, 6, 9], which recognize distinct TAAs, were kindly provides by Prof. Colnaghi, National Cancer Institute, Milan (MBr1, MOv19) and Dr. Bast, Duke University, Durham, USA (OC-125), or obtained commercially (B72.3, B6.2) from Sorin Biomedica (Saluggia VC, Italy). MAbs to keratin A and keratin B were obtained from Ortho Diagnostic [4] (Table 1). These MAbs are not expressed at detectable levels by nonepithelial cells. They were employed either as purified antibodies [8] or as culture supernatants.

Immunoperoxidase. Cytospins and cultured cells were stained with MAbs using an indirect avidin-biotin complex (ABC) immunoperoxidase method with commercially available reagents (Immunocolor kit, Sorin Biomedical). Incubation with the primary reagent lasted for 18 h. The immunoperoxidase enzymatic activity was detected using 3-amino-9-ethylcarbazole (AEC) as chromogenic substrate. The slides were rinsed and exposed to Mayer's hematoxylin (BDH, UK) for nuclear staining for 8-10 min and mounted in buffered glycerol. In control cell substrates the reaction with the primary antibody was omitted.

Results

The patient, a 62-year-old woman, underwent right radical mastectomy for a breast adenocarcinoma, stage $T_2N_0M_0$ (May 1976). After 10 years a left pleural effusion was diagnosed by chest X-rays and CT scan (September 1986). The fluid was evaluated by conventional morphological techniques and by immunocytochemical techniques employing a panel of selected MAbs directed to different TAAs. Cytological examination only revealed mostly small lymphocytes and isolated epithelial-like cells of an uncertain nature. Immunocytochemical examination was inconclusive because of the paucity of cells displaying clear-cut reactivity with MAbs. We submitted the fluid to a short-term culture. Within 2 days clusters of epithelial-like cells could be observed; these were cultured further (Fig. 1). These cultured cells were tested with the panel of MAbs and showed reactivity with MAbs to keratins A and B, thus confirming their epithelial nature, no reac-

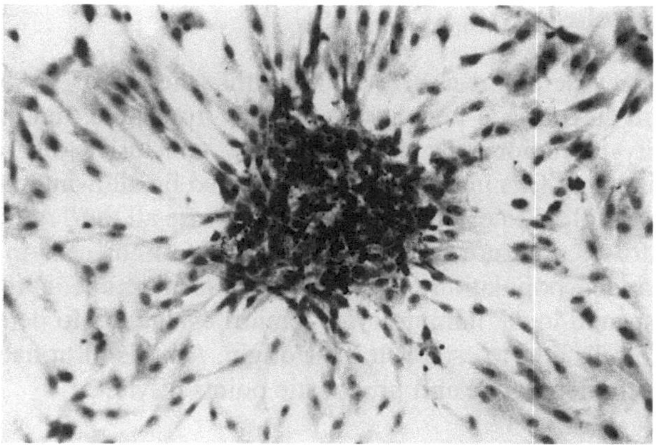

Fig. 1. Epithelial-like cells cultured from a pleural effusion on a micro chamber culture slide. ×232

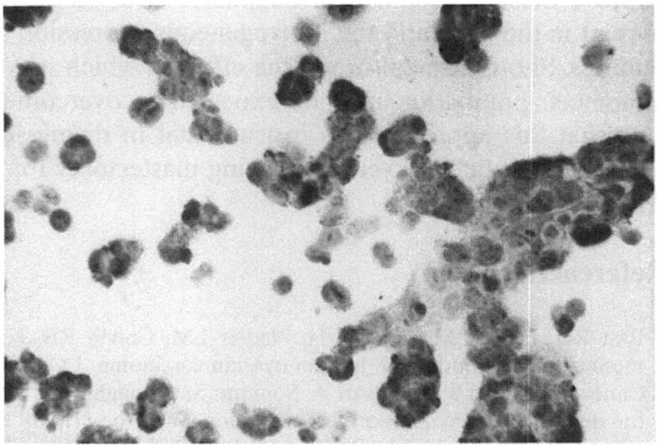

Fig. 2. Avidin-biotin immunoperoxidase staining with the MAb MBr1 on cells cultured from a pleural effusion. ×182

tion with MAbs MOv19 and OC-125, which are directed to ovarian carcinoma, and homogenous reactivity with MAbs B72.3, B6.2, and MBr1, which are directed to breast tumor associated antigens, suggesting their derivation from this transformed epithelium (Fig. 2).

On the basis of these findings the patient was treated as follows (October 1986): 5-fluorouracyl 900 mg on the 1st and 8th days, cyclofosfamide 900 mg on the 1st day, adriamycin 90 mg on 1 day every 4 weeks. After a total of six courses chest X-rays showed complete regression of the pleural effusion. No new metastatic sites were found. The patient is still disease free.

Conclusions

This case report clearly confirms that the use of MAbs can significantly improve the diagnosis of neoplastic effusions from solid tumors in cases in which the paucity of tumor cells does not allow conclusive diagnostic immunocytochemical findings. This improvement is achieved by allowing the neoplastic cell population in the fluid to expand in culture. Because all the TAAs recognized by our panel of MoAbs are known to be expressed by tumor cells in short-term culture, the expanded population clearly provides a more adequate substrate for attempting a "second-look" immunocytochemical evaluation to detect the presence of neoplastic cells and to establish their origin. All this information is clearly relevant from both therapeutic and prognostic points of view.

Summary

Although monoclonal antibodies to tumor-associated antigens (TAA) can greatly improve the cytodiagnosis of neoplastic effusions, the paucity of the tumor cells present in the fluid and the heterogeneous expression of TAA may cause negative findings. Short-term culture of the effusion which may result in enrichment of the tumor cell population may be expected to overcome this limitation. We report here that this approach was instrumental in diagnosing neoplastic effusion in a 62-year-old patient 10 years following mastectomy for breast adenocarcinoma.

References

1. Bast RO, Feeney M, Lazarus H, Nadler LM, Colvih RB, Knapp RC (1981) Reactivity of a monoclonal antibody with human ovarian carcinoma. J Clin Invest 68: 1331-1337
2. Canevari S, Fossati G, Balsari A, Sonnino S, Colnaghi MI (1983) Immunochemical analysis of the determinant recognized by a monoclonal antibody (MBR1) which specifically binds to human mammary epithelial cells. Cancer Res 43: 1301-1305
3. Colcher D, Horan P, Nuti M, Schlom J (1981) A spectrum of monoclonal antibodies reactive with human mammary tumors. Proc Natl Acad Sci USA 78: 3199-3203
4. Gown AM, Vogel AM (1985) Monoclonal antibodies to human intermediate filament proteins III. Analysis of tumors. Am J Clin Pathol 84: 413-422
5. Koss LG (1979) Diagnostic cytology and its histopathologic bases, vol 2. Lippincott, Philadelphia, pp 911-967
6. Miotti S, Canevari S, Menard S, Mezzanzanica D, Porro G, Regazzoni M, Pupa SM, Tagliabue E, Colnaghi MI (1987) Characterization of human ovarian carcinoma associated antigen defined by novel monoclonal antibody with tumor restricted specificity. Int J Cancer 39: 297-301
7. Mottolese M, Venturo I, Perrone Donnorso R, Gallo Curcio C, Rinaldi M, Natali PG (1988) Use of selected combinations of monoclonal antibodies to tumor associated antigens in the diagnosis of neoplastic effusions of unknown origin. Eur J Cancer Clin Oncol (in press)
8. Russo C, Callegaro L, Lanza E, Ferone S (1983) Purification of IgG monoclonal antibody by caprylic acid precipitation. J Immunol Methods 65: 269-271
9. Thor A, Ohuchi N, Szpak CA, Johnston WW, Schlom J (1986) Distribution of oncofetal antigen tumor-associated glycoprotein-72 defined by monoclonal antibody B72.3. Cancer Res 46: 3118-3124

Immunocytochemical Diagnosis of Neoplastic Effusions of Unknown Origin Employing Selected Combinations of Monoclonal Antibodies to Tumor-Associated Antigens*

M. Mottolese, I. Venturo, M. Rinaldi, C. Gallo Curcio, R. Perrone Donnorso, and P. G. Natali

Introduction

Malignant cells in metastatic effusions from solid tumors can easily be detected because of their morphologic features, but their histological type can rarely be defined [4]. The recent availability of an increasing number of monoclonal antibodies (MAbs) to tumor-associated antigens (TAA) may offer a powerful diagnostic complement to conventional morphologic diagnosis for the identification of the organ of origin of the primary tumor. In the present study we evaluated whether the use of selected combinations of MAbs to TAAs which lack detectable reactivity with mesothelial cells may provide useful information in the cytodiagnosis of effusions in patients with unknown primary malignancy. The results of this study show that a panel of selected MAbs is a valuable adjunct to routine cytopathology since it enables identification of the site of primary tumors in 87% of the cases studied.

Material and Methods

Patients. The 35 patients with various malignancies evaluated in this study were admitted to the Regina Elena Cancer Institute because of the appearance of either pleural or peritoneal effusions during the clinical course of a previously diagnosed malignancy. A further 38 cases were referred with a diagnosis of neoplastic effusion of unknown origin. The first group of patients received no treatment after 3-6 months and the second group received no therapy.

Cytospin Preparations. Pleural and peritoneal effusions, collected under sterile conditions, were centrifuged and resuspended after three washings with Hanks balanced salt solution (HBSS, Gibco, Paisley, UK) at a concentration of 1×10^6 cells/ml. Cytospins were done using a Shandon cytocentrifuge (Shandon, Runcorn, UK) and fixed in 95° alcohol for conventional Papanicolau staining and in cold absolute acetone for 10 min for immunocytochemical evaluation. When cytospins were not processed immediately they were stored at −20 °C, there being no loss of immunological reactivity for at least 4 months.

* Supported by CNR PF "Oncologia" no. 870155944, by AIRC, and by Tecnobiomedica

Monoclonal and Polyclonal Antibodies. MAbs for this study were chosen on the basis of two serological properties: they should recognize TAAs which are not coordinately expressed by malignant cells; they should not be expressed at detectable levels by mesothelial cells.

The reagents were kindly provided by different investigators (MBr1, MOv19, Prof. M. I. Colnaghi National Cancer Institute Milan, Italy; OC-125 Dr. Bast, Duke University, Durham, USA; KS1/4, Dr. Walker, Scripps Clinic and Research Foundation, La Jolla, CA, USA) or obtained commercially (B72.3, B6.2 Sorin Biomedica Saluggia VC Italy) (Table 1) [1, 3, 5, 7]. Fluorescein-labeled Fab_2 fraction of a goat anti-mouse immunoglobulin serum was obtained from Sorin Biomedica. The antiserum was used after absorption with human AB Rh^+ red blood cells and with insolubilized pooled normal human plasma [6].

Indirect Immunofluorescence Assay. Cellular specimens underwent indirect immunofluorescent staining with MAbs under standard conditions; they were mounted in 50% buffered glycerol and observed with a Leitz Orthoplan microscope equipped with epiillumination and phase contrast. In control cytospins the incubation with primary MAbs was omitted. The immunocytochemical diagnosis was subsequently compared with histopathological and clinical findings when these became available.

Results

The initial part of this study was performed on samples of pleural and peritoneal effusions from 35 patients who had an established diagnosis of tumor and who had received no treatment for at least 3 months.

The immunocytochemical findings of this analysis are summarized in Table 2. Almost all effusions tested reacted with MAb B72.3. The other reagents, when considered individually, each showed a preferential reactivity with a particular tumor type, but none of them except for MOv19 appeared strictly "tumor specific" with these cell substrates. On the basis of these results, distinct diagnostic patterns of reactivity could be identified (Table 3).

Once these serological patterns of reactivity were defined, 38 cases of pleural and peritoneal effusions from patients with unknown primary tumors were sub-

Table 1. Monoclonal antibodies used for the immunocytochemical diagnosis of effusions

Antibody	Isotype	Antigen	Molecular weight (daltons)	Major reactive tumor	References
B72.3	IgG1	Glycoprotein	$>10^6$	Adenocarcinomas	[3]
B6.2	IgG1	Glycoprotein	9×10^4	Breast carcinoma	[3]
MBr1	IgM	Glycolipid		Breast carcinoma	[2]
OC-125	IgG1	Glycoprotein	0.5×10^6	Ovarian carcinoma	[1]
MOv19	IgG2a	Glycoprotein	3.8×10^4	Ovarian carcinoma	[5]
KS1/4	IgG2a	Glycoprotein	4×10^4	Lung carcinoma	[7]

Table 2. Immunocytochemical diagnosis of neoplastic effusions from patients with known primary tumors employing a panel of monoclonal antibodies to tumor-associated antigens

Primary tumor	Number of patients	Reactivity with MAbs to TAA	
		(+)	(−)
Breast carcinoma	10	B72.3, B6.2, MBr1	MOv19, OC-125, KS1/4
	3	B72.3, B6.2, KS1/4	MBr1, MOv19, OC-125
	3	B72.3, B6.2, MBr1, OC-125	MOv19, KS1/4
Ovarian carcinoma	10	B72.3, OC-125, MOv19	B6.2, MBr1, KS1/4
	3	B72.3, OC-125, MOv19, MBr1	B6.2, KS1/4
	1	MOv19, OC-125, KS1/4	B72.3, MBr1
Lung carcinoma	4	B72.3, OC-125, KS1/4	B6.2, MBr1, MOv19
	1	OC-125, KS1/4	B72.3, B6.2, MBr1, MOv19

Table 3. Immunodiagnostic patterns of reactivity with monoclonal antibodies to tumor-associated antigens

Immunodiagnostic combinations with MAbs to TAAs	Immunocytochemical diagnosis
B72.3$^+$, B6.2$^+$, MBr1$^+$, OC-125$^-$, MOv19$^-$	Breast carcinoma
B72.3$^+$, MOv19$^+$, OC-125$^+$, B6.2$^-$, MBr1$^-$	Ovarian Carcinoma
B72.3$^+$, OC-125$^+$, KS1/4$^+$, MOv19$^-$, MBr1$^-$	Lung carcinoma

Table 4. Immunocytochemical definition of the tumor type of neoplastic effusions from 38 patients with unknown primary tumors employing selected combinations of monoclonal antibodies to tumor-associated antigens

Immunocytochemical diagnosis	Concordance with histological and/or clinical diagnosis
9 breast carcinoma	8/9
16 ovarian carcinoma	16/16
11 lung carcinoma	9/11
2 undefined carcinoma	1 lung and 1 ovarian carcinoma
Total tentative diagnoses 36/38 (95%)	Total concordant diagnoses 33/38 (87.5%)

* Conventional cytological techniques could indentify *only* neoplastic cells in 80% of the cases

mitted to the same immunocytochemical analysis. The results were compared with conventional histopathological or clinical findings (Table 4). The data presented show that the site of origin of the primary tumor could be indicated in 36 out of 38 patients (95%). This immunocytochemical diagnosis was subsequently confirmed histopathologically or clinically in 33 of the 38 cases.

Conclusions

In the present study we have shown that immunocytochemical analysis of metastatic effusions from solid tumors can provide a powerful diagnostic tool for identifying the site of origin of the primary tumor in those patients with no previous history of malignant disease. This can be achieved by using selected combinations of MAbs directed at various TAAs and lacking detectable reactivity to mesothelial cells. The use of more than one reagent is necessary either because of the lack of a restricted tumor specificity of the majority of MAbs available or because of the heterogeneous expression of TAAs recognized by these reagents. Our results demonstrate that the test, which is based on objective criteria of interpretation, is easy to perform and has provided a correct immunocytochemical diagnosis in 87% of the cases studied.

Summary

Conventional cytological techniques in most cases enable the identification of malignant cells in effusions from solid tumors, but rarely allow determination of the origin of the primary neoplasia.

In order to evaluate whether immunocytochemical techniques could be applied in this context we examined 35 well-defined metastatic effusions employing a panel of monoclonal antibodies to tumor-associated antigens. On the basis of this information we defined the antigenic phenotype of 38 effusions deriving from tumors of unknown origin (87.5% of the cases). These data demonstrate that selected combinations of monoclonal antibodies can be used as a powerful adjunct to traditional cytodiagnostic techniques in identifying the site of primary tumor origin.

References

1. Bast RC, Feeney M, Lazarus H, Nadler LM, Colvin RB, Knapp RC (1981) Reactivity of a monoclonal antibody with human ovarian carcinoma. J Clin Invest 68: 1331-1337
2. Canevari S, Fossati G, Balsari A, Sonnino S, Colnaghi MI (1983) Immunochemical analysis of the determinant recognized by a monoclonal antibody (MBr1) which specifically binds to human mammary epithelial cells. Cancer Res 43: 1301-1305
3. Colcher D, Horan-Hand P, Nuti M, Schlom J (1981) A spectrum of monoclonal antibodies reactive with human mammary tumors. Proc Natl Acad Sci USA 78: 3199-3203
4. Koss LG (1983) Examination of effusions (pleural, ascitic, and pericardial fluids). In: Wied GL, Koss LG, Reagan JW (eds) Compendium on diagnostic cytology. Tutorials of cytology, Chicago, pp 472-475
5. Miotti S, Canevari S, Menard S, Mezzanzanica D, Porro G, Regazzoni M, Pupa SM, Tagliabue E, Colnaghi MI (1987) Characterization of human ovarian carcinoma associated antigen defined by novel monoclonal antibodies with tumor restricted specificity. Int J Cancer 39: 297-303
6. Russo C, Callegaro L, Lanza E, Ferrone S (1983) Purification of IgG monoclonal antibody by caprylic acid precipitation. J Immunol Methods 65: 269-271
7. Varky NM, Reisfeld RA, Walker LE (1984) Antigens associated with a human lung adenocarcinoma defined by monoclonal antibodies. Cancer Res 44: 681-687

Immunocytochemistry of Malignant Pleural Effusions in Breast Cancer

J. Guzman, K. J. Bross, P. Deuscher, F. Grunert, S. von Kleist, and U. Costabel

Introduction

The cytological diagnosis of malignant pleural effusions due to breast cancer may be difficult, particularly in patients with severe hemorrhagic effusions or with inflammation where the percentage of tumor cells in the pleural spaces may be very low. In addition, invasive ductal carcinoma of the breast with a scirrhous pattern sometimes shows isolated small tumor cells which cannot easily be distinguished from other cell types included in effusions such as macrophages or mesothelial cells unless clusters of tumor cells are formed.

Because monoclonal antibodies for the detection of surface or intracytoplasmic antigens on neoplastic cells have now become available, the aim of this study was to investigate whether immunocytochemical analysis of malignant pleural effusions due to breast cancer would increase the diagnostic efficacy compared with conventional staining preparations.

Material and Methods

We investigated 30 histologically confirmed malignant pleural effusions due to invasive ductal breast cancer. The cells of the pleural effusions were analyzed (a) by conventional cytological staining and preparation techniques, and (b) immunocytochemically using a peroxidase-antiperoxidase technique with diaminobenzidine as substrate (Bross et al. 1978). The immunocytochemical analysis was performed on adhesive glass slides (Superior, Bad Mergentheim, FRG) which permit 12 different immunoreactions to occur in parallel on the same glass slide using small amounts of antibodies and cells (see Fig. 1; Bross et al. 1978; Guzman et al. 1988).

For immunocytochemical analysis, mononuclear cells from 10 ml fresh pleural fluid were separated by Ficoll Hypaque centrifugation in EDTA-coated tubes. In order to facilitate the identification of macrophages, cells were incubated with latex beads before the immunostaining was performed.

The panel of monoclonal antibodies used in this study is listed in Table 1. To evaluate the reaction, the slides were viewed with a light microscope. Positively reacting cells were clearly recognized by dark brown circular staining of the intact cell membrane (Fig. 2). No intracytoplasmic reaction occurred with this method.

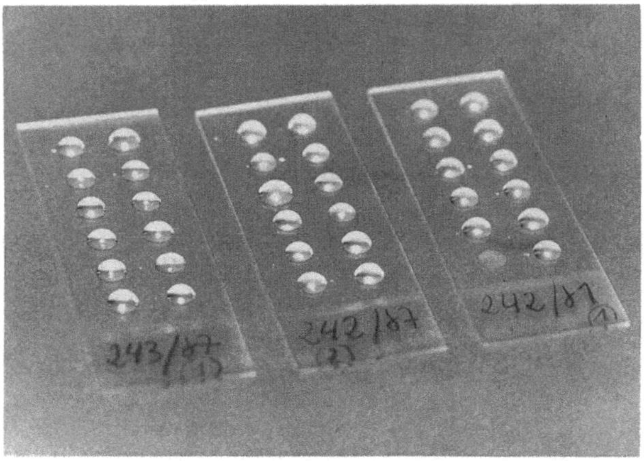

Fig. 1. Adhesive glass slides during the incubation of each reaction area with 10 μl antibody solution

Table 1. Monoclonal antibodies used in this study

Name (target)	Supplier	Dilution
CEA 3-13 (carcinoembryonal antigen)	Immunbiologie, Freiburg	1:50
CEA 5-1 (carcinoembryonal antigen)	Immunbiologie, Freiburg	1:50
HEA-125 (epithelial antigen)	Progen, Heidelberg	1:20
EMA (epithelial membrane antigen)	Dakopatts, Hamburg	1:100
Sam 2 (epithelial antigen)	Not commercially available	1:5
Sam 10 (epithelial antigen)	Not commercially available	1:5
Leu M1 (granulocyte antigen)	Becton-Dickinson, Heidelberg	1:100
My4 (monocyte/macrophage antigen)	Coulter, Florida	1:100
HLA class 1 (leukocyte antigen class 1)	Hybritech, San Diego	1:100

No nuclear counterstaining was used and endogeneous peroxidase activity was not inhibited.

Results

By conventional cytological analysis, malignant cells were detected in 13 cases, in six cases the presence of tumor cells was suspected, and in 11 cases no malignant cells were found.

By immunocytochemical analysis malignant cells were found in all 30 cases (Guzman et al. 1988). In all cases tumor cells reacted positively with EMA, HEA-125, Sam 2, Sam 10, and Leu M1. Only in 21 cases did tumor cells react positively with CEA 3-13 and CEA 5-1. There was no difference in the staining properties of tumor cells using these two different CEA antibodies. The proportion of tumor cells in the various specimens that reacted positively with the antibodies to the tumor markers ranged from 10% to 95%.

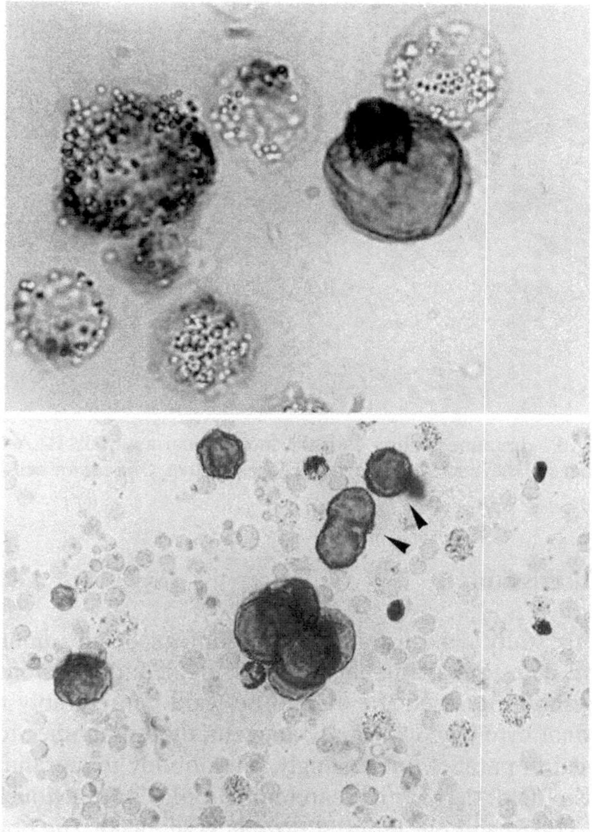

a

b

Fig. 2. **a** Immunostaining with the monoclonal antibody CEA 3-13 showing one tumor cell with a positive cell membrane reaction. The macrophages did not stain and show ingested latex beads as well as a weak endogeneous cytoplasmic peroxidase reaction. × 670. **b** Immunostaining with the monoclonal antibody EMA showing group of positively reacting tumor cells. Mesothelial cells *(arrow)* are also positive with a hairy reaction pattern. × 270

A few mesothelial cells also reacted positively with EMA, HEA-125, Sam 2, and Sam 10 but they showed a hairy and spiky reaction pattern which was clearly different from that of tumor cells, where there was linear immunostaining of the cell membrane (Fig. 2b). Macrophages never reacted with antibodies to these markers and could be identified by the phagocytosed latex beads (Fig. 2a) and the positive reaction with My4. The endogeneous peroxidase activity of macrophages did not interfere with the analysis of surface markers (Fig. 2a). A certain percentage of tumor cells from each specimen (10%-90%) did not react with HLA-1. Lymphocytes, macrophages, mesothelial cells and granulocytes always reacted positively with this antibody (Fig. 3).

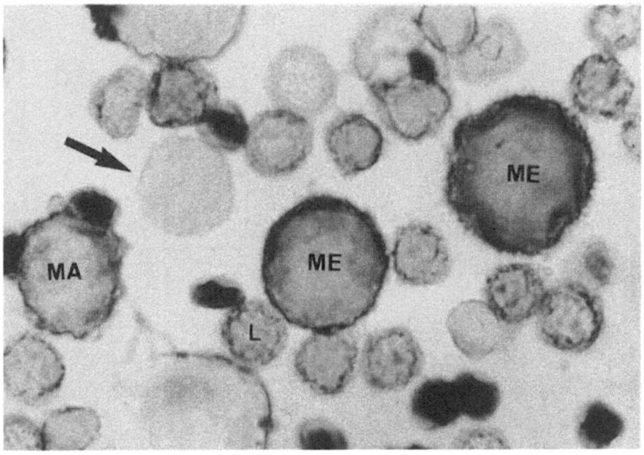

Fig. 3. Immunostaining with the monoclonal antibody HLA-1. Mesothelial cells *(ME)*, macrophages *(MA)* and lymphocytes *(L)* are positive. One tumor cell *(arrow)* is negative. ×670

Discussion

This study demonstrated the usefulness of an immunocytochemical method for the detection of malignant cells in pleural effusions due to breast cancer, this method increased the diagnostic yield considerably in comparison with conventional cytology. Our study suggests that antibodies to several markers should be used in parallel. Interestingly, an antibody to a granulocyte marker (Leu M1) can also help to identify carcinoma cells, as previously reported by Sheibani et al. (1986) and our group (Guzman et al. 1988). The observed partial or total loss of reactivity to HLA-1 only on tumor cells can be of additional help in the detection of tumor cells. In contrast to the results of Walts et al. (1987), in all cases in this study some mesothelial cells were found to react with EMA. The reaction pattern of the surfaces of mesothelial cells with their profuse and long microvilli was, however, different from that of tumor cells which showed linear surface staining.

In conclusion, we have found a markedly improved method for immunocytological diagnosis of malignant pleural effusions due to breast cancer, which gives a greater percentage of positive results than conventional cytology.

References

Bross KJ, Pangalis GA, Staatz C (1978) Demonstration of cell surface antigens and their antibodies by the peroxidase-antiperoxidase method. Transplantation 25: 331-334

Guzman J, Costabel U, Bross KJ, Wiehle U, Grunert F, Schaefer HE (1988) The value of the immunoperoxidase slide assay in the diagnosis of malignant pleural effusions in breast cancer. Acta Cytol (Baltimore) 32: 188-192

Sheibani K, Battifora H, Burke J, Rappaport H (1986) Leu M1 antigen in human neoplasms. Am J Surg Pathol 10: 227-236

Walts AE, Said JW, Shintaku IP (1986) Epithelial membrane antigen in the cytodiagnosis of effusions and aspirates. Diagn Cytopathol 3: 41-49

Monoclonal Antibodies OC125 and OV632 for the Diagnosis of Ovarian Cancer: A Comparative Immunocytochemical Study

I. A. Koelma, G. J. van Steenis, and G. J. Fleuren

Introduction

In the management of patients with ovarian cancer cytological examination of peritoneal fluid or of washings collected at the primary exploration and during a 'second look' procedure plays an important role [5]. It is important, therefore, to identify even small numbers of malignant cells. The distinction between tumor cells and mesothelial cells collected during these procedures can be very difficult, especially after a few courses of chemotherapy [1, 11]. Immunocytochemical methods of identifying malignant cells and of distinguishing them from (reactive) mesothelial cells would be of help. In the last few years many antibodies to ovarian tumor markers have been developed (for review see [2]). In the comparative study reported here we evaluated two commercially available monoclonal antibodies, OC125 and OV632, both directed against antigens occurring primarily on nonmucinous epithelial ovarian tumors [3, 6, 9], and established the value of these two antibodies for clinical cytology.

Material and Methods

Peritoneal fluids and washings were obtained by aspiration. The fluids were immediately centrifuged for 5 min at 1500 rpm, and smears and cytospin preparations were made, air dried, and Giemsa stained for screening. The preparations were classified according to the standard cytological criteria by two pathologists. Preparations containing malignant cells from patients with known ovarian carcinoma ($n=35$) and nonmalignant preparations ($n=20$) were included in the study. The nonmalignant preparations contained (reactive) mesothelial cells and inflammatory cells. The indirect immunoperoxidase technique was applied to cytospin preparations as described elsewhere for cryostat sections [7, 9]. Monoclonal antibodies were used in their optimal dilutions, 1:100 for OC125 (CIS Laboratories, France) and 1:10 for OV632 (Sanbio, Uden, The Netherlands). The peroxidase label was revealed with 3-amino-9-ethyl-carbazole. Replacement of the first antibody by phosphate-buffered saline served as a negative control, and sections of a serous cystadenocarcinoma of the ovary served as a positive control.

Sensitivity and specificity of the tests performed in this study were calculated as follows: Sensitivity equals the true-positives divided by the sum of true-positives and false-negatives. Specificity equals the true-negatives divided by the sum of true-negatives and false-positives.

Results

The results are listed in Table 1. In the majority of cases, malignant ovarian tumor cells in peritoneal fluids and washings reacted with OC125 (32 out of 35 cases; Fig. 1). The staining reaction was mainly membranous with a cytoplasmic blush. OV632 reacted with malignant cells in 34 of 35 fluids in a more cytoplasmic staining pattern (Fig. 2). In many cases mesothelial cells showed CA125 reactivity (16 of 20 nonmalignant fluids; Fig. 3). The staining pattern of the mesothelial cells was often indistinguishable from that of tumor cells in malignant fluids. OV632 showed no reactivity in the nonmalignant fluids.

The results of sensitivity and specificity calculations are shown in Table 1. OC125 was less sensitive than OV632 with respect to the recognition of ovarian tumor cells (0.91 vs. 0.97), and also less specific for the distinction between ovarian tumor cells and mesothelial cells. The specificity of OV632 in this respect was 1.00 and of OC125 0.20.

Table 1. Results of Immunocytochemical analysis of malignant and nonmalignant peritoneal effusions of patients with ovarian cancer

	OC125	OV632
Diagnosis[a]		
Malignant	32/35	34/35
Nonmalignant	16/20	0/20
Sensitivity	0.91	0.97
Specificity	0.20	1.00

[a] *n* positive/ *n* total

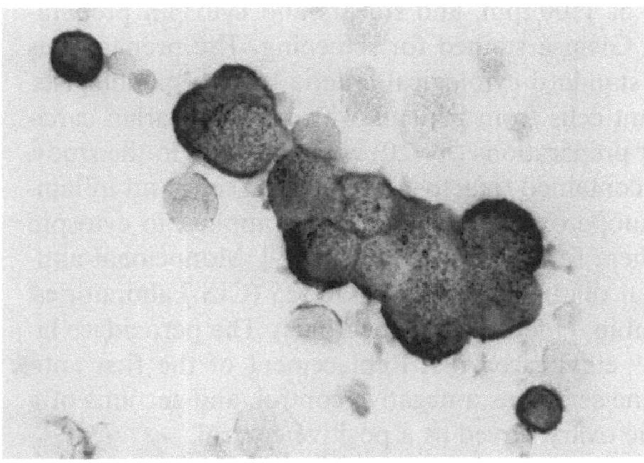

Fig. 1. Malignant cells in a peritoneal washing of a patient with serous cystadenocarcinoma of the ovary showing CA125 expression. Immunoperoxidase, ×400

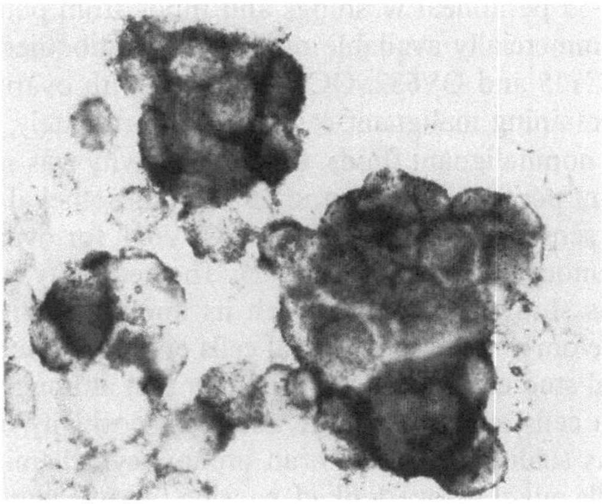

Fig. 2. OV632 positivity in clusters of malignant cells in the peritoneal effusion of a patient with ovarian cancer. Immunoperoxidase, ×400

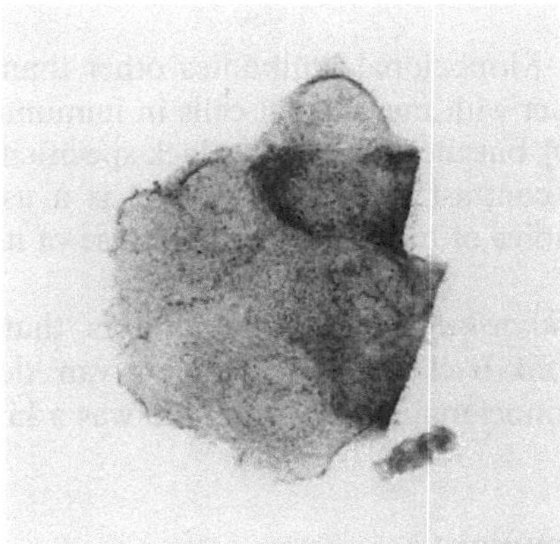

Fig. 3. OC125 immunoreactivity in a sheet of mesothelial cells found in a nonmalignant peritoneal effusion. Immunoperoxidase, ×400

Discussion

The cytological examination of peritoneal fluid, peritoneal washings, and free fluid from the pouch of Douglas has become part of the staging protocol recommended by the International Union Against Cancer for patients with ovarian tumors [5]. This kind of material is often difficult to interpret because of the contamination with blood and detached mesothelium. Furthermore, the mesothelium may show marked atypia as a result of previous operations, intraperitoneal tumor destruction by chemotherapy, or reaction to intracavitary administration of drugs [1, 11]. In histology monoclonal antibodies are widely used to identify tumor cells and to distinguish carcinoma from mesothelium. Studies of large series of cytological preparations, classified by the standard criteria for malignancy, are necessary before a monoclonal antibody can be applied as a diagnostic tool in everyday cytology. This study presents the results of an immunocytochemical investigation of 55 peritoneal washings and fluids from patients with ovarian cancer with two commercially available monoclonal antibodies raised against ovarian carcinoma, OC125 and OV632. OC125 reacted with ovarian tumor cells in 91% of 35 fluids containing malignant cells, and unfortunately, also with mesothelial cells in 80% of nonmalignant fluids. OV632 positivity was seen in tumor cells in 98% of malignant peritoneal fluids and never in mesothelial cells. These results indicate a higher sensitivity and specificity of OV632 for ovarian carcinoma than OC125 in immunocytochemistry (Table 1). The reactivity of OC125 with mesothelial cells in this study is consistent with its known reactivity with coelomic epithelium and coelom-derived mesothelial cells of the peritoneum [6]. From the immunohistological studies with OC125 it is clear that although more than 75% of serous carcinoma cells reacted with the antibody, positivity of 100% of tumor cells in a section was seldom seen [9]. In an immunocytochemical study these nonreacting tumor cells might be overlooked, causing false-negative results and lowering the sensitivity of OC125 for tumor cells in cytological examinations. It is not clear whether the antigenic heterogeneity of ovarian neoplasms is due to shedding of the antigen, the absence of the antigen in a certain population of tumor cells, or there being too little antigen for it to be detected with immunohistochemical methods [2, 4].

Monoclonal antibodies other than OV632 have been described that did not react with mesothelial cells in immunocytological studies, like CEA [8] and B72.3 [10], but these antibodies lack specificity for ovarian carcinoma. We conclude that, in contrast to OC125, OV632 is a useful diagnostic tool in immunocytological studies of ovarian cancer because of its sensitivity and specificity.

Acknowledgements. The authors thank Trudy van Leeuwen-Herberts, Colette Blonk-Beckers, and Lambert van den Broek for expert technical assistance. Monoclonal antibody OC125 was a kind gift from CIS Laboratories, France.

Summary

Two commercially available monoclonal antibodies raised against ovarian carcinoma were used in an immunocytochemical study of peritoneal fluid from patients with ovarian cancer. The study involved OC125, directed against the antigen CA125, and OV632. Cytospin preparations of peritoneal fluid from ovarian cancer patients that contained malignant cells ($n=35$) or only reactive mesothelial cells ($n=20$) were studied using an indirect immunoperoxidase technique for the presence of the antigens. OC125 recognized the majority of ovarian tumor cells (32 of 35 fluids), but CA125 was also expressed by reactive mesothelial cells in 16 of 20 cases. OV632 reacted in 34 of 35 fluids with ovarian tumor cells, but never with mesothelial cells. The sensitivity of the antibodies for ovarian cancer in this test was 0.91 (OC125), and 0.97 (OV632); the specificity was calculated to be 0.20 (OC125) and 1.00 (OV632). We conclude that because of the reactivity of OC125 with mesothelial cells OV632 is preferred for application in immunocytology.

References

1. Covell JL, Carry JB, Feldman PS (1985) Peritoneal washings in ovarian tumors. Potential sources of error in cytologic diagnosis. Acta Cytol (Baltimore) 29: 310-316
2. Fleuren GJ (1987) Ovarian cancer. In: Ruiter DJ, Fleuren GJ, Warnaar SO (eds) Application of monoclonal antibodies in tumor pathology. Nijhoff, Dordrecht, pp 237-254
3. Fleuren GJ, Coerkamp EG, Nap M, Broek van de LCJM, Warnaar SO (1987) Immunohistological characterization of a monoclonal antibody (OV632) against epithelial ovarian carcinomas. Virch Arch [A] 410: 481-486
4. Fleuren GJ, Nap M, Aalders JG, Trimbos JBMZ, Bruijn de HWA (1987) Explanation of the limited correlation between tumor CA125 content and serum CA125 antigen levels in patients with ovarian tumors. Cancer 60: 2437-2442
5. Hermanek P, Sobin LH (eds) (1987) TNM classification of malignant tumours, 4th edn. Springer, Berlin Heidelberg New York
6. Kabawat SE, Bast RC, Bhan AK, Welch WR, Knapp RC, Colvin RB (1983) Tissue distribution of a coelomic-epithelium-related antigen recognized by the monoclonal antibody OC125. Int J Gynecol Pathol 2: 275-285
7. Koelma IA, Nap M, Huitema S, Krom RAF, Houthoff HJ (1986) Hepatocellular carcinoma, adenoma and focal nodular hyperplasia, a comparative histopathologic study with immunohistochemical parameters. Arch Pathol Lab Med 110: 1035-1040
8. Koelma IA, Nap M, Eggink HF, van der Kloet M, Fleuren GJ (1987) Immunocytochemistry of CEA and OV632 in malignant and reactive effusions. 15th annual congress of the International Society for Oncodevelopmental Biology and Medicine, August 1987, Quebec, Canada
9. Koelma IA, Nap M, Rodenburg CJ, Fleuren GJ (1987) The value of tumour marker CA125 in surgical pathology. Histopathology 11: 287-294
10. Martin SE, Moshiri S, Thor A, Vilasi V, Chu EW, Schlom J (1986) Identification of adenocarcinoma in cytospin preparations of effusions using monoclonal antibody B72.3. Am J Clin Pathol 86: 10-18
11. Ravinsky E (1986) Cytology of peritoneal washings in gynecologic patients. Diagnostic criteria and pitfalls. Acta Cytol (Baltimore) 30: 8-16

Immunocytological Features of Malignant Ascites in Ovarian Carcinoma

M. Hilgarth, J. Guzman, K. J. Bross, and U. Costabel

The aim of this study was to determine whether immunocytological investigations of malignant ascites in adenocarcinoma of the ovary with monoclonal antibodies against cell surface antigens would increase the diagnostic efficacy compared with conventional cytological studies.

We studied 35 peritoneal effusions from patients with malignant ascites due to papillary serous adenocarcinoma of the ovary. Four cases showed a mucinous component. For conventional cytological investigations, air-dried smears prepared after centrifugation of the native peritoneal fluid were stained with May-Grünwald-Giemsa stain. For immunocytological studies the adhesive glass slide method was used (Bross et al. 1978; Guzman et al. 1988 a, b). The monoclonal antibodies (MAbs) CEA3-13 against carcinoembryonic antigen (Institut für Immunbiologie, Freiburg), EMA against epithelial membrane antigen (Dakopatt), and BMA-120 against endothelium antigen (Behring) were used. Positively reacting cells were recognized by a dark brown circular staining of the cell membrane (Fig. 1).

In all cases, tumor cells reacting positively with EMA and BMA-120 were present in a proportion ranging from 30% to 95%. In addition, 95% of mesothelial

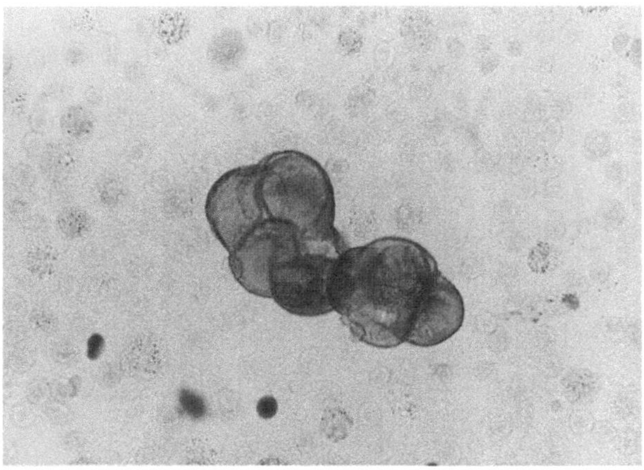

Fig. 1. Cell surface immunoreaction with the monoclonal antibody EMA showing a group of positively reacting tumor cells from papillary serous ovarian carcinoma, ×270

cells were stained with BMA-120. About 5% of the mesothelial cells also reacted with EMA (Fig. 2), but the reaction pattern was different from that of tumor cells. Whereas mesothelial cells lay isolated or in small regular groups of two to four cells of similar size, the tumor cells formed larger clusters of differently sized cells. A positive reaction of more than 40% of the tumor cells with CEA was only observed in the four cases (Fig. 3) with a mucinous component. Twenty-two cases were CEA negative, and in nine cases only a minority (5%–10%) of the tumor cells showed CEA positivity. Mesothelial cells did not react with CEA.

By conventional cytological examination, seven cases were false-negative, one case was suspicious, and 27 cases were positive for tumor cells.

In conclusion, immunocytological studies of malignant ascites can increase the diagnostic yield, as compared with conventional cytological studies. EMA and

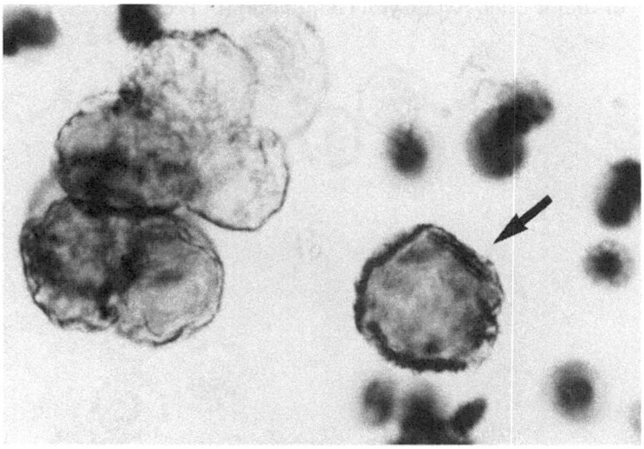

Fig. 2. Cell surface immunoreaction with the monoclonal antibody BMA-120 showing a group of positively reacting tumor cells and one positively reacting mesothelial cell *(arrow)*, ×670

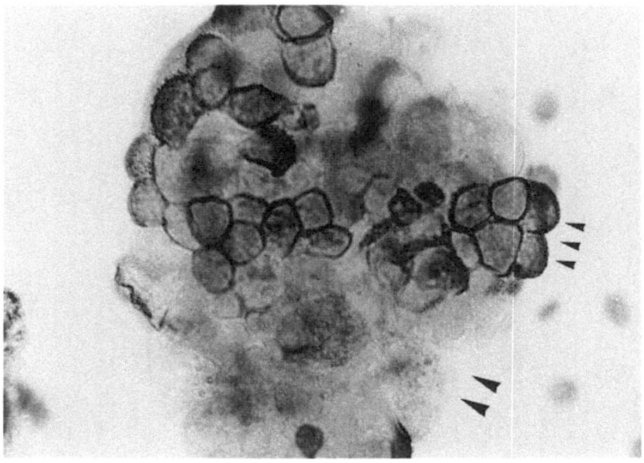

Fig. 3. Cell surface immunoreaction with CEA3-13. Only some tumor cells *(small arrowheads)* show a positive reaction. The majority of the tumor cells *(large arrowheads)* are negative, ×270

BMA-120 are useful antibodies for the recognition of tumor cells of serous papillary adenocarcinoma of the ovary, whereas CEA is not. The fact that BMA-120 (endothelial antigen) reacted with tumor cells and with the majority of mesothelial cells may support the hypothesis of a mesothelial origin of this type of tumor.

References

Bross KJ, Pangalis GA, Staatz C (1978) Demonstration of cell surface antigens and their antibodies by the peroxidase antiperoxidase method. Transplantation 25: 331-334

Guzman J, Costabel U, Bross KJ, Wiehle U, Grunert F, Schaefer HE (1988 a) The value of the immunoperoxidase slide assay in the diagnosis of malignant pleural effusions in breast cancer. Acta Cytol (Baltimore) 32: 188-192

Guzman J, Hilgarth M, Bross KJ, Ross A, Wiehle U, Kresin V, Grunert F, von Kleist S (1988 b) Malignant ascites of serous papillary ovarial adenocarcinoma. An immunocytochemistry study of tumor cells. Acta Cytol (Baltimore) (in press)

VIII. Electron Microscopy

Introduction

W. Olszewski

Recent advances within the field of immunocytochemistry have made it necessary to reconsider the use and value of electron microscopy in cytodiagnostic work.

Looking through the programs of previous European congresses on cytology one can find that almost from the beginning there have been presentations on the usefulness of electron microscopy in cytodiagnosis. Above all, there were numerous presentations and lectures on the value of electron microscopy in the evaluation of effusions and cerebro-spinal fluids (CSF) using both transmission and scanning electron microscopy techniques. The introduction and subsequent wide spread use of fine-needle aspiration biopsy again drew cytopathologist's attention to electron microscopy as a method which could help to solve some diagnostic problems. This method was, for technical reasons, not used in practical desquamative cytology.

As is well known, early researchers in electron microscopy were disappointed - the magic ultrastructural marker that could easily separate malignant tumors from benign ones was not found. On the other hand, it must be stressed that electron microscopy broadened our understanding of numerous problems in cytopathology and pathobiology, e.g., the histogenesis of numerous tumors was clarified. For many reasons, both technical and financial, electron microscopy has never become routinely used in cytopathology laboratories. The main task of cytopathology is, after all, the selection of patients at risk, rather than the fast and simple selection of patients for other diagnostic methods, e.g., surgical pathology.

In spite of the elaboration of simple and, by electron microscopic standards, fast and inexpensive methods, this technique still can only be used in cytopathology or diagnostic cytology in very selected cases. With the advances in our knowledge of tumor biology we are becoming increasingly aware of the limitations of pure morphology, particularly at the light microscopic level but of course at the electron microscopic level as well.

At the moment we are at the start of the era of immunocytochemistry, the method which we hope will fulfill most of our expectations in diagnostic work. In spite of the overwhelming enthusiasm and optimism, immunocytochemistry is not without its problems. It is enough to mention fixation, background staining, cross-reactions, and unspecific staining. From time to time, unfortunately more and more frequently, we also realise that the diagnostic specificities of the new antibodies, both poly- and monoclonals, are not as high as they might be.

It is worth mentioning that we also faced many technical problems 20 or 25 years ago when introducing electron microscopy to pathology and in particular to cytopathology.

In this part, some contributions compare the results obtained with electron microscopy with those achieved by immunocytochemistry methods. Maybe, at least for the time being, there is a place for both methods in solving the problems we face in our everyday routine diagnostic work.

Transmission Electron Microscopy of Cytological Specimens

J. M. Nesland, R. Holm, and J. V. Johannessen

Techniques

Fine needle aspiration is a rapid and convenient way of obtaining tumour cells for both cytological smears and electron microscopy. Specimens from fine needle aspirates have to be immediately fixed, for example in a cacodylate-buffered mixture of 1% glutaraldehyde and 4% formaldehyde (McDowell and Trump 1976), postfixed in buffered 1% osmium tetroxide and dehydrated in graded ethanol and propylene oxide. After infiltration in an Epon/Araldite mixture the blocks are polymerized at 60°-80 °C. Semithin sections are cut with glass knives, stained with toluidine blue and used for light microscopical orientation. Ultrathin sections are cut with diamond knives, contrasted with uranyl acetate and lead citrate and examined under the electron microscope.

Serous fluids must be handled slightly differently from fine needle aspirates. We recommend centrifuging 50 ml fluid for 10 min and pipetting out the supernatant. The remaining pellet is processed for electron microscopy as described above. If abundant blood is present, the serous fluid should be filtered before centrifugation.

Conventional preparation procedures usually require several days, but rapid preparation techniques are also available for electron microscopy (Nesland et al. 1982), enabling the pathologist to examine ultrathin sections within a few hours. When a metastasis from an unknown primary tumour is examined, rapid ultrastructural examination may simplify the search for the primary lesion.

Immunoelectron microscopy is a technically difficult method that combines the high resolution of the electron microscope with the specificity of good monoclonal and purified polyclonal antibodies. One way to avoid destruction of antigens during preparation is to process only one part of the specimen for routine electron microscopy and the other part for cryoultramicrotomy and gold labelling.

Application

The characterization of cells in pleural and peritoneal effusions is a common challenge to the cytologist. Reactive mesothelial cells can be very pleomorphic and resemble malignant cells, and it may even be difficult to tell whether cells present in an effusion are of a mesothelial or an epithelial nature. Histochemical (Ehya 1986; Sterrett et al. 1987) and immunocytochemical (Walts et al. 1983; Mènard et al.

1985; Pinto et al. 1987; Mason et al. 1987; Silverman et al. 1987; Kyrkou et al. 1985; Permanetter and Wiesinger 1987; Ehya 1986) staining methods have been applied to solve these problems, and staining for keratins, carcinoembryonic antigen (CEA) and epithelial membrane antigen (EMA) are particularly popular in routine cytology. The immunostaining results must, however, be interpreted with care.

Although mesothelial cells more often express high molecular weight keratin filaments, keratin filaments may be present in adenocarcinoma cells as well. On the other hand, the CEA and EMA immunoreactivity is usually strong in adenocarcinomas, but even mesotheliomas may present these antigens (Silverman et al. 1987; Mason et al. 1987; Pinto et al. 1987).

Mesothelial cells are usually periodic acid Schiff (PAS)-positive due to their glycogen content, and become PAS-negative after diastase treatment. A negative mucicarmine reaction and positive Alcian blue staining that becomes negative after hyaluronidase treatment are also common hallmarks of mesotheliomas (Ehya 1986). The ultrastructural features of mesothelial cells have been thoroughly studied (Bolen and Thorning 1980; Bewtra and Greer 1983; Herrerra and Wilkerson 1985). Epithelial cells are linked by desmosomes and tight junctions and have irregular nuclei with large nucleoli and deep invaginations of the nuclear membrane. Their abundant cytofilaments occasionally present the typical appearance of tonofilament bundles (Figs. 1, 2). Aggregates of glycogen rosettes are also quite

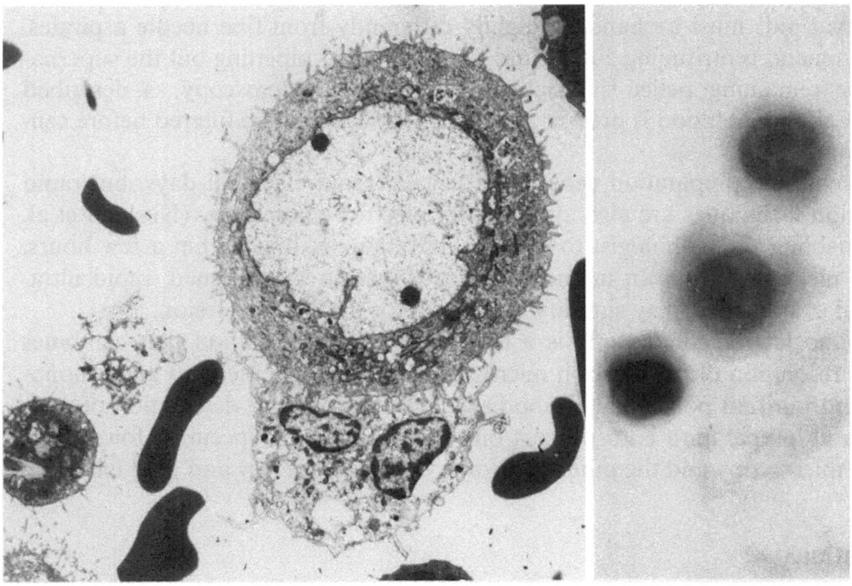

Fig. 1. Electron micrograph of a mesothelial cell in a case of malignant mesothelioma. The irregular nucleus with light-stained chromatin is centrally located. Bundles of filaments are present in the cytoplasm. The cell is equipped with numerous microvilli on the surface. A granulocyte is attached to the mesothelial cell. Uranyl acetate and lead citrate (UA/LC), ×3000. *Inset,* light micrograph of neoplastic cells in the cytoplasmic specimen prepared from left-side pleural fluid in a 62-year-old man. Papanicolaou, ×560

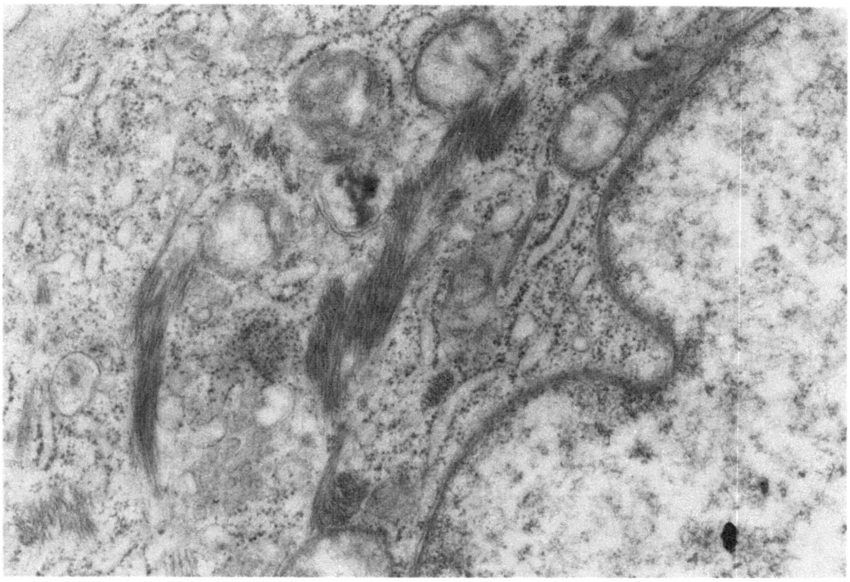

Fig. 2. Bundles of cytofilaments in the perinuclear region. UA/LC, ×26400

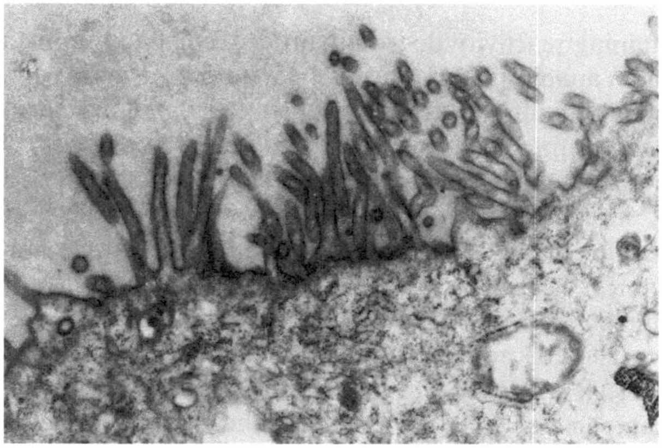

Fig. 3. Detail of mesothelioma cell with abundant microvilli on the surface. UA/LC, ×22000

common in the cytoplasm. Of major diagnostic importance is the presence of abundant, long, slender surface microvilli (Fig. 3).

Adenocarcinoma cells do not have similarly well developed microvilli (Fig. 4) and they usually show mucin vacuoles, excretory granules or intercellular or intracellular lumens. The presence of premelanosomes betrays a metastatic amelanotic malignant melanoma, but usually this tumour is more easily diagnosed with immunostaining. This reveals positivity for protein S-100 and vimentin, and negativity for keratins and carcinoembryonic antigen (Pinto 1986). However, protein S-100

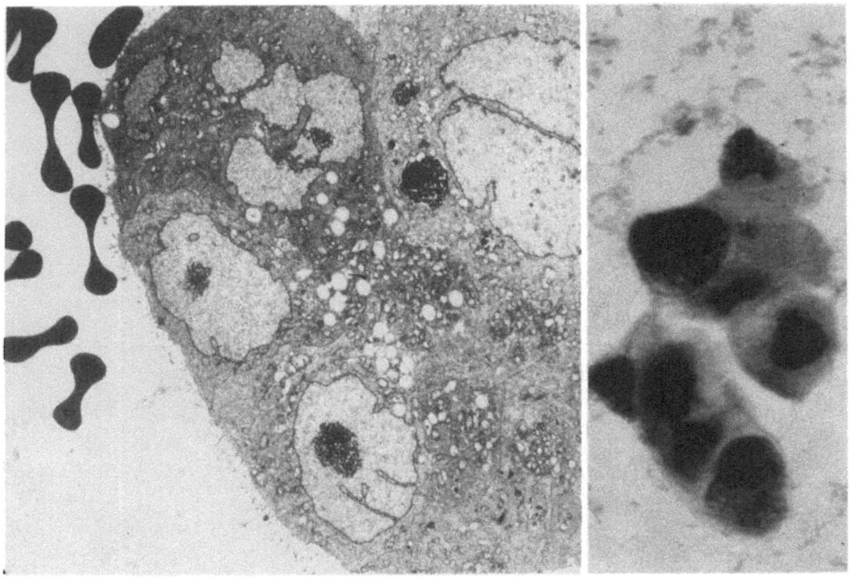

Fig. 4. Group of cells linked by desmosomes and giving an epithelial appearance. Irregular microvilli are present on the surface. UA/LC, ×22000. *Inset,* Neoplastic cells in peritoneal effusion of a 56-year-old woman. Papanicolaou, ×560

staining reactivity does not suffice to diagnose a metastatic malignant melanoma in an anaplastic tumour in a lymph node, as even carcinomas may express protein S-100 immunoreactivity. On the other hand, the needle-happy cytologist may encounter premelanosomes in clear cell sarcomas of tendon sheats, and keratin immunoreactivity may be found in epitheloid and synovial sarcomas.

Macrophages are usually easy to identify by electron microscopy. They have irregular nuclei and abundant cytoplasm loaded with lysosomes and phagolysosomes. They have irregular filipodia and they are never linked together by cell junctions.

There is no pathognomonic ultrastructural sign of malignancy in mesothelial cells. However, if the nuclei are irregular in size, with deep indentations, large, compact nucleoli, and abundant nuclear bodies, malignant behaviour of the mesothelial cells should be suspected. Boon and coworkers (1984) found cytoplasmic lipid vacuoles to be a common feature of malignant mesothelial cells, helping to distinguish them from their benign counterparts and from adenocarcinoma cells.

Electron microscopy is particularly important in three main areas of tumour pathology: small cell tumours, spindle cell tumours and anaplastic tumours (Nesland et al. 1986). Ewing's sarcoma, alveolar rhabdomyosarcoma, neuroblastoma, lymphoma and small cell carcinoma are all included in the group of malignant small round cell tumours. The cytological separation of these tumours is often extremely difficult. The major diagnostic ultrastructural features are shown in Table 1. Ewing's sarcoma does not display any specific diagnostic ultrastructural or immunocytochemical features, but the lack of cytoplasmic features like myofilaments and neuroendocrine granules as well as negative immunostaining for all

Table 1. Ultrastructural features of small round cell tumours

	Cell junctions	Glycogen	Myofilaments	Neuroendocrine granules, Neurotubuli
Ewing's sarcoma	+	+,+++	−	−
Neuroblastoma	+	++,−	−	+++
Alveolar rhabdomyosarcoma	+,−	−,+	+,+++	−
Lymphoma	−	(+),−	−	−
Carcinoma	+,+++	−	−	−,+

−, not present; (+), +, ++, +++, present, increasing amounts

common markers like myosin, desmin, neuron-specific enolase and protein S-100 helps to settle the diagnosis. Ewing's sarcoma is also vimentin-positive. A lymphoma is characterized by positive immunostaining for the common leucocyte antigen and other specific lymphoid cell markers.

Extraskeletal Ewing's sarcomas behave slightly differently from the classical skeletal tumours and often express features more in common with primitive neuroectodermal tumours (Schmidt et al. 1982; Mierau et al. 1985).

Neuroblastomas are usually positive for both protein S-100 and neuron-specific enolase but are often more easily diagnosed by the electron microscopical demonstration of small cytoplasmic electrondense granules of the neuroendocrine type.

Myofilaments are often difficult to find in electron microscopical studies of alveolar rhabdomyosarcomas but, when present, the diagnosis is unquestionable. Alveolar rhabdomyosarcomas are usually more easily recognized by immunostaining for myoglobin, desmin, Z-protein, myosin and vimentin.

Small round cell carcinomas can arise in all epithelial organs and both electron microscopy and immunocytochemistry may be needed to settle the diagnosis (Fig. 5).

Spindle cell tumours include a long list of tumours such as fibrosarcoma, malignant fibrous histiocytoma, leiomyosarcoma, rhabdomyosarcoma, synovial sarcoma, malignant Schwannoma, liposarcoma, haemangiopericytoma, angiosarcoma, mesothelioma, malignant melanoma and spindle cell carcinoma. Kindblom and coworkers (1986) have recently published a thorough study documenting the benefit of combined cytological and ultrastructural examinations in the preoperative diagnosis of soft tissue tumours. This is an extremely difficult group of tumours to classify (Fig. 6). It can also be extremely difficult to decide whether an anaplastic tumour is a carcinoma, a melanoma or even a lymphoma (Fig. 7). Some tumours remain unclassified even after careful ultrastructural examination and staining with a whole battery of antisera.

Electron microscopy may not only be useful in determining the type of tumour present, it may even reveal cell heterogeneity in one and the same tumour. We have become increasingly aware that multidirectional differentiation is a common feature of tumours. An adenocarcinoma may, for example, express neuroendocrine differentiation or squamous cell differentiation, and a medullary thyroid carcinoma may contain thyroglobulin in addition to calcitonin and other hormones

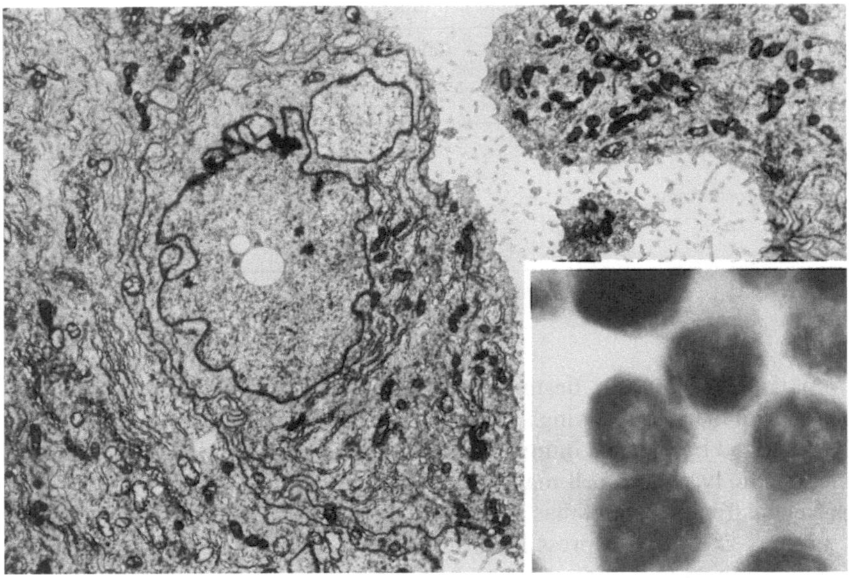

Fig. 5. Metastasis from papillary carcinoma of the thyroid. Note the nuclei with irregular contours and light-stained chromatin. Small microvilli are present in the apical parts of the cells. UA/LC, × 22000. *Inset*, Giemsa-stained cytological specimen from fine needle aspirate of an enlarged lymph node in the neck of a middle-aged woman. Giemsa, × 560

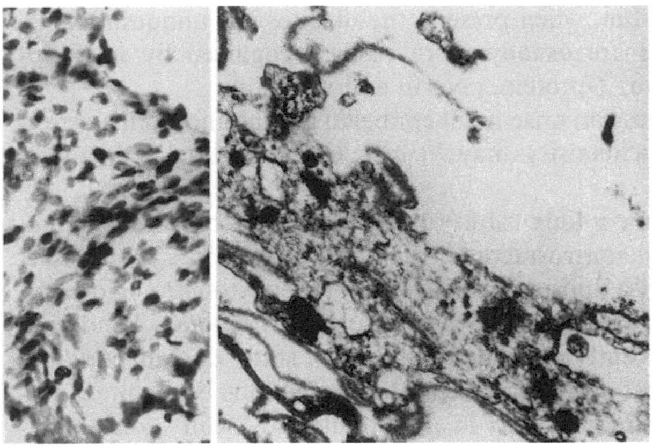

Fig. 6. Detail of neoplastic cells showing cytoplasmic extensions and external lamina material. The findings support the diagnosis of a malignant Schwannoma. UA/LC, × 15680. *Inset*, Fine needle aspirate of a tumour in the thigh of a 42-year-old man. Giemsa, × 200

(Holm et al. 1987). About one-third of all breast carcinomas express neuroendocrine features and electron microscopy reveals characteristic neuroendocrine granules in cells with exocrine features (Nesland et al. 1985). This combination is also demonstrated by immunocytochemistry using double staining with two sets of primary antibodies.

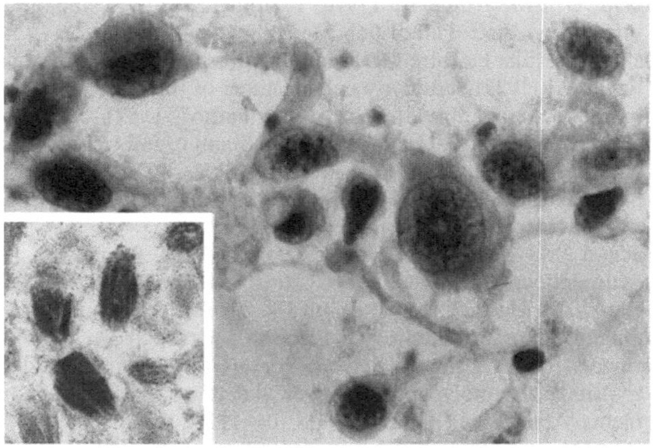

Fig. 7. Anaplastic cells present in the fine needle aspirate of an enlarged lymph node in an elderly woman. Melanin pigment is not present. Giemsa, ×400. *Inset,* Premelanosomes are present in cytoplasm of some of the tumour cells. UA/LC, ×28100

Conclusion

Recent advances in immunocytochemisrty have made it necessary to reconsider the use of electron microscopy in diagnostic cytology work. Some problems are definitely more easily solved by immunocytochemistry than by electron microscopy (EM), but unspecific staining, unspecific antibodies and destruction of antigenic properties may cause diagnostic headaches. Immunocytochemistry supplements but cannot fully replace electron microscopy in diagnostic cytology.

References

Bewtra C, Greer KP (1985) Ultrastructural studies of cells in body cavity effusions. Acta Cytol (Baltimore) 29: 226-238

Bolen JW, Thorning D (1980) Mesotheliomas. A light- and electron-microscopical study concerning histogenetic relationship between the epithelial and the mesenchymal variants. Am J Surg Pathol 4: 451-463

Boon ME, Veldhuizen RW, Ruinaard C, Snieders MW, Kwee WS (1984) Qualitative distinctive differences between the vacuoles of mesothelioma cells and of cells from metastatic carcinoma exfoliated in pleural fluid. Acta Cytol (Baltimore) 28: 443-449

Ehya H (1986) The cytologic diagnosis of mesothelioma. Semin Diagn Pathol 3: 196-203

Herrera GA, Wilkerson JA (1985) Ultrastructural studies of malignant cells in fluids. Diagn Cytopathol 1: 272-285

Holm R, Sobrinho-Simões M, Nesland JM, Sambade C, Johannessen JV (1987) Medullary thyroid carcinoma with thyroglobulin immunoreactivity. A special entity? Lab Invest 57: 258-268

Kindblom L-G, Walaas L, Widèhn S (1986) Ultrastructural studies in the preoperative cytologic diagnosis of soft tissue tumors. Semin Diagn Pathol 3: 317-344

Kyrkou KA, Iatridis SG, Athanassiadou PP, Emmanouilidou AG, Athanassiadis PP (1985) Detection of benign or malignant origin of ascites with combined indirect immunoperoxidase assays of carcinoembryonic antigen and lysozyme. Acta Cytol (Baltimore) 29: 57-62

Mason MR, Bedrossian CWM, Fahey CA (1987) Value of immunocytochemistry in the study of malignant effusions. Diagn Cytopathol 3: 215-221

McDowell EM, Trump BF (1976) Histologic fixatives suitable for diagnostic light and electron microscopy. Arch Pathol Lab Med 100: 405-414

Mènard S, Rilke F, Torre GD, Mariani-Constantini R, Regazzoni M, Tabliabue E, Alasio L, Colnaghi MI (1985) Sensitivity enhancement of the cytologic detection of cancer cells in effusions by monoclonal antibodies. Am J Clin Pathol 83: 571-576

Mierau GW, Berry PJ, Orsini EN (1985) Small round cell neoplasm: can electron microscopy and immunohistochemical studies accurately classify them? Ultrastruct Pathol 9: 99-111

Nesland JM, Millonig G, Wilson A, Johannessen JV (1982) Rapid techniques in diagnostic electron microscopy. Ultrastruct Pathol 3: 295-300

Nesland JM, Memoli VA, Holm R, Gould VE, Johannessen JV (1985) Breast carcinomas with neuroendocrine differentiation. Ultrastruct Pathol 8: 225-240

Nesland JM, Holm R, Johannessen JV (1986) Electron microscopy and immunocytochemistry in diagnostic tumor pathology. J Electron Microsc Techn 4: 130-142

Permanetter U, Wiesinger H (1987) Immunohistochemical study of lysozyme, alpha$_1$-anti-chymotrypsin, tissue polypeptide antigen, keratin and carcinoembryonic antigen in effusion sediments. Acta Cytol (Baltimore) 31: 104-112

Pinto MM (1986) An immunoperoxidase study of S-100 protein in neoplastic cells in serous effusions. Use as a marker for melanoma. Acta Cytol (Baltimore) 30: 240-244

Pinto MM, Bernstein LH, Brogan DA, Criscuolo EM (1987) Carcinoembryonic antigen in effusions. A diagnostic adjunct to cytology. Acta Cytol (Baltimore) 31: 113-118

Schmidt D, Mackay B, Ayala AG (1982) Ewing's sarcoma with neuroblastoma-like features. Ultrastruct Pathol 3: 143-151

Silverman JF, Nance K, Phillips B, Norris HT (1987) The use of immunoperoxidase panels for the cytologic diagnosis of malignancy in serous effusions. Diagn Cytopathol 3: 134-140

Sterrett GF, Whitaker D, Shilkin KB,, Walters MNI (1987) Fine needle aspiration cytology of malignant mesothelioma. Acta Cytol (Baltimore) 31: 185-193

Walts AE, Said JW, Banks-Schlegel S, (1983) Keratin and carcinoembryonic antigen in exfoliated mesothelial and malignant cells: an immunoperoxidase study. Am J Clin Pathol 80: 671-676

Ultrastructural Localization of Epithelial Membrane Antigen and Carcinoembryonic Antigen in Malignant Mesothelial and Epithelial Cells

T. H. van der Kwast, V. D. Vuzevski, M. A. Versnel, M. Delahaye, A. de Jong, and H. C. Hoogsteden

Introduction

Cytologic examination of pleural, ascitic, or pericardial effusions is mostly performed in order to establish the presence of malignancy. Reactive mesothelial cells cannot always be distinguished from malignant epithelial or mesothelial cells merely on the basis of morphological criteria. This diagnostic problem is even more severe when tumor cells are present in very low numbers. The recent advance of immunocytochemistry in cytologic diagnostics is due to the production of a number of monoclonal antibodies which might prove useful in increasing the sensitivity and specificity of cytologic diagnosis. Amongst others, several polyclonal and monoclonal antibodies directed against human milk fat globules (HMFG) or epithelial membrane antigen (EMA) have been generated and used to demonstrate the presence of (malignant) epithelial or mesothelial cells in serous effusions (To et al. 1981). Polyclonal and monoclonal antibodies were also raised to the frequently expressed tumor marker carcinoembryonic antigen (CEA). It has been claimed that the use of CEA-specific antibodies may greatly add to the diagnostic sensitivity in cytologic examination of serous effusions (Sehested et al. 1983).

Distinguishing cytologically between epithelial and mesothelial malignancy has similarly proved to be difficult. The expression of CEA by tumor cells may be helpful in differentiating adenocarcinomas from malignant mesotheliomas, since the latter tumors appear to lack CEA in the majority of cases (Said et al. 1983; Marshall et al. 1984). Unfortunately, a proportion of adenocarcinomas also lack this marker (Sehested et al. 1983). Electron microscopy (EM) permits the identification of subcellular features highly specific for mesothelial cells, distinguishing them from adenocarcinoma cells. Characteristics specific for malignancy, however, are generally not detectable by this technique (Murad 1973; Warhol et al. 1982). Although EM alone is not sufficient for conclusive diagnosis of malignant mesothelioma, immunoelectron microscopy might offer a valuable adjunct in the diagnosis of mesothelial malignancy in serous fluids (Van der Kwast et al. 1988). Here we report on the ultrastructural localization of EMA and CEA in a number of malignant mesotheliomas and adenocarcinomas.

Material and Methods

Material. Material for cytologic, immunocytochemical, and electron and immunoelectron microscopic examination was obtained from patients known to have malignant mesothelioma. In all cases the clinical diagnosis of malignant mesothelioma was confirmed by histological examination and/or EM. In addition, effusions from two patients with an adenocarcinoma of the breast and from one patient with a pulmonary adenocarcinoma were investigated.

Methods. Histochemical, immunocytochemical, and electron and immunoelectron microscopic examinations were performed essentially according to procedures described previously (Van der Kwast et al. 1988). In some postembedding immunoelectron microscopy experiments the immunogold labeling step was enhanced by microwave irradiation in a Miele magnetron oven at 150 W (Zondervan et al. 1988).

Results

Cytology and Immunocytochemistry

Cytologic and histochemical staining reactions performed on smears of serous effusions of 25 patients with clinical, histological, or ultrastructural evidence of malignant mesothelioma were always consistent with the latter diagnosis. In the first case of metastasized breast cancer the effusion contained several clusters of tumor cells while in the other case large numbers of single tumor cells were present. The pleural fluid of the patient with pulmonary adenocarcinoma contained several clusters of tumor cells which morphologically resembled mesothelial tumor cells. Of 25 cases of malignant mesothelioma, 23 showed EMA-positive tumor cells in the smear preparations. In only one case of malignant mesothelioma were a few weakly CEA-positive tumor cells observed. The two fluids containing breast cancer tumor cells and the case of a pulmonary adenocarcinoma were EMA-positive. In the smears of the pulmonary adenocarcinoma and in one of the two cases of breast cancer CEA-positive cells were present.

Electron Microscopy

The plasma membrane of the malignant mesothelioma cells showed long sinuous microvilli and the cytoplasm inconsistently included microfilaments located around the nucleus. A few lipid droplets, lysosomes, mitochondria, and free ribosomes could be seen as well as sometimes large amounts of glycogen granules. The nucleus was irregularly indented and contained a prominent nucleolus. Pulmonary adenocarcinoma cells, in contrast, included numerous mitochondria and secretory vacuoles of various sizes. The mammary adenocarcinoma cells occasionally contained intracytoplasmic lumina and only a limited number of secretory vacuoles.

Immunoelectron Microscopy

Malignant Mesotheliomas. Postembedding immunogold labeling for EMA resulted in heavy labeling of the outer cytoplasmic membranes of the long villi of cells with ultrastructural features of mesothelial cells. The staining for EMA revealed a remarkable discontinuous pattern (Fig. 1). It was striking that the nonvillous surfaces of these malignant mesothelial cells only sporadically showed gold particles. No significant labeling was detectable in the cytoplasm of the tumor cells. Preembedding immunoelectron microscopy yielded essentially the same results. In these Epon-embedded preparations slender, branched villi, glycogen granules, fat droplets, bundles of tonofilaments, and desmosomes could be identified more easily than in the Lowicryl-embedded specimens. Postembedding immunoelectron microscopic examination of pleural fluids containing large numbers of reactive mesothelial cells did not reveal significant numbers of gold depositions, although villous structures were present (Fig. 2).

Since no convincing light microscopic immunoperoxidase staining of mesothelioma cells was found with the CEA-specific antibody, no attempt was made to study its ultrastructural localization in the tumor cells.

Adenocarcinomas. The two effusions with breast cancer tumor cells and the fluid containing pulmonary adenocarcinoma cells were examined using postembedding immunoelectron microscopy with EMA-specific antibody. The cytoplasmic mem-

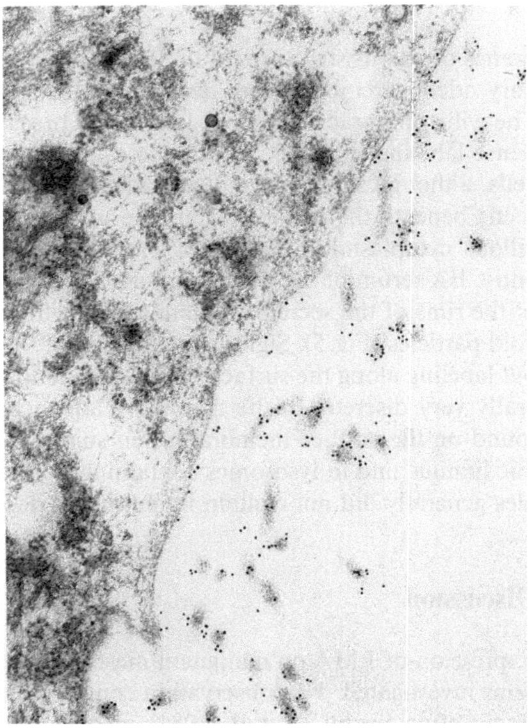

Fig. 1. Malignant mesothelioma cell, embedded in Lowicryl and labeled for EMA. Intense immunogold labeling of the long, slender villi is noted, ×27600

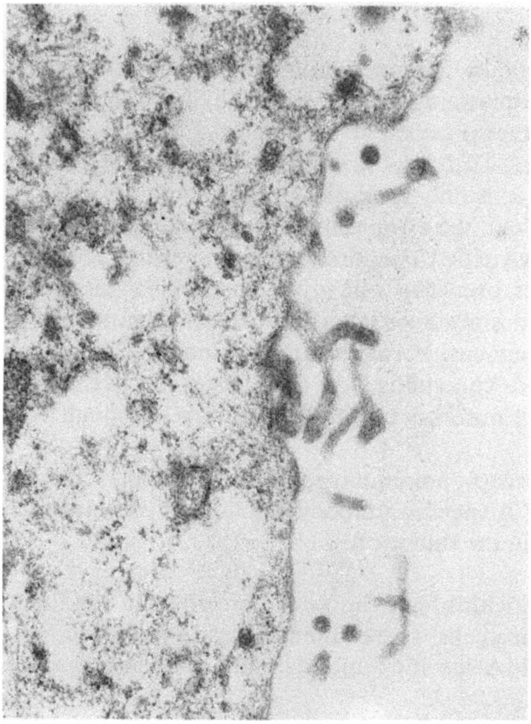

Fig. 2. Reactive mesothelial cell, embedded in Lowicryl. No significant immunogold labeling of villi is seen after incubation with EMA-specific antibody. ×27600

branes of villous structures of tumor cells of both breast cancers and of the pulmonary adenocarcinoma were heavily labeled with EMA-specific antibody (Fig. 3). The villi of intracytoplasmic lumina of breast cancer tumor cells also showed dense labeling (Fig. 4). Cytoplasmic staining was largely absent from these tumor cells, although in the pulmonary adenocarcinoma cells cytoplasmic vacuoles directly beneath the surface membrane were occasionally labeled. Labeling of nonvillous cytoplasmic membranes was focally present. Immunogold staining with anti-CEA serum of the pulmonary adenocarcinoma revealed rather dense labeling of the rims of the secretory vacuoles while their contents only contained sporadic gold particles (Fig. 5). Significant labeling of the cytoplasmic membranes occurred but labeling along the surface of the long slender villi of these tumor cells was generally very discrete. In the mammary adenocarcinoma cells, CEA reactivity was found on the surface membranes, on some villi including those of intracytoplasmic lumina, and in lysosomes containing fibrillar material (Fig. 6). Secretory vacuoles generally did not contain immunogold deposits.

Discussion

Expression of EMA on malignant mesothelial cells was found in 90% of the effusions investigated. This observation confirms the results of an earlier study on two serous effusions by To et al. (1981), who found one of two cases to be positive. In

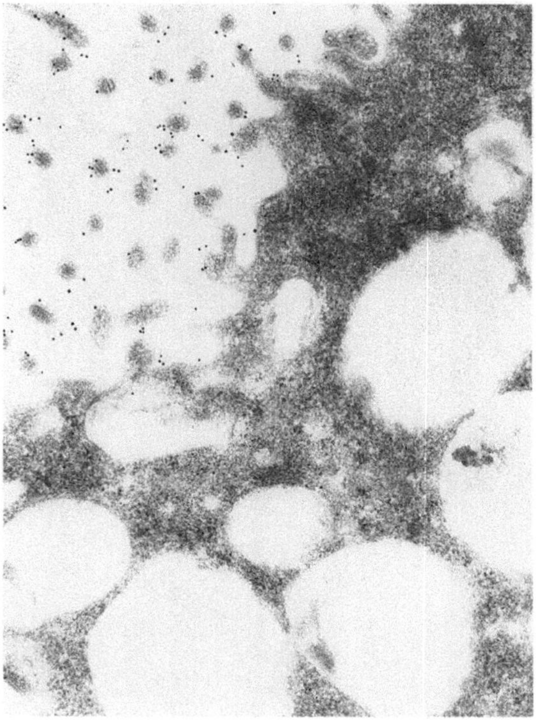

Fig. 3. Pulmonary adenocarcinoma cell, embedded in Lowicryl and labeled for EMA. Intense immunogold labeling of the villi is present, while secretory vacuoles do not show significant numbers of gold particles. ×28000

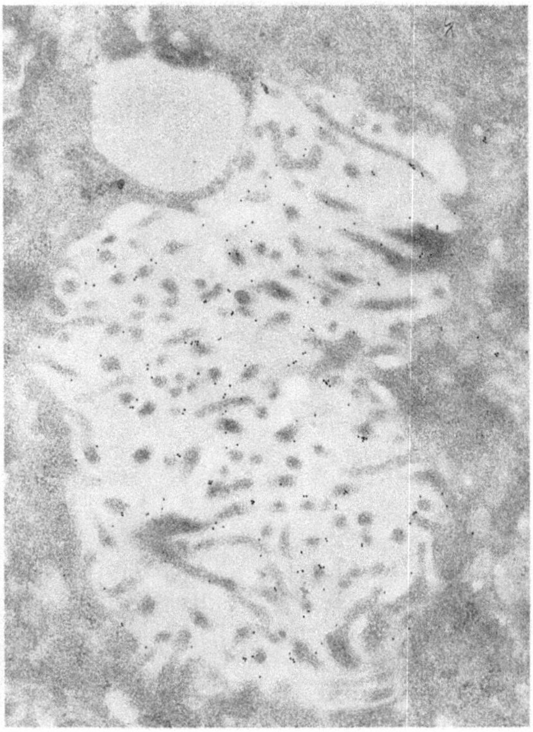

Fig. 4. Intracytoplasmic lumen of Lowicryl-embedded breast tumor cell, labeled for EMA. Intense immunogold labeling of the numerous villi is seen. ×22000

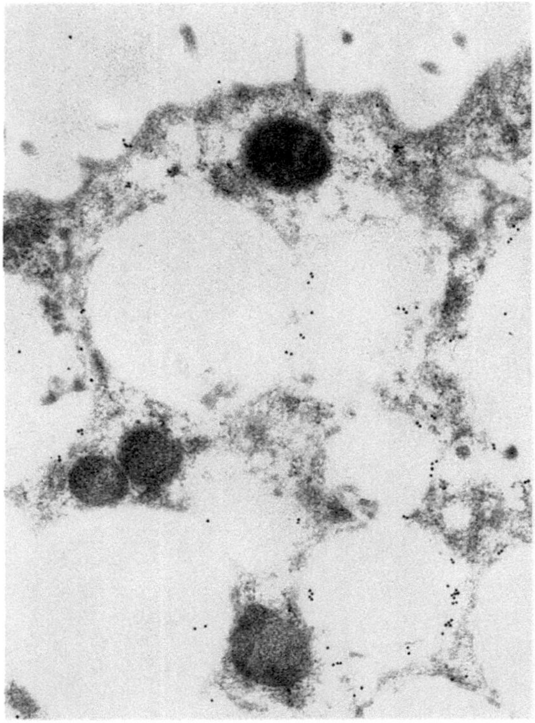

Fig. 5. Pulmonary adenocarcinoma cell, embedded in Lowicryl and labeled for CEA. Gold depositions are mainly present at the margins of the secretory vacuoles. ×26400

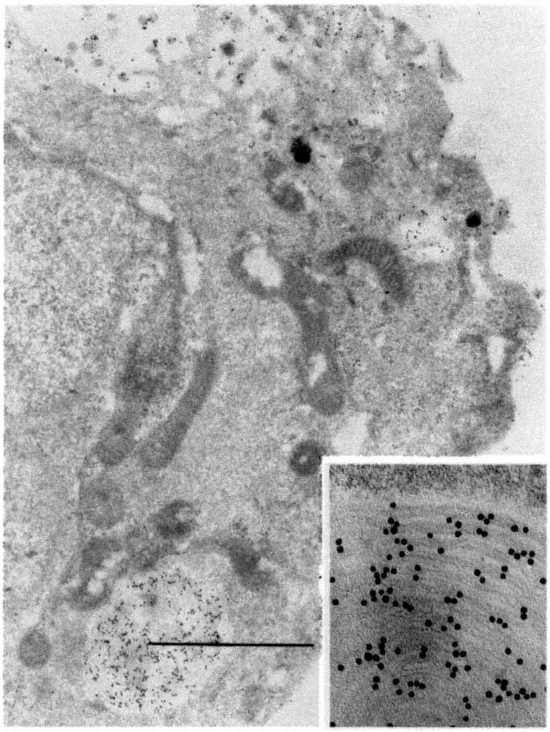

Fig. 6. Lowicryl-embedded breast tumor cell, labeled for CEA. Some immunogold labeling of the surface membrane and intense labeling of a cytoplasmic organelle with fibrillar membraneous contents (see insert, ×80000) is present. ×18000

contrast, Battifora (1984) found the only cytologic case he tested negative. Staining for HMFG glycoproteins or EMA on paraffin-embedded malignant mesothelioma tissue has yielded contradictory data. Some authors have reported that EMA-specific antibodies can be used to distinguish between adenocarcinomas and malignant mesotheliomas since no immunostaining of the latter was observed (Battifora and Kopinski 1985). The use of different EMA- or HMFG-specific antibodies might partially explain these discrepant findings. Parallel testing of a panel of HMFG-specific mouse monoclonal antibodies on a number of normal tissues and antibody-blocking experiments have indicated that at least nine different epitopes can be distinguished (Hilkens et al. 1984). The EMA-specific antibody used in our study (Gatter et al. 1984) did not in most cases react with activated mesothelial cells, as demonstrated previously by Van der Kwast et al. (1988). Since not all EMA- or HMFG-specific antibodies lack reactivity with benign mesothelial cells (Hilkens et al. 1984), the reliability of EMA as an indicator of malignancy in serous effusions depends on the epitope specificity of the antibody used. In most studies no expression or very faint expression of CEA is reported in malignant mesotheliomas (Corson and Pinkus 1982; Whitaker et al. 1984). Only in a few papers was it claimed that malignant mesotheliomas could express strong reactivity with anti-CEA sera (Warhol 1984), but in general reactivity with CEA-specific antibodies is considered to favor the diagnosis of carcinoma. Discrepant results might have resulted from different specificities of antisera or monoclonal CEA-specific antibodies. Conventionally produced polyclonal anti-CEA sera generally show cross-reactivity with nonspecific crossreactive antigen (NCA), biliary glycoprotein I, and blood-group antigens, which may result in false-positive results (Nap et al. 1983). With our absorbed anti-CEA serum none of the 25 cases of malignant mesothelioma tested gave a strong immunostaining reaction. This confirms the notion that CEA-positivity of a tumor virtually excludes the diagnosis of malignant mesothelioma. Although most ultrastructural criteria such as presence of numerous slender villi or glycogen granules are not fully specific for malignant mesothelioma, EM is often used as an ancillary test for differentiation from adenocarcinoma (Warhol et al. 1982). As the value of EM in distinguishing benign and malignant mesothelial cells is doubtful, EMA may be used as a marker of malignancy to serve this purpose.

Preembedding immunoelectron microscopy in particular gave excellent morphological results when combined with specific immunogold labeling. From postembedding immunoelectron microscopic examinations it was apparent that both in the three cases of adenocarcinoma and in the mesotheliomas preferential staining of the villi with anti-EMA occurred (Fig. 1, 3, 4). EMA labeling of cytoplasmic organelles of these adenocarcinoma cells was limited to a few vacuoles immediately below the outer surface of the tumor cells. This observation is in agreement with the data from Hilkens et al. (1984) and Beckford et al. (1985), but in contrast with the results of Major et al. (1987). Using a preembedding immunoperoxidase technique and HMFG-specific antibody, these latter authors found labeling of cytoplasmic vesicles in mammary adenocarcinoma cells, while benign mammary epithelial cells showed exclusively membrane labeling. Obviously, these discrepant results may be explained by differences in technique and in the epitope specificity of the antibodies used. In contrast to the intense EMA-positivity of the villous membranes of the pulmonary and mammary adenocarcinoma cells, these structures were only focally CEA reactive. The nonvillous membranes of the pulmonary and mammary adenocarcinoma cells demonstrated focal, though significant, CEA reactivity (Figs. 5, 6), while membranes of the secretory vacuoles were strongly CEA reactive only in the pulmonary adenocarcinoma cells. Extracellular positivity for CEA has also been described for colonic and gastric cancer cells, while mammary adenocarcinoma cells were shown to express CEA preferentially in electron-dense organelles (Wolf et al. 1986). This latter finding was confirmed in the one case of breast cancer tested in our study (Fig. 6).

Conclusions

Ultrastructurally, expression of EMA by malignant epithelial as well as by malignant mesothelial cells was observed to occur preferentially on the cytoplasmic membranes of villous structures. In contrast, secretory vacuoles were not labeled by EMA-specific antibody. A more heterogeneous picture emerged from studying expression of CEA in two cases of adenocarcinoma. In both the pulmonary and mammary adenocarcinoma cells CEA was found on the surface membranes, while in pulmonary adenocarcinoma cells intense labeling of secretory vacuoles occurred.

References

Battifora H (1984) Recent progress in the immunohistochemistry of solid tumors. Semin Diagn Pathol 1: 251-271

Battifora H, Koprinski MI (1985) Distinction of mesothelioma from adenocarcinoma: an immunohistochemical approach. Cancer 55: 1679-1685

Beckford U, Barbatis C, Beesley JE, Linsell JC, Chantler SM (1985) Localisation of Ca and HMFG2 antigens in breast tissue by immunoperoxidase, immunofluorescence, and immunoelectron microscopy. J Clin Pathol 38: 512-520

Corson JM, Pinkus GB (1982) Mesothelioma: profile of keratin proteins and carcinoembryonic antigen. An immunoperoxidase study of 20 cases and comparison with pulmonary adenocarcinomas. Am J Pathol 108: 80-88

Gatter KC, Alcock C, Herget A, Pulford KA, Heyderman E, Taylor-Papadimitriou J, Stein H, Mason DY (1984) The differential diagnosis of routinely processed anaplastic tumors using monoclonal antibodies. Am J Clin Pathol 82: 33-43

Hilkens J, Buijs F, Hilgers J, Hageman P, Calafat J, Sonnenberg A, Van der Valk M (1984) Monoclonal antibodies against human milk-fat globule membranes detecting differentiation antigens of the mammary gland and its tumors. Int J Cancer 34: 197-206

Major P, Lavallee M, Minassian H, Kovac P, Wang N-S (1987) Ultrastructural localization of a breast tumor-associated antigen. J Histochem Cytochem 35: 375-379

Marshall RJ, Herbert A, Braye SG, Jones DB (1984) Use of antibodies to carcinoembryonic antigen and human milk fat globule to distinguish carcinoma, mesothelioma, and reactive mesothelium. J Clin Pathol 37: 1215-1221

Murad TM (1973) Electron microscopic studies of cells in pleural and peritoneal effusions. Acta Cytol (Baltimore) 17: 401-409

Nap M, Ten Hoor KA, Fleuren G-J (1983) Cross-reactivity with normal antigens in commercial anti-CEA sera, used for immunohistology. The need for tissue controls and absorptions. Am J Clin Pathol 79: 25-31

Said JW, Nash G, Tepper G, Banks-Schlegel S (1983) Keratin proteins and carcinoembryonic antigen in lung carcinoma: an immunoperoxidase study of fifty-four cases, with ultrastructural correlations. Hum Pathol 14: 70-76

Sehested M, Ralblejaer E, Rasmussen J (1983) Immunoperoxidase demonstration of carcinoembryonic antigen in pleural and peritoneal effusions. Acta Cytol 27: 124-127

To A, Coleman DV, Dearnaley DP, Ormerod MG, Stelle K, Neville AM (1981) Use of antisera to epithelial membrane antigen for the cytodiagnosis of malignancy in serous effusions. J Clin Pathol 34: 1326-1332

Van der Kwast TH, Versnel M, Delahaye M, De Jong A, Zondervan PE, Hoogsteden H (1988) Expression of epithelial membrane antigen on malignant mesothelioma cells. An immunocytochemical and immunoelectron microscopic study. Acta Cytol (Baltimore) 32: 169-174

Warhol MJ (1984) The ultrastructural localization of keratin proteins and carcinoembryonic antigen in malignant mesotheliomas. Am J Pathol 116: 385-390

Warhol MJ, Hickey WF, Corson JM (1982) Malignant mesothelioma: ultrastructural distinction from adenocarcinoma. Am J Surg Pathol 6: 307-314

Whitaker D, Sterrett GF, Shilkin KB (1982) Detection of tissue CEA-like substance as aid in the differential diagnosis of malignant mesothelioma. Pathology 14: 255-258

Wolf B, Schwinde A. Thompson J, Von Kleist S (1986) Ultrastructural localization of carcinoembryonic antigen (CEA) on ultrathin cryosections of human tumours. Anticancer Res 6: 91-96

Zondervan PE, De Jong A, Sorber CWJ, Kok LP, De Bruijn WC, van der Kwast TH (1988) Microwave-stimulated incubation in immunoelectron microscopy. A quantitative study. Histochem J (in press)

On the Fine Structural Differential Diagnosis of Primary and Metastatic Lymph Node Tumors with Special Reference to the Nucleus

I. Valkov

Introduction

A great deal of research on the use of transmission electron microscopy (TEM) for tumor typing, e.g., for revealing tumor histogenesis, has occurred in the last 20 years and many publications, including general reviews, have appeared (Bernhard 1969; Ghadially 1975; Reichlin et al. 1981; Krustev et al. 1982; Denk 1987). As a rule, the use of cytoplasmic features for diagnostic purposes has been stressed.

Discrimination between some kinds of non-Hodgkin's malignant lymphomas[1] (NHML) and anaplastic and poorly differentiated epithelial tumor metastases is one of the main points in lymph node diagnostics. According to our experience, routine cytological examination (fine needle aspiration biopsy and imprints) enables almost 89% of malignancies of to be discovered. In 11.2% of true-positive cases differentiation between primary and secondary (metastatic) tumors was difficult, even at a histological level, and additional techniques including immunohistochemistry and electron microscopy had to be applied (Valkov et al. 1987).

Material

The present investigation deals with samples from 61 lymph node biopsies examined using TEM. There were 29 cases of NHML, 12 of them being of low grade malignancy and 17 of high grade (Table 1). The metastatic tumors examined are listed in Table 2. Table 3 indicates the most essential types of lymphoma and metastic tumor that must be differentiated.

Methods

TEM was used as part of a five-step system (Table 4) being applied in our laboratory to improve the diagnosis of lymph node lesions. Semithin sections were used as an additional but precise histological method. The different types of ribonucleo-

[1] Terminology used here for malignant lymphomas is according to the new Kiel classification (4th workshop of the European lymphoma study group, Brussels, October 1986).

Table 1. Non-Hodgkin's malignant lymphomas

Chronic lymphatic leukemia (atypical)	6
Immunocytic lymphoma	4
Centroblastic/centrocytic lymphoma	2
Centroblastic lymphoma	1
Immunoblastic lymphoma	4
Lymphoblastic lymphoma	8
T-large cell pleomorphic lymphoma	1
Giant cell lymphoma	2
Histiocytic sarcoma	1
Total	29

Table 2. Metastatic tumors

Small cell anaplastic cancer	6
Poorly differentiated squamous cell cancer	6
Poorly differentiated adenocarcinoma (including adenosquamous cancer)	16
Neuroblastoma	4
Total	32

Table 3. Essential differential diagnoses

Chronic lymphocytic leukemia (atypical)	Anaplastic small cell cancer
Centroblastic malignant lymphoma	Neuroblastoma
Lymphoblastic malignant lymphoma	
Centroblastic/centrocytic malignant lymphoma	Poorly differentiated squamous cell cancer
Histiocytic sarcoma	
Immunocytic malignant lymphoma	Poorly differentiated adenocarcinoma
Immunoblastic malignant lymphoma	
Giant cell malignant lymphoma	
T-large cell pleomorphic malignant lymphoma	Anaplastic large cell cancer

Table 4. Five-step system for lymph node examination

1	Screening of suspected cases (express cytology - Diff-Quick)
2	Routine cytology (Pappenheim)
3	Histology (hematoxylin and eosin, V.Gieson, periodic acid-Schiff, Giemsa, Gömöry; semithin sections - methylen blue)
4	Immunomorphology (monoclonal and monospecific antibodies, immunofluorescence - imprints, frozen sections; peroxidase antiperoxidase - dab smears, paraffin blocks)
5	Transmission electron microscopy (1,6-glutaraldehyde, Sörensen buffer pH 7.3, osmium tetroxide, Durcupan, uranyl acetate, lead citrate)

protein structures in 640 random nuclear profiles from 19 biopsy specimens (9 with NHML and 10 with cancer metastases) were simply counted.

Results and Discussion

In small cell tumors the most frequent problem in cytological examinations was differentiation between lymphoblastic malignant lymphoma and anaplastic small cell cancer metastases (Fig. 1). A tendency for the cells to gather in groups and the coarser chromatin pattern in cancer cells helped the discrimination but were not sufficient to allow a definitive decision. Even with histological examination using semithin sections the difference was not clear in some cases (Fig. 2). Using TEM we looked for the classic cytoplasmic and cell membrane features of cell differentiation that indicate the histogenesis: desmosomal contacts between cells, the cytoplasm rich in polyribosomes in immunoblastic lymphomas, the abundance of rough endoplasmic reticulum cysternae in adenocarcinomas but also in immunocytic lymphomas, secretory granules in apudomas, melanosomes and premelanosomes in achromatic melanomas, etc.

The nuclear pattern of tumor cells was the subject of special attention, in particular the peculiarities of the main nuclear ribonucleoprotein structures: nucleolus, nuclear bodies, and perichromatin granules.

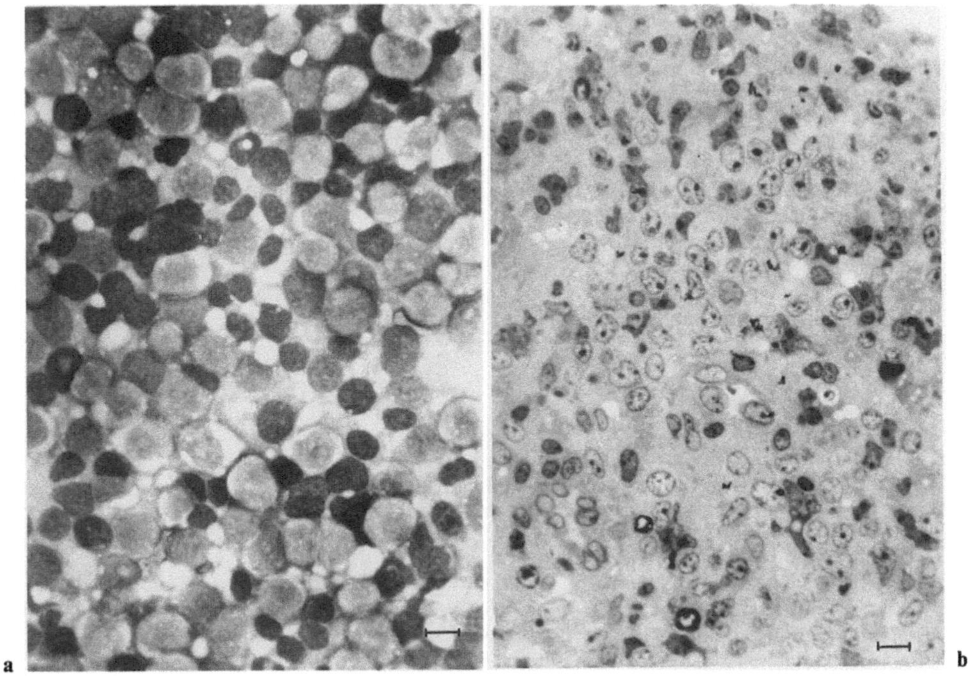

Fig. 1 a, b. B-cell lymphoblastic malignant lymphoma, **a** Imprint (Giemsa); *bar*, 10 μm. **b** Semithin section (methylene blue); *bar*, 10 μm

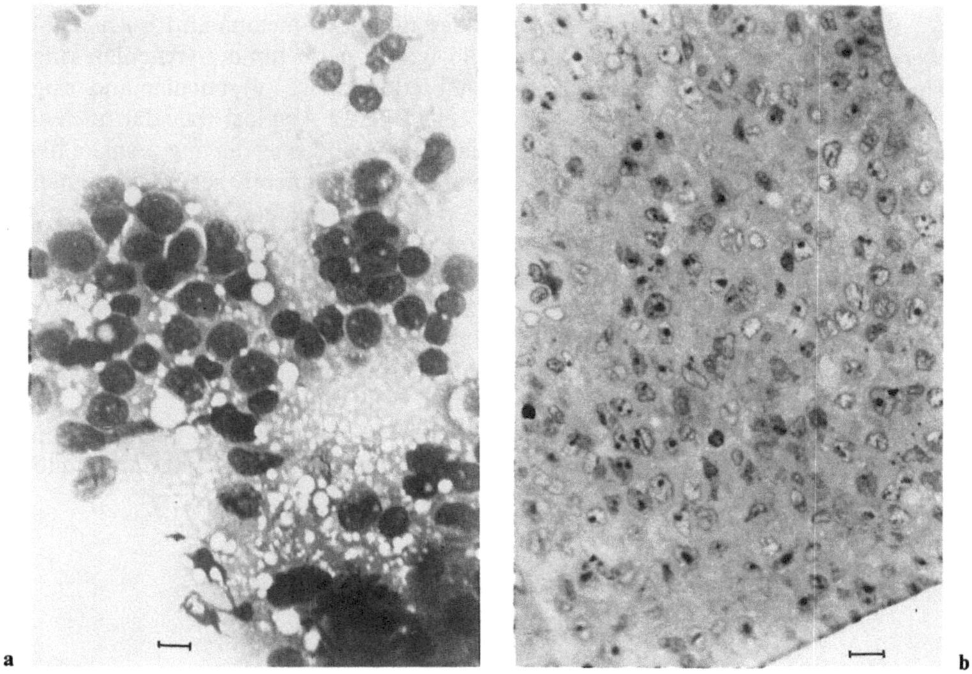

Fig. 2a, b. Small cell anaplastic lung cancer metastasis. **a** Imprint (Giemsa); *bar,* 10 μm. **b** Semithin section (methylene blue); *bar,* 10 μm

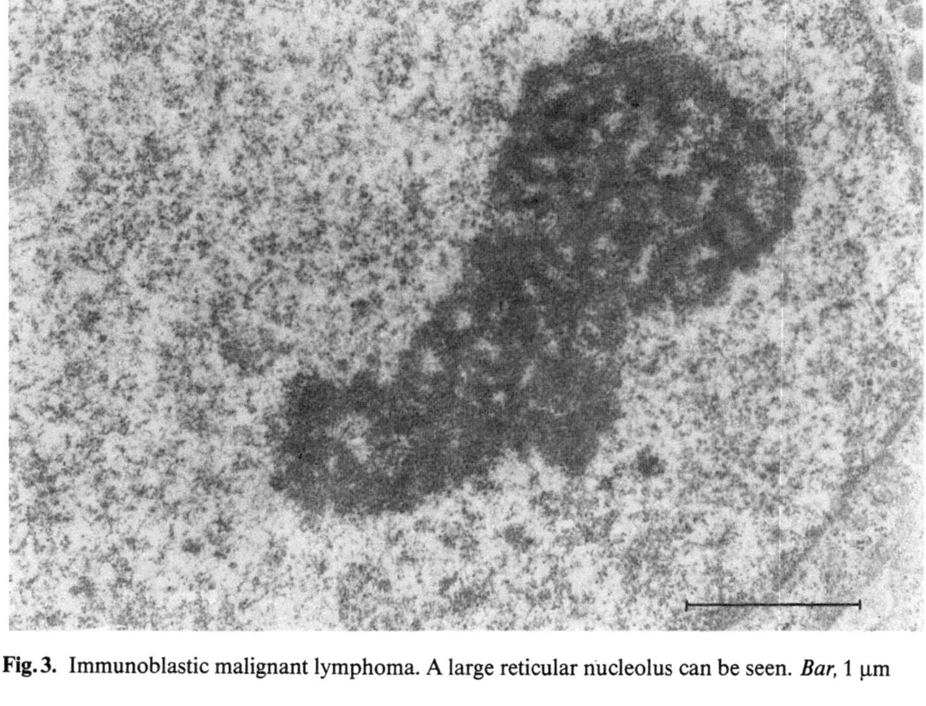

Fig. 3. Immunoblastic malignant lymphoma. A large reticular nucleolus can be seen. *Bar,* 1 μm

Regarding the nucleolus, all the types described by Smetana and Bush (1974) were seen in both primary and metastatic lymph node tumors (reticular, ring-shaped, compact, and micronucleolus). A predominance of reticular and ring-shaped nucleoli was observed in NHML cells (Fig. 3). Atypical reticular nucleoli such as spotted ones (described by Bernhard 1969) and ones having a snake-like nucleolonema (first described by us) were also more frequently seen in malignant lymphoma cells (Fig. 4). A submembranous position of reticular nucleoli in centroblasts was also noticed in cases with centroblastic/centrocytic malignant lymphoma (Fig. 5). Large compact nucleoli sometimes accompanied by deep nuclear invaginations (up to pseudomultinucleation; Fig. 6) were found mostly in metastatic tumor cells. Two types of atypical nucleoli (pseudoannular and multichamber, previously observed in embryonic and cancer cells; Valkov et al. 1982, 1984) were seen in a small percentage of metastatic cancer cells (Fig. 4).

Nuclear bodies were very often found in two examined groups of tumors, but in NHML cells they were found to be disposed singly and in metastatic tumor cells they were sometimes arranged in chains (Fig. 7).

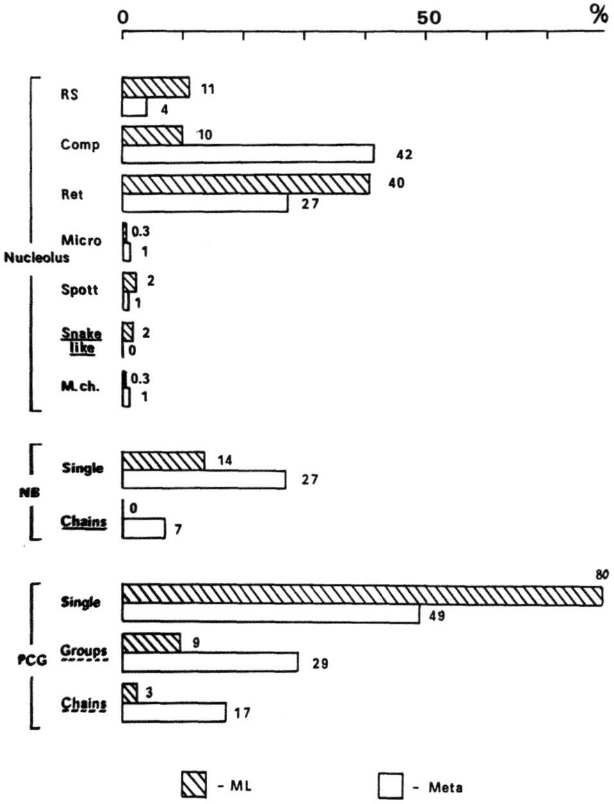

Fig. 4. Percentage occurrence of different kinds of ribonucleoprotein structures in nuclear profiles. *RS,* ring shaped; *Comp,* compact; *Ret,* reticular; *Micro,* micronucleolus; *Spott,* spotted; *Snake like,* with snake-like nucleolonema: *M. ch.*, multichamber; *ML,* malignant lymphomas; *NB,* nuclear body; *PCG,* perichromatin granules; ▨, malignant lymphoma; □, metastases

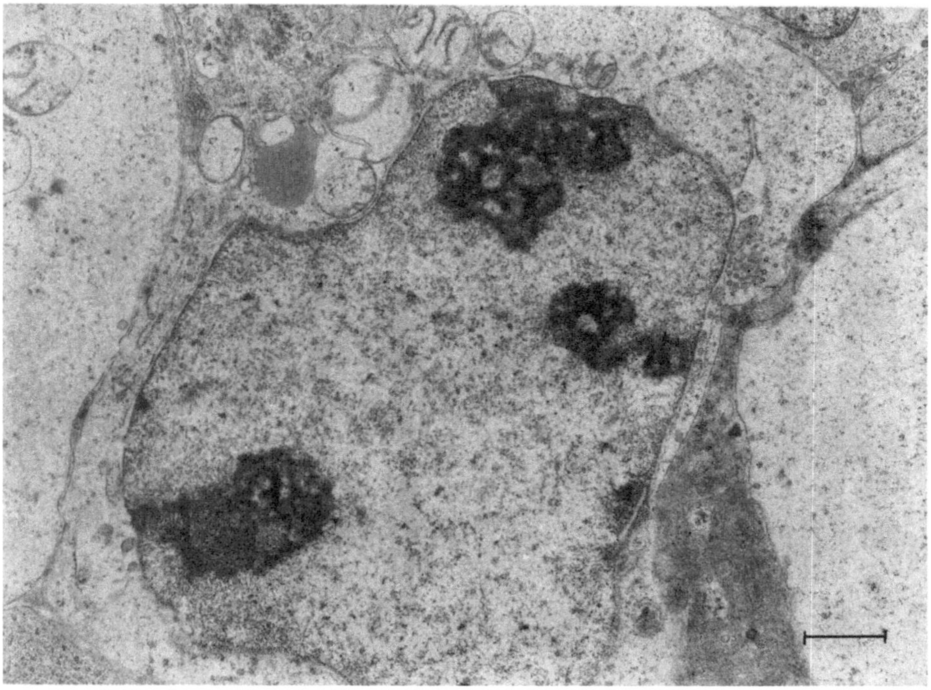

Fig. 5. Centroblastic/centrocytic malignant lymphoma. The centroblast has three large submembranous reticular nucleoli. *Bar*, 1 μm

Perichromatin granules were pleomorphic in most cells in both NHML and metastatic tumors; this pattern has been accepted as a malignancy marker (Valkov and Vassilev 1977). The granules were more frequently found to be gathered in clusters or in chains (Fig. 8) in metastatic tumor cells than in malignant lymphoma cells. The above-mentioned data are summarized in Fig. 4.

Many investigations concerned with the biological role of the nuclear ribonucleoprotein structures have suggested that their morphological appearances are connected with different steps of RNA metabolism. It has been accepted that ring-shaped nucleoli are less active in rRNA synthesis than reticular or compact ones as result of the different relationships between the fibrillar centers and the nucleolonema (Goessens 1984, Jordan 1984). Indeed, ring-shaped nucleoli are more characteristic of low grade malignant lymphomas, especially lymphocytic ones, than of high grade ones, where reticular nucleoli are the most common type. The function of nuclear bodies has not yet been clarified, but it has been suggested by Dupuy-Coin and Bouteille (1975) that they are involved in nucleocytoplasmic exchange of proteins. On this basis we have suggested that disturbances of this exchange in cancer cells may be give rise to the arrangement of the nuclear bodies in chains, this being an expression of pathologic functioning of the nucleocytoplasmic pathway (Valkov et al. 1984). The appearance of pleomorphic perichromatin granules in tumor cells could be related to disturbances in the synthesis or processing of rRNA and mRNA. Similar but not identical pleomorphisms and the

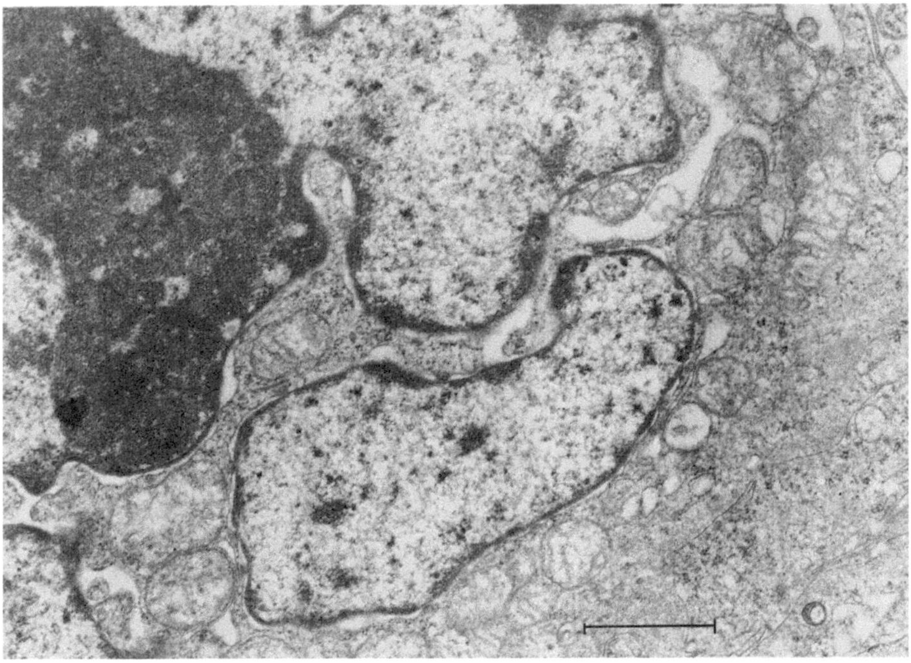

Fig. 6. Poorly differentiated squamous cell cancer. Deep nuclear invaginations give the appearance of pseudomultinucleation and there is a large compact nucleolus attached to the nuclear membrane. *Bar,* 1 μm

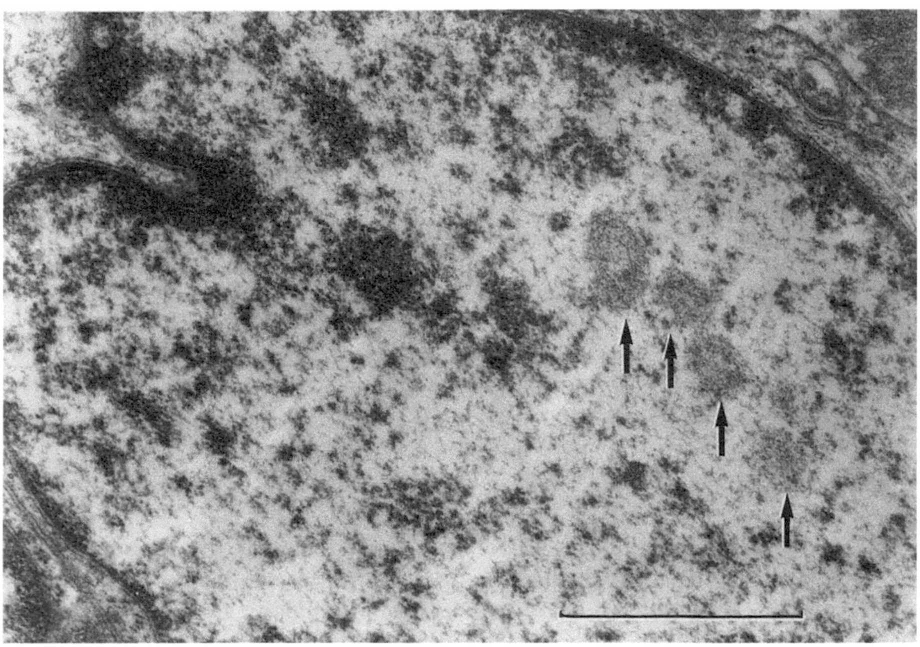

Fig. 7. Anaplastic small cell cancer. There is a chain of simple nuclear bodies *(arrow heads)*. *Bar,* 1 μm

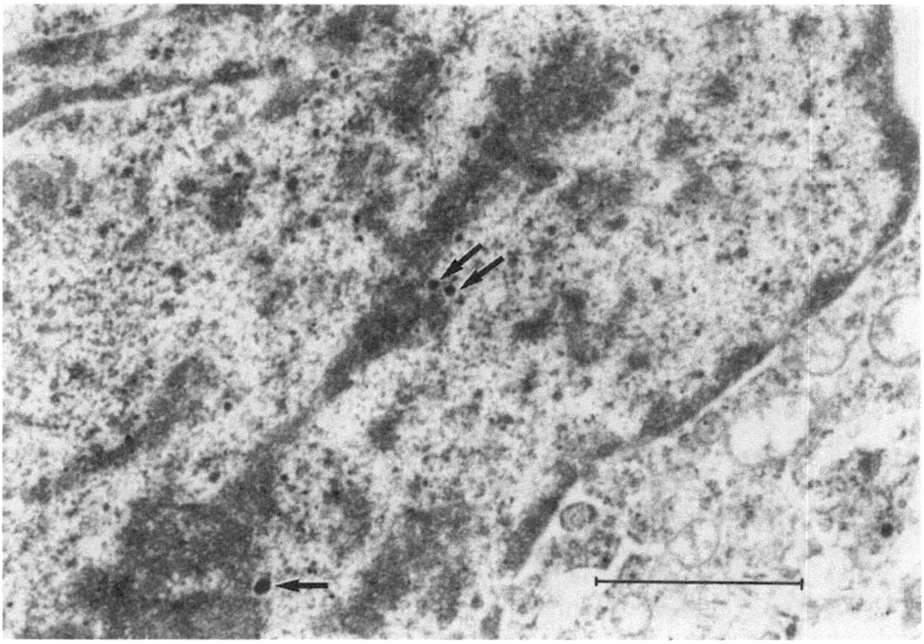

Fig. 8. Poorly differentiated adenocarcinoma. Many pleomorphic perichromatin granules *(arrow heads)* are clustered in the middle. *Bar,* 1 μm

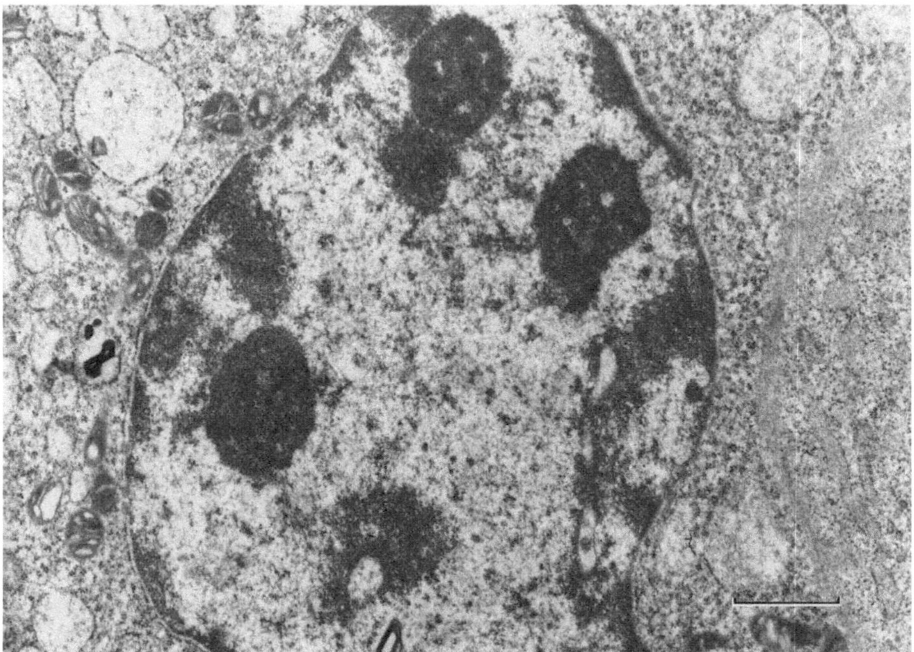

Fig. 9. Immunoblastic malignant lymphoma. Myelin-like figures can be seen both in mitochondria and in the nucleus. *Bar,* 1 μm

appearance of perichromatin-like granules have been shown to occur with treatment with cyclohexamide, which is an agent that blocks RNA synthesis (Moyne et al. 1977). According to the hypothesis of Cervera et al. (1983), perichromatin granules play a role in hnRNA processing and their accumulation is related to the block in the transport of hnRNA (Cervera et al. 1985). In any case, it is probable that the degree of disturbance of RNA metabolism is not identical in different kinds of tumor cells as perichromatic granules occur singly in some and gathered in groups or chains in others. A certain predominance of groups and chains was noticed in cancer cells, especially in adenocarcinoma cells. Evidently this is one diagnostic feature with some value.

Some unusual nuclear inclusions evidently originating from the cytoplasm could be also helpful in diagnosis. The presence of cytokeratin or tubular inclusions (Krustev et al. 1982) has been shown to be characteristic of cancer cells. In rare cases in the material presented here we observed nuclear inclusions such as myelin-like figures identical to those in the mitochondria and areas of fine fibrillar material located under nuclear membrane (Fig. 9). The latter was similar to the contents of the rough endoplasmic reticulum cysternae. The above-mentioned inclusions were seen in one case of immunoblastic lymphoma and in one case of immunocytic malignant lymphoma.

In conclusion it must be pointed out that the use of TEM is a valuable method for discrimination between NHML and anaplastic tumor metastases in the lymph nodes in difficult cases. Careful examination of the nucleus may be helpful and examination using TEM should be part of the complex of ultrastructural investigative techniques used to reveal the tumor histogenesis.

References

Bernhard W (1969) Ultrastructure of cancer cells. In: Lima de Faria A (ed) Handbook of molecular cytology. North-Holland, Amsterdam, pp 688-715

Cervera J, Almar M, Martinez A, Renau-Piqueras J (1983) Nuclear alterations induced by cadmium chloride and L-canavaline in HeLa S3 cells. Accumulation of perichromatin granules. J Ultrastruct Res 82: 241-263

Cervera J, Baguena-Cervellera R, Martinez A (1985) The effect of zink chloride on the ribonucleoprotein structures in Hep2 cells. Accumulation of perichromatin granules. J Ultrastruct Res 93: 129-137

Denk H (1987) Immunological heterogeneity of malignant tumors. Pathol Res Pract 182: 483

Dupuy-Coin A-M, Bouteille M (1975) Protein renewal in nuclear bodies, as studied by quantitative ultrastructural autoradiography. Exp Cell Res 90: 111-118

Ghadially FN (1975) Ultrastructural pathology of the cell. Butterworths, London, pp 1-91

Goessens G (1984) Nucleolar structure. Int Rev Cytol 87: 107-158

Jordan EG (1984) Nucleolar nomenclature. J Cell Sci 67: 217-220

Krustev L, Valkov I, Raitchev R, Prokopanov C (1982) Ultrastructural basis of pathology (in Bulgarian). Medizina i Fizkultura, Sofia, Bulgaria

Moyne G, Nash E, Puvion E (1977) Perichromatin granules in isolated rat hepatocytes treated with cortisol and cyclohexamide. Biol Cell 30: 5-16

Reichlin NT, David H, Lapis K (1981) Ultrastructure of tumors in man (in Russian). Meditsina, Moskva, pp 13-20

Smetana K, Bush H (1974) The nucleolus and nucleolar DNA. In: Bush H (ed) The cell nucleus. Academic, New York, pp 75-147

Valkov I, Vassilev N (1977) Comparative morphologic and morphometric study on the perichromatin granules in normal and tumor lymph node cells. C R Acad Bulg Sci 30: 1347-1350
Valkov I, Radeva V, Roussev G (1982) Nuclear ultrastructure of the newt's neuroepithelial cells after treatment with embryoactive substances. Verh Anat Ges 76: 237-238
Valkov I, Topov J, Vassilev N (1984) Nuclear ribonucleoprotein structures in malignancy. Proceedings, 13th European congress on cytology, Varna, Bulgaria, Sept 27-29, p 204
Valkov I, Popov A, Goranova I (1987) On the complex approach in the differential diagnosis between primary and metastatic lymph node tumors. Pathol Res Pract 182 (4): 570-571

Immunologic Reactions on Cell Surfaces Investigated with Scanning and Transmission Electron Microscopy Using Replicas and Ultrathin Sections

D. Hoder, H.-J. Scholman, A.-M. Lajous-Petter, and S. Blümcke

Introduction

The aim of any morphologic preparation is selectively to present cells and specific structural elements. Immunohisto- and immunocytochemical methods introduced in recent years have led to great progress in this area. Nevertheless, many problems remain unsolved; for instance the antibodies thus far developed only enable cells to be classified to their tissue (e.g., epithelial, nerve, muscle, mesenchymal) and not to their organ of origin. This deficiency is hindering research on malignant tumors. With light microscopy it is often not possible to distinguish benign and malign cells by direct immunocytochemical methods and the developmental stages of various kinds of cells cannot be determined accurately enough.

It is possible that ultrastructural features could be found which, through immunologic reactions, could enable demonstration of characteristics that facilitate cell differentiation. For these reasons, the surfaces of outer membranes of epithelial cells, lymphocytes, and leukocytes were studied with different immunoelectronmicroscopic procedures and the results of these methods were compared.

Material and Methods

Blood lymphocytes were taken from patients with chronic lymphatic leukemia, blood leukocytes from patients with chronic myelocytic leukemia, urine sediments from material delivered to our institute, and oral mucous membrane swabs from the authors.

The primary antibodies used were the monoclonal antibodies (MAbs) EMA (Dako) against epithelial membrane antigen, LC (Dako) against common leukocytic antigen, and Pan-B (Dako) against B-lymphocyte antigen, each diluted 1:20. The secondary antibodies were rabbit anti-Mouse IgG (RaMIgG; Dako) and home-made protein A-gold (12 nm goldcitrate; see Slot and Geuze 1981), and an immunogold reagent (5 nm Auroprobe EM G5, Janssen).

Cell separation was followed by cytocentrifugation of the cells (Shandon Cytospin 2, 700 rpm, 5 min) on coated cover-slips or carbon chips (Abbott tissue adhesive 1% or alcian blue 1%), fixation in 3% paraformaldehyde plus 0.5% glutaraldehyde in 0.1 *M* phosphate-buffered saline (PBS; 5 min), incubation with the primary MAb (EMA, LC, or Pan-B, 16 h, 4 °C) and the secondary label either with

RaMIgG and 12 nm protein A-gold in succession (1 h each, 20 °C) or directly with 5 nm Auroprobe (1 h, 20 °C). After each preparatory step the samples were thoroughly washed with 0.1 *M* PBS (pH 7.3), taking care not to allow them to dry out. After dehydration with graded alcohols, scanning electron microscope (SEM) specimens and replicas were critical-point dried, coated with gold (SEM) or carbon (energy dispersive analysis of X-rays, EDX), and double coated with platinum or carbon respectively; further treatment of the replicas was according to Hohenberg and Mannweiler (1980). The transmission electron microscope (TEM) preparations were postfixed with 2% OsO_4, embedded in Araldite, and contrasted with uranyl acetate and lead citrate. For further details of the preparation process see Lajous-Petter et al. (1986). The electron microscopes used were an EM 10 B (Zeiss) and a Stereoscan 250 (Cambridge).

Results

Cells were concentrated by means of the cytocentrifuge to a circular area of about 5 mm diameter in the middle of the cover-slips or carbon chips. Single-layered coats with a generally homogeneous distribution of cells were obtained without any difficulty by dilution of the cell suspensions. TEM and SEM observations at high magnification showed good conservation of cells after the cytocentrifugation (Figs. 1-3).

TEMs of ultrathin sections and replicas and SEMs are suitable for demonstration of immunologic reactions on cell surfaces. These methods permit direct spatially resolved classification the products of reactions with cell membrane structures. Quantification of immunologic reactions is possible by counting the number

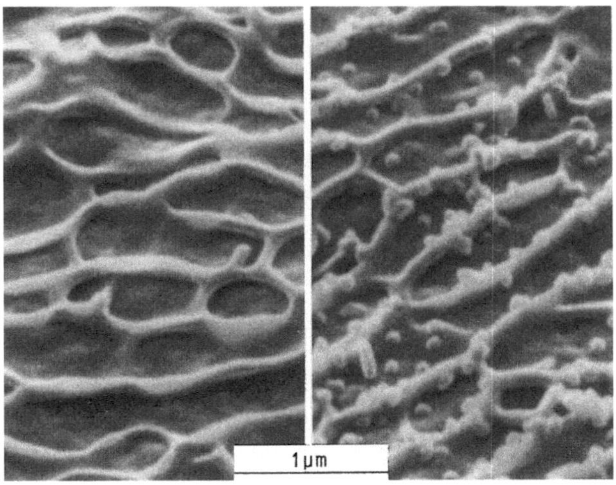

Fig. 1. Scanning electron micrographs of oral mucosa epithelial cells. Distribution of the enlarged gold particles. Labelling: **left** no primary antibody, 5 nm immunogold, silver enhancement; **right** antibody against epithelial membrane antigen (EMA), 5 nm immunogold, silver enhancement

of gold particles per unit area. On samples with gold coating, gold particles of down to 12 nm diameter could be discerned in SEMs. Even smaller gold particles could have been used, but they would have to have been enlarged later by the silver enhancement method (Fig. 1).

Our material showed important variations in staining density. Variations were noticed between preparations from the same cell types as well as between different kinds of cells. Specific patterns were not be observed. However, an accumulation of immunologic reaction products close to the extended membrane protrusions

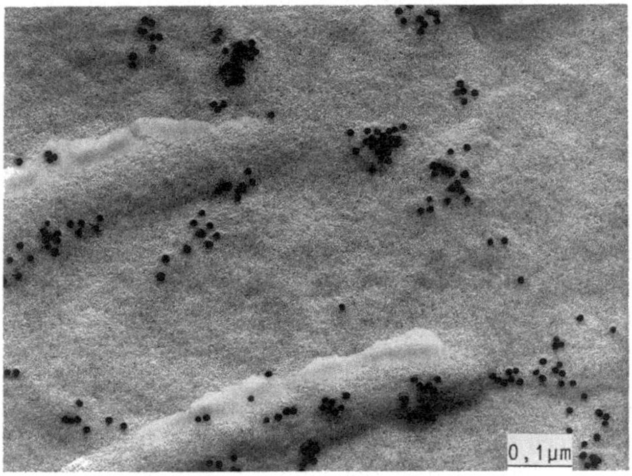

Fig. 2. Transmission electron micrograph of an oral mucosa epithelial cell (replica). Accumulation of gold particles on extensive and small cell membrance protrusions. Labelling: EMA, rabbit anti-mouse IgG, protein A-gold 12 nm

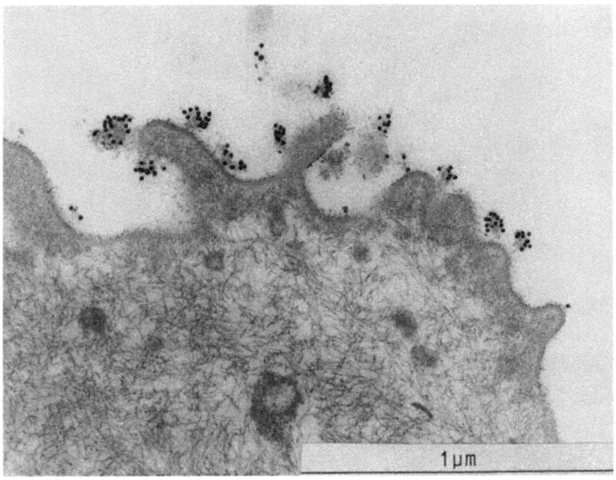

Fig. 3. Transmission electron micrograph of an oral mucosa epithelial cell (ultrathin section). Localization of single reaction sites and their relation to membrane structures. Labelling: EMA, rabbit anti-mouse IgG, protein A-gold 12 nm

was frequently seen (Figs. 1-3). When the quoted incubation periods were used, no disturbing secondary reactions occurred.

The specificity of gold accumulations was checked by negative controls, i.e. by leaving out the primary antibody (Fig. 1 A). A further control in SEM was given by the EDX method. This method enables uncoated samples and carbon-vapor coated samples to be tested to determine whether structures which appeared to be gold particles really did contain gold (spot analysis). For samples with silver enhancement and gold coating, the silver content of the structures was tested similarly. Comparisons of these methods led to the conclusion that, in comparison with SEM, TEM study of replicas (Fig. 2) only allows visualization of considerably smaller field, but offers clearer localization of the labeled areas and better differentiation of membrane structures. Localisation of individual immunologic reactions and the relation to the membrane structures, however, can best be determined with ultrathin sections using TEM.

Discussion

The concentration of cells on a small area through the use of the cytocentrifuge requires only small quantities of expensive immune sera for incubation and offers the possibility of rapid examination of cells. The single-layered state of the cells ensures even access of the immunologic reagent to cell structures and guarantees good observation of all cells. Given that the fine structure of cells is maintained even after cytocentrifugation, this preparation method can be regarded as highly advantageous and can be recommended for the similar electron microscopic examinations.

Points in cells where an immunologic reaction has occurred are labeled by gold particles. However, in quantitative estimations of reactions, it must be taken into account that only some of the potential reaction points on the cell are labeled. One of the reasons for this is steric hindrance between the gold particles during the binding process. The closer the possible reaction points are together and the bigger the gold particles are, the less binding occurs (Horisberger 1979). For homogeneous, highly sensitive labeling small gold particles of 3-5 nm diameter are recommended. Such particles, however, can no longer to be distinguished by SEM. An acceptable compromise between high sensitivity and good visibility is the silver enhancement method; the gold particles are enlarged after the labeling process by silver accumulations. Depending on the period over which the silver reagent acts, different sizes of particles can be obtained. On cells from the oral mucous membrane which are marked with EMA and protein A-gold, the gold particles mainly assemble on and around the membrane ruffles. This distribution pattern, however, is not specific for this type of cell, as gold is also mainly attached in the area of extended cell membrane protrusions in lymphocytes and leukocytes.

If the EDX method is used as a control, the required specificity is obtained, but this requires plenty of time. With SEM a quick specificity control during the observation of the gold binding can be obtained by determining the proportion of backscattered and secondary electrons (De Harven et al. 1984).

Of the methods used, the replica technique is of special importance as the gold particles can be clearly distinguished, small gold particles can be used, and extremely fine details of the cell surface are shown. For example, only the replica micrograph shows that gold particles are to be found not only on the larger cell protuberances, but also in between them and on smaller circular elevations of the cell membranes.

In summary, it can be stated that any of the methods described here is suitable for demonstrating immunologic reactions. The methods, however, are complementary and should not be used as alternatives.

Summary

Suspensions of cells from human body fluids (blood lymphocytes, blood leukocytes, urinary sediments) and from oral mucosa smears were incubated with monoclonal antibodies (EMA, LC, Pan-B) and protein A-gold and prepared for SEM and TEM. Cell surface replicas were also made up for TEM. Comparative examination of these methods led to the following conclusion: SEM is well suited to quick presentation of immunologic reaction products but the capacity to resolve fine structures is limited. The labeling specificity can be increased by employing small particles of colloidal gold and subsequently the silver enhancement technique. In contrast, the replica enables a more precise study of surface localization and allows better quantification of the immunologic reaction points. Conventional TEM clearly demonstrates the relation between surface markers and membrane structures.

Acknowledgements. We wish to thank Mrs. D. Hildebrandt for her support during preparation of the manuscript and Mrs. R. Tiburtius for the photographic work.

References

De Harven E, Leung R, Christensen H (1984) A novel approach for scanning electron microscopy of colloidal-gold-labeled cell surfaces. J Cell Biol 99: 53-57

Hohenberg H, Mannweiler K (1980) Semi-automatic washing device for simultaneous cleaning of surface replicas under identical conditions. Mikroskopie 36: 145-154

Horisberger M (1979) Evaluation of colloidal gold as a cytochemical marker for transmission and scanning electron microscopy. Biol Cell 36: 253-258

Lajous-Petter A-M, Scholman H-J, Hoder D, Blümcke S (1986) Präparative Gesichtspunkte bei der elektronenmikroskopischen Darstellung von immunologischen Reaktionen an Zelloberflächen. Beitr elektronenmikroskop Direktabb Oberfl 19: 399-404

Slot JW, Geute HJ (1981) Sizing of protein A-colloidal gold probes for immunoelectron microscopy. J Cell Biol 90: 533-536

Ultrastructure of Human Breast Carcinoma from a Pleural Effusion*

S. Scarpa, A. Modesti, G. D'Orazi, F. Massari, and M. Matzeu

Introduction

The fluid aspirated from pleural effusions contains various cells including lymphocytes, polymorphonuclear leukocytes, macrophages, mesothelial cells, fibroblasts, and a few tumor cells in neoplastic patients (Hofmann et al. 1984). The cause of the effusion is often unknown and the tumoral component is often insufficient to make a certain cytologic diagnosis. We propose that the cellular component should be analyzed ultrastructurally to characterize the cell histotypes. Furthermore, the few tumor cells present in the fluid can be plated in vitro for a short-term cell culture, offering a valid source of fresh living cells to study better the specific causes of the initial pleural effusion.

Material and Methods

Case Report. The patient had had a mastectomy for breast carcinoma 5 years before. She had recently had a relapsing bilateral pleural effusion of unknown cause. Cytologic examination, evaluated by conventional smear and hematoxylin and eosin stain, showed the presence of uncertain elements.

Tissue Culture. Cells were pelleted from the pleural effusion by centrifugation and washed several times in Hanks' balanced salt solution. They were then plated in RPMI 1640 medium enriched with 10% fetal calf serum and antibiotics. Many epithelial adherent cell colonies with high growth rates were visible after a few days.

Electron Microscopy. Cells were pelleted from the fluid, fixed in 2.5% glutaraldehyde, postfixed in osmium tetroxide, and embedded in Epon 812. Ultrathin sections were stained with uranyl acetate and lead citrate.

Immunofluorescence. Either adherent cells in culture or cells detached with EDTA were fixed for 10 min with cold absolute ethanol and incubated with the specific primary antibody for 1 h. Cells were then washed with phosphate-buffered saline

* This work was partially supported by an Italian CNR Grant (PFO-CNR 87.01360.44).

(PBS) and incubated with a fluorescein-conjugated second antibody diluted 1:20 for 1 h. Antibodies to extracellular matrix proteins were polyclonal, from rabbit; all the others were monoclonal (Sanbio).

Results

The cellular component of the pleural effusion was analyzed by electron microscopy: the majority of cells were lymphocytes, macrophages and erythrocytes, but few cells with constant morphologic characteristics - secretory granules, intermediate filaments (keratin-like), many desmosomes, and intracytoplasmic lumen-like formations (vacuoles; Fig. 1) - were present.

The cells from the fluid were plated in vitro and many epithelial adherent cell colonies with high growth rates spread as monolayers to confluency; they then packed tightly together spontaneously forming domes (Fig. 2).

The ultrastructural features of cells in culture were similar to those already seen in some cells from the initial pleural effusion: they appeared covered by abundant microvilli and were joined by well-developed desmosomes. Short bundles of intermediate filaments (cytokeratin) were present in the cytoplasm (Fig. 3).

The short-term culture was further characterized by immunofluorescent staining, as summarized in Table 1. Cells expressed basement membrane components (laminin and type IV collagen) but were negative for type V collagen and fibronectin; of all five intermediate filaments, only keratin was positive. Cells were also

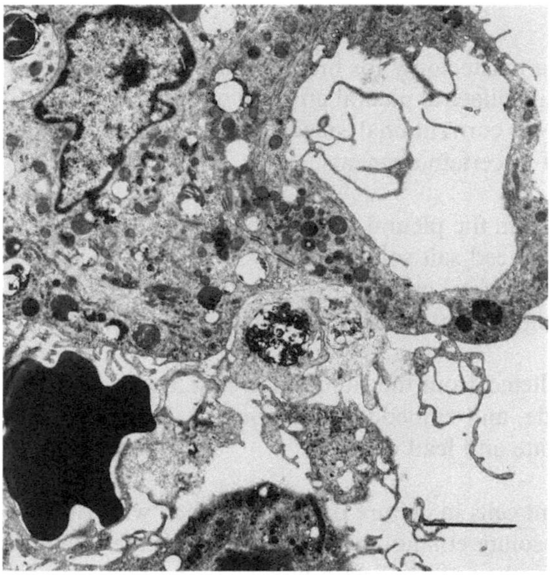

Fig. 1. Ultrastructural appearance of the cellular components of a pleural effusion. An erythrocyte is visible and an epithelial-like cell with microvillae, many desmosomes, and cytokeratin. *Bar,* 4 µm

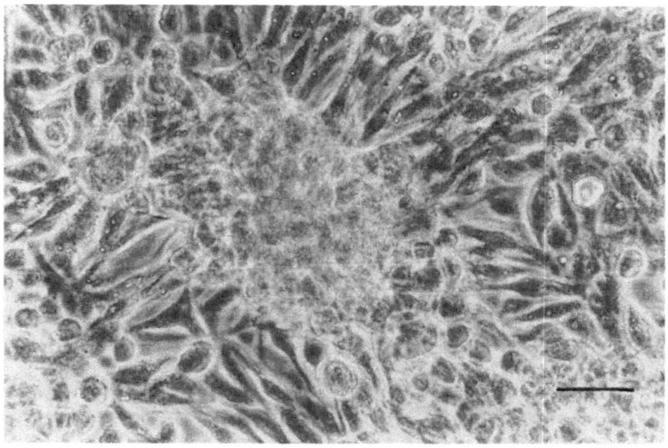

Fig. 2. Short-term cell culture from a pleural effusion. Monolayer cells form a typical dome. *Bar,* 15 μm

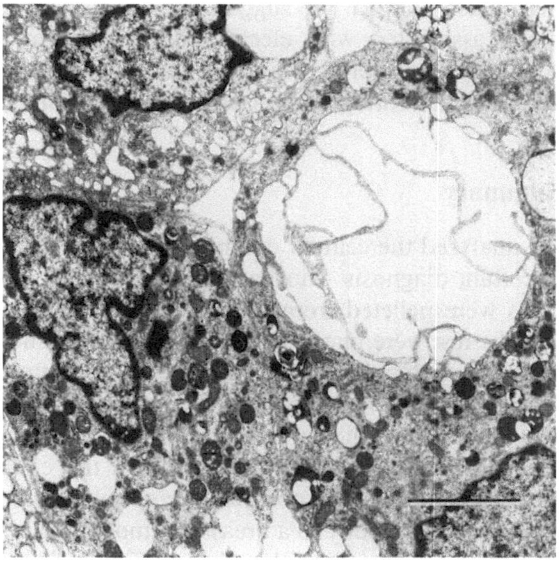

Fig. 3. Electron micrograph of cultured cells from a pleural effusion. Tumor cells are packed tightly, expressing epithelial markers: microvillae, desmosomes, cytokeratin. *Bar,* 4 μm

Table 1. Immunofluorescence of short-term cell culture

Extracellular matrix	Laminin	Collagen IV	Fibronectin	Collagen V	
	+	+	–	–	
Intermediate filaments	GFAP*	Neurofilaments	Keratin	Vimentin	Desmin
	–	–	+	–	–
Tumor-related antigens	1-80 (BC)	CEA	HLA		
	+	+	+		

* GFAP, Glial fibrillary acidic protein

stained by monoclonal antibodies to CEA, HLA, and breast carcinoma related antigen (1-80).

Discussion

The ultrastructural markers typical for epithelial cells such as microvilli, desmosomes, and cytokeratin (Nesland et al. 1987; Rizk and El-Rakhawy 1981; Osborn and Weber 1983) were found in the cells from this pleural effusion of unknown cause. This allowed characterization of the cell histotype and suggested a metastatic origin from the previous primary breast carcinoma. The high plating efficiency of these cells in vitro confirmed their malignant nature. Furthermore, the analysis of the short-term cell culture confirmed the presence of epithelial markers, demonstrated both by electron microscopy and by immunofluorescent staining, cells being positive for basement membrane components (Birembaut et al. 1985) and keratin. The diagnosis of recurrent breast carcinoma was finally confirmed by positive fluorescent staining with a monoclonal antibody specific to a breast carcinoma related antigen. We suggest that the utilization of the short-term culture technique associated with electron microscopy and immunocytochemical staining should be considered as an additional diagnostic tool in order to make a firm diagnosis when interpretation of cytologic specimens is uncertain.

Summary

We analyzed the cellular components of a pleural effusion from a patient with an uncertain diagnosis who had a mastectomy for breast carcinoma 5 years before. Cells were pelleted from the fluid: some were analyzed by electron microscopy and the rest were plated in vitro. An adherent short-term cell culture was obtained and was characterized ultrastructurally and immunologically. Of the various cell types visible by electron microscopy in the initial pleural effusion, few were present with epithelial features (desmosomes and cytokeratin) and the same ultrastructural characteristics were found in the cultured cells. Immunofluorescence analysis confirmed the positivity for epithelial markers and showed staining with a monoclonal antibody to a breast carcinoma related antigen. All these data suggested a metastatic origin of the pleural effusion from the previous primary breast carcinoma.

References

Birembaut P, Caron Y, Adnet JJ (1985) Usefulness of basement membrane markers in tumoural pathology. J Pathol 145: 283-296

Hofmann V, Berens ME, Früh U (1984) Analysis of malignant effusions by cellular composition, proliferation kinetics, and in vitro clonogenicity. Recent Results Cancer Res 94: 35-40

Nesland JM, Holm R, Lunde S, Johannessen JV (1987) Diagnostic problems in breast pathology: the benefit of ultrastructural and immunocytochemical analysis. Ultrastruct Pathol 11: 293-311

Osborn M, Weber K (1983) Tumor diagnosis by intermediate filament typing: a novel tool for surgical pathology. Lab Invest 48: 372-394

Rizk NN, El-Rakhawy MT (1981) Tissue culture and scanning electron microscopy of breast carcinoma cell line MCF-7 from pleural effusion. Acta Anat (Basel) 109: 70-74

Electron Microscopy in Childhood Tumors

M. Us-Krašovec

Introduction

The management of childhood malignancies continues to improve. There is increased survival for many tumors and in some cases patients may even be cured. However, the therapeutic regimes may have significant adverse effects resulting in a variety of psychological and physical sequelae. Accurate morphologic diagnosis is therefore mandatory before embarking on a definitive treatment course. A pathologist today has not only to solve the basic dilemma of whether the tumor is benign or malignant but also to assess its histogenesis and differentiation. To achieve this, he often has to employ all the available modern diagnostic techniques: electron microscopy, immunohistochemistry, cytogenetics, etc. [18, 34].

Electron Microscopy

Electron microscopy (EM) is a valuable adjunct to routine techniques in tumor pathology. A number of investigators have demonstrated its particular value in the differential diagnosis of small round cell tumors of childhood. Neuroblastoma, Ewing's sarcoma, rhabdomyosarcoma, and malignant lymphoma are most often considered as members of this group. Their basic diagnostic ultrastructural characteristics are features found in less differentiated cells. Some of the most important features in neuroblastoma cells are the neurosecretory, 50- to 200-nm, dense core granules and cytoplasmic extensions. These cytoplasmic extensions resemble neural processes and contain microtubules and filaments that are often surrounded by neurosecretory granules. Primitive cell-to-cell attachments and abortive synaptic junctions are often observed [27, 31]. Ewing's sarcoma is composed of primitive cells with conspicious aggregations of cytoplasmic glycogen. They have few cytoplasmic organelles and primitive cell-to-cell attachments [25]. Rhabdomyosarcoma cells can only be distinguished with EM when the features of developing skeletal muscle are found. Parallel arrangements of thin (actin) and thick (myosin) filaments in bundles with Z bands are of diagnostic relevance [21, 26, 28]. Bundzen wrote that "more than one specific feature of the sarcomere in an appropriate malignant setting is sufficient for ultrastructural diagnosis of rhabdosarcoma" [11]. Others think that thick (myosin) filaments surrounded by chains of ribosomes may be considered as the minimal evidence on which a diagnosis can be made [26].

Myosarcoma cells may have basal laminae and junctions, and often contain abundant glycogen. Lymphoma cells lack cell-to-cell attachments or other objective ultrastructural diagnostic characteristics.

Fine Needle Aspiration Biopsy

Morphologic diagnosis can be made preoperatively by fine needle aspiration biopsy (FNAB). This method, less traumatic than open biopsy, is especially suitable in high-risk pediatric patients who are too debilitated to withstand surgery or in whom surgery is not a primary treatment modality. Knowing the many diagnostic problems which small round cell tumors pose in surgical biopsies, it would seem unlikely at first sight that a histogenetic classification of these tumors could be made by cell sample examination alone. Both the manner of sampling and the sample itself, if only one criterion is used, i.e., cell morphology, would theoretically decrease the diagnostic accuracy of the method. Experience of the conventional FNAB technique in the diagnosis of pediatric tumors is limited by the low incidence of the cancers themselves, but mostly by the reluctance to use this method. Yet, the few reported results are suprisingly good. Taylor and Nunez were able to classify correctly 16 out of 19 tumors. In our laboratory, a correct preoperative classification was made in 74% of a group of 46 children with small round cell tumors [20, 33, 36, 37]. Correct diagnosis in histological terms is possible when cells in aspirates show characteristic morphologic features or cell arrangements. A precise diagnosis cannot be made when tumors show poorly differentiated cells without any specific diagnostic signs though their individual morphologies are not identical.

Electron Microscopic Examination of Fine Needle Aspirates

As ultrastructural diagnostic features are mainly confined to the cytoplasm and, to a much lesser extent, to the tissue architecture, EM examination of fine needle aspirates should have the same value as in surgical pathology. Several cytopathologists have demonstrated that cell samples can be processed for EM. Methods for routine (2-3 days) and quick (5-6 h) processing of aspirates have been developed [2, 10, 13, 19, 22, 29, 32, 38].

The quality of thin Epon sections is high. Aspiration samples contain single cells, small cell clusters, and small tissue fragments which can be optimally fixed. The structures of the cells and cytoplasmic organelles are usually well preserved. The samples allow the study not only of the interior of single cells but also of their surfaces, cell borders, and cell contact, and even of the relationships of cells to the stroma. According to Kindblom the morphologic preservation of small (less than 0.6 mm) tissue fragments found in such samples is superior to that obtained from large postoperative specimens [22].

In the majority of aspiration biopsy series it has been found that EM provides additional information and enables conclusive preoperative diagnosis in more than one-third of examined cases [29, 38]. EM studies have been conducted mostly on tumors in adults. Only a few articles were published by Akhtar on EM contributions to preoperative diagnosis of small round cell tumors of childhood. His experience, gained through EM study of cell samples of 16 neuroblastomas, 12 Ewing's sarcomas, 29 malignant lymphomas, and 13 miscellaneous malignant tumors of childhood indicates that EM can be an effective adjunct in the preoperative diagnosis of these lesions [3, 4, 5].

Immunohistochemistry

In tissue diagnosis, apart from morphology, immunohistochemistry is used to facilitate the identification and classification of poorly differentiated tumors. At present we have a large and rapidly proliferating panel of polyclonal and monoclonal antibodies raised against different tumor-associated antigens [9, 24]. A variety of markers is available which can be used to differentiate and classify malignant lymphomas [30]. Antibodies to intermediate filament proteins, which can distinguish cells of muscular, epithelial, glial, and neuronal origin, are useful in the differential diagnosis of small blue cell tumors [7, 15, 16, 35]. In an attampt to increase the frequency of specific preoperative diagnosis, immunological techniques have been introduced into cytopathology. The results show that this method can improve the accuracy of cytopathologic diagnosis by providing additional objective diagnostic criteria [8, 12, 14, 17].

Conclusions

In our experience, as well as that of other cytopathologists, technical and sampling problems may be encountered in EM examination and in immunostaining of aspiration biopsy samples. Technical problems are caused by the small amount of cells, their further loss during processing, and by excessive admixture with erythrocytes. Owing to this about 50% of submitted samples proved unsuitable for EM. In the face of improved techniques for selective concentration of cells, up to one-third of the specimens are still inadequate for EM [6, 23, 29, 38]. In a small sample, the representative, i.e., more mature, cells may not be present. So, we were able to find the EM diagnostic pattern of rhabdomyosarcoma in only three out of ten children presumed to have rhabdosarcoma by light microscopy.

Some tumor cells have a small amount of cytoplasm which is often ill-defined. In cell samples of poorly differentiated tumors naked nuclei may prevail due to the high vulnerability of the cytoplasm. Such samples preclude the identification of ultrastructural features as well as an immunological reaction.

The minute aspiration biopsy samples usually prevent the use of both techniques simultaneously. As yet it is not possible to define the precise role or priority of either EM or immunodiagnosis in the preoperative cytopathologic evaluation of particular tumors of childhood.

References

1. Akhtar M, Ashraf MA, Owen E, Bakry M (1980) A simple method for processing fine-needle aspiration biopsy specimens for electron microscopy. J Clin Pathol 33: 1214-1216
2. Akhtar M, Ashraf A, Owen EW (1981) Application of electron microscopy in the interpretation of fine-needle aspiration biopsies. Cancer 48: 2458-2463
3. Akhtar M, Ali M, Sabbah R (1985) Aspiration cytology of Ewing's sarcoma. Light and electron microscopic correlations. Cancer 56: 2051-2060
4. Akhtar M, Ali M, Sabbah R et al. (1985) Fine-needle aspiration biopsy (FNAB) diagnosis of round cell malignant tumors of childhood: a combined light and electron microscopic approach. Cancer 55: 1805-1817
5. Akhtar M, Ali M, Sabbah RS et al. (1986) Aspiration cytology of neuroblastoma. Light and electron microscopic correlations. Cancer 57: 797-803
6. Akhtar M, Bakry M, Nash EJ (1986) An improved technic for processing aspiration biopsy for electron microscopy. Am J Clin Pathol 85: 57-60
7. Altmannsberger M, Osborn M, Treuner J, Hölscher A, Weber K, Schauer A (1982) Diagnosis of human childhood rhabdomyosarcoma by antibodies to demin. The structural protein of muscle specific intermediate filaments. Virchows Arch [B] 39: 203-215
8. Altmannsberger M, Osborn M, Droese M, Weber K, Schauer A (1984) Diagnostic value of intermediate filaments antibodies in clinical cytology. Klin Wochenschr 62: 114-123
9. Baumal R, Khan HJ, Bailey D et al. (1984) The value of immunohistochemistry in increasing diagnostic precision of indifferentiated tumors by surgical pathologists. Histochem J 16: 1061-1078
10. Berkman WA, Chowdhury L, Brow NL, Padleckas R (1983) Value of electron microscopy in cytologic diagnosis of fine-needle biopsy. Am J Roentgenol 140: 1253-1258
11. Bundtzen JL, Norback DH (1982) The ultrastructure of poorly differentiated rhabdomyosarcoma: a case report and literature review. Hum Pathol 13: 301-313
12. Chess Q, Hajdu SI (1986) The role of immunoperoxidase staining in diagnostic cytology. Acta Cytol (Baltimore) 30: 1-7
13. Collins VP, Ivarsson B (1981) Tumor classification by electron microscopy of fine needle aspiration biopsy material. Acta Pathol Microbiol Scand [A] 89: 103-105
14. Dalquen P (1987) Immunocytochemical cytodiagnosis of tumors. In: Seifert G, Hübner K (eds) Pathology of cell receptors and tumor markers. Fisher, Stuttgart, pp 196-201
15. De Jong ASH, Kessel-van Vark M, Albus-Lutter CE (1985) Skeletal muscle actin as tumor marker in the diagnosis of rhabdomyosarcoma in childhood. Am J Surg Pathol 9: 467-474
16. Denk H, Krepler R, Artlieb U et al. (1983) Proteins of intermediate filaments. Am J Pathol 110: 193-208
17. Domagala W, Lubinski J, Weber K, Osborn M (1986) Intermediate filament typing of tumor cells in fine needle aspirates by means of monoclonal antibodies. Acta Cytol (Baltimore) 30: 214-224
18. Favara BE (1976) Diagnostic methods in pediatric oncology. Pediatr Clin North Am 23: 55-65
19. Hagelquist E (1978) Light and electron microscopy studies on material obtained by fine needle biopsy. Acta Otolaryngol [Suppl] (Stockh) 354: 5-75
20. Jereb B, Us-Krašovec M, Jereb M (1978) Thin needle biopsy of solid tumors in children. Med Pediatr Oncol 4: 213-220
21. Kahn HJ, Yeger H, Kassim D et al. (1983) Immunohistochemical and electron microscopic assessment of childhood rhabdomyosarcoma. Cancer 51: 1897-1903
22. Kindblom LG (1983) Light and electron microscopic examination of embeded fine-needle aspiration biopsy specimens in the preoperative diagnosis of soft tissue and bone tumors. Cancer 51: 2264-2277
23. Lazzaro AU (1983) Improved preparation of fine needle aspiration biopsies for transmission electron microscopy. Pathology 15: 399-402
24. Leong AS (1986) Antibody probes in the diagnosis of anaplastic tumors. I. malignant round cell tumors. Pathology 18: 296-305
25. Llombart-Bosch A, Blache R, Peydro-oaya A (1978) Ultrastructural study of 28 cases of Ewing's sarcoma: typical and atypical forms. Cancer 41: 1362-1373

26. Mierau GW, Favara BE (1980) Rhabdomyosarcoma in children. Ultrastructural study of 31 cases. Cancer 46: 2035-2040
27. Misugi I, Misugi N, Newton WA (1968) Fine structure of neuroblastoma, ganglioneuroblastoma and pheochromocytoma. Arch Pathol 86: 160-169
28. Morales AR, Fine G, Horn RC (1972) Rhabdomyosarcoma: an ultrastructural appraisal. Pathol Annu 7: 81-106
29. Nordgren H, Akerman M (1982) Electron microscopy of fine needle aspiration biopsy from soft tissue tumors. Acta Cytol (Baltimore) 26: 179-188
30. Picker JL, Weiss LM, Medeiros LJ et al. (1987) Immunophenotypic criteria for the diagnosis of non-Hodgkin's lymphoma. Am J Pathol 128: 181-201
31. Romansky SG, Crocker DW, Shaw KNF (1978) Ultrastructural studies of neuroblastoma. Cancer 42: 2392-2398
32. Sehested M, Francis P, Hainau B (1983) Electron microscopy of trans-thoracic fine needle aspiration biopsies. Acta Pathol Microbiol Scand [A] 91: 457-461
33. Taylor SR, Nunez C (1984) Fine-needle aspiration biopsy in a pediatric population. Cancer 54: 1449-1453
34. Triche TJ, Askin FB, Kissane JM (1986) Neuroblastoma. Ewing's sarcoma, and the differential diagnosis of small-round-blue cell tumors. In: Finegold MS (ed) Major problems in pathology: pathology of neoplasia in children and adolescents. Saunders, Philadelphia, pp 145-195
35. Tsokos M, Linnoila RI, Chandra RS, Triche JT (1984) Neuronspecific enolase in the diagnosis of neuroblastoma and other small, round-cell tumors in children. Hum Pathol 15: 575-584
36. Us-Krašovec M, Golouh R (1983) Cytology of small round-cell tumors of childhood (Abstract). Pathol Res Pract 178: 169
37. Valkov I, Bojikin B (1987) Fine-needle aspiration biopsy of abdominal and retroperitoneal tumors in infants and children. Diagn Cytopathol 3: 129-133
38. Wills EJ, Carr S, Philips J (1987) Electron microscopy in the diagnosis of percutaneous fine needle aspiration specimens. Ultrastruct Pathol 11: 361-387

IX. Reactive Cellular Changes

Introduction

M. Us-Krašovec

One of the fundamental characteristics of living organism is their ability to react to stimuli in order to maintain structural and functional integrity. These stimuli may be physiological or pathological and may originate from the internal or external environment. When the compensatory mechanisms of the cell are insufficient cell injury and cell death may result.

The variety of cellular reactions depends on the type, duration and severity of the stimulus as well as on the type and adaptability of the cell. Reactive cellular changes we encounter daily in our practices as they represent the basis of disease. Morphological changes in cell structures occur in retrogressive as well as in hyperplastic and neoplastic responses to injury. The range of morphological changes of the cell is limited and similar alterations can be observed with a variety of noxious stimuli. Cellular manifestations of increased biological activity in a variety of reactive lesions are characterized by significant nuclear enlargement and the presence of prominent nucleoli. These may be confused with similar features of cancer cells. To avoid diagnostic errors it is essential to bear in mind that only a combination of morphological features may provide a characteristic diagnostic pattern. The recognition of such a pattern supported by clinical and other relevant data enable a correct cytopathological diagnosis.

The reactive cellular changes of urothelium, mesothelium, soft tissue, lymph nodes, and thyroid glands, which are the subjects of this section, are only some representative examples of the wide variety of cellular reactions found in different organs to different stimuli.

Reactive Cellular Changes in Soft Tissue

M. Åkerman

Introduction

Preoperative fine needle aspiration and cytodiagnosis (aspiration cytology) of soft tissue tumors is being used with increasing frequency [1, 2]. Aspiration cytology (AC) of soft tissue tumors, usually to differentiate benign lesions or tumors from sarcomas, has so far shown itself to have a diagnostic accuracy comparable to that of AC of various types of epithelial tumors [1, 2].

Cytodiagnosis of soft tissue lesions may, however, be inaccurate and both false-negative and false-positive diagnoses regarding benignancy and malignancy have been reported [1]. One reason for false-positive cytodiagnosis is the difficulty in distinguishing between benign reactive cellular changes in mesenchymal tissue and true sarcoma cells.

The main reactive changes to be aware of are the reactive fibroblasts, inflammatory reactive changes in fatty tissue, and regenerating striated muscle fibers.

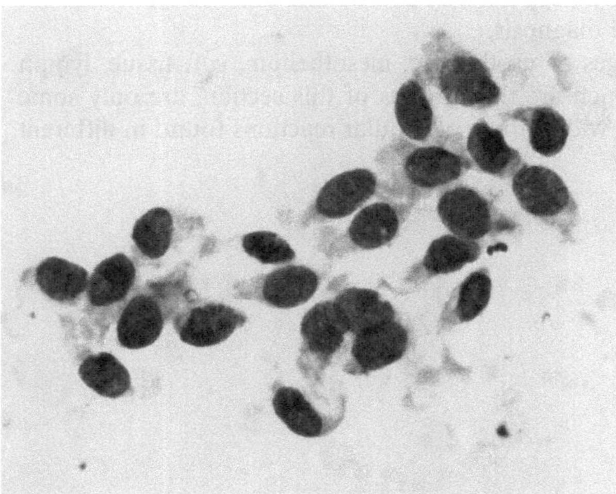

Fig. 1. Normal fibroblasts. Uniform elongated cells with rounded or ovoid nuclei and moderate amount of cytoplasm. May-Grünwald-Giemsa, ×580

Fibroblasts

In aspirate smears normal fibroblasts appear as spindle-shaped cells with slender cytoplasm. The cytoplasmic borders may be indistinct. The nuclei are ovoid or rounded with regular chromatin and small nucleoli, if any. The fibroblasts are seen either as dissociated cells or in small clusters of loosely attached cells. Naked nuclei devoid of cytoplasm are not uncommon (Fig. 1).

Reactive fibroblasts show, irrespective of etiology, wide variation in size and shape. The cells are fusiform with more or less elongated cytoplasmic processes, and may be polyhedral or triangular. They may show angulated cytoplasmic extensions. The nuclei vary in size and may be rounded, ovoid, or spindle-shaped, the chromatin structure may be irregular, and the nucleoli are prominent (Fig. 2). Binucleated cells are not uncommon. A typical example of the variegated appearance of reactive fibroblasts is seen in aspirates from nodular fasciitis [3].

Fatty Tissue

Normal fatty tissue in aspirate smears is seen as small fragments or clusters of large fat cells with abundant univacuolated cytoplasm and small, regular, dark nuclei. In larger fragments a discrete network of branching slender capillaries is observed. Dissociated fat cells are uncommon. In reactive conditions, posttraumatic or inflammatory, the fatty tissue fragments may show a myxoid-like background, the capillary network is more pronounced, the fatty vacuoles are smaller, and the fragments are more cellular than normal because of a fibroblastic and histiocytic reaction. Between fragments histiocytes with vacuolated or foamy cytoplasm appear.

Muscle Fibers

Regenerating striated muscle fibers are seen in fine needle aspirates as multinucleated cells of varying sizes and shapes. Their cytoplasm is dense, eosinophilic in hematoxylin and eosin, and dark blue in May-Grünwald-Giemsa stained smears.

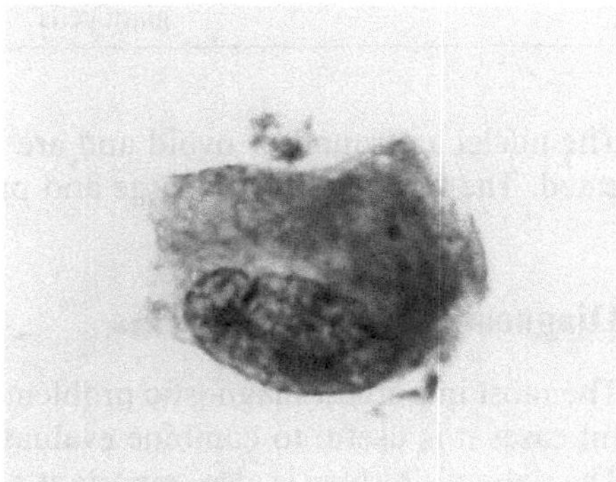

Fig. 2. A reactive fibroblast. A polyhedral cell with abundant cytoplasm. Irregular chromatin structure and prominent nucleolus. Hematoxylin and eosin, ×900

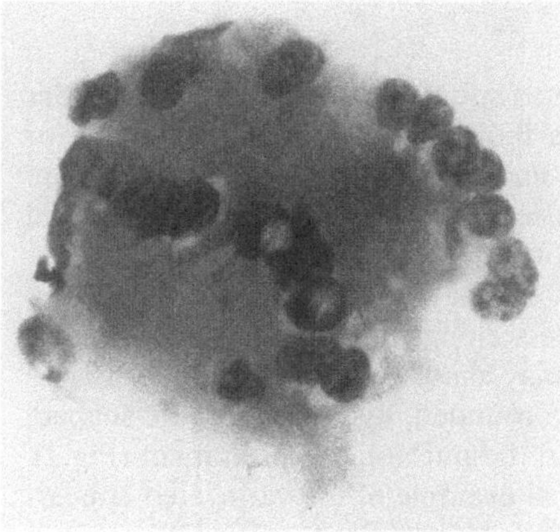

Fig. 3. Regenerating striated muscle fiber appearing as a multinucleated giant cell. Nuclei in rows and eccentrically located. May-Grünwald-Giemsa, ×540

Table 1. Benign soft tissue lesions with reactive cellular changes which might be misinterpreted as malignant in fine needle aspirate smears

Benign lesion	Cytologic findings	Wrong diagnosis
Pseudosarcomatous proliferative lesions in soft tissue (nodular fasciitis, proliferative fasciitis, proliferative myositis)	Reactive fibroblasts	"Spindle cell sarcoma," polymorphous sarcoma
Posttraumatic reaction in fibrous tissue	Reactive fibroblasts	"Spindle cell sarcoma," polymorphous sarcoma
Reactive changes in fatty tissue	Increased vascularity, myxoid degeneration, increased cellularity, histiocytic reaction	Myxoid liposarcoma, highly differentiated lipoma-like liposarcoma
Regenerating striated muscle fibers in posttraumatic conditions, in benign tumors infiltrating striated muscle, and in malignant tumors other than sarcoma infiltrating striated muscle	"Muscle giant cells" of varying sizes and shapes	Sarcoma with polymorphous giant cells
Pseudomalignant myositis ossificans	Reactive fibroblasts, "muscle giant cells"	Sarcoma
Infantile fibromatosis colli	Reactive fibroblasts, "muscle giant cells"	Sarcoma

The nuclei are round or ovoid and are often arranged in rows or eccentrically located. The nucleoli may be large and prominent (Fig. 3).

Diagnostic Problems

The most important diagnostic problems are presented in Table 1. In many doubtful cases it is useful to combine evaluation of clinical and of cytological findings. The patient's history is also important as many patients go in for sports and athletics and have a history of trauma. Pseudosarcomatous lesions of soft tissue often grow rapidly and tenderness is not uncommon [4]. The pseudomalignant myositis ossificans has a rather typical appearance upon radiological investigation [5].

References

1. Åkerman M, Rydholm A, Persson BM (1985) Aspiration cytology of soft tissue tumors. The 10-year experience at an Orthopedic Oncology Center. Acta Orthop Scand 56: 407-412
2. Layfield L, Anders KH, Glasgow BJ, Mirra JM (1986) Fine needle aspiration of primary soft tissue lesions. Arch Pathol Lab Med 110: 420-424
3. Dahl I, Åkerman M (1981) Nodular fasciitis. A correlative cytologic and histologic study of 13 cases. Acta Cytol (Baltimore) 3: 215-223
4. Enzinger FM, Weiss SW (1983) Soft tissue tumors. Mosby, St Louis, pp 15-31
5. Lagier R, Cox JN (1975) Pseudomalignant myositis ossificans. A pathological study of eight cases. Hum Pathol 6: 653-665

Reactive Changes of the Mesothelium

P. Pfitzer, G. W. Lipke, W. Hoeltermann, and E. Schlotmann-Höller

Reactive cellular changes of mesothelial cells are still a very embarrassing problem, not only for newcomers to cytology but also for more experienced cytologists. The borderline between reactive changes and the highly differentiated mesothelioma is the best-known example of these difficulties. All new techniques (electron microscopy, tissue culture, morphometry, cytochemistry, and monoclonal antibodies) have been used to try to solve this problem. However, it is still only partly solved.

One of the reasons for the difficulties is our limited knowledge of normal mesothelium in situ. It is limited by the anatomical condition itself since many details of a monolayer such as details of the mesothelium cannot be seen in histological cross sections. We therefore tried to develop techniques allowing a view of the surface of the mesothelium using various technical approaches. The first was a stripping technique which allows larger mesothelial pieces to be studied and we applied it first to the peritoneal mesothelium of fifteen 3- to 4-week-old NMRI mice (Cremerius 1984; Pfitzer 1984). The main steps in the procedure are treatment of unfixed material with 0.3% silver nitrate solution for 30 s, fixation in ethanol for 30 min, treatment with ether, air-drying, and stripping of the mesothelial layer with Tesafilm (hydrated cellulose) which is transferred to slides, cleansed with water and xylene, and finally stained with May-Grünwald-Giemsa stain (MGG).

In a second series of 52 6-month-old guinea pigs used for routine tests of tuberculosis (with negative results), mesothelium from various locations was investigated (Lipke 1987). For this material a second technique that involved freezing with carbon dioxide after silver nitrate had to be applied since the stripping technique worked only with peritoneal mesothelium supported by the abdominal muscles and not with visceral pleura.

The technique resulted in cells having silver-stained borders and demonstrated the surprisingly irregular arrangement of the cells (Fig. 1). They show more or less invaginated margins and not just the theoretically expected regular pavements.

There are reserves of membrane presumably necessary for the changes of shape needed to allow for movement of the underlying organs. The arrangement of the nuclei may be either irregular or surprisingly regular. In the latter case the nuclei are found in a line along their long axes; the direction of this line is independent of the direction of the long axis of the cell and may even be at a right angle to it. What influences such arrangements can only be speculated upon.

Sometimes two nuclei are in positions which seem the consequence of a re-

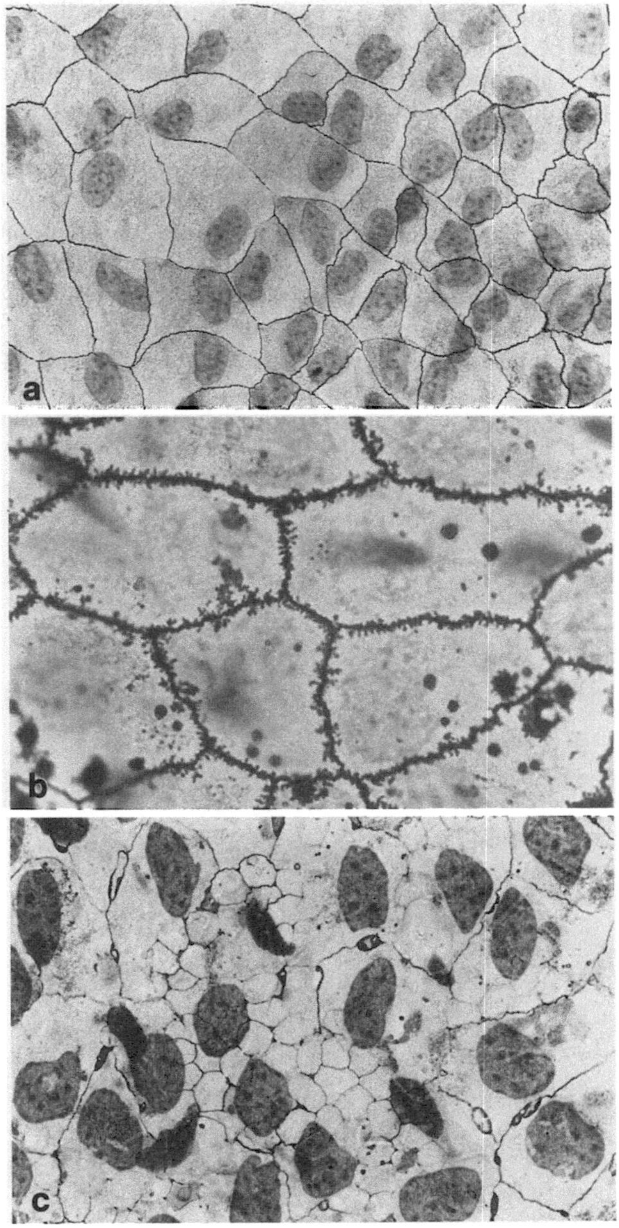

Fig. 1 a–c. Various examples of mesothelial tissue of guinea pigs. **a** Parietal pleura, 1 month old, ×400. **b** Peritoneum, 6 months old, ×560. **c** Peritoneum, 3 months old, ×560. Silver impregnation/MGG

cently finished mitotic division and indeed cells in various stages of mitosis can be found. The cells may be isolated or grouped in clusters which seem to represent regions of mesothelial growth.

Some problems arise from the silver staining of cell borders, which may be sharply delineated, flushy, or even missing. Sometimes globular structures such as pinocytic vesicles appear near the borders or in the cytoplasm. Occasionally pores can be identified.

The appearances of the silver-stained cell borders were independent of the technique used. In some specimens a meshwork of silver-stained structures replaced in parts the solid borderlines, and occasionally nuclei overlapping with cell borders demonstrated that only parts of the plasmalemma were stained. In addition, no cells undergoing mitosis were found in 6-month-old animals. In a younger control (30 days old), however, between 0.2% and 0.9% of cells were undergoing mitotic divisions in the various locations investigated (Fig. 2).

Beneath the mesothelial layer a loose meshwork of fibroblasts and fibrillar structures with some lymphocytes and eosinophils is found, at least in guinea pigs after tuberculosis testing (Fig. 3). In the case of costal pleura there is adipose tissue underneath this mesenchymal net.

Naturally we wanted to extend these investigations to human material. Our

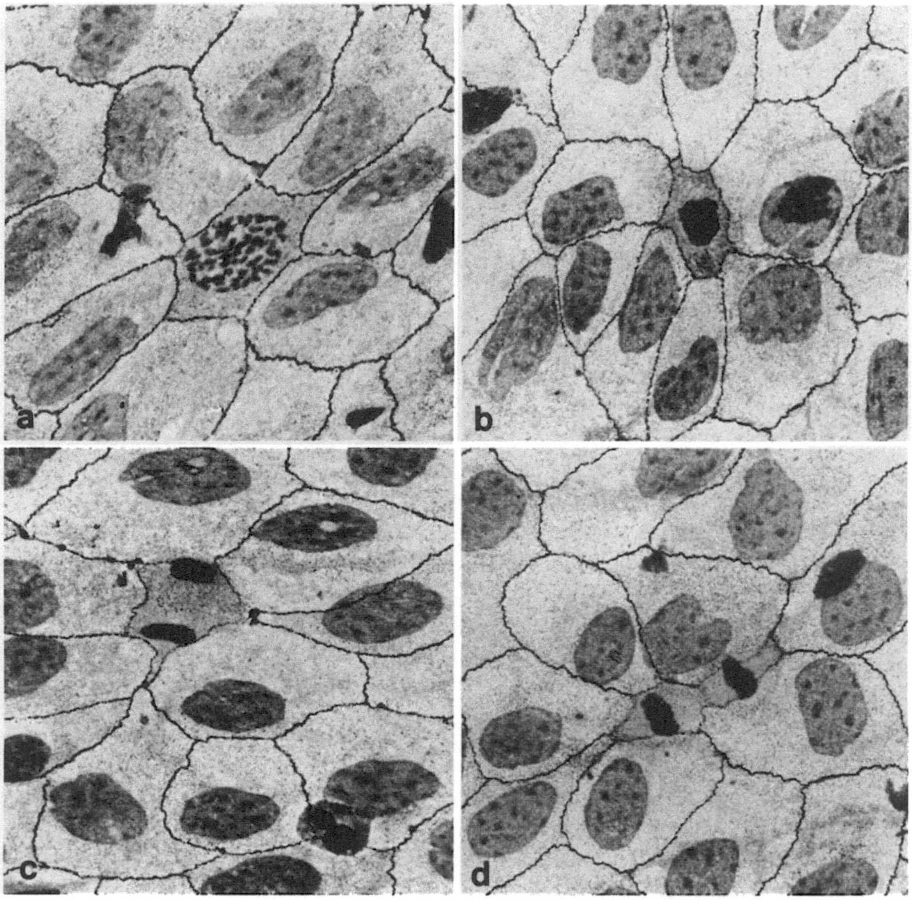

Fig. 2 a–d. Various stages of mitosis and mesothelial cell division in peritoneal mesothelium of a 1-month-old guinea pig. Silver/MGG, ×600

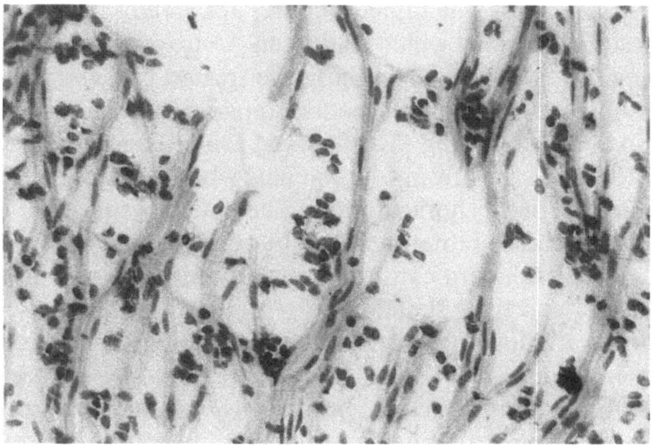

Fig. 3. Submesothelial layer of the peritoneum from a 6-month-old guinea pig. Silver/MGG, ×170

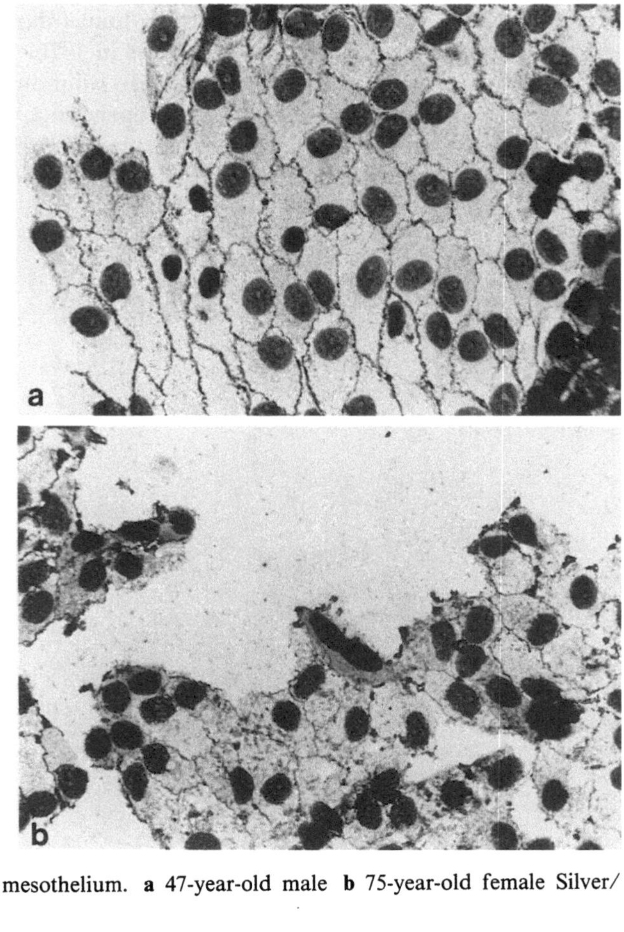

Fig. 4 a, b. Human peritoneal mesothelium. **a** 47-year-old male **b** 75-year-old female Silver/MGG, ×500

first series was a complete failure, apparently because of the vulnerability of the mesothelial cells, which had been destroyed by the palpating hands of the surgeon. To check this hypothesis we treated five guinea pigs accordingly and made scanning electron microscopic preparations. They showed a thick irregular fibrinous coat. Knowing this, we tried the freezing technique once more using small untouched strips from four patients who had had abdominal operations and seven who underwent thoracotomies. The patients' ages were between 45 and 95; younger individuals are not included in our series.

The results are consistent with those from guinea pigs as regards the cellular borders and their stainability by silver impregnation (Fig. 4). However, there was some variation in nuclear size, there being some larger nuclei with two or four times the normal DNA content as measured by scanning cytophotometry of Feulgen-stained preparations. This means that some polyploid nuclei are present in situ. Also, some multinucleated cells similar to those found in effusions were also found in situ, and there were indications that local hyperplastic reactions had occurred.

Proliferations in effusions were our main object of study in a second series of investigations on reactive mesothelial changes, concentrating on a sort of "experimental" ascites: continuous ambulatory peritoneal dialysis (CAPD). This method was introduced by Popovich and co-workers in 1976-1978 and consists of lavage of the peritoneal cavity with 2000 ml glucose solution over 4-5 h. It can be performed by the patients themselves. This independence of the patient from machinery naturally demands that the bags be changed carefully. Infections are continuous danger. We first received the material in the cytology laboratory when it was

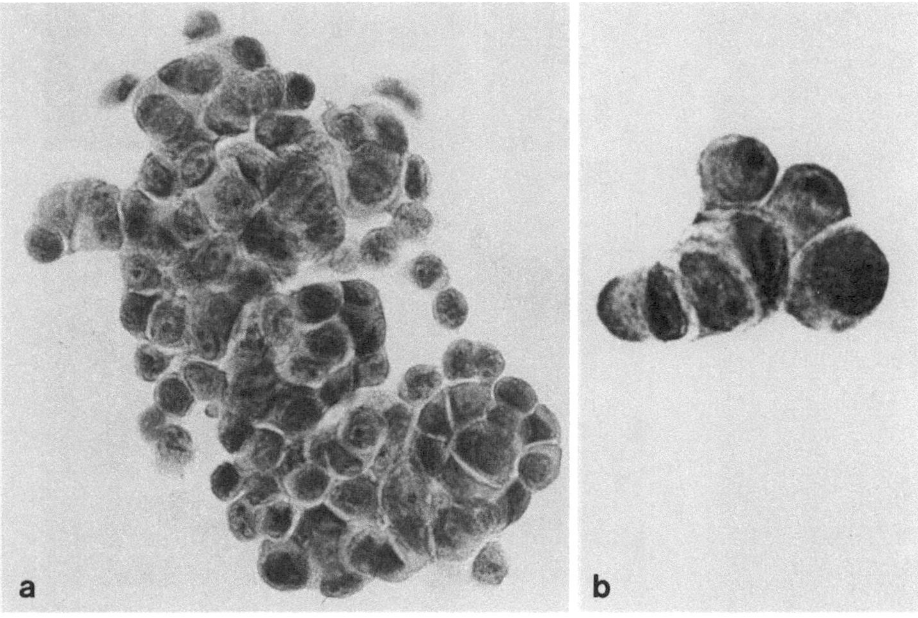

Fig. 5 a, b. Mesothelial proliferations after CAPD. **a** Papanicolaou, ×690. **b** MGG, ×1000

sent to us for confirmation of suspected eosinophilic inflammations; we found various degrees of general inflammation in 85% of the 362 specimens. The inflammations did not occur continuously during the study period but there was a certain increase in chronic form with time. Eosinophilic reactions were observed sporadically in 27 of the 32 patients (84%).

The mesothelial reactions which occurred in 39% of the specimens are especially interesting from the cytologic point of view (Fig. 5). Some of them appear highly suggestive of malignancy because of the variability of the clusters and some of the single cells. However, if their numbers are considered in relation to the 2000 ml of fluid they are rather rare. However, the number of atypical, often polyploid mitoses found is rather high (7.5% of the CAPD specimens; Fig. 6). Such mitoses are occasionally found in routine smears with high mesothelial cell populations, but in CAPD fluid they appear isolated because of the low cellularity.

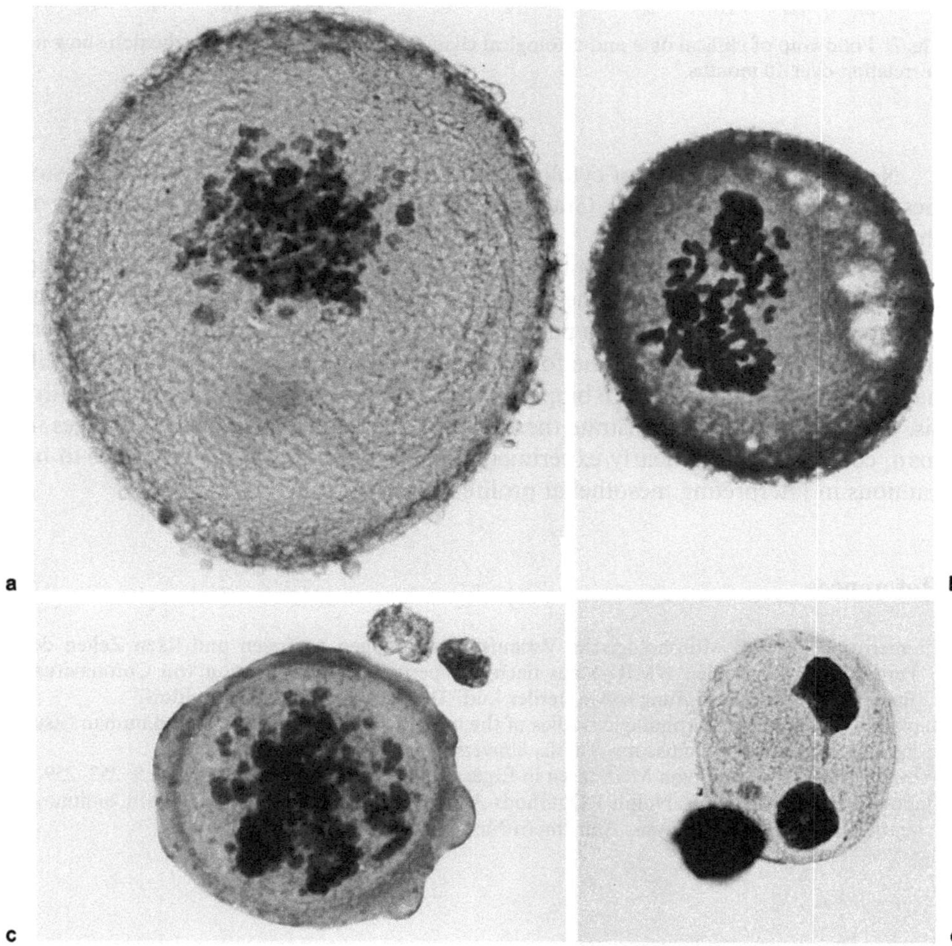

Fig. 6 a–d. Atypical mitoses and their consequences. **d** Interrupted chromosomal bridge. **a, d** Papanicolaou **b, c** MGG. ×1100

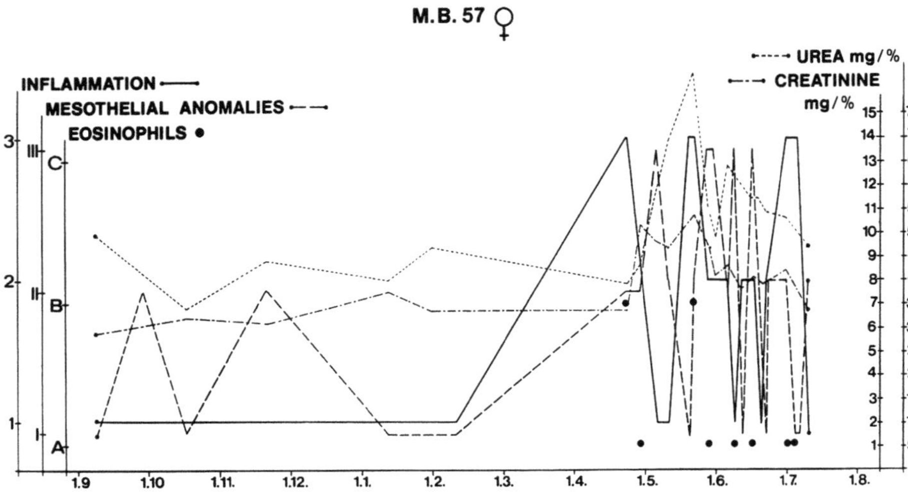

Fig. 7. Follow-up of clinical data and cytological changes in a semiquantitave evaluation show no correlation over 10 months

No relevant correlation of cytological results with the clinical data or the histories of the patients was found. Graphs (Fig. 7) demonstrate only the irregularity of the changes in all 362 samples.

The most interesting results of the cytological examination of the CAPD fluid are certainly the quality of the mesothelial changes and the relatively high number of atypical mitoses found after a period of only 4-6 h. This is certainly a very short time in comparison with the time for which a spontaneous peritoneal effusion will have existed before a puncture biopsy is taken by the clinician for cytologic analysis. However, it does demonstrate the extent of the reactive mesothelial changes in man, controlled under nearly experimental conditions, and it is a reminder to be cautious in interpreting mesothelial proliferation.

References

Cremerius PG (1984) Morphologische Verlaufsbeobachtungen an freien und fixen Zellen des Peritoneums der weißen NMRI-Maus nach intraperitonealer Applikation von Coronavirus-, Eiweiß- oder Kochsalzlösung sowie steriler Luft. Thesis, University of Düsseldorf

Lipke GW (1987) Cytomorphologic studies of the mesothelium in situ. Animal and human tissue in light and electron microscopy. Thesis, University of Düsseldorf

Pfitzer P (1984) Nachweis von Metastasen in Ergüssen der Körperhöhlen. Pathologe 5: 252-259

Popovich RP, Moncrief JF, Nolph KD, Ghods AJ, Twardowski ZJ, Pyle WK (1978) Continuous ambulatory peritoneal dialysis. Ann Intern Med 88: 449-456

Reactive Cellular Changes: Pitfalls in Fine-Needle Aspirates from Reactive Lymph Nodes

P. van Heerde

Introduction

Because of differences in antigenic stimulation, the lymph node cells found in fine-needle aspirates may vary considerably in terms of cellular composition, the size and shape of the cells, and other morphological aspects. Various agents can stimulate a lymph gland, including micro-organisms, drugs, tissue antigens, auto-antigens, and other, sometimes unknown, factors.

Lymph node cells must reach a high position in the "top ten" of the "tricky cells" that successfully try to mislead our cytological eyes. Stimulated and accumulated lymph node cells, i.e., lymphocytes, histiocytes, and a variety of derived cells, may greatly resemble neoplastic cells [1, 2]. "Overdiagnosis" in the sense of false-positive cytological conclusions usually leads to serious consequences for the patient, but exact figures about overdiagnosis are difficult to find in the literature. There are some publications on false-positive histological reports; in an old publication [6], 35% of the referred cases initially supposed to be Hodgkin's disease (on the base of histomorphology) were in fact reactive lymph nodes! In more recent reports [7] the cytological figures are less alarming.

Awareness of the possible pitfalls, strict diagnostic criteria, proper handling of the aspirate specimen, and sensible interpretation of the clinical data reduce the number of overdiagnoses substantially.

This article gives a short overview of the reactive changes that can simulate neoplastic lesions, illustrated with some examples. Lymph node hyperplasias may resemble not only Hodgkin's disease, non-Hodgkin's lymphoma, and related disorders, but also others such as metastatic lesions [2]. The reverse may also happen: malignant cells may mimic reactive lymphoid cells, especially in low-grade non-Hodgkin's lymphomas [7].

Cellular changes leading to lymphoid hyperplasia may occur in different compartments of a lymph gland [2]:

1. B-cell compartment: follicles, reaction centres, medullary cords (plasma cells)
2. T-cell compartment: paracortical area
3. Histiocytes and derived cells
   a) Sinus system: histiocytes/macrophages
   b) Other histiocyte-derived cells: interdigitating cells in the paracortical area, epitheloid cells

In many instances, a combined reaction of T cells, B cells, and histiocytes occurs on antigen stimulation.

Follicular Reaction

A follicular reaction often occurs in so-called aspecific conditions with unknown cause. Furthermore, it is regularly found in toxoplasmosis (often together with small granulomas) and many other infectious diseases. In secondary syphilis, initial stages of the acquired immunodeficiency syndrome (AIDS), and rheumatoid arthritis a very pronounced follicular reaction may be present.

Cytologically a more or less heterogeneous cell population is seen: centroblasts (syn. large non-cleaved follicular centre cells), centrocytes (mainly small cleaved follicular centre cells), some plasma cells and starry sky macrophages, in a background of small round lymphocytes (see Fig. 1). A combination of centroblasts, centrocytes and lymphocytes with starry sky macrophages is strongly indicative of a benign follicular proliferation [7], whereas the same cell types without starry sky macrophages may indicate a follicular centre cell malignant lymphoma! In rheumatoid arthritis, binucleated Sternberg-Reed-like cells occasionally occur that may lead to a misdiagnosis of Hodgkin's disease (see Fig. 2).

Clinical data including the age of the patient are important. In dubious cases the fine-needle aspiration is repeated and should include immune typing of aspirated cells in suspension.

Paracortical Reaction

Fine-needle aspirates from a highly stimulated paracortical T-cell area in the lymph gland are a well-known problem in diagnostic cytology. In such cases an extensive accumulation of immunoblastic cells may mimic a malignant lymphoma

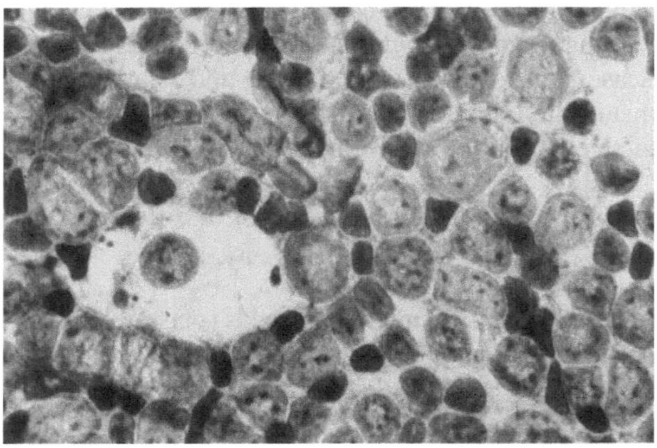

Fig. 1. Lymphoid hyperplasia of follicular centre cell type. Starry sky macrophage surrounded by centroblasts, centrocytes and lymphocytes. Giemsa, ×500

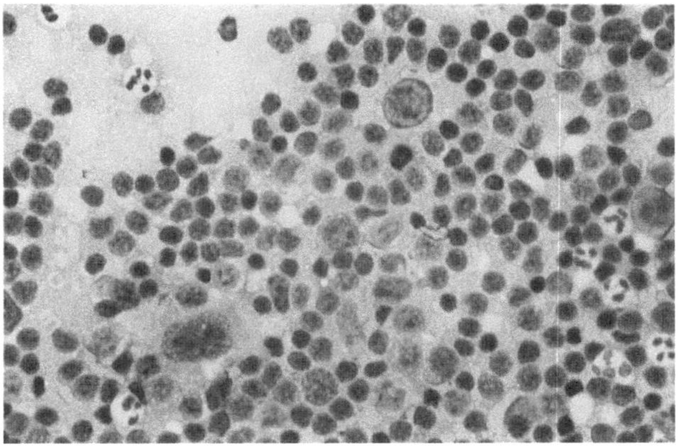

Fig. 2. Rheumatoid arthritis. Large binucleated lymphoid cell in the lower left corner. Giemsa, ×500

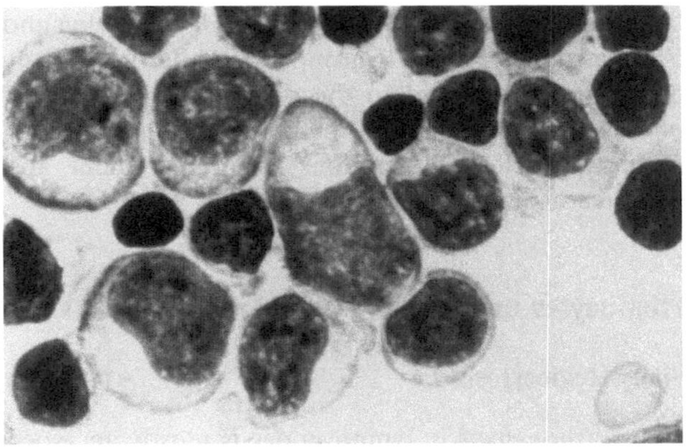

Fig. 3. Infectious mononucleosis. Large immunoblastic cells often predominate. Giemsa, ×1000

[1, 2, 7]. Binucleate Sternberg-Reed-like cells in such cases may lead to an erroneous diagnosis of Hodgkin's disease [1, 2]. In such cases immunological investigation with antibodies to markers of B cells, T cells and Hodgkin cells such as leu M-1 [1] may add substantial information; however, aspecific reactions may occur.

Paracortical reactions are characterized cytologically by the presence of immunoblastic cells, plasma cells in different stages of maturation, in a background of small lymphocytes (see Fig. 3). A range of plasmacytoid cells from immunoblastic cells to mature plasma cells is strongly indicative of benign paracortical proliferation. These reactions may occur in viral infections (especially in infectious mononucleosis), in sensitivity reactions due to drugs (hydantoin and others), with vaccination, with transplants, etc.

In fine-needle aspirates from chronically stimulated paracortical areas a large number of small to medium-sized T cells may be present, characterized by irregu-

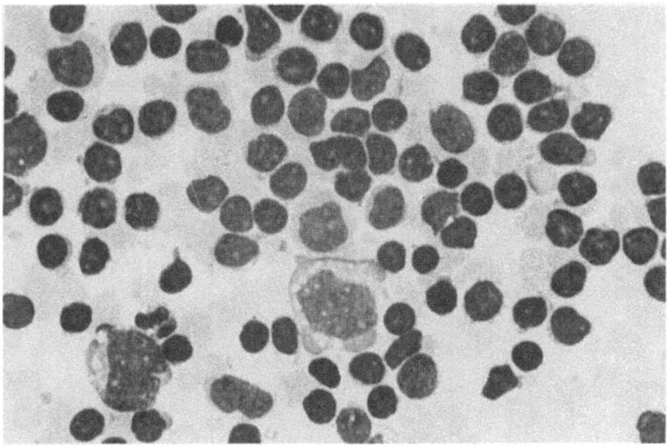

Fig. 4. Chronically stimulated paracortical T cells: irregularly outlined hyperchromatic nuclei with a single pale nucleolus. Two immunoblastic cells. Giemsa, ×500

lar hyperchromatic nuclei, a single pale nucleolus and a moderate amount of pale cytoplasm (see Fig. 4). If grouped together in large monotonous fields, these cells may erroneously lead to the diagnosis of malignant lymphoma. In T-cell lymphomas, however, nuclear hyperchromatism and anisonucleosis are more pronounced, nuclear indentations are more evident, and the nucleoli are mostly inconspicuous in the small cells [7].

Histiocytes and Other Derived Cells

Sinus Histiocytosis

Sinus histiocytosis in lymph nodes is mostly an aspecific sign, often occurring together with T- and/or B-cell accumulation. Even in extensive sinus histiocytosis lymphoid cells may predominate in the fine-needle aspirates. Histiocytes may resemble micrometastases, e.g. from melanoma, breast carcinoma or renal carcinoma. Careful examination, however, reveals features of macrophages in some of these cells: abundant foamy cytoplasm, signs of phagocytosis, monocytoid nuclei, reticular (retiform) chromatin, no distinct cell cohesion and a specific immunohistochemical pattern [8] (see Fig. 5). In addition, if metastatic lymph nodes become palpable, the major part of the lymph node is occupied by tumour tissue, so micrometastases seldom occur in palpable lymph nodes.

Another example of accumulation of benign histiocytes within the lymph nodes sinuses is so-called sinus histiocytosis with massive lymphadenopathy [5]. This clinically mimics a malignant lymphoma, with greatly enlarged lymph nodes mostly in the neck. Its etiology is unknown and it occurs mostly in children. Characteristic for this lesion is the presence of a large number of histiocytes, occasionally with large nuclei and nucleoli, which phagocytose (emperipolesis) lymphocytes and other cells extensively (see Fig. 6); it is a benign lesion with clinically and sometimes cytologically an alarming picture.

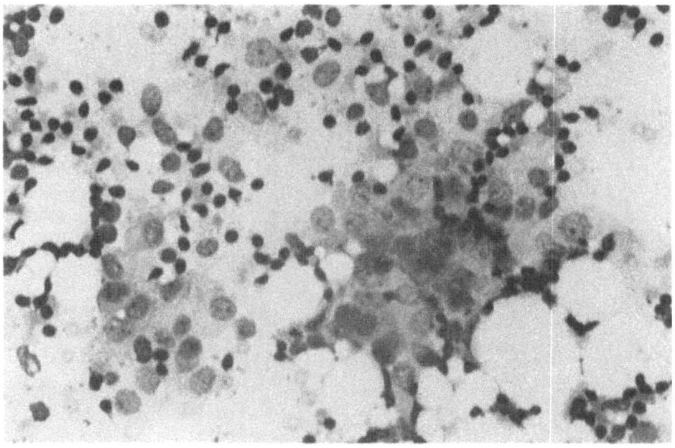

Fig. 5. Sinus histiocytosis. Clustered and dissociated histiocytes with abundant cytoplasm, pale reticular chromatin and small nucleoli. Giemsa, ×500

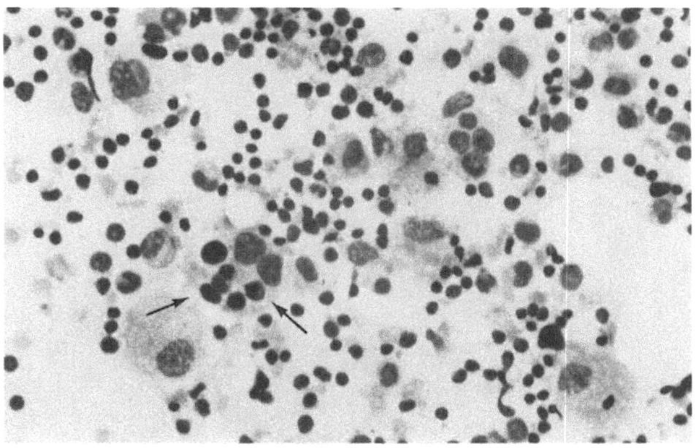

Fig. 6. "Sinus histiocytosis with massive lymphadenopathy." Large histiocytes. Phagocytosis of lymphocytes and plasma cells *(arrows)*. Giemsa, ×500

The so-called virus-associated haemophagocytic syndrome in immunosuppressed patients is another rare example of an accumulation of benign, normal-looking histiocytes; however, this is not limited to the lymph nodes and there is striking erythrophagocytosis [3, 4].

In malignant histiocytosis, in contrast to benign histiocytosis, cellular and nuclear atypia are very obvious, there being characteristic ring-shaped nuclei, the so-called "window-nuclei" [8].

Other Histiocyte-Derived Cells

Granulomatous lesions may occur in many pathological conditions and may be caused by numerous agents. Cat scratch disease is a relatively frequently occurring example of an accumulation of epitheloid cells and histiocytes, characteristically

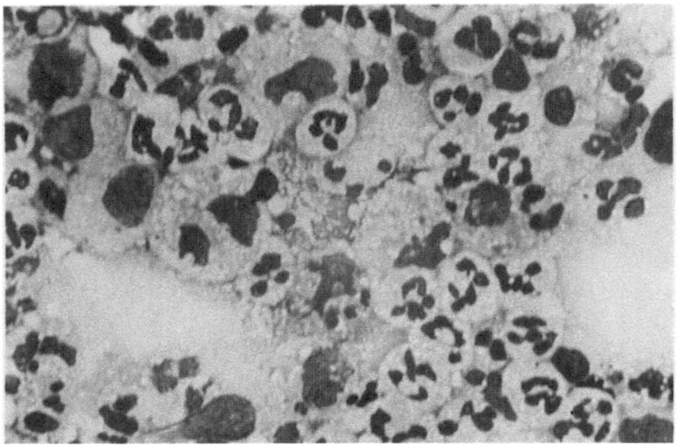

Fig. 7. Cat scratch disease. Neutrophils, some lymphocytes and stimulated histiocytes. Giemsa, ×500

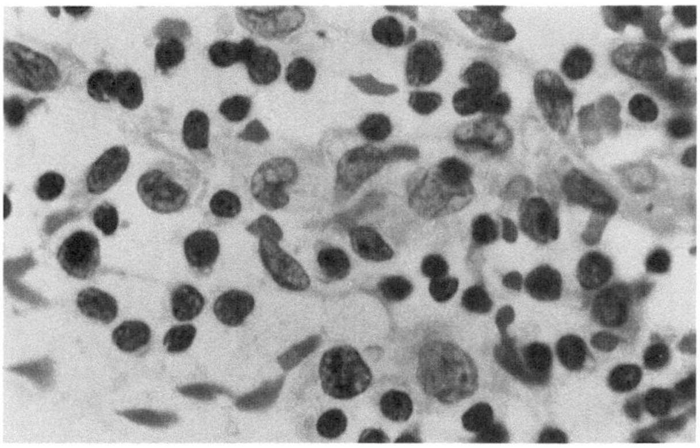

Fig. 8. Dermatopathic lymphadenopathy. Lymphocytes and interdigitating cells with pale folded nuclei and long slender protrusions of the cytoplasm. Papanicolaou, ×500

admixed with neutrophils. Clustered histiocytes may resemble a metastatic lesion (see Fig. 7).

Accumulation of interdigitating (reticulum) cells is seen in dermatopathic lymphadenopathy. This commonly causes no striking diagnostic problems; only in cases with large collections of interdigitating cells may a malignant proliferation be erroneously assumed. The characteristic morphology of interdigitating cells is demonstrated in Fig. 8: pale chromatin, folded nuclei and pale, ill-defined cytoplasm with long slender protrusions.

Summary and Conclusions

In conclusion, benign cellular changes in aspirated lymphoid cells may cause considerable diagnostic problems. Clinical features are important for the differential diagnosis. In a fast-growing tender lymph node in a young person - despite an alarming cytological picture - a benign lesion must always be kept in mind until the contrary is proven, whereas lymph node enlargement in older people must be considered malignant - despite an "innocent" cytological picture in a substantial number of cases - until the contrary is proven.

Attention must be paid to whether (small) malignant lymphoid clones are present in reactive extranodal lesions, the so-called "pseudo-pseudolymphomas", such as occurs in some cases of Sjögren's disease in the parotid gland, lymphoid hyperplasia of the stomach, etc.

The diagnostic problems can be diminished or solved in most instances by good clinical information, smears of high quality, awareness of the existence of the various aspects of lymphoid cells, using strict criteria for the diagnosis, and immunocytochemical or other extra diagnostic methods.

References

1. Childs CC, Parham DM, Berard CW (1987) Infectious mononucleosis. The spectrum of morphologic changes simulating lymphoma in lymph nodes and tonsils. Am J Surg Pathol 11 (2): 122-132
2. Dorfman RF, Warnke R (1974) Lymphadenopathy simulating the malignant lymphomas. Hum Pathol 5: 519-550
3. McKenna RW, Risdall RJ, Brunning RD (1981) Virus associated hemophagocytic syndrome. Hum Pathol 12: 395-398
4. Risdall RJ, McKenna RW, Nesbit ME, Krivit W, Balfour HH, Simmons RL, Brunning RD (1979) Virus associated hemophagocytic syndrome. Cancer 44: 993-1002
5. Rosai J, Dorfman RF (1969) Sinus histiocytosis with massive lymphadenopathy. Arch Pathol 87: 63-70
6. Symmers WSC (1968) Survey of the eventual diagnosis in 600 cases referred for a second histological opinion after an initial biopsy diagnosis of Hodgkin's disease. J Clin Pathol 21: 650-653
7. Van Heerde P, Go DMDS, Koolman-Schellekens MA, Peterse JL (1984) Cytodiagnosis of non-Hodgkin's lymphoma; a morphological analysis of 215 biopsy proven cases. Virchows Arch [A] 403: 213-233
8. Van Heerde P, Feltkamp CA, Hart AAM, Somers R, van Unnik JAM, Vroom TM (1984) Malignant histiocytosis and related tumors. A clinicopathologic study of 42 cases using cytological, histochemical and ultrastructural parameters. Hematol Oncol 2: 13-33

Evaluation of Fine-Needle Aspiration Biopsy Findings in Treatment of Grave's Disease

G. Kyriakos, M. Ioannidou, and G. Tolis

Introduction

Fine-needle aspiration biopsy (FNAB) of the thyroid is now accepted as a standard procedure in the diagnosis of thyroid diseases, although the cytologist must be experienced and the physician must sample from more than one area of the mass under consideration. The anatomical position of the thyroid gland allows an easy and safe approach to this area. The clinician (endocrinologist) also benefits from this diagnostic method, by being able to evaluate clinically the biopsy findings of FNAB and thus follow the correct course of treatment for a particular patient.

To date FNAB has been very useful in the differential diagnosis of benign and malignant nodules and in supporting the clinical diagnosis of subacute and autoimmune thyroiditis [1]. This study deals with the usefulness of FNAB in assessing the effect of antithyroid medication upon thyroid cytoarchitecture and function and in the prediction of relapse.

Material and Methods

Thirty female patients, aged 30–50 years, with established Grave's disease (increased radioactive iodine uptake, blunted thyrotropin-releasing hormone test, elevated triiodothyronine and thyroxine) were evaluated. Following the evaluation they were treated with thionamides at an initial dose of 40–60 mg per day. This dose was gradually diminished at a rate that depended on the results of laboratory tests and the clinical picture. Fine-needle aspirates were taken from all of them at the acute phase of the disease prior to and at various intervals during therapy, namely in the 4th, 8th, 12th and 18th months.

Results

The cytological smears made before therapy revealed in all 30 patients an absence of colloid and thyrocytes with abundant, finely granular, "fragile" cytoplasm and large peripheral vacuoles (flaming cytoplasm) on a bloody background (Fig. 1). After 4 months of therapy and when the patients had been rendered euthyroid, the

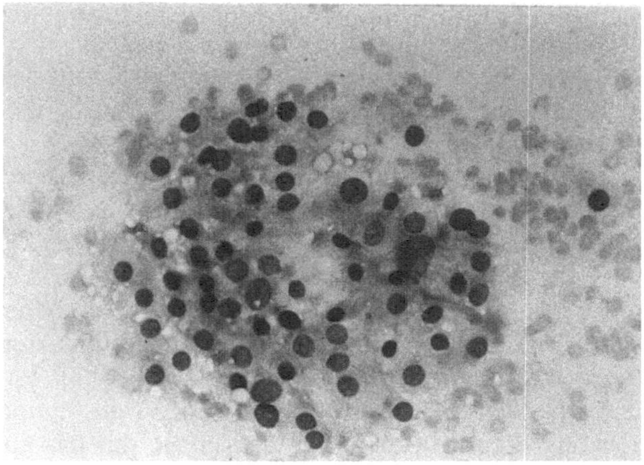

Fig. 1. Hyperthyroidism: thyrocytes with the "flaming" cytoplasm. May-Grünwald-Giemsa, ×300

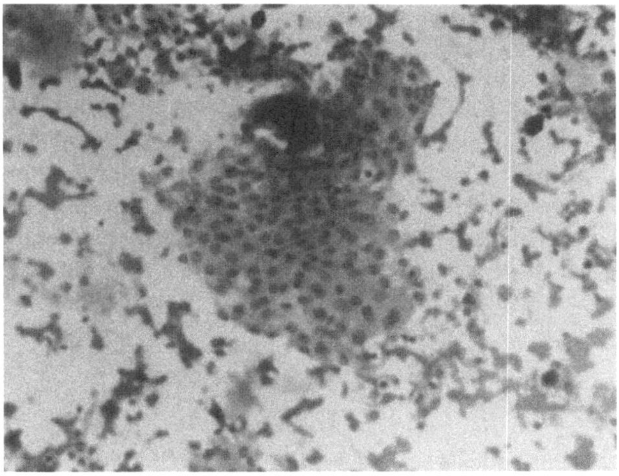

Fig. 2. An aggregation of benign thyrocytes without "abnormal" findings. May-Grünwald-Giemsa, ×170

findings from the smears in 22 patients (73%) were the same as before therapy. In eight patients (27%), however, the picture was now consistent with a non-toxic goiter, i.e., the thyrocytes were normal and did not exhibit granular cytoplasm and peripheral vacuoles (Fig. 2).

Eight months after therapy 20 patients (66%) had a smear consistent with hyperthyroidism, whereas in the rest of the patients (34%) the smear was consistent with a normally functioning thyroid. Twelve months after therapy 18 patients (60%) still had a smear consistent with hyperthyroidism, whereas 12 patients (40%) had a smear consistent with a nontoxic goiter. By the 18th month 13 patients (42%) had a "normal" smear.

Within 1 year of discontinuation of therapy 18 patients showed clinical and biochemical hyperthyroidism. These patients included the 17 in whom the smears had always been consistent with hyperthyroidism despite normal thyroxine and triiodothyronine levels; in contrast, only one of the patients with "euthyroid" smears relapsed.

Discussion

In patients receiving treatment for Grave's disease the prognosis is of great interest. It is well known that relapse or long-term remission rates are exceedingly variable in such patients after therapy is stopped [3]. To date, despite numerous attempts by well-known researchers working on thyroid disease to find a method providing accurate prognostic assessment, this has not been achieved [4]. So far, the tendency to relapse has been correlated with markers such as CD_4/CD_8, IgG, and antibodies against eye muscles or thyroid-stimulating hormone receptors (thyroid-stimulating immunoglobulin, TSI; thyroid-stimulating immunoantibody, TSIab). These examinations are cumbersome and are performed only in highly specialized research units; they are thus not available for the practising endocrinologist [2, 3, 5, 6].

From studies using these methods it has been reported that the titre of TSI starts to fall after 46 months of therapy [5, 6]. In this respect, it is of interest to note that our results demonstrate a normalization of cytological features after 4 months of treatment. Our additional observation that even after 18 months of therapy and despite clinical and biochemical euthyroidism, the results of FNAB can predict which patients will relapse, makes FNAB of potentially great value. It will be of great interest to correlate the results of FNAB with circulating markers in a larger sample of such patients.

Conclusion

FNAB provides a cheap, simple and quick laboratory test which can establish the diagnosis of hyperthyroidism. After 8 months of therapy with thionamides, results of FNAB help to predict the relapse of Grave's disease.

In our opinion, use of FNAB in patients with Grave's disease receiving treatment with thionamides may lead endocrinologists to seek a more specific mode of therapy.

Summary

Relapse is not unknown in Grave's disease after discontinuation of medical treatment; prediction of this phenomenon has been attempted using circulating serum markers. Our study deals with cytological features determined after fine-needle aspiration biopsy (FNAB) prior to and during therapy with thionamides. Thirty female patients aged 40 ± 10 years with active Grave's disease and ophthalmopathy were evaluated. FNAB was performed prior to and 4, 8, 12 and 18 months after the initiation of therapy. After the 2nd month of therapy all patients became eurythroid and remained so throughout the study period. Despite this, the cytological picture varied among patients. Thus, prior to therapy all patients had "normal" cytological findings which subsequently disappeared in approximately one-third of patients. After the 18th month, treatment was discontinued. All of the 17 (100%) patients who continued to show cytological abnormalities until the 18th month relapsed, but only one of 13 patients (8%) with normal cytological results showed clinical exacerbation. Persistence of abnormal cytology by the 8th month of therapy with thionamides allows one safely to predict relapse of the disease process after the discontinuation of medical therapy. These findings, if confirmed in larger samples, may indicate that treatment using drugs should be discontinued and that more specific modes of treatment should be sought.

References

1. Willens J-S, Löwhagen T (1981) Fine needle aspiration cytology in thyroid disease. Clinics in endocrinology and metabolism. Pathology and Management of Thyroid Disease p 247
2. Duron F (1982) Maladie de Basedow et immunoglobulines stimulant la thyroide. Encyclopédie Medico-Chirurgiale. Glandes endocrines Vol 1, Paris p 1
3. Duron F (1987) Action immunosuppressive des antithyroidiens de synthèse au cour de la maladie de Basedow. Encyclopédie Medico-Chirurgiale. Glandes endocrines Vol 1, Paris p 2
4. Walfish P, Strawbridge H, Rosen H (1985) Management implications of routine needle biopsy of hyperfunctioning thyroide nodules. Surgery 98: 1179-1188
5. Madek A, Laurent M, Lorey Y, Le Querrier F (1984) Thyroid stimulating antibodies: an aid to the strategy of treatment of Grave's disease? Clin Endocrinol (Oxford) 21: 247-255
6. Allannic H (1987) Exploration de l' autoimmunité thyroidienne en practique courante. La Revue Française d'Endocrinologie Clinique, Nutrition et Métabolisme 28: No 4-5

Cytological Monitoring of Human Hepatic Allografts

J. A. Young and R. M. Kirby

Introduction

Deterioration of hepatic function following orthotopic liver transplantation may be due to rejection, infection, ischaemia or biliary disorder [5]. Because of the risks associated with frequently repeated histological biopsy [3], a study was carried out to assess the feasibility of using cytological techniques to confirm or exclude the presence of cellular rejection. Fine-needle aspiration (FNA) is a well-established procedure for monitoring renal allograft dysfunction [1] but only very limited investigation of its possible role following liver transplantation has previously been undertaken [6–8].

Materials and Methods

Material

Control Specimens. Ten aspirates were obtained at laparotomy from patients who were being investigated for hepatic dysfunction but who were not part of the transplant programme, and eight were collected from excised livers removed prior to transplantation.

Transplant Specimens. 198 specimens from 36 grafts in 30 patients were examined. There were 24 baseline FNAs taken at the time of operation and 83 follow-up FNAs, 71 of which were percutaneous and 11 of which were obtained at laparotomy. These were supplemented by 92 lavage specimens taken from Menghini needles following biopsy. Histological samples were obtained at the same time as the cytological specimens on 103 occasions.

Methods

Methodology was based on that described by Hayry and von Willebrand [2] for renal allografts. A 21-gauge needle was used for FNA and laparotomy and a Chiba 'skinny' needle or a 23-gauge, 6-inch spinal needle (Steriseal) for percutaneous aspiration. FNAs were collected into 5 ml heparinised transplant medium (modified Eagle minimum essential medium with foetal bovine serum at 60 m/l). A similar volume of the same medium was used to flush the Menghini needles for the lavage samples. Differential cell counts were performed on the FNAs and washings and also on peripheral blood specimens. The blood values were subtracted from the hepatic values to obtain the total corrected increment (TCI) of inflammatory cells in the graft [2]. A TCI of 3.5 or above was considered indicative of rejection. At the end of the study the cytological results were compared with the final diagnosis based on histological biopsy, biochemical findings and subsequent clinical history.

Results

The mean TCI together with standard deviation and minimum and maximum values for controls (18), baseline specimens (24) and follow-up-specimens (174) are shown in Table 1. Simultaneously histological biopsy, available on 103 occasions, showed rejection in 73 instances and no evidence of rejection in 30. In the remaining episodes (71) without histological biopsy, retrospective clinical assessment indicated probable rejection on 17 occasions and that rejection was unlikely on 54. In total there were 21 false-negative cytological results and 11 false-positives, as summarised in Table 2. This gives an overall sensitivity of 76.7%, specificity of 86.9%, positive predictive value of 86.3%, and negative predictive value of 77.7%. Of the cytological specimens, 145 were obtained in the first 2 months after transplantation. These results are summarised in Table 3 and give a sensitivity of 81.3%, specificity of 90.0%, positive predictive value of 89.7% and negative predictive value of 81.8%. The overall accuracy in the first 2 months was 85.5%. This fell to 62.1% thereafter ($p = 0.006$, χ^2).

Table 1. Total corrected increments (TCI)

Group	*n*	TCI	SD	Min.	Max.
Controls	18	0.96	1.26	0	4.9
Baseline	24	1.15	1.16	0	3.5
Rejection (histological control)	73	5.15	2.44	0.2	11.2
Rejection (without histological control)	17	4.96	1.99	2.3	8.4
Rejection (total)	90	5.11	2.35	0.2	11.2
No rejection (histological control)	30	3.08	1.99	0.3	8.5
No rejection (without histological control)	54	1.83	1.50	0	7.3
No rejection (total)	84	2.27	1.79	0	8.5

Table 2. Results of cytological examination, all specimens

Cytology	Histology/retrospective diagnosis	
	Rejection	No rejection
Rejection	69	11
No rejection	21	73

$p < 0.001$ (χ^2 with Yates' correction)

Table 3. Results of cytological examination, first 2 months

Cytology	Histology/retrospective diagnosis	
	Rejection	No rejection
Rejection	61	7
No rejection	14	63

$p < 0.0001$ (χ^2 with Yates' correction)

Discussion

This study allowed a full assessment of the simplicity, safety, acceptability, accuracy and reproducibility of cytological monitoring of replacement livers. There was no reason to suspect that FNA was responsible for any trauma resulting in haemorrhage or pneumothorax and the procedure was well tolerated by patients.

Despite the good separation between the TCI values of the rejecting and non-rejecting groups there were a substantial number of false-negative and false-positive cytological results, although less in the first 2 months after grafting. In this early period, where cytological accuracy was 85.5%, it was found that FNA was clinically helpful as a procedure complementary to histological biopsy in the management of patients. However, after 2 months, when accuracy fell to 62.1%, it was felt that cytological monitoring was too unreliable to be clinically useful. Unfortunately it is just those patients still requiring investigation after 2 months that are most difficult to manage and in whom laboratory techniques are required to supplement clinical observations. The straightforward patient will have been discharged by this time and the majority of patients who die will have done so in the first 30 days after transplantation. It is therefore concluded, in the Birmingham liver transplant programme, that histological biopsy remains the 'gold standard' for the diagnosis of rejection [4] and that, apart from its limited value in the first 2 months, cytological monitoring is insufficiently sensitive or specific to contribute to patient management.

Summary

198 fine-needle aspirates and Menghini needle lavage specimens were obtained from 36 hepatic allografts in 30 patients. These were examined to determine the presence or absence of cellular rejection. Overall accuracy in the first 2 months following transplantation was 85.5% but this fell to 62.1% thereafter. It was concluded that cytological monitoring was insufficiently reliable for patients management.

Acknowledgments. We thank Dr. S. G. Hubscher, Mr. P. McMaster, Dr. E. Elias and Mr. W. Cuthbertson for clinical and technical assistance.

References

1. Hayry P (1985) Transplant aspiration cytology. Quality control analysis in 1984. Transplant Proc 17: 2100-2105
2. Hayry P, von Willebrand E (1981) Practical guidelines for fine needle biopsy of human renal allografts. Ann Clin Res 13: 288-306
3. Hegarty JE, Williams R (1984) Liver biopsy: techniques, clinical applications and complications. Br Med J 288: 1254-1256
4. Hubscher SG, Clements D, Elias E, McMaster P (1985) Biopsy findings in cases of rejection of liver allografts. J Clin Pathol 38: 1366-1373
5. Kirby RM, McMaster P, Clements D, et al. (1987) Orthotopic liver transplantation - postoperative complications and their management. Br J Surg 74: 3-11
6. Lautenschlager I, Hockerstedt K, Ahonen J, et al. (1987) Cellular characteristics of liver allograft rejection. Transplant Proc 19: 2485-2486
7. Vogel W, Fagen EA, Bomford A, et al. (1986) Wash off liver cytology: a complementary diagnostic tool to liver biopsy. J Clin Pathol 39: 449-452
8. Vogel W, Margreiter R, Schmalzl F, Judmaler G (1984) Preliminary results with fine needle aspiration biopsy in liver grafts. Transplant Proc 16: 1240-1242

X. Histology versus Cytology, Cell Collection and Staining

Cytology versus Histology in Carcinoma in Situ of the Breast and the Role of Fine-Needle Aspiration Cytology in the Treatment of Carcinoma in Situ

K. Lindholm

Introduction

The value of fine-needle aspiration cytology (FNAC) in the triple diagnosis of palpable breast lesions is well recognized and the value of FNAC in non-palpable mammographically revealed breast lesions has been documented in the last few years. The role of FNAC in the treatment of carcinoma in situ of the breast, however, needs to be discussed.

Discussion

Carcinoma in situ of the breast is described as two distinct entities - lobular carcinoma in situ (LCIS) and ductal carcinoma in situ (DCIS). LCIS is easily and reproducibly diagnosed in sections by today's histopathologists. LCIS seems to be common - in autopsy material LCIS is diagnosed in about 8% of women over 40 years of age. Retrospective studies have shown that 20% of women with unrecognized LCIS developed invasive cancer after an average of 17 years. However, the cancers developed as often in the ipsilateral as in the contralateral breast and as often into infiltrating ductal cancers as into infiltrating lobular cancers.

LCIS cannot be diagnosed preoperatively. Clinically it seldom gives parenchymal condensation or a palpable mass and mammographically it has no pattern. Consequently, the histological diagnosis of LCIS is an accidental one and surgery is performed for other reasons.

With this knowledge as a base a conservative treatment protocol with follow-up after accidental histological diagnosis of LCIS has now been decided upon in most places, but FNAC is hardly mentioned in this context. Can LCIS be diagnosed cytologically? Many experienced cytologists know of one or two cases reported as atypia, suggestive, or maybe even as cancer where subsequent biopsy revealed that the cellular material might correspond to LCIS (Fig. 1) but where on pure morphological grounds there was little in favour of a precise cancer diagnosis. I doubt that modern methods like DNA measurements or pure morphometry will solve the problem, although image analysis with pattern recognition could be of some help.

In my opinion FNAC has no place in the primary diagnosis of LCIS as there are no guiding methods for aspiration and the cytological picture does not permit a precise diagnosis. However, FNAC might have a place in the follow-up of LCIS since invasive cancer or DCIS may develop.

DCIS is a variable entity - variably and unreproducibly diagnosed by even trained pathologists. DCIS seems to be common. In autopsy material about 10% of 65-year-old women have clinically unknown DCIS. DCIS is a risk factor for invasive cancer but the natural history of DCIS is less well known than that of LCIS as cases with atypical ductal proliferations have seldom escaped further treatment after the diagnostic biopsy. It is calculated that about 50% of untreated cases develop invasive cancer after an average of 7 years. The cancer appears in the ipsilateral breast and is of the ductal type. This knowledge of DCIS has resulted in proposed treatment protocols which are conservative but more aggressive than those for LCIS. Segmental resections up to simple mastectomy followed only by clinical investigation and mammography, or randomized trials where segmental resections are followed by radiotherapy or follow-up alone have been suggested. In this context FNAC is mentioned in the Swedish protocol but not in the ones suggested by other European countries.

Could DCIS be diagnosed preoperatively? Widespread DCIS could give a palpable mass or parenchymal condensation but mammography is the method of choice. The typical calcifications in the intraductal lesions are not only the basis for the radiological diagnosis but also a superb guidance for FNAC.

There are two problems in cytological diagnosis of DCIS. The first is the above-mentioned variable cytological picture, which varies from the obvious malignant large cell comedo carcinoma in situ with its necrosis to the well-differentiated small cell cribriform lesions and well-differentiated papillary lesions (see Figs. 2-5).

The other problem is whether the tumor is only in situ or in situ and invasive. The occurrence of necrosis, large flat epithelial sheets, papillary structures, dissociated, rounded, degenerated single cells or small epithelial fragments and macrophages are all cytological details which suggest an in situ lesion, but invasion can never be ruled out. There are hallmarks for infiltration (Fig. 6) which should be looked for and reported. In my opinion, however, the final diagnosis of in situ or in situ and invasive cancer is a histopathological one which the surgeons must be made aware of.

FNAC has a definite place in the primary diagnosis and the treatment of DCIS. As there are guiding methods for aspiration, cytological cancer diagnosis is possible and, together with mammography, gives a sound basis for treatment.

The use of modern techniques such as DNA-measurements, possibly other markers, morphometry, and image analysis on cytological samples might also permit subtyping of DCIS into prognostic groups which could allow a more differentiated treatment.

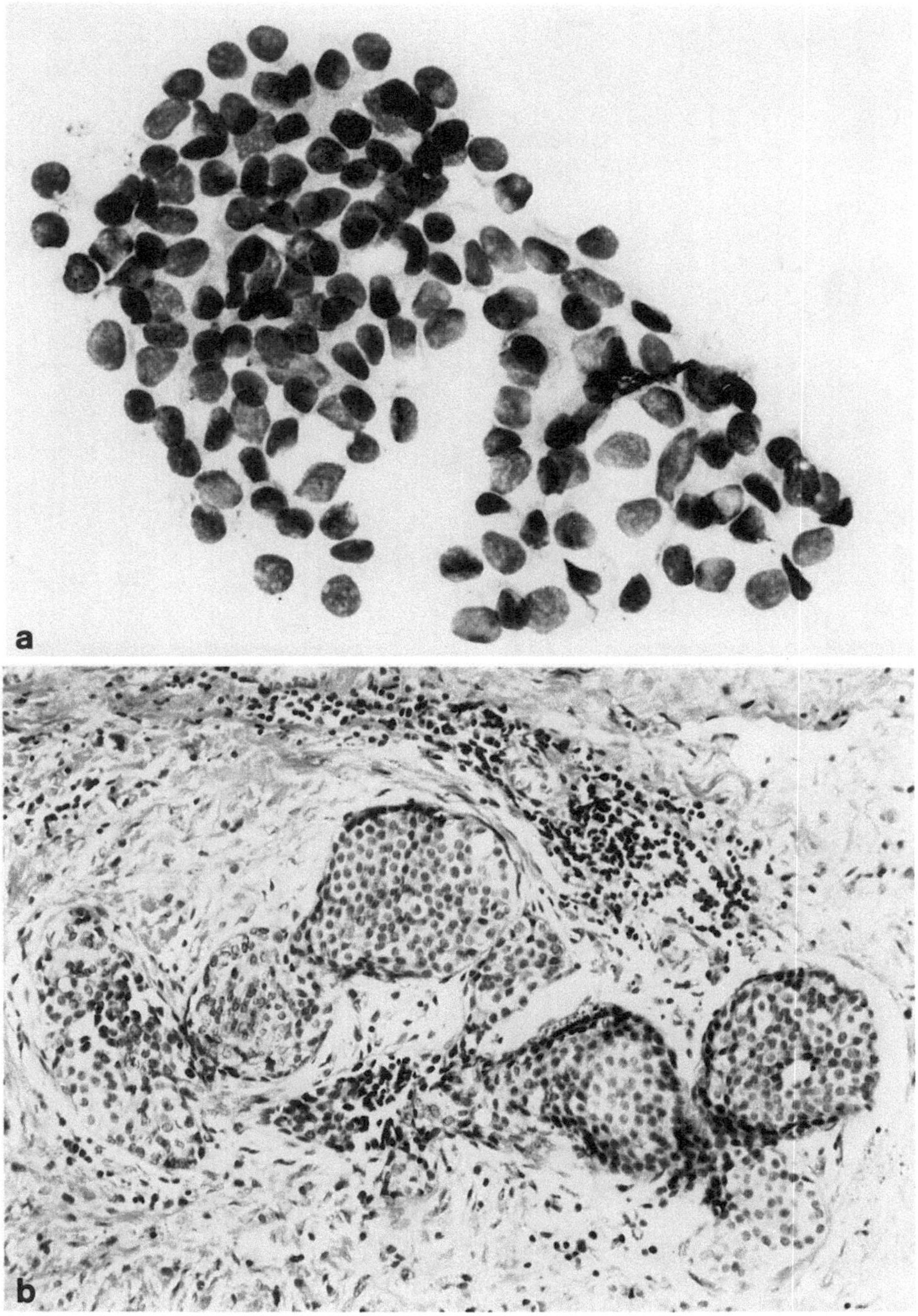

Fig. 1a, b. LCIS. **a** FNAC reported as showing equivocal atypia - mild nuclear atypia, lack of myoepithelial cells, slightly atypical cellular arrangement. $\times 470$. **b** Corresponding histology showed LCIS and diffusely infiltrating lobular cancer in fibrous tissue. $\times 180$

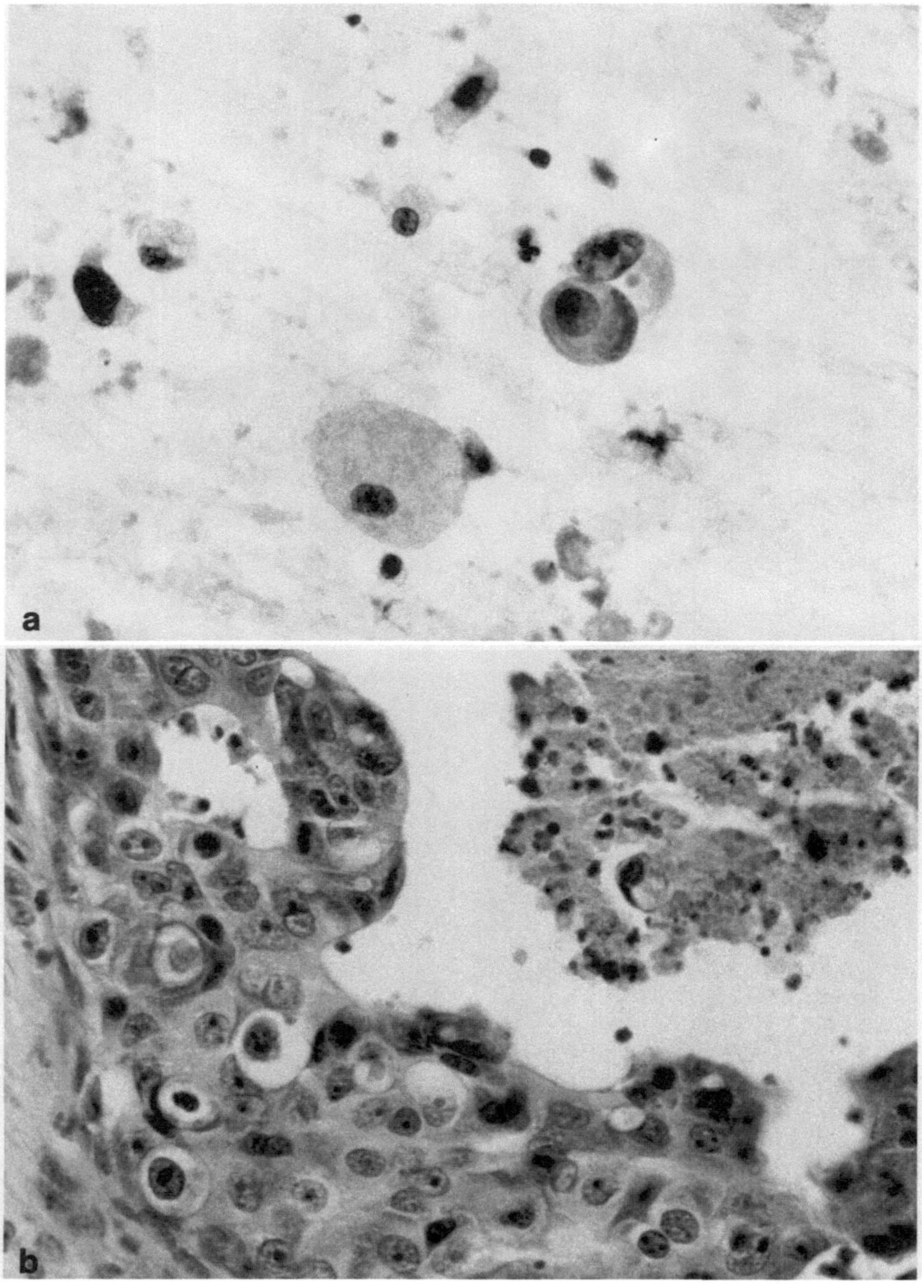

Fig. 2a, b. DCIS, comedo type. **a** FNAC reported as showing malignancy, consistent with DCIS - an obviously malignant cell group, pycnotic cell nuclei and a benign macrophage. × 580. **b** Histology showed widespread DCIS. × 360

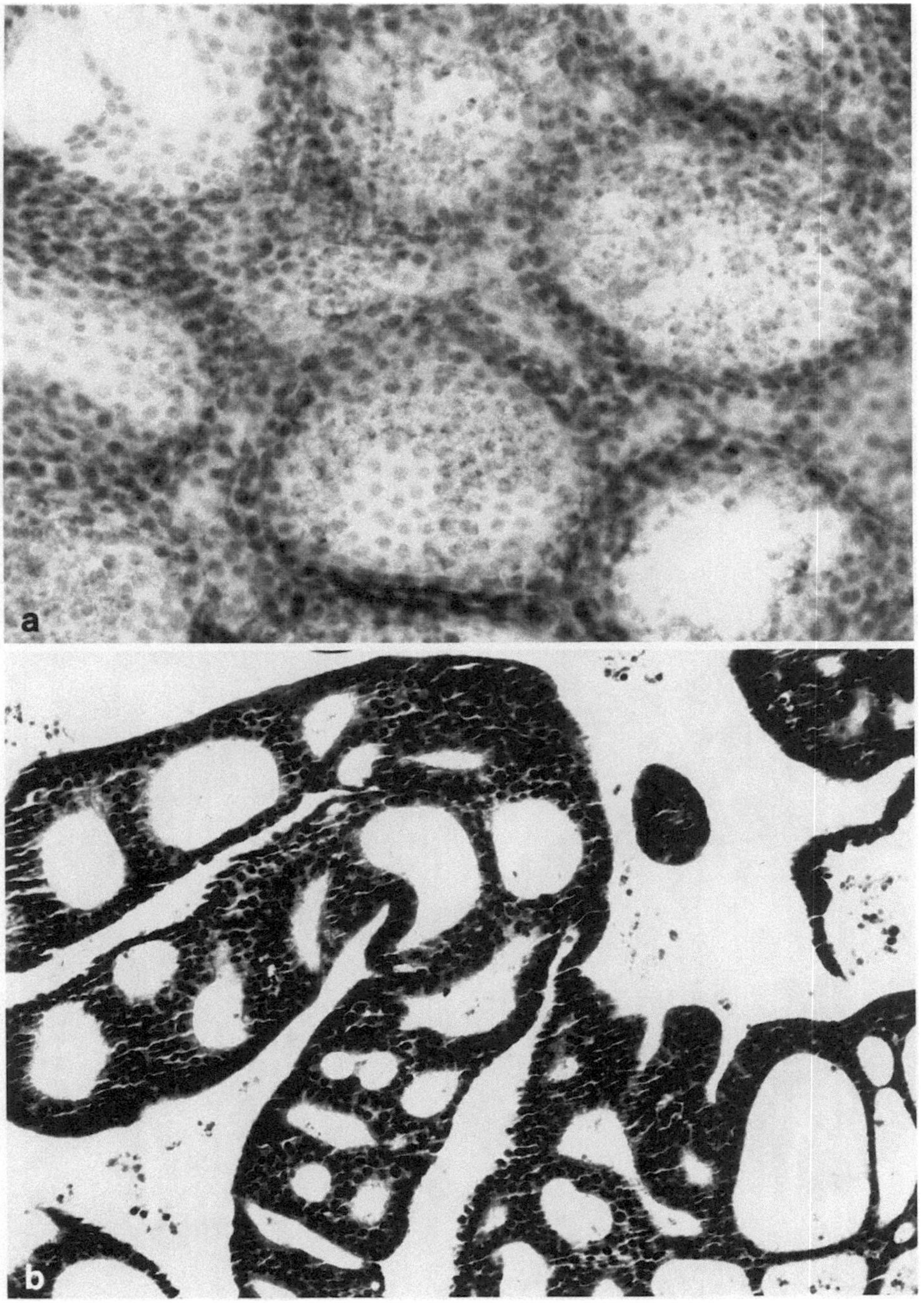

Fig. 3 a, b. DCIS, cribriform type. **a** FNAC reported as suggesting DCIS - one of several fragments with atypical cribriform pattern and necrotic dust in and around the fragments, no nuclear atypia. × 180. **b** Histology showed cystically dilated ducts with cribriform DCIS. × 180

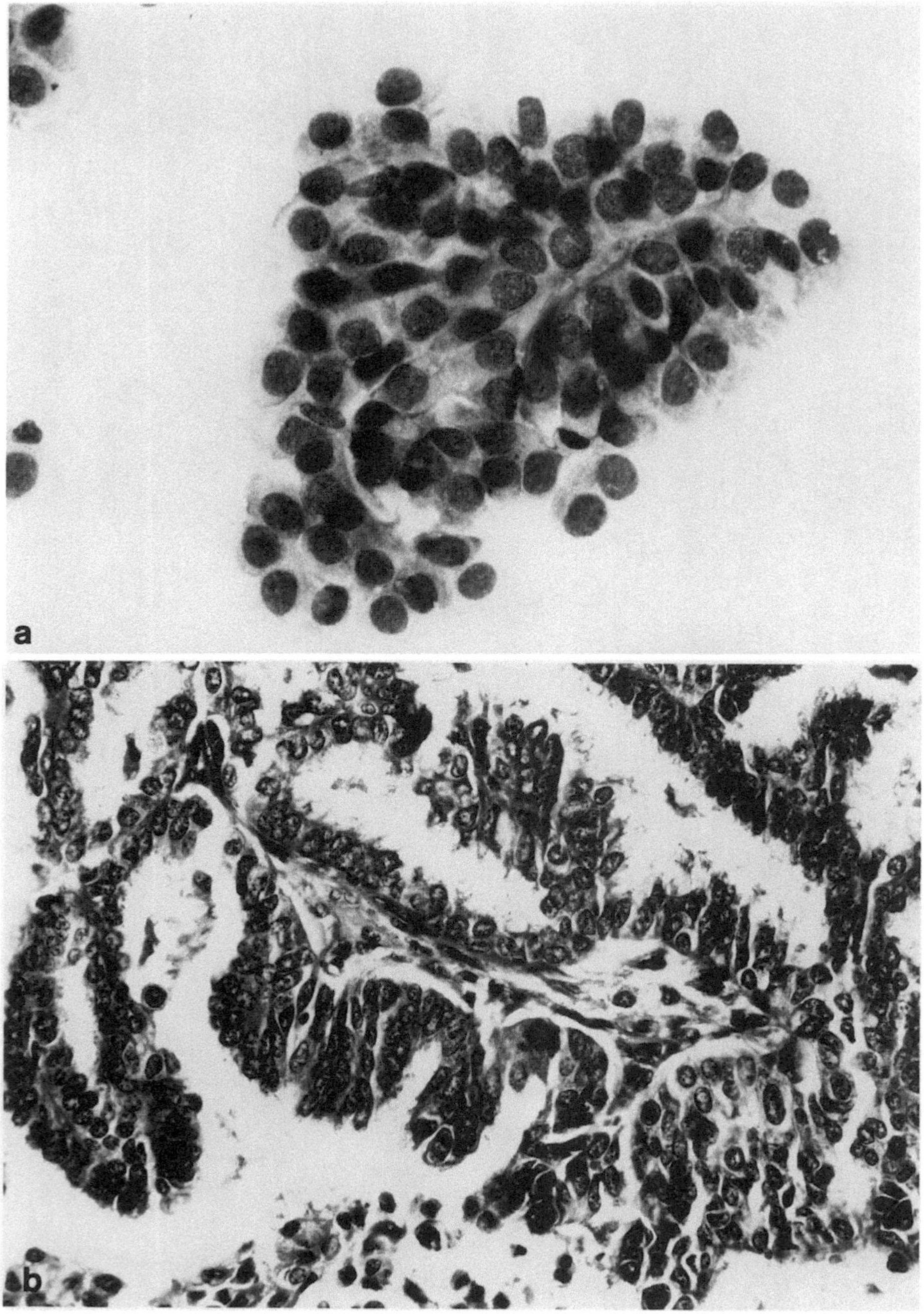

Fig. 4 a, b. DCIS, papillary type. **a** FNAC reported as showing malignancy consistent with well differentiated intracystic or intraductal cancer - cohesive fragment with central delicate fibrous stalk, palisading of the nuclei, no nuclear atypia (in other parts of the same smear dissociation and nuclear atypia). × 470. **b** Histology reported as showing intracystic cancer. × 360

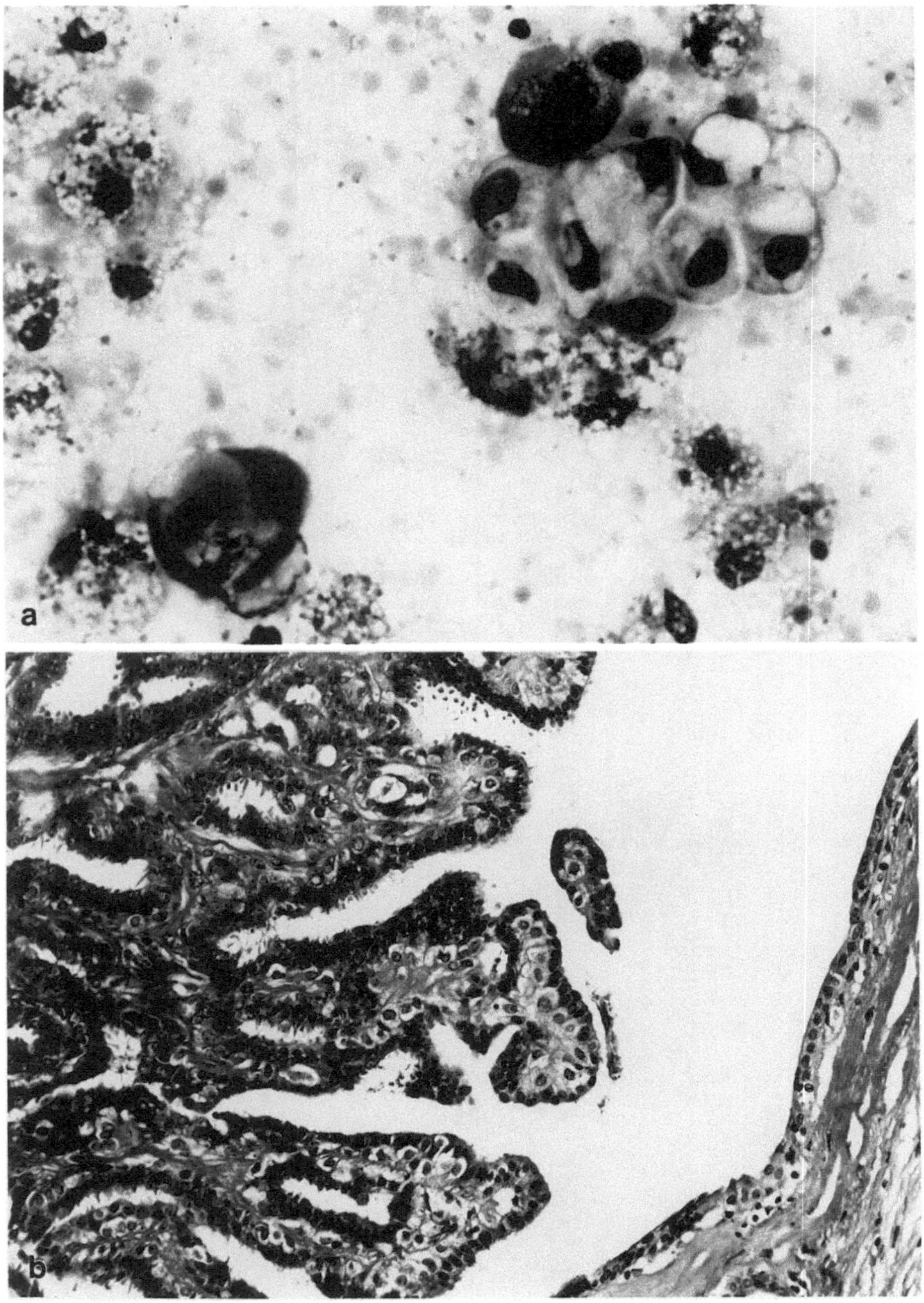

Fig. 5 a, b. Intraductal papilloma. **a** FNAC reported as being suspicious for DCIS - overdiagnosis due to the degenerative changes in the nuclei of the apocrine epithelium; note also the hemorrhagic background and pigment-loaded macrophages. × 470. **b** Histology. × 180

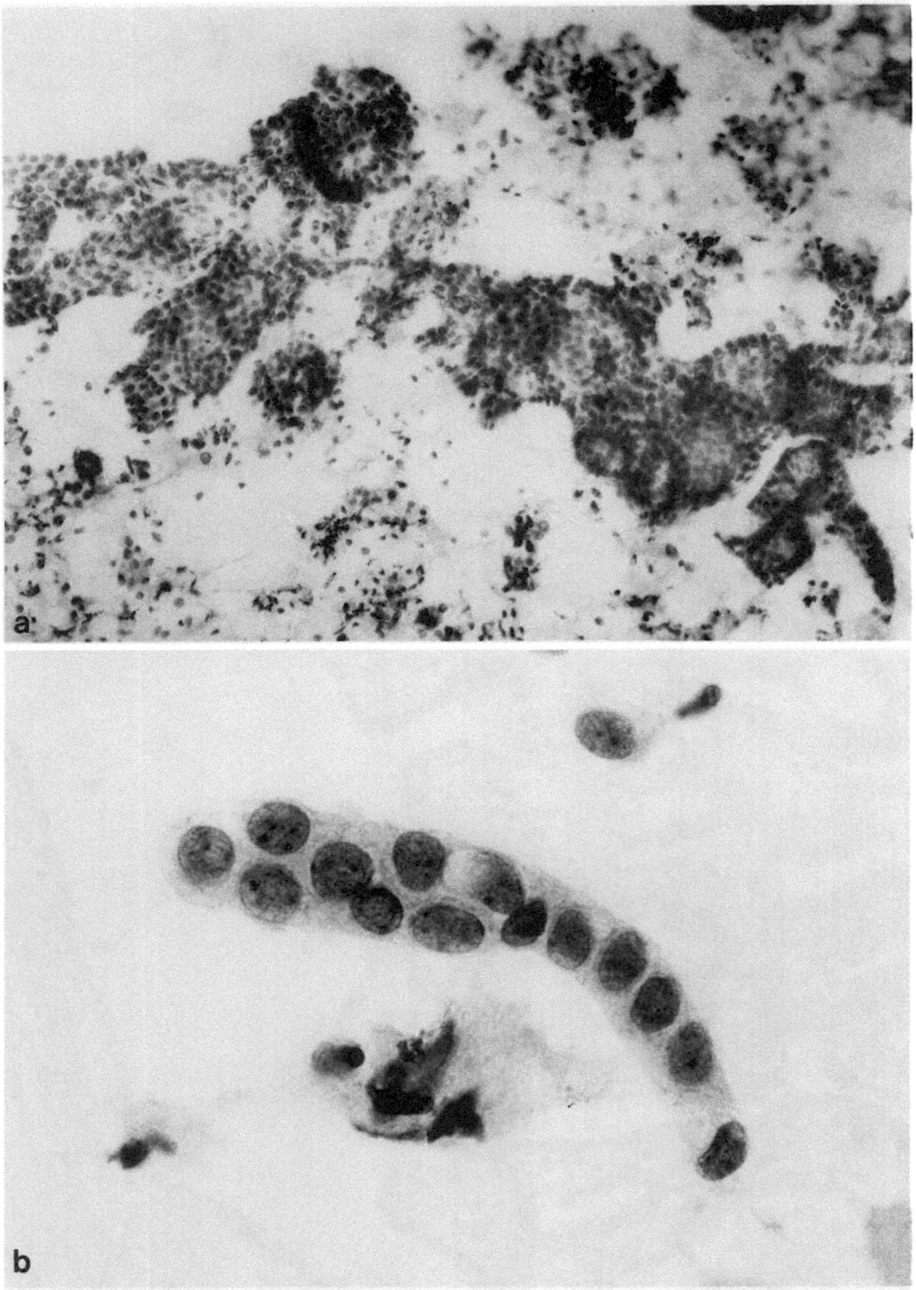

Fig. 6 a, b. FNAC of infiltrating mostly well-differentiated ductal cancer with surrounding cribriform DCIS. **a** Low magnification with large cohesive fragments probably from the DCIS but in a dirty background with fibrous tufts not typical of an in situ lesion. × 180. **b** High magnification - atypical epithelial cord with a small fibrous tuft both suggesting infiltrating cancer. × 900

Summary

FNAC is of no value in the primary diagnosis of LCIS but should be included in the follow-up treatment protocol as LCIS is not only a risk factor for recurrent LCIS but also for invasive cancer and DCIS. FNAC has a definite role in the primary diagnosis and the treatment of DCIS. When discussing the role of FNAC in this context the problem of the variable and unreproducible diagnosis of DCIS even by trained pathologists must be recognized, as must the fact that cytology can suggest the occurrence of DCIS but can never rule out invasion.

Comparison Between Fine-Needle Aspiration Cytology and Core Biopsy Histology in Prostatic Carcinoma

J. A. M. Beekhuis-Brussee, A. A. B. Lycklama à Nijeholt, J. L. Bruins, A. E. J. L. Kramer, F. Eulderink, H. J. H. M. van Krieken, G. J. van Steenis, and P. W. Arentz

Introduction

Ferguson (1930) introduced a method for transperineal needle puncture and aspiration of tissue from prostatic nodules in which carcinoma was suspected. This method was adapted by Franzén et al. (1960) to fine-needle aspiration biopsy (FNAB) for cytological examination. The latter authors used a specially designed flexible needle and a needle guide that makes a transrectal approach possible. They demonstrated a high accuracy and a low morbidity using this technique. Advantages of FNAB over core biopsy are precise puncture of small suspicious nodules, a less painful procedure, and a low complication rate (Esposti et al. 1975; Melograna et al. 1982; Zattoni et al. 1983; Carter et al. 1986; Bodner et al. 1987).

In Europe, FNAB of the prostate has become a valuable diagnostic method. The largest series were presented by Lin et al. (1979) who reported on 1280 FNABs and by Esposti (1966) who reported on 1110 cases. The limited number of reports from North America indicates that this technique is not yet widespread there (Hosking et al. 1983; Chodak et al. 1986). Uncertainty about diagnostic accuracy is probably the major reason (Hosking et al. 1983; Carter et al. 1986; Ljung et al. 1986), but lack of experience in the technique leading to inadequate sampling or inadequate judgement also may play a role (Lin et al. 1979; Melograna et al. 1982; Hosking et al. 1983; Zattoni et al. 1983; Carter et al. 1986; Chodak et al. 1986; DeBellis and Schumann 1986; Ljung et al. 1986; Schned et al. 1986; Bodner et al. 1987). The present study compares retrospectively the results of FNAB with the corresponding histological results obtained by core biopsy in patients in whom prostatic carcinoma was suspected.

Material and Methods

From 1 January 1980 till 31 December 1986, 826 patients with abnormal results of rectal examinations suggesting prostatic carcinoma were studied. Biopsies were taken in every case of even minimal suspicion. All biopsies were performed in the lithotomy position. Nearly all patients were outpatients. No anesthesia was used and the patients did not have bowel preparations. All patients received antibiotic prophylaxis (Bactrim) for 5 days, starting 2 days before the procedure. The transrectal FNAB was followed directly by the transrectal core biopsy. With both tech-

Table 1. Definition of test values

$$\text{Sensitivity} = \frac{\text{true positive}}{\text{true positive} + \text{false negative}}$$

$$\text{Specificity} = \frac{\text{true negative}}{\text{true negative} + \text{false positive}}$$

$$\text{Correlation} = \frac{\text{true positive} + \text{true negative}}{\text{total number of specimens except insufficient samples}}$$

niques the nodule and the contralateral side of the prostate were punctured. This resulted in two to four biopsies for both the aspiration and the core methods. The biopsies were performed routinely by the urologist or the urological resident in charge of the patient.

The FNAB was performed as described previously by Franzén et al. (1960; cf. Melograna et al. 1982). At least two smears were prepared immediately after each needle passage. The slides were air-dried and stained with May-Grünwald-Giemsa stain. All smears were prepared and read by the same cytotechnologist (J. A. M. B.-B.). The final report on the aspirate was authorized by the cytopathologist, who was not informed about the results of the prostatic core biopsy. The cytological specimen was classified according to the Papanicolaou system into one of five categories. Papanicolaou classes 1, 2, and 3 were considered as negative for malignant cells and classes 4 and 5 as positive.

The transrectal core biopsy was also performed without anesthesia. A disposable Travenol or Tru-cut needle was used to obtain the specimen. The left index finger guided the tip of the needle towards the biopsy region. The biopsy was repeated if the obtained core of tissue was unsatisfactory. After immediate fixing with formalin the specimen was stained with hematoxylin and eosin. The core biopsies were read by a pathologist. The tissue was classified as benign or malignant. The value of FNAB in comparison to core biopsy histological examination was expressed in terms of sensitivity, specificity and correlation (Table 1). The sensitivity is the fraction of positive results of histological tests that are confirmed by cytological tests and the specificity is the fraction of negative results of histological tests that are confirmed by cytological tests (Ljung et al. 1986).

In a group of patients with a discrepancy between the cytological and histological findings, data from the medical history were available that could provide the final diagnosis. Repeated biopsies, histological tests of surgically acquired or autopsy material, indications of disease progression by rectal palpation or by blood phosphatase levels, and bone scan data were considered. The cytological and histological preparations were then reexamined, without knowledge of the primary or the final diagnosis.

Cytological Appearance and Pitfalls

Cytological studies of benign prostate tissue show a honeycomb-shaped monolayer with small monomorphic nuclei and tiny nucleoli (Fig. 1 a), whereas the malignant prostate shows polymorphic overlapping nuclei and multiple prominent mac-

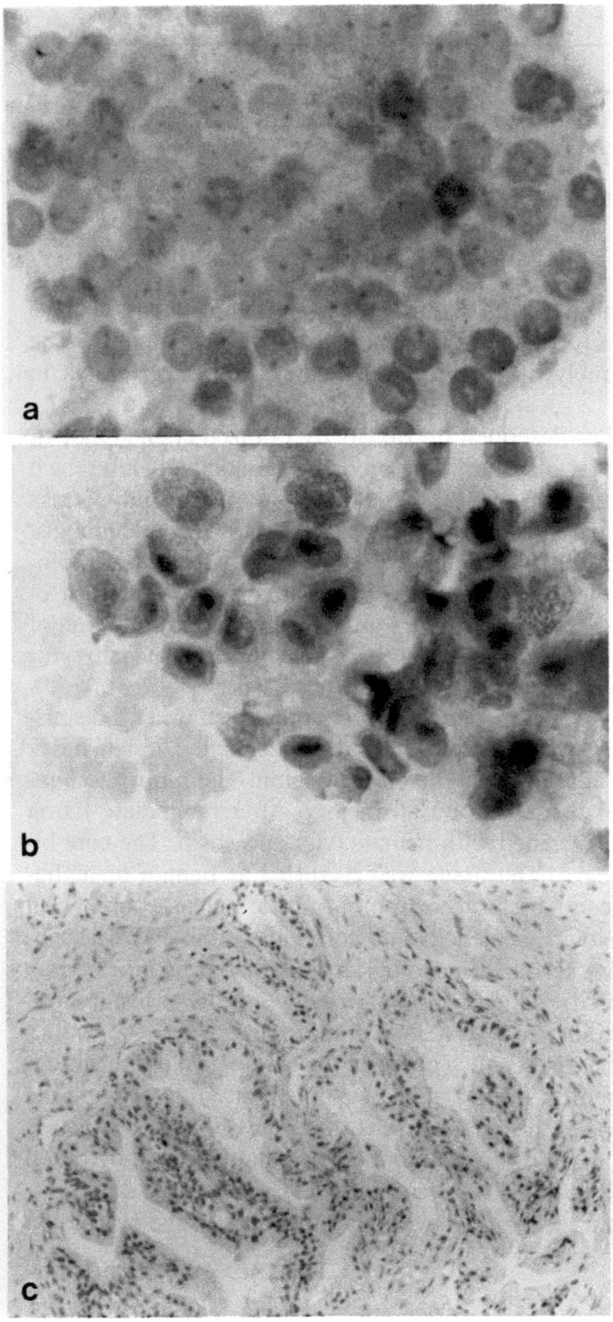

Fig. 1 a–d. FNAB cytology and core biopsy histology of the prostate. **a** Cytology benign **b** Cytology malignant **c** Histology benign **d** Histology malignant

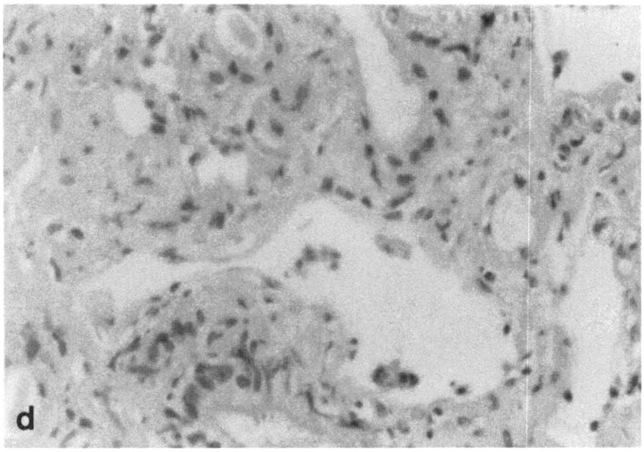

Table 2. Pitfalls of prostatic fine-needle aspiration biopsy cytology

False-positive results
1. Rectum epithelial cells
2. Urothelial cells
3. Seminal vesical cells
4. Severe inflammation with histiocytic giant cells
False-negative results
1. Unnoticed insufficient or inadequate material
2. Well differentiated adenocarcinoma

ronucleoli (Fig. 1 b). In Fig. 1 c, 1 d the corresponding histological preparations are shown.

Cytological studies, however, are prone to error (Table 2). False-positive results can occur in the prostatic FNAB when severe inflammation with histiocytic giant cells is present (Fig. 2) or when the technique is applied inadequately and rectum epithelial cells, urothelial cells, or seminal vesical cells are aspirated. False-negative results can occur in cases of well-differentiated adenocarcinoma of the prostate, or if there is unnoticed insufficiency of material or inadequate sampling that misses the suspect nodule. Differentiation between well-differentiated adenocarcinoma (Fig. 3 a) and nodular glandular hyperplasia (Fig. 3 b) can be extremely difficult (Maier et al. 1984).

Results

The initial comparison of cytological and histological results showed a sensitivity of 89.7%, a specificity of 81.5%, and a correlation of 84.4%. The combined results were as follows: 27.4% of cases were positive for malignancy, 46.0% were negative for malignancy, and 13.5% showed a discrepancy; in 13.6% of the cases inadequate or insufficient material was collected (Table 3).

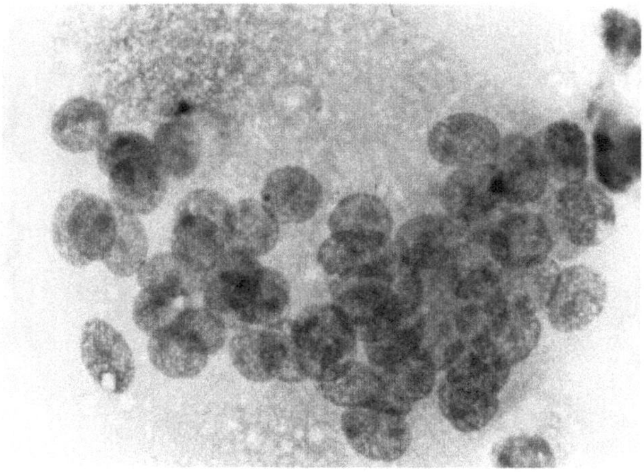

Fig. 2. False-positive FNAB cytology of the prostate: histiocytic giant cells

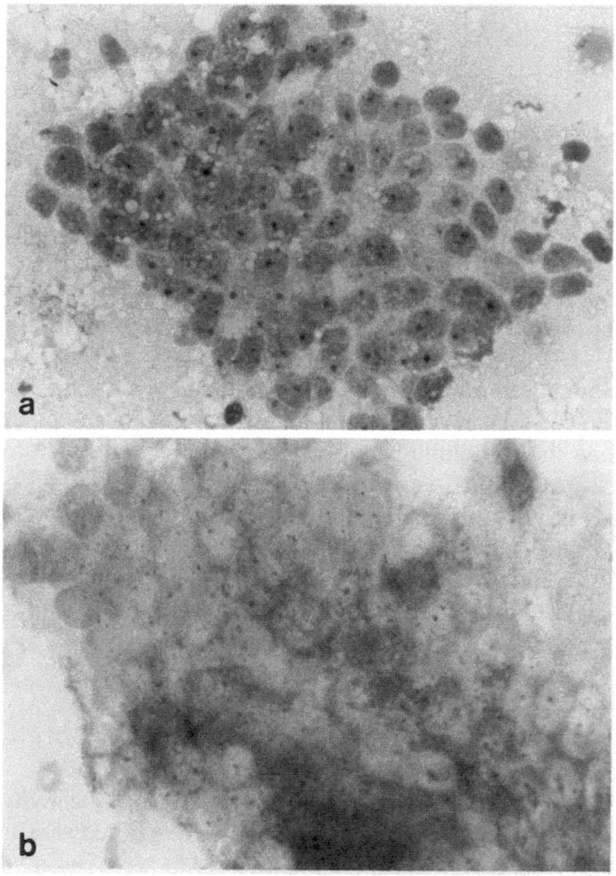

Fig. 3 a, b. Possible false-negative FNAB cytology of the prostate. **a** Well differentiated adenocarcinoma **b** Nodular glandular hyperplasia

Follow-up data were available from 84 of the 112 patients with discrepant diagnoses. Further histological tests or clinical parameters enabled a final diagnosis to be made in 62 patients (Table 4). In 22 patients no final diagnosis could be made. Malignancy was present in 32 patients, while 30 patients showed no signs of malignancy on follow-up.

Up to the present, 38 cytological and 37 histological preparations of 41 discrepant cases have been revised. This has led to 15 changes in the initial cytological diagnoses and six in the initial histological diagnoses (Table 5).

Comparison, after correction of the 21 revised cases, between cytological and histological results shows that the cytological tests had a sensitivity and a specificity of 90.8% and 85.3%, respectively. The correlation then reached 87.2% (Table 6).

In 91 cases there still existed a discrepancy between results of cytological and histological tests. In 41 of these patients the final diagnosis had been confirmed

Table 3. Correlation between initial cytological and histological findings ($n=826$)

Histology	Cytology							
	Positive		Negative		Insufficient		Total H	
	(n)	(%)	*(n)*	(%)	*(n)*	(%)	*(n)*	(%)
Positive	226	27.4	26	3.1	22	2.7	274	33.2
Negative	86	10.4	380	46.0	68	8.2	534	64.6
Insufficient	4	0.5	10	1.2	4	0.5	18	2.2
Total C	316	38.3	416	50.3	94	11.4	826	

C, cytology; H, histology

Table 4. Parameters determining a final diagnosis ($n=69$)

Parameter	*(n)*
Histology	
Repeat biopsy (if positive)	21[a]
Surgery	
Transurethral prostatectomy	19[b]
Radical prostatectomy	1
Millin prostatectomy	3
Lymph node dissection	3
Autopsy	2
Clinical	
Progression of disease	
Rectal examination	16
Alkaline phosphatase	4
Acid phosphatase	11
Bone scan	7

[a] $n=14$: primary diagnosis C+H−; repeated histology H+; $n=1$: primary diagnosis C−H+; repeated cytology C+.

[b] $n=3$: histological examination of resected material: benign in case of malignancy from an earlier biopsy.

Table 5. Discrepancy group with follow-up. Revision of deviant preparations ($n = 41$)

	Initial deviance	Revision
Cytology	29 C+H−	12 C+ 13 C−* 1 Co* 3 not found
	12 C−H+	11 C− 1 C+*
Histology	12 H+C−	7 H+ 2 H−* 3 not found
	29 H−C+	24 H− 4 Ho* 1 not found

Symbols: +, malignancy; −, no indications for malignancy o, insufficient/inadequate material *, change of initial reading

Table 6. Correlation between cytological and histological findings after revisions ($n = 826$)

Histology	Cytology							
	Positive		Negative		Insufficient		Total H	
	(n)	(%)	*(n)*	(%)	*(n)*	(%)	*(n)*	(%)
Positive	227	27.4	23	2.8	22	2.7	272	32.9
Negative	68	8.2	395	47.8	69	8.4	532	64.4
Insufficient	8	1.0	10	1.2	4	0.5	22	2.7
Total C	303	36.6	428	51.8	95	11.6	826	

C, cytology; H, histology

Table 7. Sensitivity and specificity of cytology and histology in discrepant readings for patients with a final diagnosis ($n = 41$)

	Malignant ($n = 29$)		Benign ($n = 12$)		Sensitivity	Specificity
	+	−	+	−	(%)	(%)
Cytology	20	9	9	3	69	25
Histology	9	20	3	9	31	75

Table 8. Sensitivity and specificity of FNAB and core biopsy histology with respect to existing disease

	Malignant ($n = 256$)		Benign ($n = 407$)		Sensitivity	Specificity
	+	−	+	−	(%)	(%)
Cytology	247	9	9	398	96.5	97.8
Histology	236	20	3	404	92.2	99.3

Table 9. Studies of fine-needle aspiration biopsy cytology of the prostate

Study	Biopsies (*n*)	Inadequate cytology (%)	Cyto-histological correlation (%)	Sensitivity[a] (%)	Specificity[a] (%)
Melograna et al. (1982)	87	15	81	79	93
Hosking et al. (1983)	74	?	81	75	84
Zattoni et al. (1983)	195	6	96	98	92
Maier et al. (1984)	433	4.2	78.8	81.9	85.0
Carter et al. (1986)	110	6	82	92	73
Bodner et al. (1987)	158	2	95	94	96
Urology, Leiden					
Previous study (de Voogt 1979)	294	4.1	94.0	97.9	89.9
Present study					
Initial	826	11.4	84.4	89.7	81.5
Revised data	826	11.6	87.2	90.8	85.3

[a] Compared to histological diagnosis.

(see Table 4). For this group of patients it was then possible to compare the cytological and histological findings with the final diagnosis. Sensitivity and specificity could be given with respect to the final diagnosis of malignancy (29 patients) or benign pathology (12 patients; Table 7).

An attempt can now be made to define an absolute sensitivity and specificity of both FNAB and core biopsy histology with regard to the existing disease. If it is taken for granted that the correlated results in Table 6 (227 positive and 395 negative) reflect 227 malignant cases and 395 benign cases, these can be added to the data from Table 7 for the final diagnosis in 41 discrepant cases. The results are shown in Table 8; the FNAB has a sensitivity of 96.5% and a specificity of 97.8%, while core biopsy histology has a sensitivity of 92.2% and a specificity of 99.3%.

Discussion

The value of FNAB cytology in the detection of prostatic carcinoma has been compared with core biopsy histology by a number of authors (Table 9). In general, the data presented here on a far greater number of biopsies, compare well with those from the other groups, although in our material both sensitivity and specificity seem to be lower than optimal. Also, the percentage of insufficient or inadequate FNABs in this material (11.6%) is clearly higher. Both results are probably explainable by the relative inexperience of the physician who collected the samples in our routine setting. Indeed it has been demonstrated that the fraction of inadequate sampling is diminished with growing technical experience (de Voogt 1979; Hosking et al. 1983; Carter et al. 1986; DeBellis and Schumann 1986; Schned et al. 1986). On the other hand, however, a repeat FNAB is feasible because the technique is relatively painless and harmless and can be performed rapidly (Hosking et al. 1983; Bodner et al. 1987). If applied in the same way to the approximately 12% of patients in whom punctures had been unsuccessful in the first session, and if this was again unsuccessful in 12% of them, this would diminish the failure rate to about 1.4% overall (Maier et al. 1984). Because less material is needed to perform the cytologic examination than the histologic one, the FNAB cytology may be expected to be more sensitive to clinically proven malignancy than the core biopsy histology is (Table 5; Ljung et al. 1986). This will be the case especially when (multiple) small suspect nodules are detected (Bodner et al. 1987). Because of the specialized technique of prostatic cytology and its pitfalls (Table 2), only an experienced cytopathologist should interpret the smears (Melograna et al. 1982; Bodner et al. 1987).

Repeat FNAB's are indicated in the case of insufficient or inadequate samples. A negative diagnosis in a patient with suspected cancer can be revised using the original preparation (Carter et al. 1986) or may warrant a repeat procedure (Melograna et al. 1982; Zattoni et al. 1983; Maier et al. 1984; Chodak et al. 1986). Core biopsy histology can then be performed to confirm or reject a subsequently negative cytology. Even then, however, a small fraction of prostatic carcinoma may be missed (Table 4; Bodner et al. 1987).

In this study, FNAB cytology and core biopsy histology showed correlated results in 87.2% of the cases, with a sensitivity of 90.8% and a specificity of 85.3% of cytology as compared to histology. The higher sensitivity of FNAB cytology in cases of clinically proven malignancy (Table 8; Hosking et al. 1983; Zattoni et al. 1983; Chodak et al. 1986; Ljung et al. 1986; Bodner et al. 1987), the relative ease of the technique, and the lower burden placed on the patient indicate that FNAB cytology of the prostate has a firm place in the diagnosis of prostatic carcinoma.

References

Bodner DR, Hampel N, Maksem JA, Suarez M, Resnick MI (1987) Aspiration biopsy of the prostate. World J Urol 5: 62-64

Carter HB, Riehle RA, Koizumi JH, Amberson J, Vaughan ED (1986) Fine needle aspiration of the abnormal prostate: a cytohistological correlation. J Urol 135: 294-298

Chodak GW, Steinberg GD, Bibbo M, Wied G, Strauss II FS, Vogelzang NJ, Schoenberg HW (1986) The role of transrectal aspiration biopsy in the diagnosis of prostatic cancer. J Urol 135: 299-302

DeBellis C, Schumann GB (1986) Prostatic aspiration cytology in the university teaching environment (Abstract). Acta Cytol (Baltimore) 30: 591-592

de Voogt HJ (1979) Urologische cytologie (in Dutch). Ned Tijdschr Geneeskd 123: 530-535

Esposti PL (1966) Cytologic diagnosis of prostatic tumours with the aid of transrectal aspiration biopsy. A critical review of 1110 cases and a report of morphologic and cytochemical studies. Acta Cytol (Baltimore) 10: 182-186

Esposti PL, Elman A, Norlén H (1975) Complications of transrectal aspiration biopsy of the prostate. Scand J Urol Nephrol 9: 208-213

Ferguson RS (1930) Prostatic neoplasm: their diagnosis by needle puncture and aspiration. Am J Surg 9: 507-510

Franzén S, Giertz G, Zajicek J (1960) Cytological diagnosis of prostatic tumours by transrectal aspiration biopsy: a preliminary report. Br J Urol 32: 193-196

Hosking DH, Paraskevas M, Hellsten OR, Ramsey EW (1983) The cytological diagnosis of prostatic carcinoma by transrectal fine needle aspiration. J Urol 129: 998-1000
Lin BPC, Davies WEL, Harmata PA (1979) Prostatic aspiration cytology. Pathology 11: 607-614
Ljung B-M, Cherrie R, Kaufman JJ (1986) Fine needle aspiration biopsy of the prostate gland: a study of 103 cases with histological follow-up. J Urol 135: 955-958
Maier U, Czerwenka K, Neuhold N (1984) The accuracy of transrectal aspiration biopsy of the prostate: an analysis of 452 cases. Prostate 5: 147-151
Melograna F, Oertel YC, Kwart AM (1982) Prospective controlled assessment of fine-needle prostatic aspiration. Urology 19: 47-51
Schned AR, Jordan AG, Castle WN, Richardson J, Selikowitz SM, Sullivan MM (1986) Fine needle aspiration and core biopsy of the prostate: a retrospective correlative study (Abstract). Acta Cytol (Baltimore) 30: 592
Zattoni F, Pagano F, Rebuffi A, Costantin G (1983) Transrectal thin-needle aspiration biopsy of prostate: four years' experience. Urology 22: 69-72

Technical Advances in Cell Collection as Applied to Prostatic Cytology

M. Devonec, P. Mouriquand, E. Piaton, M. Monsallier, P. Perrin, and J. Mouriquand

The diagnostic value of fine-needle aspiration biopsy of the prostate (FNAP) has been evaluated by many authors [1, 2] over the past few years to investigate whether this simple technique could challenge prostatic core biopsy (PCB) and even replace it in the diagnosis of prostatic cancer. These studies have demonstrated that FNAP has a good specificity but a lower sensitivity than PCB, even in series realized under apparently favorable conditions, i.e., with a high rate of prostatic cancer and with clinically palpable tumors. This explains why PCB has, up to now, remained the best clinical diagnostic reference, to which any new technique has to be matched.
The aims of our study were to:

1. Evaluate the feasibility of using the FNAP technique in a urology department in which there was no previous experience in this field
2. Compare the value of FNAP and PCB in the diagnosis of prostatic cancer
3. Compare the diagnostic value of FNAP against PCB when compared to a third technique, i.e., transurethral resection (TUR) chips taken as a histological reference

Material and Methods

FNAP was performed in 307 patients (group 1) between August 1986 and August 1987; among them 193 patients (group 2) also underwent a PCB. FNAP and PCB were performed either on a patient with a suspect prostate during rectal or ultrasound examination, or systematically while the patient was under general anesthesia prior to TUR of the prostate in 105 patients (group 3) with urinary obstructive symptoms.

FNAP was carried out either with a spinal needle (in 149 patients) or a Wescott (B-D) needle (in 158 patients); the transperineal route was used in 74 cases and the transrectal route in 233 cases. Transperineal puncture was ultrasonographically guided by an endorectal probe and was carried out when a suspect area, mainly hypoechoic and located in the peripheral zone, was detected by ultrasound [5], whether rectal examination was normal or showed a palpable nodule. Transperineal puncture guided with ultrasound requires local anesthesia but allows the puncture to be performed with great precision. Transrectal puncture was digitally guided. The procedure does not require local anesthesia; however, precision of the puncture is highly related to the skill of the urologist.

The FNAP technique used in this study is original and can be summarized as follows [3]: A one-way stopcock is connected to a 60-cc syringe and turned to the off position. The plunger is then vigorously pulled back and remains automatically locked in this position by a specially designed bolt previously adapted to the plunger. Thus, a high vacuum is created and maintained inside the syringe. After the prostate has been punctured, cell aspiration takes place automatically when the stopcock is turned to the on position. The puncture is then easy to perform and is very precise; the procedure avoids uncontrolled movements of the hand which are usually necessary to create vacuum with other techniques. Once the aspirate has been judged to be satisfactory, the stopcock is turned to the off position; the tip of the needle is then retrieved from the prostate and immediately immersed in the fixative solution. As soon as the stopcock is turned to the on position, the cellular material inside the lumen of the needle is automatically aspirated into the cylinder of the syringe, and in this way automatically fixed. In this study we used Fixfrot fixative solution (Biolyon) and Papanicolaou's staining technique [5]. PCB was only carried out transperineally, either with finger guidance (in 119 patients) or with ultrasonic guidance (in 74 patients).

Results

Among 307 cytologic samples collected in 1 year, 232 samples were evaluable. Samples it was not possible to evaluate were bloody, acellular, or badly fixed.

Cytology and PCB results were correlated for 193 patients (Table 1). The prevalence of prostatic cancer in group 2 was 19% (37/193). In this group 58% (22/37) of the cancers were non palpable, i.e. with a normal, adenomatous, or firm prostate but without any palpable nodule on rectal examination. The 15 remaining cancers were palpable on rectal examination. Among 37 malignant biopsies (PCB), 26 had a malignant FNAP. Thus, FNAP sensitivity was 70% (26/37), when PCB was considered as the standard. Among 156 benign biopsies (PCB), 140 had a benign FNAP and 16 had a positive result. Thus FNAP sensitivity was 90% (140/150) with the same standard. The results of FNAP and PCB were similar in 166 patients (85%) and divergent in 27 patients (14%).

Then we compared the results of FNAP and PCB with those from examination of TUR chips in the 105 patients of group 3, who had a FNAP and a PCB prior to TUR of the prostate (Table 2). When the TUR was positive, both the

Table 1. Correlation between FNAP and PCB results in 193 patients (group 2)

	PCB+	PCB−	Total
FNAP+	26	16	42
FNAP−	11	140	151
Total	37	156	193

Table 2. Correlation between TUR and FNAP plus PCB results in 105 patients (group 3)

FNAP/ PCB results	TUR+	TUR−	Total
FNAP+ PCB+	10	3	13
FNAP− PCB−	7	74	81
FNAP+ PCB−	3	5	8
FNAP− PCB+	2	1	3
Total	22	83	105

FNAP and the PCB were positive in ten patients, the FNAP or the PCB was positive in five patients, and the FNAP and the PCB were negative in seven patients. When the TUR was negative, both the FNAP and the PCB were negative in 74 patients, the FNAP or the PCB was positive in six patients, and both the FNAP and the PCB were positive in three patients.

Discussion

It was not possible to interpret 75 cytologic punctures; there are four factors which explain this high rate (75/307: 24%) of samples which could not be evaluated.

1. With a skilled urologist, the quality of the samples is improved: this was clearly demonstrated in our study, where newly trained urologists lacking experience with FNAP had the highest rate of non significant samples at the beginning of the study.
2. The size of the tumor: most of the FNAP carried out under ultrasonographic control were performed on nonpalpable hypoechoic areas less than 10 mm in their largest dimension.
3. Technical failure: red cell clumps inside the lumen of the needle have led to uninterpretable samples (acellular or sparsely cellular). This can easily be prevented with the use of heparin solution in the syringe, needle, and stopcock prior to puncture.
4. The quality of the fixative solution also plays an important role: the rate of specimens which could not be evaluated decreased significantly to 8% after we had changed our fixative solution and introduced material heparinization in our last 70 cases in a more recent study.

FNAP specificity was good in our study, at 90%, which is comparable to the performance recorded in the literature. However, our sensitivity was only 70%. This may be explained by the characteristics of our prostatic cancer group; the prevalence of prostatic cancer was low: only 19% (37/193) as compared to a much higher prevalence in other series [3, 4], which might partly explain the better score in these studies. Secondly, the size of prostatic cancers diagnosed in our series was small; indeed 58% of the cancers diagnosed were nonpalpable on rectal examination.

Finally, in 27 cases the FNAP results contradicted those of the PCB. FNAP was negative in 11 cases while PCB was positive. In these cases, one has to admit the superiority of PCB in the diagnosis of prostatic cancer. However, when FNAP was positive in 16 patients while PCB was negative, we, like other authors, do not claim that cytology is superior to histology, since we know the difficulty of cytological interpretation in prostatic cancer and the possibility of a false-positive answer, especially with a low-grade result.

The only way to evaluate the comparative diagnostic values of FNAP and PCB would be a comparison of these two techniques to a third one in which the diagnostic accuracy were not questionable: the only known unequivocal method is careful examination of step sections of the entire prostate. Careful examination of all TUR chips represents another possibility, which although much less accurate than the latter, offers the advantage of feasibility; this was the main objective of our study, although limited to only 105 patients.

Using TUR chip examination as a histological reference in group 3 (Table 3), modified the diagnostic sensitivity of both FNAP and PCB:

- FNAP sensitivity improved, going up to 95%, whereas it was only 70% in group 2 (193 patients), where FNAP was matched against PCB which was taken as the histological reference.
- FNAP demonstrated a better sensitivity than PCB: 95% versus 72% respectively with TUR as the histological reference in group 3.

These results underline the importance of taking into consideration a reference when evaluating the diagnostic performance of a technique. Finally, if the diagnostic reference included PCB and TUR taken together, the score of each of these

Table 3. Comparison of the diagnostic sensitivity of one single technique or two techniques according to the histological reference

Histological reference	Diagnostic sensitivity							
	FNAP		PCB		TUR		FNAP plus PCB	
	(n)	(%)	*(n)*	(%)	*(n)*	(%)	*(n)*	(%)
TUR (22 cases)	21	95	16	72	22	100	–	
TUR plus PCB (26 cases)	21	80	16	61	21	84	24	92
TUR plus PCB plus FNAP (31 cases)	21	67	16	51	22	70	24	77

When the diagnostic reference was TUR, 22 cancers were diagnosed; when the diagnostic reference was TUR plus PCB, 26 cancers were diagnosed; when the reference was TUR plus PCB and FNAP: 31 cancers were diagnosed.

techniques changed again and appeared to be slightly lower, still giving a better score for FNAP, 80%, versus 61% for PCB.

We do not conclude that FNAP sensitivity is better than that of PCB, but we emphasize that the diagnostic score can be optimized by the addition of FNAP and PCB: it improves to 92% or 77% according to the diagnostic reference, TUR plus PCB or TUR plus PCB plus FNAP, respectively, which leads to the conclusion that these techniques are complementary.

No complications have been recorded with FNAP technique in this study, whereas we experienced a morbidity rate of 10% in a previous study of 271 PCBs [6]. This advantage coupled with an excellent diagnostic sensitivity makes FNAP a very attractive tool for the diagnosis of prostatic cancer.

Summary

Fine needle aspiration of the prostate (FNAP) was performed in 307 patients. Among them 193 also underwent a prostatic core biopsy (PCB) and 105 a PCB and a transurethral resection (TUR). The prevalence of prostatic cancer was 19% (37/193) and the tumors were small: 58% (22/37) of the cancers were nonpalpable on rectal examination. The diagnostic values of FNAP and PCB were compared. The sensitivity of FNAP was 70%, while its specificity was 90%, when PCB was considered as the histological reference. However, when TUR was taken as the histological reference, FNAP sensitivity was better than that of PCB (95% versus 72%). These results show the importance of the diagnostic reference when evaluating the diagnostic performance of a technique. The best results were obtained with the combination of FNAP and PCB, with 92% sensitivity. FNAP and PCB are complementary techniques in the diagnosis of prostatic cancer.

References

1. Carter HB, Riehle RA, Koizumi JH, Amberson J, Vaughan ED (1986) Fine needle aspiration of abnormal prostate: cytological correlation. J Urol 135: 294-298
2. Chodak GW, Steinberg GD, Bibbo M, Wied G, Straus FS, Vogelzang NJ, Schoenberg HW (1986) Role of transrectal aspiration biopsy in diagnosis of prostatic cancer. J Urol 135: 299-303
3. Devonec M, Mouriquand P, Perrin P (1987) A new technique for fine needle aspiration biopsy of solid tumors. Proceedings of the XVth European Congress of Cytology, Sept 1987, Baden-Baden
4. Lee F, Littrupp J, McLeary RD, Kumusaka GH, Borlaza GS, McHugh TA, Soiderer MH, Roil D (1987) Needle aspiration and core biopsy of prostate cancer: comparative evaluation with biplanar transrectal US guidance. Radiology 163: 515-520
5. Mouriquand P, Mouriquand J, Devonec M, Delorme E, Perrin P (1987) L'apport du Polyéthylène Glycol dans la ponction cytologique de prostate. J Urol (Paris) 93: 73-75
6. Perrin P, Mouriquand P, Delorme E, Grisard M, Dutrieux-Berger N, Dassonville F (1986) Valeur de la ponction biopsie prostatique dans le diagnostic des cancers prostatiques: à propos de 271 observations. J Urol (Paris) 92: 277-281

Transthoracic Fine-Needle Aspiration Cytology Guided by Fluoroscopy of Pulmonary Masses: Analysis of 460 Cases

G. Ciarmiello, A. Dal Pozzo, C. Sartori, I. Di Natale, G. Chiara, and G. Di Falco

Introduction

Transthoracic fine-needle aspiration is now in widespread use as a method for obtaining deep material for microscopic diagnosis; it is particularly useful for diagnosis in patients with suspected cancer [5]. In the last decade it was shown that the method, if properly applied, gives excellent results because of its high diagnostic accuracy and the low incidence of complications [3, 4]. However, this method is utilized in only a few centers in Italy, where it has been described as "excellent but unfamiliar."

The purpose of the present study was the evaluation of the results obtained with transthoracic fine-needle aspiration cytology in patients with pulmonary masses which could not be diagnosed using other methods.

Materials and Methods

Aspiration biopsies were performed by a surgeon under fluoroscopic control (biplane fluoroscopy) in the majority of cases, and by use of ultrasound (real time) in the cases of pleural adjacent mass and in Pancoast's tumors. We used a 22 or 23 gauge needle, choosing the perpendicular route. Part of the material obtained was immediately fixed in 95% ethanol for subsequent cytologic staining, and part was stained using the May-Grünwald-Giemsa technique and viewed by the cytologist in the radiology room. We thus reduced the specimens which were inadequate for evaluation. There were 460 individuals in the study, who were observed from January 1982 to December 1987. Of those in the study, 20 (4%) were outpatients; 365 were men and 95 women; most of the patients were 40-74 years old.

Indications

Any undiagnosed lung mass is a basic indication for needle aspiration biopsy; this lesion can be symptomatic (cough, fever or hemoptysis) or discovered incidentally in a totally asymptomatic patient; it can also be a metastatic lesion of unknown origin. The clear demonstration of the nature of a lung lesion by FNAC allows appropriate and efficient management of the patient.

Contraindications

Serious chronic obstructive or restrictive pulmonary disease ($FeV_1 = 1$ liter), contralateral pneumonectomy, serious bleeding diathesis, a suspected echinococcus cyst, or lack of cooperation are contraindications. In interstitial pneumopathy the FNAC is not performed because the method gives results which are usually inadequate for diagnosis [2].

Technique

The biopsy was performed without any premedication or fasting. We used a fluoroscopy that allows monitoring of the procedure in the frontal and lateral projections by rotating a tube and an image intensifier (Siemens Arkoscop Pandoros-Optimatik). The patient lay prone or supine depending whether an anterior or posterior approach was planned; in rare cases, we anesthetized the skin and the tissues down to the pleura with 2% Xylocaine. Breath-holding was limited when the needle was crossing the pleura and shallow breathing was encouraged for the rest of the time. The perpendicular route of the needle was controlled under direct fluoroscopy; when the lesion was reached, the lesion motions, in lateral control, demonstrated the correct position of the needle. The material was obtained by aspiration of a constant suction syringe connected to the needle. After 1 h we checked all the patients through standard chest radiography.

Complications

There were no deaths. Pneumothorax was the most common lesion, occurring in 12% of the cases, of which only a few required a chest tube. Hemoptysis, which was usually minimal, occurred in about 10% of patients; transient pleuritic chest pain occurred occasionally; we did not observe subcutaneous emphysema, but there was one case of large bloody pleural effusion, which was treated by thoracocentesis.

Results

Of 460 consecutive FNACs performed, in 432 a diagnosis of benign or malignant lesion was indicated; 28 (6%) samples were estimated inadequate for diagnosis. Of the 432 cases, 320 (74%) were estimated to be malignant lesions, comprising me-

Table 1. Cytohistologic confirmation

	True	False			%
		(n)	(%)		
Negatives	43	5	10	Sensitivity	96,6
Positives	145	1	0.6	Specificity	97,5
	188	6			

Table 2. Cytologic typing confirmed by histology

	Cytologic diagnosis	Histologic diagnosis	Correlation (%)
Adenoca	32	23	72.0
Large cell ca.	21	16	77.4
Squamous cell ca.	34	28	82.3
Small cell ca.	30	30	100
Lymphoma	3	3	100
	120	100	

tastatic lesions, and 112 (26%) were benign lesions. Among these cases we obtained 194 histologic confirmations (Table 1): there were 43 true negatives and 5 (10%) false negatives; there were 145 true positives and 1 (0.6%) false positive (a case of pseudolymphoma).

Among the 145 malignant lesions, in 25 not differentiated cases we were just able to exclude small cell carcinoma. In 120 cases the tumor was typed cytologically; the cytohistologic correlation is shown in Table 2.

Discussion

The sensitivity of 0.96 was higher than that reported in other studies. The diagnosis of malignancy is reliable since the predictive value for the positive results is 0.99, with one false positive in a pseudolymphoma. The diagnostic reliability of cytologic typing of primary malignant tumors with histologic confirmation is 0.83; the value of results reported in the literature varies from 0.65 [3] to 0.89 [1]. In conclusion we think that this method, inexpensive, quick, and absolutely less invasive than a thoracotomy, looks like the better to assess the nature of selected pulmonary lesions, before surgery.

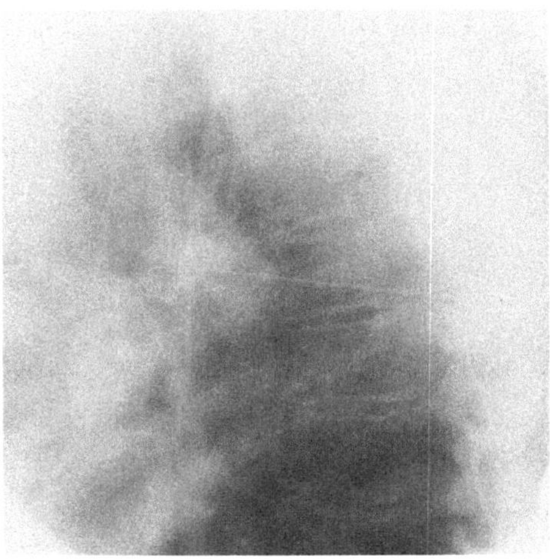

Fig. 1. Nodular pulmonary lesion tested by fine-needle biopsy

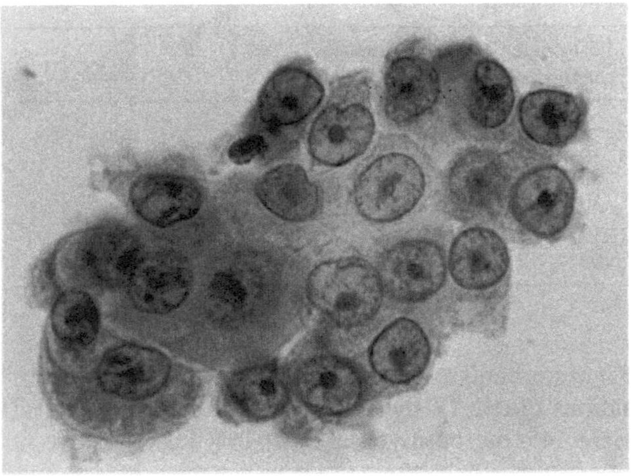

Fig. 2. Microscopic photography of a cytologic smear (fine-needle aspirate) of adenocarcinoma. H & E, × 1000

References

1. Michel RP, Lushpihan A, Nisar Ahmed M (1983) Pathologic findings of transthoracic needle aspiration in the diagnosis of localized pulmonary lesions. Cancer 51: 1663-1672
2. Montali G, Croce F, Ravetto C, Ghiringhelli G, Solbiati L (1983) Agobiopsia polmonare percutanea per aspirazione con ago sottile. Radiol Med (Torino) 69: 559-565
3. Pilotti S, Rilke, F, Gribaudi G, Damascelli B (1982) Fine needle aspiration biopsy; cytology of primary and metastatic pulmonary tumors. Acta Cytol (Baltimore) 26: 661-666
4. Sinner WN (1979) Pulmonary neoplasm diagnosed with transthoracic needle biopsy. Cancer 43: 1533-1540
5. Tao LC, Pearson FG, Delarue NC, Langer B, Sanders DE (1980) Percutaneous fine needle aspiration biopsy: 1. Its value to clinical practice. Cancer 45: 1480-1485

Fine-Needle Aspirates of Nonpalpable Tumor Masses Guided by Computerized Tomography: Definition of Their Primary or Metastatic Nature Using Immunocytochemical Methods*

M. Mottolese, M. Rinaldi, I. Venturo, C. Gallo Curcio,
R. Perrone Donnorso, and P. G. Natali

Introduction

Fine-needle aspiration biopsies sampled under computerized tomography (CT-FNA) is increasingly performed for the cytodiagnosis of nonpalpable tumor masses [5]. While the neoplastic nature of the lesions can be recognized in most instances by morphologic criteria, whether the neoplasia is primary or metastatic can rarely be established. The use of monoclonal antibodies (MoAbs) to tumor-associated antigens (TAAs) in this context has resulted only in a limited improvement because most of the TAAs recognized are not strictly tumor specific and are endowed with variable degrees of heterogeneous expression [12]. In the present study, we have explored whether, instead of this, the use of combinations of MoAbs, selected on the basis of an extensive immunohistochemical testing, may be of potential value for: (a) the diagnosis of primary cancer, and (b) the differentiation between a primary or metastatic lesion. The data presented in this chapter demonstrate that this methodologic approach can significantly improve the cytodiagnosis of CT-FNAs of nonpalpable tumor masses with an accuracy of 84%.

Materials and Methods

Patients

A total of 51 patients (42 males and 9 females) admitted to the Regina Elena Cancer Institute were referred for CT-FNA biopsy because of radiographic appearance of one single, or multiple pulmonary or abdominal lesions, at varying times, following treatment for various malignancies or for a tumor of unknown origin.

Biopsies Procedure and Cellular Sampling

Tumor biopsies were done with a 22-gauge disposable needle under CT guidance. Cellular specimens were immediately smeared onto glass slides and fixed in 95% alcohol for conventional Papanicolaou staining or in cold acetone for 10 min for

* Supported by CNR PF "Oncologia" No. 870155944, by AIRC, and by Tecnobiomedica

immunocytochemical tests. Surgical biopsies of different tumors were frozen in liquid nitrogen and stored at 20 °C until use. Also, 4 μm cryostat sections were obtained and fixed for 10 min in absolute acetone for immunohistochemical tests. These fixation and storage conditions maintained the serological reactivity of the specimens for at least 4 months.

Monoclonal Antibodies

MoAbs to TAAs employed in this study were chosen on the basis of two main serological characteristics: (a) they were to react only with transformed cells, and (b) their reactivity was to be of limited heterogeneity among tumor cells. MoAbs to various TAAs were kindly provided by different investigators (MBR1, MOv19, OC-125, KS1/4, D612, RC38) or obtained commercially (B72.3, B6.2 Sorin Biomedica Saluggia Vc, Italy; Table 1). [1-3, 7, 10, 14].

Immunoperoxidase

Tissue specimens were stained using an indirect avidin-biotin complex (ABC) immunoperoxidase method with commercially available reagents (Immucolor Kit, Sorin Biomedica) [4]. Slides were incubated overnight with MoAbs at 4 °C in a humid chamber. The protein concentration of the primary antibody ranged from 25 to 50 μg/ml in Hanks balanced salt solution (HBSS) containing 1% bovine serum albumin (Sigma, St. Louis, Mo, USA). The enzymatic activity was developed using 3-amino-9-ethylcarbazole (AEC) as chromogenic substrate. Slides were counterstained with Mayer's hematoxylin (BDH Chemicals Ltd, Pool, England) and mounted in buffered glycerol. Immunocytochemical diagnosis, evaluated independently by at least two investigators, was subsequently compared with histopathological and/or clinical data.

Results

The data in Table 2 demonstrate that the combinations of MoAbs with the different TAAs we selected, when used in immunoenzymatic tests, were able to identify through their positive and/or negative reactivity, tumors of different origin.

Table 1. Monoclonal antibodies used for the immunocytochemical diagnosis of CT-FNA

Antibody	Isotype	Antigen	Molecular weight	Major reactive tumor (carcinomas)	References
B72.3	IgG1	Glycopr.	$>10^6$	Adenocarcinomas	[1-4, 7, 10, 13, 14]
B6.2	IgG1	Glycopr.	9×10^4	Breast	[3]
MBr1	IgM	Glycolip.		Breast	[2]
OC-125	IgG1	Glycopr.	0.5×10^6	Ovarian	[1]
MOv19	IgG2a	Glycopr.	3.8×10^4	Ovarian	[7]
KS1/4	IgG2a	Glycopr.	4×10^4	Lung	[14]
D612	IgG2a	Glycopr.	$>2\times10^5$	Colonic	[12]
RC38	IgG1	Glycopr.		Renal	[10]

Table 2. Immunodiagnostic patterns of reactivity of neoplastic tissue with monoclonal antibodies to tumor-associated antigens

B72.3	B6.2	MBr1	MOv19	OC-125	KS1/4	D612	RC38	Immunocytochemical diagnosis (carcinomas)
+	+	+	−	−	ns	−	−	Breast
+	−	−	+	+	ns	−	−	Ovarian
+	−	−	−	+	+	−	−	Lung
+	−	−	−	−	−	+	−	Colonic adenocarcinoma
+	−	−	−	−	−	−	+	Renal

ns, not significant

Table 3. Definition of primary or metastatic nature of CT-FNA of nonpalpable tumor masses using selected combinations of monoclonal antibodies to TAA

Site of CT-FNA		Patients (*n*)	Immunocytochemical diagnosis (carcinomas)	Concordance with histologic and/or clinical diagnosis
Lung		40	25 Primary lung	25/25
			4 Metastatic colonic	4/4
			1 Metastatic breast	1/1
			1 Metastatic ovarian	1/1
			1 Metastatic renal	1/1
			8 Undefined	6 Primary pulmonary carcinoma 2 Met. larynx carcinoma
Liver		3	1 Metastatic ovarian	1/1
			1 Metastatic breast	1/1
			1 Metastatic lung	1/1
Pelvis	Lymph nodes	3	3 Metastatic colonic	3/3
	Intestinal tumor	1	1 Metastatic ovarian	1/1
	Supraren. Gl.	2	2 Metastatic lung	2/2
	Lymph nodes	1	2 Metastatic renal	1/1
	Lymph nodes	1	1 Undefined	1 bladder carcinoma
				Total diagnosis 42/51 (83%)*

* It was possible to reach a tentative diagnosis cytomorphologically in a minority of cases (24%)

Among the MoAbs used it should be underlined that MoAb B72.3 detects an antigen shared by most adenocarcinomas [13] while the reagents MOv19, D612, and RC38 have a restricted serological reactivity with ovarian, gastrointestinal, and renal carcinoma respectively.

On the basis of the results shown in Table 2, we examined immunocytochemically a total of 51 CT-FNAs sampled from pulmonary or abdominal tumor masses. As summarized in Table 3, 25 out of 40 CT-FNA of single or multiple lung lesions could be defined as primary lung tumors. This diagnosis was based on two parameters: firstly the reactivity of the sampled cells with at least two MoAbs identifying lung TAAs (B72.3, OC-125, and KS1/4) and secondly their lack of reactivity with the other MoAbs (Fig. 1a–b). The 15 CT-FNA which appeared clearly morphologically malignant, but failed to react with MoAbs directed to lung TAAs, were tested with MoAbs chosen on the basis of available clinical data. As a result of this additional testing, four of these pulmonary lesions could be diagnosed as metastatic (from a colonic adenocarcinoma, one from a breast carcinoma, one from an ovarian carcinoma, one from a renal carcinoma), while in eight instances the isolated expression of the pancarcinoma antigen TAG-72 recognized by MoAb B72.3 did not allow any diagnosis beyond that of "undefined carcinoma." Conventional morphologic analysis of the same specimens could establish the primary or metastatic origin of the pulmonary malignant tumor in about 24% of the instances.

Abdominal CT-FNA

The immunocytochemical analysis of three hepatic CT-FNA allowed the identification of the metastatic origin of these lesions indicating the ovarian, mammary,

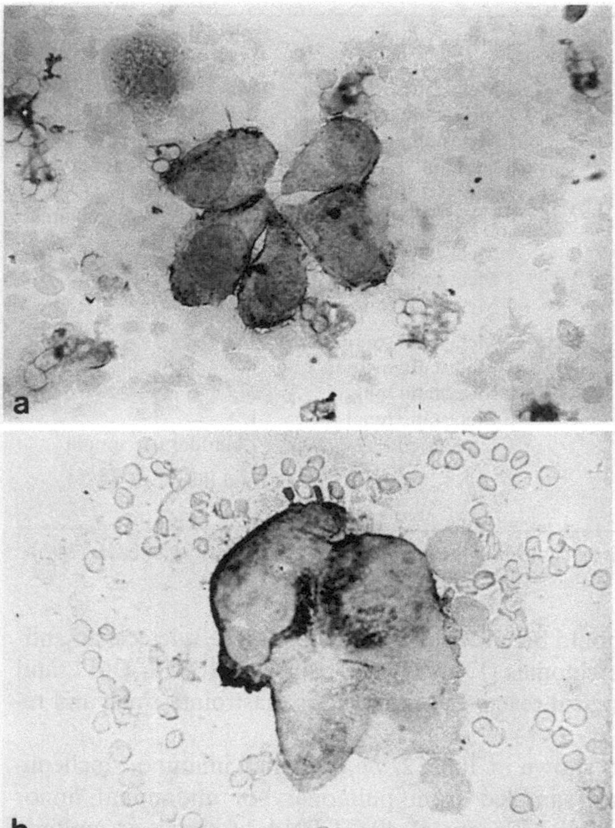

Fig. 1a–b. Biotin-avidin immunoperoxidase staining of CT-FNA of a primary lung tumor. Cellular specimens display a homogeneous reactivity with MoAb KS1/4 (**a**) and MoAb B72.3 (**b**)

and pulmonary origin of the metastasis. In seven out of eight CT-FNAs of nonpalpable pelvic tumor masses, through immunologic diagnosis we were able to define the metastatic origin of the lesions identifying the origin of the primary neoplasia in 87% of the cases (Table 3).

Discussion

A number of malignancies (i. e., mammary, renal, and gastrointestinal) are characterized by the property of metastatizing to the lung or in abdominal districts during their clinical course producing radiological images of single or multiple lesions [9, 11]. This feature may raise, depending on the clinical situation, a number of diagnostic questions such as: (a) Is the mass a primitive tumor? (b) Is the lesion a metastasis from a previously treated cancer? (c) Is this a metastasis from a new primary tumor?

This differential diagnosis, which is obviously relevant from the therapeutic as well as prognostic points of view, has recently been improved by the possibility of sampling tissue for cytodiagnosis by FNA under CT guidance. Despite this improvement, this still conventional cytology can, in most instances, because of the paucity of the tissue sampled, only discriminate between benign and malignant cells, while it may be severely limited in establishing the primitive or metastatic nature of the lesion, or in identifying the organ of origin of the metastasis. The use of MoAbs to TAAs or cytoskeletal components in immunocytochemical assays is expected to complement current cytodiagnosis by adding both serological specificity and objectivity in the interpretation of the findings. Immunohistochemical tests which employ either a single MoAb or a panel of antibodies have in fact been described [6]. Using these methods the attempt to distinguish between metastatic focus of colonic, mammary, and ovarian origin, and primary lung neoplasms on cryostat sections was not significantly successful. We have recently shown that the use of a combination of MoAbs to TAAs can significantly improve the diagnostic accuracy of cytodiagnosis of neoplastic effusions as well as FNAs [8]. The panel of the reagents chosen in these studies was such that the detrimental effect of the heterogeneity of expression of the TAAs was minimized, and that it was possible for both positive and negative findings to complement each other.

Some limitations of the assay should be underlined. A primary lung cancer was more accurately diagnosed in male patients. In females the cross-reactivity of MoAbs KS1/4 and OC-125 with breast and ovarian carcinoma provide a questionable immunocytochemical diagnosis. In this study we have shown that with a limited number of monoclonal reagents, the accuracy of current cytopathology can be greatly improved, identifying the primary or metastatic origin of the neoplasia in 84% of cases, answering the diagnostic questions which are raised by the appearance of pulmonary or abdominal masses.

Summary

The possibility of performing fine-needle aspiration (FNA) biopsies of nonpalpable masses under the guidance of computerized tomography (CT-FNA) has significantly increased the diagnostic use of FNA cytology. In order to evaluate whether immunocytochemical methods employing selected panels of monoclonal antibodies (MoAbs) to tumor-associated antigens (TAA) can improve the diagnostic potential of FNAs we have evaluated 51 CT-FNAs. Of 40 pulmonary, three hepatic, and 11 pelvic CT-FNAs, the primary or metastatic nature of the lesions could be established in 84% of the cases. The present data demonstrate that the use of immunocytochemical methods can significantly improve the diagnostic accuracy of cytology and expand the application of CT-FNAs.

References

1. Bast RC, Feeney M, Lazarus H, Nadler LM, Colvin RB, Knapp RC (1981) Reactivity of a monoclonal antibody with human mammary carcinoma. J Clin Invest 68: 1331-1337
2. Canevari S, Fossati G, Balsari A, Sonnino S, Colnaghi MI (1983) Immunochemical analysis of the determinant recognized by a monoclonal antibody (MBr1) which specifically binds to human mammary epithelial cells. Cancer Res 43: 1301-1305
3. Colcher D, Horan-Hand P, Nuti M, Schlom J (1981) A spectrum of monoclonal antibodies reactive with human mammary tumors. Proc Natl Acad Sci USA 78: 3199-3203
4. Hsu SM, Raine L, Fanger H (1981) Use of avidin-biotin complex (ABC) in immunoperoxidase technique: a comparison between ABC and unlabeled antibody (PAP) procedures. J Histochem Cytochem 29: 577-580
5. Kaminsky DB (1981) Aspiration biopsy for the community hospital. Mason, Paris
6. Meijer CJLM, Mullink SC, Henzen-Logmans SC (1987) Practical application of monoclonal antibodies to diagnostic tumorpathology and some future perspectives. In: Ruiter DJ, Fleuren GJ, Warnaar SO (eds) Application of MoAbs in tumor pathology. Nijhoff, Dordrecht
7. Miotti S, Canevari S, Menard S, Mezzanzanica D, Porro G, Regazzoni M, Pupa SM, Tagliabue E, Colnaghi MI (1987) Characterization of human ovarian carcinoma associated antigens defined by novel monoclonal antibodies with tumor restricted specificity. Int J Cancer 39: 297-303
8. Mottolese M, Venturo I, Perrone Donnorso R, Gallo Curcio C, Rinaldi M, Natali PG (1988) Use of selected combinations of MoAbs to TAA in the diagnosis of neoplastic effusions of unknown origin. Eur J Cancer Clin Oncol (in press)
9. Nystrom JS, Winer JM, Wolf RM et al. (1979) Identifying the primary site in metastatic cancer of unknown origin. JAMA 241: 381-383
10. Oosterwijk E, Ruiter DJ, Wakka JK, Huiskens JW, Jonas U, Fleuren GJ, Zwartendijk J, Hoedemaeker P, Warnaar SO (1986) Immunohistochemical analysis of monoclonal antibodies to renal antigens. Application in the diagnosis of renal cell carcinoma. Am J Pathol 123: 301-309
11. Ruiter DJ (1986) Differential diagnosis of metastatic lesions in the lung. In: McVie SG, Bakker W, Wagenaar SS, Carney D (eds) Clinical and experimental pathology of lung cancer. Nijhoff, Dordrecht
12. Schlom J, Weeks MO (1985) Potential clinical utility of monoclonal antibodies in the management of human carcinomas. In: DeVita VT, Hellman S, Rosemberg SA (eds) Important advances in oncology. Lippincott, Philadelphia, pp 170-192
13. Thor A, Ohuchi N, Szpak CA, Johnston WW, Schlom J (1986) Distribution of oncofetal antigen tumor associated glycoprotein-72 defined by MoAb B72.3. Cancer Res 46: 3118-3124
14. Varky NM, Reisfeld RA, Walker LE (1984) Antigens associated with human lung adenocarcinoma defined by monoclonal antibodies. Cancer Res 44: 681-687

Fine-Needle Aspiration in Ophthalmic Pathology

P. Boccato and E. Midena

Introduction

Over the past few years fine-needle aspiration (FNA) has gained relatively high acceptance in ophthalmology [9]. With this technique solid or cystic masses of the orbit, lid, or globe can be promptly and inexpensively investigated, greatly increasing the cytodiagnostic possibilities available to the clinical ophthalmologist.

In comparison with classical surgical biopsy techniques, FNA is atraumatic and currently performed on an outpatient basis, except for intraocular lesions. Concerning orbital and periorbital tumors (or space-occupying lesions), surgical biopsy without aggressive procedures is limited to the anterior orbit. FNA allows a delicate and precise approach to anterior and posterior orbital lesions, and yields sufficient material in most cases sampled.

Even if the accuracy of clinical diagnosis of intraocular tumors has improved in recent years, there are still confusing cases [1]. Except for iris lesions [5] there is not a sufficently safe biopsy technique to be used in investigating suspected intraocular malignancies. At present, FNA is the most adequate method for sampling intraocular tumors.

This paper describes our techniques for fine-needle aspiration of orbital, periorbital, and ocular masses, the assessment of the adequacy of samples and our own experience.

Principles of the Technique

Fine-needle aspiration of 82 orbital (and periorbital: lids and lacrimal gland) and 39 intraocular tumors was performed. Enough material for cytologic diagnosis was yielded from 115 lesions. We used the following techniques of aspiration, depending on the localization of the tumor: (a) in the orbit or (b) in the eyeball.

In palpable (anterior) lesions the material was obtained by:

- Simply introducing a fine needle into the tumor mass without aspiration [11]
- Using a butterfly needle [3], inserted into the lesion by the operator's hand, which rested comfortably on the patient's face (Fig. 1)
- Using a needle connected to the syringe by an extension tubing, or attached to the syringe as usual [8]

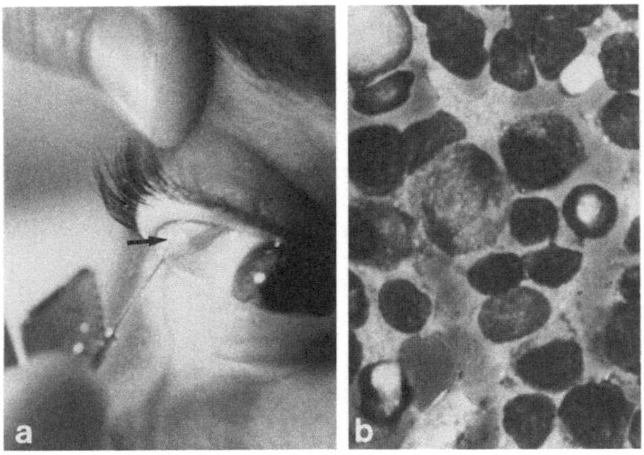

Fig. 1. a A small lacrimal gland lump *(arrow)* is sampled using a butterfly fine needle inserted into the lesion by the operator's right hand, which leans comfortably on the patient's face. **b** The corresponding cytological picture indicates with an inflammatory pseudotumor of the lacrimal gland (May-Gruenwald-Giemsa, original magnification × 360)

In nonpalpable, deep lesions the shortest and safest [7] way was chosen and the operator generally reached the target mass with the aid of ultrasonography or of CT-scan. Except when the space-occupying lesion was really anterior (ultrasonographically checked), we commonly used the needle with stylet (spinal needle) to avoid contamination by lacrimal gland tissue [6].

Intraocular lesions were reached by various methods [9]:

- For anterior segment lesions, the needle was inserted through the limbus, opposite the mass
- For posterior segment lesions, the *pars plana* approach opposite the tumor was usually chosen
- If the tumor was in the ciliary body and the patient was aphakic, the cornea provided a possible insertion point

Recently a modified technique of FNA has been proposed in cases of suspected, posterior-confined, retinoblastoma [10]. The cornea is perforated (1-2 mm inside the limbus) and the ophthalmologist proceeds directly through the iris root and zonula into the (suspected) tumor.

In order to assess the adequacy of samples obtained we used a rapid staining technique [2] and microscopic observation was performed in a room near the operating theater (Fig. 3 A). A recently introduced procedure is the observation of air-dried, unstained smears under a microscope with the condenser lowered, the screen diaphragm narrowed, and the additional lens excluded, i.e., under high-refringence conditions (Fig. 2). This can be done while waiting to examine other rapid-stained smears or may sometimes be sufficient on its own. This microscopic observation "in black and white" [4] of aspirated material allows us: (a) to evaluate the adequacy of a sample, (b) to avoid performing other aspirations when the material

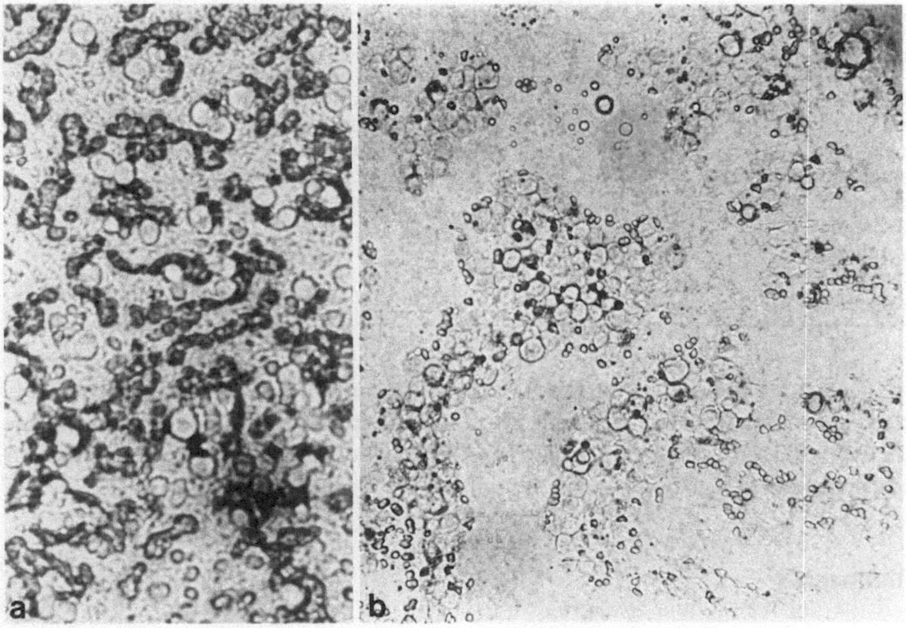

Fig. 2. **a** FNA of an orbital mass (plasmocytoma) in a 82-year-old woman: isolated, round, monomorphic cells (at a higher magnification plasmocytoid in type) intermingled with erythrocytes are clearly visible. **b** Cytocentrifuge preparation from an aqueous tap, performed in a subject (65-year-old woman) affected by an adenocarcinoma of the nonpigmented epithelium of the ciliary body: large clusters of neoplastic cells, epitheliomorphic in type are identifiable. In these air-dried, unstained smears (**a**, × 330; **b**, × 200) the sampling was judged adequate (so-called cytology in black and white)

collected is adequate, and (c) to perform other aspirations in selected cases, especially when needed to provide additional material for special staining or immunocytochemistry.

Results

Table 1 summarizes the results from 121 patients in whom FNA of ophthalmic lesions was performed. In six cases the orbital specimen was inadequate for interpretation because of the fibrous consistency of the tumor, or its localization in the orbital apex. Our samples from intraocular lesions were not inadequate, and we believe that this was mainly due to the use of rapid staining in order to assess the adequacy of the samples.

We observed only a few cases of subconjunctival hemorrhage after orbital FNA. Some orbital hematomas are reported in the literature: they all resolved spontaneously, without sequelae. During FNA of intraocular malignancies, using the transvitreal approach, we observed the appearance of a tiny vitreous hemorrhage. It commonly cleared spontaneously.

Table 1. Summary of the cytological findings

Lids, lacrimal gland, and orbit		Intraocular	
Mucocele	6	Nevus (iris)	2
Cysts	20	Adenocarcinoma	1
Angioma	7	Melanoma	23
Benign mixed tumor	5	Retinoblastoma	2
Optic nerve glioma	1	Metastasis	11
Carcinoma	12		
Lymphoma	15		
Metastasis	10		
Inconclusive	6		

In expert hands, none of the patients who underwent intraocular FNA suffered any major visual disturbance. Nobody reported local dissemination of malignancy after orbital or intraocular FNA. The cytohistologic correlation made in most instances was especially encouraging [1].

Comments

Our data confirm that FNA is a useful and safe cytologic technique for obtaining a rapid and correct diagnosis of orbital and eye "tumors." Sometimes it allows the solving of difficult and complex cases (Fig. 3 B). We hope that in the future, FNA will allow ophthalmologists to define accurately not only the nature (benign, malignant, melanoma, etc.) of intraocular tumors, but also the histologic diagnosis of melanoma-type cells (Fig. 3 A).

In our experience FNA for cytologic investigation of ophthalmic-mass lesions is an important diagnostic technique with an acceptable complication rate. It gives the clinical ophthalmologist accurate diagnostic information, often preventing unnecessary, invasive biopsy or surgical procedure (metastasis, lymphomas, benign tumors). This technique must be carried out by an ophthalmologist who deals with orbital and intraocular mass lesions on a regular basis. A cytopathologist cooperating with the ophthalmologist performing the aspiration makes an ideal team.

Summary

During the past few years fine-needle aspiration (FNA) has gained relatively high acceptance in ophthalmology. With this technique, solid or cystic masses of the orbit, lid, or globe can be investigated promptly and inexpensively, greatly increasing the cytodiagnostic possibilities available to the clinical ophthalmologist.

The eye and orbit represent a complicated and delicate area to be sampled; different approaches and procedures and our own experience (since 1979) in FNA of intraocular and orbital lesions are presented. In the authors' experience, FNA for cytologic investigation of ophthalmic mass lesions is an important diagnostic technique with acceptable complication rate. This technique must be done by an

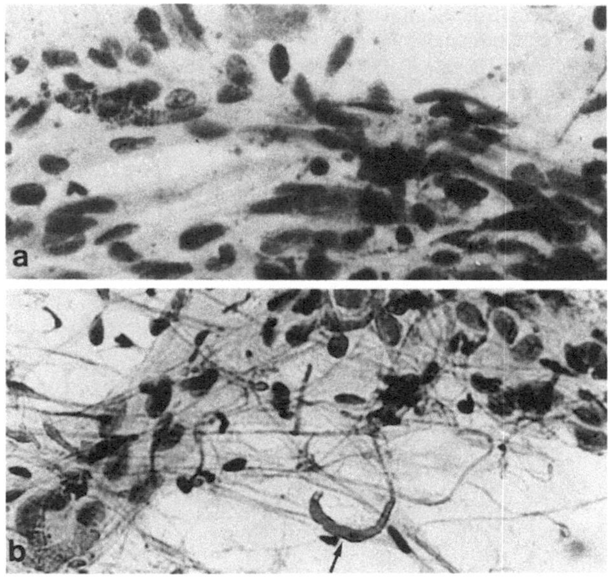

Fig. 3 **a** Fine-needle aspirate (performed in the operating room before surgery) from a neoplasm of the posterior chamber; a preoperative cytological diagnosis of spindle cell, malignant melanoma was made, according to "rapid" staining; (the same preparation restained with a standard Papanicolaou procedure is depicted; original magnification × 200); **b** FNA of an orbital mass, performed on a 9-year-old boy: the feature is that of a fibrillated matrix, formed by entangled fibrillated cytoplasmic processes; a Rosenthal fibre is also visible *(arrow)*: a cytological diagnosis (histologically confirmed) of juvenile pilocytic astrocytoma of the optic nerve was made (Papanicolaou, original magnification × 140)

ophthalmologist who deals with orbital and intraocular mass lesions on a regular basis. A cytologist cooperating with the ophthalmologist performing the aspiration makes the ideal team.

References

1. Augsburger JJ, Shields JA, Folberg R, Lang W, O'Hara BJ, Claricci JD (1985) Fine needle aspiration biopsy in the diagnosis of intraocular cancer. Ophthalmology 92: 39-49
2. Boccato P (1983) Rapid staining techniques employed in fine needle aspiration. Acta Cytol (Baltimore) 27: 82-83
3. Boccato P (1987) Fine needle aspiration biopsy of small targets. Acta Cytol (Baltimore) 31: 200-201
4. Boccato P, Briani G, Bizzaro N, d'Atri C, Vallis G, Midena E, Blandamura S, Pasini L (1987) Cytology "in black and white". Acta Cytol (Baltimore) 31: 643-645
5. Czerniak B, Woyke S, Domagala W, Krzysztolik Z (1983) Fine needle aspiration cytology of intraocular malignant melanoma. Acta Cytol (Baltimore) 27: 157-165
6. Dekker A, Johnson BL, Kennerdell JS (1984) Occurrence of benign glandular tissue in orbital aspirates. Acta Cytol (Baltimore) 28: 171-174
7. Kennerdell JS, Dubois PJ, Dekker A, Johnson BL (1980) CT-guided fine needle aspiration biopsy of orbital optic nerve tumors. Ophthalmology 87: 491-496

8. Kennerdell JS, Slamovits TL, Dekker A, Bruce LJ (1985) Orbital fine-needle aspiration biopsy. Am J Ophthalmol 99: 547-551
9. Midena E, Segato T, Piermarocchi S, Boccato P (1985) Fine needle aspiration biopsy in ophthalmology. Surv Ophthalmol 29: 410-422
10. Shields JA, Augsburger JJ, Donoso LA (1986) Recent developments related to retinoblastoma. J Pediatr Ophthalmol Strabismus 23: 148-152
11. Zaijdela A, de Maublanc MA, Schlienger P, Haye C (1986) Cytologic diagnosis of orbital and periorbital palpable tumors using fine-needle sampling without aspiration. Diagn Cytopathol 2: 17-20

Cytological Diagnosis of Orbital and Periorbital Palpable Tumors

A. Zajdela, P. Zillhardt, P. Vielh, P. Schlienger, and C. Haye

Introduction

Fine-needle sampling without aspiration has been employed by pathologists of the Institute Curie in Paris for the cytological diagnosis of orbital and periorbital palpable tumors for approximately 10 years. Because of its swiftness and reliability, fine-needle puncture allows an initial approximation of tumor diagnosis which is almost as precise as histological examination.

Material

Up to 1985, 72 patients were studied, comprising 68 adults (42 men and 26 women) between 30 and 89 years of age, and four children, aged between 4 and 12. All tumors were palpable and varied in diameter from 3 to 60 mm and caused deviation of the eye in 28 cases. The anatomical sites of the tumors are listed in Table 1. Tumors were mostly localized in the inner and outer lateral canthus and in the orbital roof (25 and 8 patients respectively). In 66 patients a subsequent histological diagnosis was performed by surgical removal or biopsy. A histological examination following biopsy of local nodal metastasis was carried out in four additional patients. In the two remaining cases, diagnosis was confirmed by the evolution of the disease.

Table 1. Orbital and periorbital tumors

Localization	No.
Supraorbital or eyebrow	5
Lacrimal gland	8
Orbital roof	8
Upper eyelid	9
Inner canthus	17
Lower eyelid	6
Orbital floor	6
Outer canthus	13
Total	72

Technique

The exact location of the palpable tumors in the orbital roof is determined with one hand over the closed desinfected eyelids, protecting the ocular globe by displacing it down from the superior rim of the orbit using one or more fingers; the tumor is simultaneously immobilized. Tumors located on the eyelid or at the inner or outer canthus are first immobilized prior to the puncture by abducting the eyelids. The operator then introduces a 25-Gauge (0.5 mm), 3-cm-long fine needle at the point of the tumor. After a few seconds in a stationary position, the needle is moved back and forth slightly in different planes of the tumor. The needle is quickly withdrawn and the procedure repeated in an another area of the tumor (Fig. 1).

Cells are collected by capillary action and expelled onto a microscope slide, using an air-filled syringe and then stained by the May-Grünwald-Giemsa technique. Using a "quick Giemsa," a diagnosis can be made within a few minutes. Simultaneously, slides for special staining or immunochemical evaluation can be prepared, as cell material is generally abundant. Punctures can also be performed in tumors located in the enucleated orbit.

Results

Of the 72 patients, 66 had a subsequent histological diagnosis performed by surgical removal or biopsy. Concordance between histologic and cytologic examination was established in 59 of these 66 cases (89.5%) with regard to malignancy and type. In three cases (4.5%), the diagnosis of malignancy was established by cytological examination, but not the type of neoplasia, which had to be done by biopsy

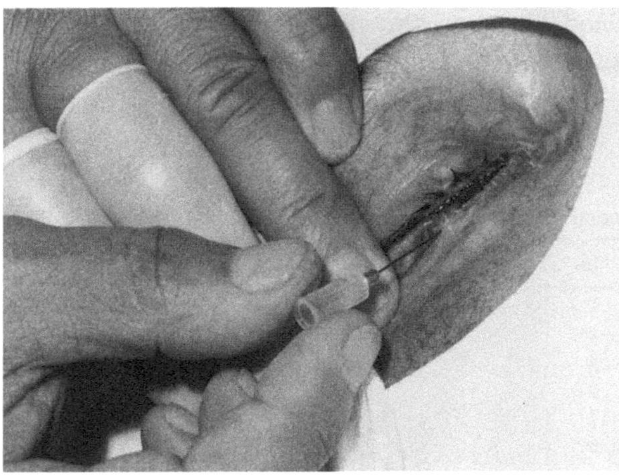

Fig. 1. Puncture of a tumor of the eyelid. The operator pulls the eyelids to immobilize the tumor and introduces a fine needle. After a few seconds in a stationary position, the needle is moved back and forth slightly in different planes of the tumor

or surgical excision (a plasmocytoma, a rhabdomyosarcoma, and a retinoblastoma).

There were three discordant results: one false benign tumor (1.5%) corresponding to a non-Hodgkin's malignant lymphoma, two false malignant tumors (3%), corresponding to a meningioma, and a reactive lymphoid hyperplasia. In one case (1.5%), the cellular material was insufficient for diagnosis but was found to be a malignant melanoma by histology.

In the other six cases malignant tumors were cytologically diagnosed and their type determined, diagnosis being confirmed biochemically for two myelomas showing a monoclonal spike, and by biopsy of local nodal metastasis for two

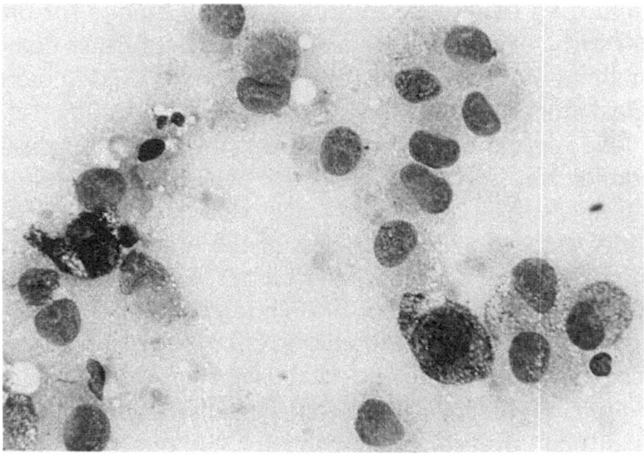

Fig. 2. Cytologic smear from malignant melanoma. Poorly cohesive malignant cells with abundant pigment both intracytoplasmic and dispersed in the background. (May-Grünwald-Giemsa stain, H & E × 200)

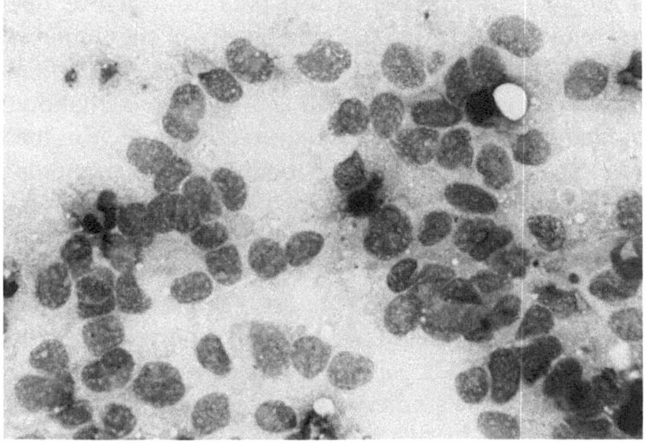

Fig. 3. Isolated malignant cells without any differentiation extracted from an undifferentiated carcinoma. (May-Grünwald-Giemsa stain, H & E × 200)

squamous cell carcinomas and two melanomas. No orbital hemorrhage or complications were observed after puncture. Figures 2 and 3 show examples of tumor cells extracted by this method.

Discussion

The great cytological variability of orbital and periorbital tumors can be easily demonstrated by this technique [11]. The diagnosis of malignancy was correctly established in 62 of the 66 cases (94%) and the type in 59 of the 66 cases (89.5%). Similar results were reported by Kennerdell et al. [5] with 92% concordant diagnoses and the same rate of discordant result as found in our study. Malberger [7] emphazised the great potential of this technique for the diagnosis of orbital adenoid cystic carcinoma. Reliability has been further demonstrated by the increased number of patients on whom this method is used in our department in which cytological analysis of 290 lesions confirmed the results [12].

The success of this method is also due to the considerable experience of our pathologists. Since 1954, they have performed over 100000 cytologic examinations by fine-needle aspiration in almost all organs [9, 10, 13]. We must also attribute the results to the close cooperation of the ophthalmologist, radiologist, and pathologist practised in different medical teams. The small, round, poorly differentiated cell tumors frequently found in children, such as rhabdomyosarcomas and retinoblastomas, are difficult to characterize [4]. In two cases in our survey, the type of such tumors could not be established by cytology. However, the difficulties encountered in diagnosis are often as great in cytology as in histology. Cytologically, it is difficult to distinguish reactive lymphoid hyperplasia and non-Hodgkin's malignant lymphoma, which may lead to misdiagnosis [1, 2, 6].

As in histopathology, the use of immunocytochemical techniques can provide substantial support in the differential diagnosis of varied and difficult cases, such as those mentioned [3]. This rapid method can provide the clinician with an important preliminary diagnosis regarding malignancy and type prior to definitive histological confirmation. These data can often render unnecessary neurosurgical intervention and taking a single biopsy of the tumor externally to confirm the cytologial diagnosis, before starting radio- or chemotherapy. This is particularly appreciated in certain cases such as lymphomas, rhabdomyosarcomas, and some inflammatory granulomas which require specific medical therapy rather than surgical intervention. Furthermore, fine-needle sampling without aspiration entails minimal cost and is painless for the patient, making hospitalization and anesthesia unnecessary. As emphasized by Midena et al. [8] in their paper focusing on fine-needle biopsy - its history, complications, and interpretation - traumatic complications are infrequent and almost never serious.

In our survey, there have been no reported adverse effects, whereas Kennerdell et al. [5] reported six cases of orbital hemorrhage that resolved spontaneously out of the 50 patients studied. The absence of aspiration in our method may explain this difference. For all these reasons, we strongly believe that fine-needle sampling without aspiration is a suitable technique for routine use by physicians and ophthalmologists.

References

1. Calle R, Zajdela A, Haye C, Schlienger P (1971) Tumeurs lymphoïdes malignes primitives de l'orbite, de l'oeil et de ses annexes. Bull Cancer (Paris) 131: 329-350
2. Calle R, Zajdela A, Haye C, Schlienger P (1975) Primary malignant lymphoid tumors of the orbit. The eye and its annexa. Eye Ear Nose Throat Mon 54: 25-35
3. Dalquen P (1987) Immunocytochemical cytodiagnosis of tumors. In: Seifert G, Hübner K (ed) Pathology of cell receptors and tumor markers. Fischer, Stuttgart, pp 196-201
4. Harms D, Schmidt H (1987) Solid tumors of childhood. Immunochemistry, diagnosis and differentiation. In: Seifert G, Hübner K (ed) Pathology of cell receptors and tumor markers. Fischer, Stuttgart, pp 182-195
5. Kennerdell JS, Dekker A, Johnson BL (1979) Fine needle aspiration biopsy. Its use in orbital tumors. Arch Ophthalmol 97: 1315-1317
6. Knowles DM, Jakobiec FA (1980) Orbital lymphoid neoplasms. A clinical pathological study of 60 patients. Cancer 46: 576-589
7. Malberger E, Gdal-On M (1985) Adenoid cystic carcinoma of the orbit diagnosed by means of aspiration cytology. Ophthalmologica 190 (2): 125-127
8. Midena E, Segato T, Piermarocchi S, Boccato P (1985) Fine needle aspiration biopsy in ophthalmology. Surv Ophthalmol 29: 410-422
9. Zajdela A, Ghossein NA, Pilleron JP et al. (1975) The value of aspiration cytology in the diagnosis of breast cancer. Experience at the Fondation Curie. Cancer 35: 499-506
10. Zajdela A, Ennuyer A, Bataïni P, Poncet P (1976) Valeur du diagnostic cytologique des adénopathies par ponction aspiration. Confrontation cyto-histologique de 1756 cas. Bull Cancer (Paris) 63 (3): 327-340
11. Zajdela A, de Maublanc MA, Schlienger P, Haye C (1986) Cytologic diagnosis of orbital and periorbital palpable tumors using fine-needle sampling without aspiration. Diagn Cytopathol 2 (1): 17-20
12. Zajdela A, Voillemot N, Zillhardt P, Schlienger P, Haye C (1987) Cytology of palpable orbital tumors and lid lesions. Ophthalmic cytology symposium, Parma, p 6
13. Zajdela A, Zillhardt P, Voillemot N (1987) Cytological diagnosis by fine needle sampling without aspiration. Cancer 59: 1201-1205

Routine and Special Staining in Cytology*

D. Wittekind and E. Schulte

Introduction

The title of this necessarily brief overview implies that there might be a difference in principle between stains which are routinely applied for "general" diagnostic purposes on the one hand and those which serve in the recognition of selected components of cells and tissues on the other. It is expected that routine stains are technically simple - sometimes they are - and furthermore, that their theoretical background be equally simple - but the theory underlying biological staining is far from being simple. As a point of fact, a general theory of staining does not exist (Horobin 1977). It is also expected that special staining differs from routine staining by its high level of sophistication. In the passages to follow we shall comment on these assumptions.

Cytologists and Their Routine Stains

Cytologists do not generally devote significant amounts of their careers to the study of the nature of staining reactions, in spite of the fact that the discipline of morphology is heavily based on the analysis of stained materials. The majority of cytologists and histologists never find the time and the energy to develop much detailed knowledge in the basic chemistry and physics of the complex dye-substrate relations. This is unfortunate since we perform better if we have a better understanding of what we are doing (Baker 1958). If cytologists were in fact acquainted at least with the basic principles of biological staining they would be in a better position to judge the quality of their staining results rationally and to improve them if necessary. To improve a staining method for widespread and frequent use, it is important that cytologists be familiar with the various ways of controlling, and hence manipulating, the different phases of staining (Rilke 1986). It is also important that they adapt the prerequisites, the conditions, and the results finally obtained to quantitative, i.e., objective measurements.

The latter point is of major importance. There is a fundamental difference between the demands on staining techniques for visual and for computer-guided cell image analysis, which must be exactly reproducible (and that means clearly),

* Supported by Bundesminister für Forschung und Technologie, Federal Republic of Germany

based on quantitatively defined dye-substrate relations. It is mainly due to the impact of modern electronic data processing on the analysis of cytologic material that the attention of morphologists in general, and of cytologists in particular, became focused on the basic mechanisms of staining - again or anew; as mentioned above, cytologists and cytotechnologists do not care too much about understanding the theoretical background of their stains. The importance of theory was noticed by the "cytologic public" only after the advent of automated high-resolution image analysis, when the latter began to establish itself as "a competitor" to the visual observer who over the decades had the monopoly in cytologic diagnosis.

The visual observer relies on the capacity of the eye-brain combination. It recognizes the cellular constituents essential for cell classification reliably yet mainly at an unconscious level (Rosenthal 1984). These constituents are fairly independent of type and intensity of color, and thus of the dyes used. The eye-brain classification combination will not fail even if an improper use of impure dyes tends to disguise cell structures rather then to reveal them (Wittekind 1985). Automated image analysis, on the other hand, has to rely entirely on measured quantities, be they geometric or photometric. The latter are evidently more specifically dye dependent than the former (Schulte 1988).

We must now ask ourselves: which cellular constituents are in fact essential for cell classification? For the practising cytologist the answer seems fairly simple: (a) the cell nuclei: mainly their size, shape, density and the arrangement of chromatin (chromatin texture); (b) the cytoplasm, its color, sometimes cytoplasmic inclusions, and sometimes merely an estimation of the important nuclear-to-cytoplasmic ratio. The state of the cytoplasmic condition usually contributes distinctly less to the recognition of malignancies, in particular, than that of the nucleus.

Taken as a whole, cytological diagnosis is guided by experience. The overall impression of cell populations is quickly transformed into a classification of the cell types seen - without the necessity to subject each single cell to careful morphological analysis. The eye-brain combination can afford this. The visual impression it can process is that of an arrangement of cell masses.

The Theorists and "Their" Stains

Theorists of staining may look at the same smears that were quickly diagnosed, i.e., classified into cell types by cytologists, from quite a different point of view: a theorist may know virtually nothing about the biological significance of the cells he also sees. Certainly he should not be interested a priori in the types of cells observed. He may ask:

(a) *Why* are cells stained at all by the dyes used?
(b) Why is not everything stained by *one* dye, or in other words, why are different dyes taken up by different substrates? For example, why in the Papanicolaou stain are cell nuclei colored by hematoxylin and the cytoplasms "counterstained" by the light green SF-eosin Y mixture?

The theorist and the cytologist agree on the *purpose* of staining, that is, to obtain *contrast*. However, whereas the theorist is concerned with the *reasons* for the contrast between one biological structure and another, differently stained structures or the unstained surroundings, the cytologist on the other hand is interested in the results of the contrast producing process - whatever its nature.

It is proposed here that closer cooperation between theorists and cytologists be established. It is to be expected that from such cooperation an improvement of the staining performance and thus an improvement of cytological diagnoses would result, which is of course in the interest of the patient. Cytologists should learn to interpret the colors of substrates in scientific terms. And the theorists should learn to interpret the substrates which they know how to stain.

Who is the Competent Theorist for the Cytologist?

It seems a surprising reality that many cytologists will be unable to answer this question. The physicochemists are the experts in charge. The fact that we, as cytologists, have to search for "our" theorists can be explained by the history of biological staining. This largely originated on a trial-and-error basis. With noteable exceptions (e.g., Paul Ehrlich's work) biological staining has been carried out empirically. With the typical cytologists' attitude to staining described above, anatomists, pathologists, and other scientists of preponderantly morphologic orientation were engaged in the achievement of contrast by using a variety of dyes under various and insufficiently controlled staining conditions. The Papanicolaou stain serves as an example. Papanicolaou, with admirable patience and empiricism, adapted the old, familiar trichrome staining technique (e.g., Masson 1929) to cytology. The physicochemistry of the Pap stain is difficult to control, and it was thus unable to fulfill the hopes and technical requirements of bioengineers engaged in constructing sophisticated, powerful, and expensive machines to classify cells which, however, escape classification for lack of strictly reproducible dye-substrate relations.

Theory and Practice of Cytological Staining - a Simplified Attempt to Merge Them with the Description of Routine Cytological Staining

In Fig. 1, the main stages of cytological staining are listed with some keywords to indicate the nature of the successive steps. We shall try here to provide the essential keywords with a brief theoretical comment, necessarily very incomplete, yet practice-related - to avoid explaining the term. Cytological practioners are asked whether they consider the following primitive "glossary of cytotechnology" as common knowledge or whether they have learned something from reading it.

The order of the keywords is in accordance with the order of events.

Fixation and Fixatives. These serve to stabilize macromolecules against bacterial (heterolysis) and intrinsic (autolysis) enzymatic attack. Alcohols (A) perform fixation by protein coagulation, aldehydes (e.g., formaldehyde) by cross-linking protein chains. Coagulation favors staining more than cross-linking. Cross-linking causes heavy shrinkage of cell films. Fixation is a highly complex interaction of fixatives with substrates.

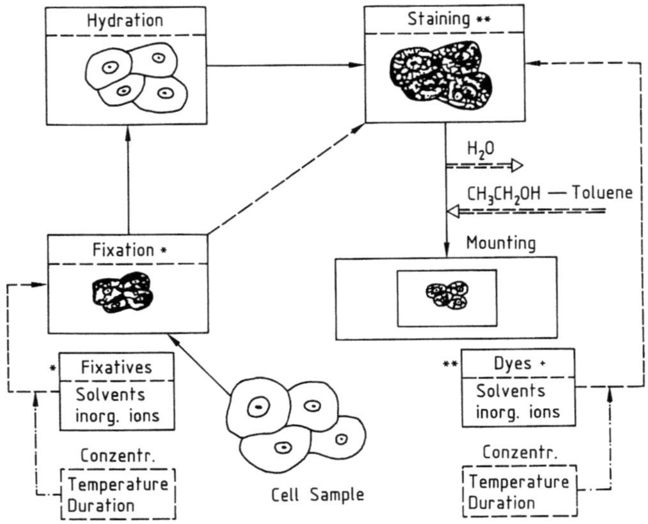

Fig. 1. The main stages and factors of cytological staining

Concentration of Fixatives. Absolute alcohol is recommended for cell films. Addition of water (70% A) destroys erythrocytes. Shrinkage by withdrawal of water through A is limited because cells stick to dry glass surfaces.

Duration of Fixation. This is not very critical for H because A leaves macromolecules chemically unchanged.

Temperature of Fixation. As is well known, room temperature is customary. Ice-cold A is said to preserve structure especially well.

Solvents and Inorganic Ions. These may influence staining results if formaldehyde and other fixatives which need solvents are used. Addition of salts may augment or impair uptake of dyes by substrates.

Hydration. This may revert shrinkage effects and is needed to wash out fixatives which are not chemically inert. The diagonal arrows in Fig. 1 refer to the infrequent cases where dyes (see below) are applied in alcohol and where the alcoholic staining solution follows alcoholic staining immediately.

Staining. Dyes dissolved in appropriate solvents can be designated as *stains.* The most important solvent is water; it has incomparable physical and chemical properties. A highly polar fluid, it has to be used when electrically charged dyes are expected to be bound to substrates that are oppositely charged. The pH of the staining solution is also important. The pH is adjusted by buffers, which contribute inorganic ions. Low pH favors uptake of anionic "acid" dyes, high pH the uptake of cationic "basic" dyes, as explained by the fundamental laws of basic chemistry. Inorganic ions may compete with equally charged dye ions for oppositely charged substrate binding sites.

Differentiation. (Keywords at the arrow below the staining stage). It is important to remove the staining solution and "excess" of bound dye from substrates, leaving only the colored substrate on the slide. Differentiation is one of the most critical steps in the staining procedure. Ethanol, an intermedium often used between an aqueous staining solution and the mounting medium, is incompatible with the latter. Benzene, toluene, or xylene are suitable intermedia between A and the apolar-mounting media. The constant problem is to remove water without detaching the dyes from the substrates. Virtually complete destaining (cationic dyes) can be achieved with acidified ethanols because protons (H^+) displace dye cations from anionic binding sites.

The stability of the dye-substrate complex largely depends on the nature of the substrates and the chemistry of the dyes. Understanding the dye-substrate relations means understanding biological staining and definetely assists in the interpretation of staining results which deviate from "standard" staining patterns. Hematoxylin (H)-chromatin complexes are especially stable and are therefore highly resistant to aqueous and alcoholic differentiation. This is important for the cytologist, irrespective of the fact that the H-chromatin complex is not yet fully understood by the physicochemist.

Dyes. Some properties of dyes have been mentioned before. Dyes used in biology are colored organic molecules with large systems of delocalized electrons (see Horobin 1982 for references). Consequently, most dyes are aromatic compounds. In Fig. 2 the structure formula of the important dye azure B is given. Cationic azure B is one of the constituent dyes of the Romanowsky-Giemsa (RG) stain, the other is eosin Y. All other dyes present in the commercial mixtures, mostly methylene blue, are superfluous. The RG stain is the only staining technique which is standardized (DIN 58995/1; International Committee for Standardization in Haematology 1984) and thus reproducible.

The chromophore groups (conjugate double bonds) impart color to dye molecules. The auxochromic substituents provide electric charge and thus the capacity to bind to "counterions" on substrates by coulombic forces. Other binding forces

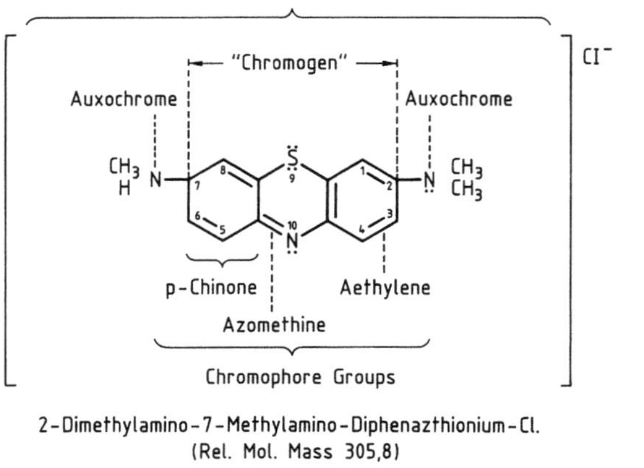

Fig. 2. Constituents of a dye: azure B

(e.g. dipole-dipole, van der Waals', H-bridge formation) are also influential, again depending on the chemical structure of the dye.

These grossly simplified and cursory comments on cytological staining describe, in condensed form, some of the basic facts underlying the theory of staining. It is a worthwhile task for cytologists to broaden their knowledge of these facts. Cytologists may soon find out that by doing this they can transform cytological staining from a tedious routine into a fascinating occupation.

Standardization of Cytological Staining

In our modern world there is a growing and justified tendency towards standardization of material, methods, and prescriptions applied in laboratories, industrial processes, and authorities. Unfortunately, insight into the necessity of standardization has not yet spread to the community of cytologists, who maintain their reluctant attitude. It is founded on the cytologists mostly erroneous conviction that the individual staining techniques they use are already optimal. In order to establish that this belief is in need of urgent correction, it is only necessary to compare ten "optimal" Pap stains from different laboratories. You will find about eight versions of an "optimal" Pap stain.

The aim of standardizing cytologic stains is to render the staining results reproducible and thus comparable. To achieve this requires first and foremost the employment of pure dyes and, additionally, the standardization of each preparatory step listed in Fig. 1. With the good will of all concerned, this can be done. An extensive discussion of the problems involved in standardization has been published elsewhere (Wittekind 1985).

Special Staining in Cytology

In the Introduction it was implied that special stains differ from routine stains by their higher level of sophistication. This statement requires some elaboration. Take immunocytochemistry and enzyme cytochemistry as examples. These techniques aim at identifying one specific constituent, e.g., an antigen (AG), in immunological staining by employing the corresponding antibody (AB). Cell nuclei and cytoplasms, i.e., the cellular components cytologists are familiar with, may or may not also be seen clearly. In some cases they can be brought out by routine poststaining.

Now, to design immunostaining techniques and to bring them to general application presupposes:

1. Familiarity with the basics of organic chemistry in general and of protein chemistry in particular.
2. Advanced knowledge of immunology, including the pertinent fields of genetics.
3. Very desirable is special knowledge of the mechanism of AG-AB interaction.

It is fairly obvious that a fair amount of theoretical knowledge is required in order to achieve meaningful results with special stains. For the majority of special stains, however, theory is fairly close to the clinical field of application. So the necessity for reverting to theory is far more obvious with regard to the special stains than to the routine stains. To use immunostaining presupposes familiarity with immunology, both theoretical and applied. But if the mechanism of AG-AB binding has to be analyzed and efforts at standardization of immunostaining are to be undertaken, it will soon become evident that the same binding forces prevail that were enumerated in the section on routine staining. Here again, cooperation with physicochemists is desirable.

References

Baker IR (1958) Principles of biological microtechnique: a study of fixation and dying. Methuen, London

DIN (1985) Normenausschuss Medizin (NaMed) in DIN e.V. Norm 58995 Teil 1 Medizinische Mikrobiologie Verfahren zur Anwendung der Farbstoffe Romanowsky Giemsa Färbung, Deutsches Institut für Normung

Horobin RW (1977) Theory of staining. In: Bancroft ID, Stevens A (eds) Theory and practice of histological techniques. Churchill Livingstone, Edinburgh, p 75

Horobin RW (1982) Histochemistry. Fischer/Butterworths, Stuttgart London

International Committee for Standardization in Haematology (1984) ICSH reference method for staining of blood and bone marrow films by azure B and eosin Y (Romanowsky Stain). Br J Haematol 57: 707-710

Masson P (1929) Some histological methods, Trichrome staining and their preliminary technique. Bull Int Assoc Med 12: 75

Papanicolaou GN (1942) A new procedure for staining vaginal smears. Science 95: 438-439

Rilke F (1986) Foreword. In: Boon ME, Kok LP (eds) Standardization and quantitation of diagnostic staining in cytology. Coulomb, Leyden

Rosenthal DL (1984) Book review. Anal Quant Cytol 6: 144

Schulte E (1988) Bildanalytische Untersuchungen über den Einfluss der Zell- und Gewebspräparation auf das Färbemuster zytologischer und histologischer Präparate des Bronchialcarcinoms. Habilitationsschrift, vorgelegt der Medizinischen Fakultät der Universität Freiburg

Wittekind D (1985) Standardization of dyes and stains for automated cell pattern recognition. Anal Quant Cytol 7: 6-30

XI. Epidemiology, Screening Programs, Early Cancer Detection

Introduction

E. Grundmann

Cancer screening programs, first introduced in Europe and the United States in the 1950s, are now being implemented in almost every country. This has not been an easy achievement since the methods were a matter of dispute and the results uncertain for quite some time. There is no doubt that the pioneering work of Papanicolaou in not only developing a simple staining method but also ensuring handy practicability has helped make it possible to establish worldwide screening programs for cervical cancer.

Efforts have been made to extend such cytological screening to other organs. Some promising results have been obtained by correct sputum sampling to screen patients at a high risk for bronchial carcinoma. Screening programs for gastrointestinal cancer, however, are still less effective.

Radiological screening has proven to be more meaningful in the case of other organs, especially the female breast. These organs have been made an object of cancer prophylactic screening, often conducted in conjunction with cytology.

Therefore this part of the book focuses on cytology of the cervix and screening for breast cancer. The experience reported here helps to refute criticism voiced in recent years. If we want to tackle these malignancies successfully, the only approach is prevention, predominantly what is called secondary prevention. It is here that cytodiagnostis plays an important role.

Let me just mention the new immunohistochemical staining procedures which may come to supplement and enlarge our conventional methods in the near future. We know that a large proportion of cervical carcinomas contain human papillomaviruses (HPV), which may be demonstrated by immunohistochemical methods, in particular those of DNA hybridization. I would even like to predict that the first results of cancer screening programs utilizing immunohistochemical HPV demonstration will be discussed at one of our very next meetings. These, however, are indeed future aspects as yet unsubstantiated by conclusive data. These chapters concentrate on current experience and criteria that can be applied reliably to presently available screening programs to evaluate efficiency.

This introduction, however brief, should not end without the grateful acknowledgement of the conscientious and never-ending work of many thousands of doctors and cytology assistants for the benefit of persons protected by truly successful programs.

Experiences with a Nationwide Cancer Screening Programme for Cervical Cancer in the Federal Republic of Germany

B.-P. Robra and B. Berghof

Mortality rates for malignant neoplasms of the uterus have decreased considerably in the Federal Republic of Germany, in particular for cohorts born since 1920 (Fig. 1). These are the cohorts with a reasonably high participation rate in the national screening programme.

Screening for cervical cancer was introduced into the statutory health insurance system of the Federal Republic of Germany in 1971. In 1982 the lower age limit was reduced from 30 to 20 years. An annual Pap smear is recommended [1-4]. Results of the screening programme are documented by the physicians and analysed centrally. In the following, we report on the 1984 results of the national screening programme (see [1] for raw data). For technical reasons, data from one service region (KV Nordrhein) were unavailable.

Almost 7 million women attend each year. The overall attendance rate is close to 30% (29.9% in 1984 and 31.1% in 1985). Annual age-specific attendance rates are 40%-45% in age groups from 25-29 to 45-49. Of the women attending, 57.8% were repeat screenees with an interval of 1 year since last screening.

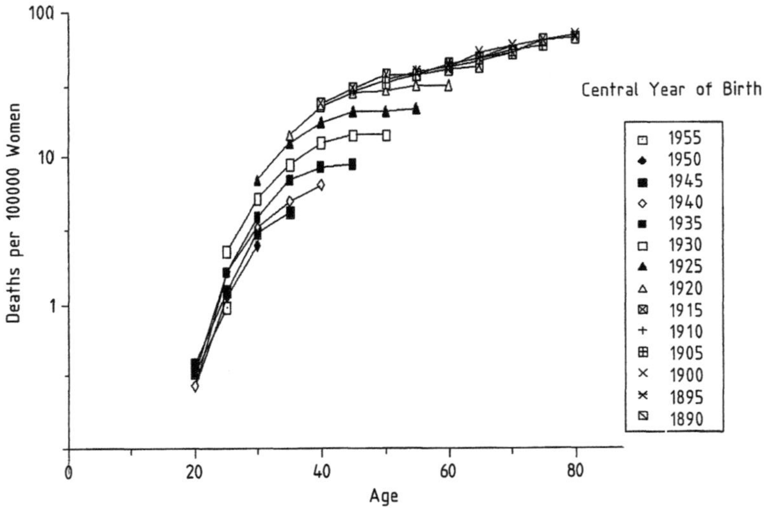

Fig. 1. Age-specific mortality rates by birth cohorts for cancer of the uterus, Federal Republic of Germany, 1955-1959 to 1980-1984

Detection rates for carcinoma in situ (CIS) and invasive cervical cancer (INCA) have been available since 1983. Crude detection rates are rather low. In 1984 2.5 CIS and 0.9 invasive cancers of the cervix per 10000 screening examinations were documented. Detection rates are age dependent. In age brackets older than 60–64 years, INCA is more frequent than CIS (Fig. 2).

Some information on individual history is included in the programme's data sheet. The number of pregnancies, for instance, turns out to be a risk factor for cervical abnormalities (Table 1). The association, however, is not very strong and would probably not be useful for targetting the programme to groups at high risk as defined by the number of pregnancies.

Women attending for the first time comprised 13% of the attendees in 1984, and 42% of the invasive cancers were seen in this group. Repeat screenees with a 1-year interval since last screening (57.8%) had 37.5% of CIS and 16.7% of INCA.

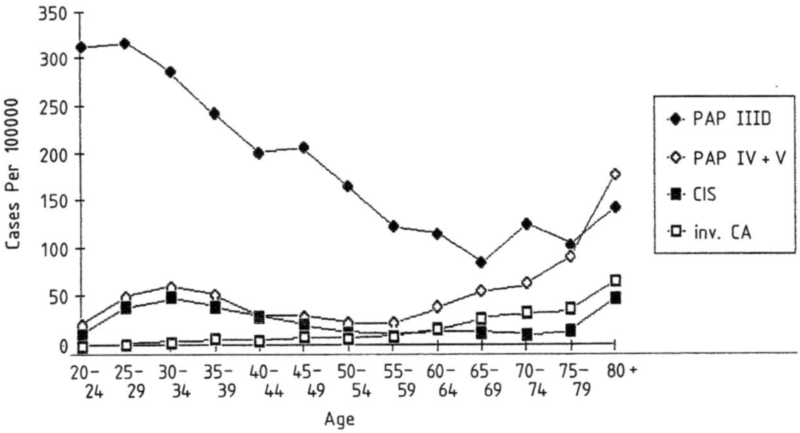

Fig. 2. Cancer screening programme in the Federal Republic of Germany. Age-specific cytological findings and detection rates 1984 (based on data from Bavaria for Pap results and from the Federal Republic (excl. North Rhine) for carcinoma in situ *(CIS)* and invasive cervical cancer *(inv. CA)*

Table 1. Odds ratios[a] for detecting cervical abnormality by number of pregnancies in Bavaria[b] in 1984

No of pregnancies	Pap IV and higher	Pap V	CIS and higher	INCA
0	1	1	1	1
1 and more	1.10	1.46	1.71	1.69
2 and more	1.10	1.37	1.20	1.73
3 and more	1.23	1.51	1.29	1.87

[a] Mantel-Haenszel odds, adjusted for age (in 10-year age groups) and interval since last screening examination.

[b] In 1984, Bavaria was the only service region documenting this information in detail.

The group of participants delaying their repeat examination for 3 or more years (10%) had 22.1% of CIS and 30.5% of INCA. Also, of all the documented cases with invasive cancer, 74.4% were 45 years or older, compared to 27.4% of the cases with CIS. In the age group 60 and above, 53% of the cases of INCA were detected in the small group of women attending for the first time (8.2% of all those over 60). Invasive carcinoma of the cervix, as documented in the national screening programme, is therefore to a large extent a problem of women in the higher age groups and of first attendees, i.e. a problem of past ommissions. Age-standardized detection rates in the programme continue to decline (Fig. 3). Hence screening efficiency (i.e. the amount of resources invested to produce one unit of health benefit) has declined as well [2, 3].

Detection rates depend on the interval since the last screening examination (Table 2). The (age-standardized) relative risk for INCA is 0.06 and for CIS 0.31 for screenees at an interval of 1 year compared to first attenders. Stated another way, a screening examination "reduces" the risk of invasive cancer 18-fold for a period of 1 year. The relative "protective" effect over a period of 2 years is 10.8-fold. In these comparisons, incidence screening rounds are compared to a prevalence screening round.

We are, unfortunately, unable to report the cumulative number of smears that

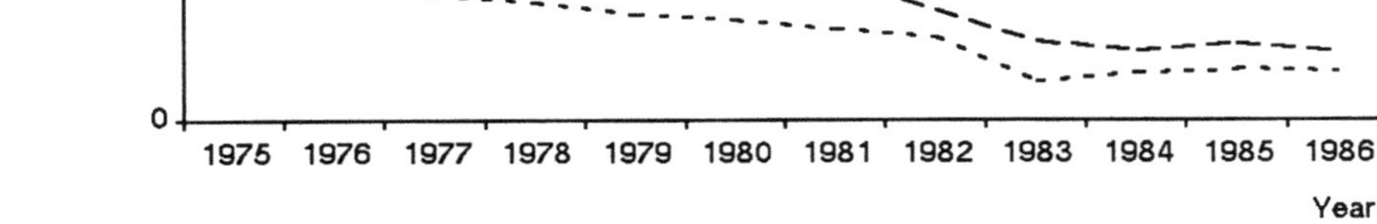

Fig. 3. Cancer screening programme in the Federal Republic of Germany. Detection rates for uterine cancer 1975-1986 (age-standardized). *Continuous line*, first examination; *large-dashed line*, repeat examination after 2 years; *small-dashed line*, repeat examination after 1 year

Table 2. Age-standardized[a] detection rates (per 100000) by interval since last screening examination, Federal Republic of Germany, 1984

	First attendance	Repeat screening examinations after		
		1 year	2 years	3 years or more
Carcinoma in situ	48	15	20	51
Invasive cervical cancer	54	3	5	23

[a] European standard population

the group of women reattending had in the past. It is reasonable to assume that the majority of them had more than one previous (negative) smear. In addition, the group of "first" attendees seems to be progressively "contaminated" by previously screened women, as indicated by the decreasing detection rate for this group in Fig. 3. With these assumptions, we compared our estimate of "relative protection" against cervical cancer with that calculated by the IARC Working Group on Cervical Cancer Screening on the basis of ten centres or countries [5]. Their estimate of relative protection for an exposure of two or more negative smears is 15.3 (95% CI 10.0-22.6) if the time since the last negative smear is 0-11 months and 11.9 (95% CI 7.5-18.3) if the time since the last negative smear is 12-23 months.

The Federal Republic of Germany was not represented in the IARC Working Group. But since our results seem to agree with international experience, we can hardly evade the conclusions of this collaborative project. The most relevant conclusion is that screening from the age of 25 with repeat smears every 3 years will bring about a reduction of about 90% in the incidence of invasive disease. "More intensive screening is unlikely to improve effectiveness more than marginally" [6].

In addition to marginal effectiveness, imperfect specificity is an argument in favour of a screening interval longer than 1 year. The specificity of a screening programme determines to a large extent the number of follow-up examinations necessary and, accordingly, the resources required as well as inconvenience and risks for healthy participants [7].

It is possible to estimate the specificity of our cervical screening programme from the national data base (using a method developed in [7]; clinical as well as cytological "suspicion of cancer" is the relevant positive test result and CIS or higher is the confirmed disease). Specificity is 99.6% overall. This value is somewhat lower than the 99.9% discussed previously [8, 9]. Specificity is particularly high in repeat screening examinations after 1 year. If we use this latter value (99.65%) and consider a screening programme offered repeatedly to asymptomatic healthy women (i.e. removing positive women from the routine programme), we arrive at the following test results (assuming independence of tests):

No of repeat examinations	Percent positive at least once	Percent positive at least twice
10	3.4	0.05
20	6.8	0.22
30	10.0	0.50

Thirty examinations can be easily reached in the life-time of a not very compliant woman on an annual programme. So even with this very high specificity a healthy woman taking part annually has a considerable chance of being positive at least once. This is acceptable only if the follow-up strategy for positive screenees is non-aggressive. The frequency of conizations, however, seems to be rather high in Germany [10].

Lynge and Poll [11] estimated the incidence of INCA in women with 2-4 negative smears at ages 30-64 as 0.479 per 10000 women years (i.e. 0.0000479 per woman per year) if the time elapsed since the last negative smear is 0-11 months. If the time elapsed is 12-23 months, the incidence is 0.615 per 10000 woman-years. Compared to this, the probability of a false positive test in our programme is 0.0035 per woman per year (i.e. 1 – specificity [7]). If we accept the Danish results as applying to our country, there will be between 57 and 73 false positives for one invasive cancer in an annual repeat screening programme (depending on how closely a 12-month interval is met). By far the most likely outcome for a compliant (i.e. repeatedly negative) participant who happens to be test positive is a false positive examination. Are we really prepared to stick to the annual Pap smear for repeatedly negative women with false positives in this order of magnitude?

We should now reshape our screening policy according to IARC recommendations. The routine interval should be 3 years after two or three smears have been negative. Systematic efforts should be made to recall symptomless women after this interval in order to improve the cytologic screening coverage of our population. Individual invitations may be sent by the respective health insurance funds, by the Associations of Health Insurance Physicians (Kassenärztliche Vereinigungen) or by the physicians in charge of primary health care. The latter should use patient encounters made for other reasons to verify the screening history of their patients and to take appropriate action.

Complete and immediate follow-up of women with positive smears has in the past been a problem in this country [12] but may have improved [13]. Nevertheless, inadequate follow-up of abnormal smears is considered to be the most important principal failure in many cervical screening programmes [14]. Resources invested here are likely to be more productive than rapid repeat screening examinations in previously negative women.

References

1. Kassenärztliche Bundesvereinigung und Spitzenverbände der Krankenkassen (eds) (1987) Krankheits-Früherkennung Krebs - Frauen und Männer - Aufbereitung und Interpretation der Untersuchungsergebnisse aus den gesetzlichen Früherkennungsmaßnahmen 1984. Deutscher Ärzte-Verlag, Köln (Wissenschaftliche Reihe des Zentralinstituts für die kassenärztliche Versorgung in der Bundesrepublik Deutschland, vol 36)
2. Schwartz FW, Brecht JG (1984) Evaluation of German cancer screening by cross-sectional data. Int J Epidemiol 13: 283-286
3. Robra B-P, Schwartz FW, Brecht JG (1985) Evaluation of the screening program for cervical cancer in the Federal Republic of Germany from an epidemiological perspective. In: Bender HG, Beck L (eds) Cancer of the uterine cervix. Fischer, Stuttgart, pp 23-30
4. Robra B-P (1986) Ergebnisse und Probleme des „gesetzlichen" Krebsfrüherkennungsprogramms in der Bundesrepublik Deutschland. In: Schrage R, Hillemanns H-G (eds) Das Krebsfrüherkennungsprogramm - gegenwärtige Situation und Ergebnisse. Fischer, Stuttgart, pp 11-23
5. IARC Working Group on Cervical Cancer Screening (1986) Summary chapter. In: Hakama M, Miller AB, Day NE (eds) Screening for cancer of the uterine cervix. International Agency for Research on Cancer, Lyon, pp 133-141
6. IARC Working Group on Cervical Cancer Screening (1986) Report: screening for cervical cancer. In: Hakama M, Miller AB, Day NE (eds) Screening for cancer of the uterine cervix. International Agency for Research on Cancer, Lyon, pp 289-290
7. Brecht JG, Robra B-P (1987) A graphic method of estimating the specificity of screening programmes from incomplete follow-up data. Meth Inform Med 26: 53-58
8. Soost H-J (1985) Possibilities of quality control in diagnostic cytology. In: Bender HG, Beck L (eds) Cancer of the uterine cervix. Fischer, Stuttgart, pp 115-121

9. Schwartz FW, Robra B-P (1985) Some considerations on screening models and their empirical application to cervical screening. In: Bender HG, Beck L (eds) Cancer of the uterine cervix. Fischer, Stuttgart, pp 15-21
10. Bender HG, Schnürch H-G (1985) Conization data in the Federal Republic of Germany. In: Bender HG, Beck L (eds) Cancer of the uterine cervix. Fischer, Stuttgart, pp 59-66
11. Lynge E, Poll P (1986) Risk of cervical cancer following negative smears in Maribo County, Denmark, 1966-1982. In: Hakama M, Miller AB, Day NE (eds) Screening for cancer of the uterine cervix. International Agency for Research on Cancer, Lyon, pp 69-86
12. Wagner D (1979) Zur Frage der Effizienz von Krebsvorsorgeuntersuchungen. Fortschr Med 97: 513-516
13. Schrage R (1986) Ergebnisse des Kollumkarzinomscreenings bei der neuen Dokumentationsweise. In: Schrage R, Hillemanns H-G (eds) Das Krebsfrüherkennungsprogramm - gegenwärtige Situation und Ergebnisse. Fischer, Stuttgart, pp 71-80
14. Chamberlain J (1986) Reasons that some screening programmes fail to control cervical cancer. In: Hakama M, Miller AB, Day NE (eds) Screening for cancer of the uterine cervix. International agency for research on cancer, Lyon, pp 161-168

Criteria for Evaluating the Screening Efficacy for Cervical Cancer: Estimated Sensitivity and Protective Effect of the Screening Program of Florence

S. Cecchini

Recently, case control studies have been employed to study the "protection" offered by Pap screening against cervical cancer. This method has been proved to be effective for evaluating screening efficacy [1, 3]. The case control study performed by our Center (CSPO), whose preliminary results will be presented here, differs from previous studies in that the information about the screening histories of patients and controls is taken from the computerized archives of the CSPO instead of being assessed by interviews. Thus, the exact data and diagnosis of every Pap test are known.

Cervical cancer screening has been conducted in the province of Florence since 1970, and computerized data are available since 1973. Moreover, the incidence of cervical cancer has been assessed by a histologic registry since 1975, and a tumor register has been in operation since 1984. The screening is based on an exocervical and an endocervical sample, stained according to conventional Papanicolaou technique. Colposcopy is performed in selected cases (about 10% of the screened population), at the suggestion of the cytopathologist or on the basis of clinical abnormalities diagnosed by the performer of the Pap test. Until 1980 screening was self-referred; since 1980 mail calls have been sent every 3 years to resident women aged 18-60. This caused the number of women examined per year to increase from 50000 (1975-1979) to 65000 (1980-1986).

All the cervical cancers in the province of Florence in the years 1982-1985 have been considered as eligible cases. Cases aged over 75 were excluded. A total of 208 cases were thus identified; 191 were interviewed (91.8%). For each case three living population controls were identified and matched by age, residence, and immigration history. Hysterectomized controls were excluded.

The Pap tests that were considered for the study had to be performed at least 6 months before the diagnosis, assuming that more recent exams could be related to the presence of symptoms. The Pap tests considered as negative were all the class I and the class II Pap tests (that is, every cytologic diagnosis slighter than mild dysplasia). For every case and control a questionair has been compiled. For dead cases a relative was interviewed. The investigated variables were: civil and socioeconomic status, place of birth, smoke, deliveries and abortions, age at first marriage, menarche, and age at first delivery. These variables have been used to adjust the relative risks (RR) estimates. In fact, a possible bias of case control studies is the supposed lower compliance of high-risk population, causing the overestimation of the protective effect of the screening. As we will see by our results (and as previously reported in an Italian case control study [7]) to adjust the RR by confounding variables does not change their estimates, in fact the adjusted odds ratios (OR) are even lower than the crude ones (Table 1).

Table 1 shows the RR of the population screened with (at least 1 Pap in the previous 10-12 years) versus the never screened. The protection offered by the Pap test (that is, the inverse of the RR) is about 7 times. The variation of the RR according to the number of Pap tests performed is analyzed in Table 2. The protection offered by only 1 Pap test is about three fold, whereas that offered by 2, 3, or more tests increases to 7 and 16.

Then, we tried to evaluate the difference in the RR according to the interval since the last Pap tests, as shown in Table 3, but it can be noticed that, whereas there is still a difference between the three groups no Pap, only 1, and 2 or more, there is no trend according to interval within the two groups studied. This fact can be easily explained. In fact, the variation of the RR according to the number of the Pap tests versus a nonscreened population is so much greater than the variation according to the interval, that this last effect can be completely masked. Thus, we chose to set on another study design, aiming to analyze the difference within the screened population according to the different rescreening histories. We utilized the case control method, applied to a cohort of women screened at least once. The main aspect of this study is that cases and controls were matched by age, residence, and year of first Pap test. Thus, we have a cohort of women who were

Table 1. Previous screening history (ever/never) tested and relative risks (odds ratios, matched data) of cases and controls

	Pap test		
	0	1 or more	Total
Cases	155	36	191
Controls	290	265	555
Crude OR	1[a]	0.19 (0.12-0.31)	
Adjusted OR	1[a]	0.14 (0.08-0.24)	

[a] Reference category

Table 2. Distribution of the number of Pap tests performed and relative risks (odds ratios, matched data) for cases and controls

	Number of Pap tests				
	0	1	2	3+	Total
Cases	155	20	9	7	191
Controls	290	87	59	119	555
Crude OR	1[a]	0.32	0.23	0.08	
CL		(0.18-0.58)	(0.10-0.49)	(0.03-0.18)	
Adjusted OR	1[a]	0.29	0.14	0.06	
CL		(0.14-0.55)	(0.05-0.30)	(0.02-0.14)	

[a] Reference category

screened in the same year and had a negative diagnosis: cases (who developed cervical cancer in the years 1980–1985), and 5 controls for each case. The different rescreening histories will be correlated to different RR. Obviously the controls must be not hysterectomized at the time of cancer diagnosis. Matching cases and controls by date of first Pap tests avoids the bias due to the inclusion of first-screen detected cases (which have infinite period of time since the previous Pap test) because these cases are excluded. Also in this study the same confounding variables were investigated and used to adjust the RR. In Table 4, the reference category is the population which has had only one negative Pap test, and it can be noticed that when two or more Pap tests had already been performed an interval longer than 5 years is less protective than a shorter one (2 times versus 5–7 times).

On the basis of the results, it seems that the protection offered by the number of Pap tests performed in the last 10 years is more important than that offered by a short interval. An interval longer than 5 years is shown to be less effective than shorter ones. Thus, the possible failure of the screening program seems to be due to a poor sensitivity (related to the number of Pap tests) rather than to fast growing cancers (an effect related to the interval since the last test).

Table 3. Distribution of the number of Pap tests and the interval since the last exam and relative risks (odds ratios, matched data) for cases and controls

	Cases	Controls	Adjusted OR (CL)
Never screened	155	290	1[a]
Only 1 Pap test in the last			
<36 months	7	26	0.34 (0.12–1.00)
36–59 months	5	15	0.34 (0.09–1.29)
60–119 months	5	26	0.18 (0.05–0.62)
>120 months	3	20	0.33 (0.09–1.22)
2 or more Pap tests in the last			
<36 months	7	97	0.07 (0.03–0.19)
36–59 months	6	39	0.13 (0.05–0.38)
60–119 months	2	37	0.06 (0.01–0.27)
>120 months	1	5	0.06 (0.02–0.74)

[a] Reference category

Table 4. Distribution of screened women according to the number of Pap tests and to the interval since the last rescreening, and relative risks (odds ratios, matched data)

	Cases	Controls	Adjusted OR (CL)
Only 1 Pap test	25	76	1[a]
2 or more Pap tests in the last			
<36 months	9	72	0.21 (0.08–0.55)
36–59 months	4	42	0.14 (0.04–0.51)
>60 months	6	30	0.47 (0.14–1.54)

[a] Reference category

As a side consideration, "compliance," which usually refers to the attendance rate, is not a very useful indicator of which part of the resident population is protected against cervical cancer. Following a proposal of Duca et al. [4], the data given in Table 4 have been analyzed as in Table 5. Relative protection (the inverse of the RR) has been calculated using as a reference category the population that we assumed to be the most protected, that is the women who have had 2 or more Pap tests and the last within the previous 5 years. Then, the screened population is standardized according to the coefficients and the ratio between protected women and resident women will be called the "protection index" of the screened population, PI:

$$PI = W_a + (0.39 \times W_b) + (0.18 \times W_c)/W_r$$

W_a is the number of women with 2 or more Pap tests, the last within 5 years; W_b is the number of women with 2 or more Pap tests, the last over 5 years; W_c is the number of women with only 1 Pap test; and W_r is the number of resident women. We give some examples of possible uses of this protection index in the following tables. In Table 6, the slight decrease in cervical cancer incidence in the province of Florence is compared to a corresponding increase in the protection index. A cohort study has shown that the decrease in incidence is due to a lower incidence in younger women, who present a higher compliance than those over 60 years. In fact, in the Tables 7 and 8 the incidence ratios in cohorts born 1926-1935 is lower

Table 5. Distribution of screened women according the number of Pap tests and to the interval since the last rescreening, and relative protection (inverse of the odds ratios, matched data)

	Cases	Controls	Relative protection
2 or more Pap tests in the last			
<5 years	13	114	1[a]
>5 years	6	30	0.39 (0.12-1.22)
only 1 Pap test	25	76	0.18 (0.07-0.45)

[a] Reference Category

Table 6. Incidence of cervical cancer in the province of Florence in the years 1975-1985, and the relative protection index

	Incidence (×100000)[a]	Protection index
1975-1976	16.43	24.8
1977-1978	16.60	33.4
1979-1980	18.72	39.9
1981-1982	14.95	43.8
1983-1984	11.07	44.7
1985	11.05	45.0

[a] Truncated 25-74 years, standardized for European population

Table 7. Incidence of cervical cancer at the age of 50-59 years in two consecutive cohorts, and relative protection index

Year of birth	Cases (*n*)	Incidence[a] (× 100000)	Protection index (%)
1916-1925	126	27.9	23.3
1926-1935	84	17.6[b]	38.4

[a] Years 1975-1985
[b] $p<0.05$

Table 8. Incidence of cervical cancer at the age of 40-49 years in two consecutive cohorts, and relative protection index

Year of birth	Cases (*n*)	Incidence[a] (× 100000)	Protection index (%)
1926-1935	80	17.3	23.3
1936-1945	50	10.4[b]	47.7[b]

[a] Years 1975-1985
[b] $p<0.05$

at the age of 50-59, than in preceeding cohorts; the same happens in women born in 1936-1945, at the age of 40-49, with regards to the older cohort.

The detection of cervical cancer is not the only goal of Pap test screening, but providing an estimate of the screening sensitivity for invasive cervical cancer should help in the planning of screening programs, especially with regard to the choice of interval between screening rounds. The estimated sensitivity of a screening program is usually calculated as the ratio between detected cancers and detected cancers plus interval cancers. Different values can thus be calculated according to the interval length, following the method proposed by Habbema [5].

In the years 1975-1985, 251516 women were screened for the first time and 183 invasive cervical cancers were diagnosed, a detection rate of 0.73‰. This population was followed up until January 1987 and all the incident cancers, detected either by further screenings or otherwise, were considered in estimating the first screening sensitvity. Table 9 reports the estimated sensitivity according to different time intervals. The interval was calculated from the month and year of the first Pap test. Since the follow-up length is variable, both the population at risk and the number of screen-detected and interval cancers vary according to interval. The shorter the interval considered, the higher the number of women exposed to the risk of developing a cancer during this interval and thus available for the study. The sensitivity values decrease as the interval since the first screening increases, being 0.91, 0.79, and 0.71 as the interval increases from 1 to 3 to 5 years.

In estimating the sensitivity of a screening program, interval cancers must not be differentiated into "false negatives" of screening and "true" interval cancers, since both kinds escape detection by the screening procedure. Nevertheless, the weight of false negatives is much greater in the sensitivity values estimated at shorter intervals, while true interval cancers increase with the length of the interval

Table 9. Sensitivity of first screening according to the length of the time interval since the test

1st screening in the years	Cancers detected by 1st screening	Interval (years)	Cancers observed in the interval	Sensitivity[a]
1975-1985	183	1	18	0.91
1975-1984	163	2	22	0.88
1975-1983	144	3	38	0.79
1975-1982	136	4	45	0.75
1975-1981	121	5	49	0.71
1975-1980	102	6	45	0.69
1975-1979	81	7	36	0.69
1975-1978	69	8	28	0.71
1975-1977	52	9	19	0.73

[a] Detected cancers/detected cancers plus interval cancers

Table 10. Further screening sensitivity according to the length of the interval since the test

Rescreening in the years	Cancers detected by rescreening	Interval (years)	Cancers observed in the interval	Sensitivity[a]
1975-1985	41	1	17	0.70
1975-1984	37	2	29	0.56
1975-1983	36	3	43	0.45

[a] Detected cancers/detected cancers plus interval cancers

considered. Therefore, we can assume that the value at 1 year (0.91) is due almost completely to the false negatives and is indicative of the sensitivity of the screening procedure, whereas the values calculated at other intervals decrease according to the increasing number of new cancers and are indicative of the sensitivity of the screening programs employing different intervals. The present data show a progressive reduction in sensitivity as the interval from the first screening grows, but the reduction is relatively small and sensitivity values of about 0.70 are calculated even after an interval of 9 years. According to a rough estimate of the detection lead time of 3.5 years (calculated as the ratio between prevalence and incidence data), one should expect a more dramatic drop of sensivity after 4-5 years. The low number of observed interval cancers can be easily explained by the protective effect of the screening against cancer, realized by diagnosis and treatment of precancerous lesions, and proved by the preliminary results of the case control study.

Moreover, a distinction between first and further screening sensitivity should be made. In fact, it was suggested that sensivitiy should decrease at further screenings, as first-screening detected cancers are most prevalent, whereas further screenings have to detect incident cases and cancers missed by previous Pap tests, and thus selected on a basis of difficult detection [2, 6]. Therefore, women undergoing second or further Pap tests have also been considered in evaluating the sensitivity of repeated screening examinations. The follow-up also varies in this population from 1 to 10 years, but due to the low number of screen-detected and interval cancers, the intervals considered for the sensitivity estimate is 1-3 years. As shown in Table 10, the sensitivity was 0.70, 0.56, and 0.45 for an interval of

1, 2 and 3 years respectively; this is considerably lower than the corresponding values for first-screening sensitivity. These data confirm the quoted hypothesis and suggest that repeated screenings are surely less cost effective than first screening.

References

1. Berrino F, Gatta G, d'Alto M, Crosignani P, Riboli E (1986) Efficacy of screening in preventing invasive cervical cancer: a case control study in Milan, Italy. In: Hakama M, Miller AB, Day NE (eds) Screening for cancer of the uterine cervix. International Agency for Research on Cancer, Lyon (IARC scientific pulications, no 76)
2. Cecchini S, Palli D, Casini A (1985) Cervical intraepithelial neoplasia III: an estimate of screening error rates and optimal screening interval. Acta Cytol (Baltimore) 29: 329-333
3. Clarke EA, Anderson TW (1979) Does screening by "pap" smears help prevent cervical cancer? Lancet II: 1-4
4. Duca P, Turolla E, Galli C (1986) Efficacia dello screening di massa mediante Pap-test: validità e utilità di un indicatore di copertura. Epidemiol Prevenzione 28: 50-53
5. Habbema JDF, van Putten DJ, van Oortmarssen GJ, van der Maas PJ (1981) Preliminaries for evaluation of breast cancer screening. Third progress report of research project. Decision making on mass screening. Dept of Public Health and Social Medicine, Erasmus University, Rotterdam
6. Habbema JDF, van Oortmarssen GJ (1982) Performance characteristics of screening tests. Clin Lab Med 2: 639-656
7. La Vecchia C, Franceschi S, de Carli A, Fasoli M, Gentile A, Tognoni G (1984) "Pap" smear and the risk of cervical neoplasia: quantitative estimates from a case-control study. Lancet II: 779-782

Screening for Cervical Cancer – The Use of Epidemiological Studies to Evaluate Screening Policies

S. M. Moss

Introduction

There is now considerable evidence that well-organised screening programmes can be successful in reducing both the mortality from cervical cancer and the incidence of invasive disease [1]. However, far less is known about the effect of different screening patterns on the risk to individual women, and it is this information which is required in order to produce recommendations for screening policies in terms of which women should be offered screening and the time intervals at which screens should be spaced.

Given the current extent of screening for cervical cancer it would not be feasible to introduce controlled trials of different screening policies. An alternative approach to estimating the effect of such policies is to use the epidemiological data available from screening programmes already in existence to estimate the risk of invasive disease associated with different screening histories. This was the approach used in a collaborative study carried out by the International Agency for Research on Cancer [2].

Risk of Invasive Disease Following a Negative Smear

The variable which requires estimation is the risk of invasive disease following a negative cervical smear, in terms of both the time elapsed since the most recent negative smear and the number of previous negative smears. Once a woman has had a positive smear, she should be followed up clinically, and whilst the completeness of follow-up of such cases is an important issue, it is not considered further here.

The definition of a negative smear requires some care. The IARC study used a definition based on that of Boyes et al. [3]. Class III smears were counted as positive. Class II or 'atypical' smears were counted as positive if there were three or more consecutive such smears, if there were two or more spanning a time interval of 10 months or more, or if a class II smear was followed either by a class III smear or by positive histology. A negative smear within 6 months of a previous negative smear was ignored to avoid double counting due to the repeat of inadequate smears.

In the first 2 or 3 years following a single negative smear, the risk of invasive

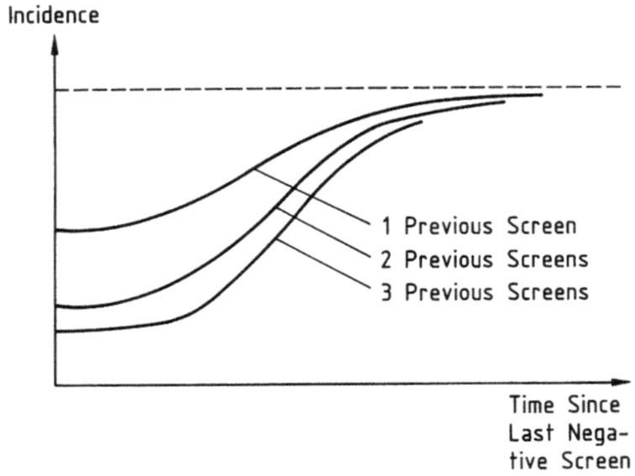

Fig. 1. Anticipated incidence following one or more negative screens, by time since last negative screen, Walter and Day [4]

disease results largely from false negatives of the screening test, and hence gives a measure of the test sensitivity. With increasing time since a negative smear, newly arising cases contribute a greater part of the incidence of invasive disease (Fig. 1). After two or more negative smears, the effect of false negatives of the screening test disappears to a great extent; the risk of invasive disease is then almost entirely due to newly arising cases, and the incidence curve with time simply reflects the cumulative distribution function of the detectable preclinical phase of the disease [4].

The IARC Collaborative Study

Description of the Study

The IARC study included data from screening programmes in eight different countries in Europe and North America. The programmes varied widely in respect of factors such as the degree of central organisation of the screening, and the extent to which a single laboratory was responsible for the cytological evaluation, which greatly facilitates the determination of screening histories. Nevertheless, an attempt was made to ensure uniform terminology, for example in the definition of a negative smear.

The aim of the study was to estimate the risk of invasive cervical cancer in successive years following 1, 2 or more negative smears. The data include both cohort and case-control studies. Cohort studies were those where the whole population in the screening programme could be identified and full screening records and cancer registration data were available. Case-control studies were originally used in situations where complete population data was not available, but were also subsequently used as a flexible and economical approach to the analysis of the data from the centrally organised programmes in Aberdeen and Iceland.

Results

The results of the study are expressed in terms of 'relative protections' - i.e. the inverse of the relative risks - and are mainly for women aged 35-64. For the cohort studies, the relative protections were calculated using either unscreened women in the population or an equivalent population prior to the introduction of screening as the baseline.

Table 1 shows the relative protection for Aberdeen with time since a single negative smear after adjustment for number of smears [5]. There were no cases in the first 12 months following a single smear, giving a relative protection of infinity. Thereafter, the relative protection falls with increasing time following a negative smear.

In general, there was greater variation between centres in the results following a single smear than in those following two or more smears, suggesting that while the sensitivity of the test is a function of the particular screening programme and varies between centres, there is a common underlying disease natural history. Table 2 shows the results following two or more smears combined for all centres by taking the geometric mean. The relative protection falls gradually with time since last smear, but some benefit remains even after 7-9 years.

Table 1. Relative protection with time since last negative smear (Aberdeen)

	Months since last negative smear (after adjustment for number of smears)						
	0-11	12-23	24-35	36-47	48-71	72-119	>120
Relative protection		6.9	3.4	1.7	1.6	1.0	1.0
95% Confidence interval		1.3-37.7	1.0-11.4	0.5-5.2	0.7-3.9	0.4-2.7	

Macgregor et al. [5]

Table 2. Summary of relative protections for all centres (geometric mean values)

Time since last negative smear (months)	Relative protection (no. of cases in parentheses)	95% Confidence intervals
0-11	15.3 (25)	10.0-22.6
12-23	11.9 (23)	7.5-18.3
24-35	8.0 (25)	5.2-11.8
36-47	5.3 (30)	3.6- 7.6
48-59	2.8 (30)	1.9- 4.0
60-71	3.6 (16)	2.1- 5.9
72-119	1.6 (6)	0.6- 3.5
120+	0.8[a] (7)	0.3- 1.6
Never screened	1.0	

[a] Based on figures from Aberdeen and Iceland only IARC Working Group [9]

Analysis by Age

In view of discussion in the literature of the possibility of a more rapidly progressive disease occurring in younger women [6, 7], the data for Aberdeen and Iceland have been analysed in two age groups, women below 50 and those 50 and over, in order to investigate possible variation in the disease natural history with age. There was no significant difference between the results for the two groups [8]. However, in this study little data was available for women aged under 35, which is the age group in which concern about possible fast-growing disease exists.

Estimates of the Effects of Different Screening Policies

These results can be used to estimate the potential effects of different screening policies in terms of the reduction in the cumulative incidence of invasive disease that can be achieved. Table 3 shows the predicted reduction in incidence for a number of different screening frequencies over the age range 35-64, together with the number of tests involved per woman over this age range. It can be seen that very little is gained by screening every year or even 2 years compared with every 3 years, and that screening every 5 years still offers considerable protection.

Two main factors must be borne in mind when considering these results. Firstly, these reductions are those that would be achieved with 100% compliance with the screening programme. Screening the whole target population every 10 years is of greater benefit than screening only a third every 3 years, and a high acceptance rate should be a major aim of any screening programme.

Secondly, any decision concerning screening policy must be based on a cost-benefit approach. Table 4 shows for a number of different screening policies the predicted reduction in incidence and the number of cases prevented per 100000 tests [9]. The most cost-effective regime is screening every 5 years starting at age 35. Screening every 3 years increases the benefit but with a reduced cost effectiveness, as does commencing screening at age 25. More frequent screening or commencing at ages below age 25 merely reduce the cost effectiveness with little or no additional benefit.

Table 3. Percentage reduction in the cumulative rate of invasive cervical cancer over the age range 35-65, with different frequencies of screening

Screening frequency	Reduction in the cumulative rate[a] (%)	Test *(n)*
1 year	93.3	30
2 years	92.5	15
3 years	91.4	10
5 years	83.9	6
10 years	64.2	3

[a] Assuming a screen occurs at age 35, and that a previous screen had been performed
IARC Working Group [9].

Table 4. Effect of different screening policies on incidence rates of cervical cancer in women aged 20-64

Screening programme	Cumulative rate/10^5 women	Reduction in incidence (%)	Tests (*n*)	Cases prevented/ 10^5 tests
No screening	1575			
Screening every 5 years				
Ages 20-64	258.6	84	9	146
Ages 25-64	287.8	82	8	161
Ages 35-64	480.9	70	6	182
Screening every 3 years				
Ages 20-64	138.9	91	15	96
Ages 25-64	161.8	90	13	109
Ages 35-64	354.9	78	10	122
Screening every year				
Ages 20-64	105.2	93	45	33

IARC Working Group [9]

References

1. Hakama M (1981) Trends in the incidence of cervical cancer in the Nordic countries. In: Magnus K (ed) Trends in cancer incidence, causes and practical implications. Hemisphere, New York
2. IARC working group on cervical cancer screening (1986) Screening for squamous cervical cancer - the duration of low risk following negative results in cervical cytology tests: introduction. In: Hakama M, Miller AB, Day NE (eds) Screening for cancer of the uterine cervix. IARC, Lyon (IARC scientific publications, no 76)
3. Boyes DA, Morrison B, Knox EG, Draper GJ, Miller AB (1982) A cohort study of cervical cancer screening in British Columbia. Clin Invest Med 5: 1-29
4. Walter SD, Day NE (1983) Estimation of the duration of a pre-clinical disease state using screening data. Am J Epidemiol 118: 865-886
5. Macgregor JE, Moss SM, Parkin DM, Day NE (1985) A case-control study of cervical cancer screening in north-east England. Br Med J 290: 1543-1546
6. Berkeley AS, Livalsi VA, Schwartz PE (1980) Advanced squamous cell carcinomas of the cervix with recent normal Papanicolaou tests. Lancet II: 375-376
7. Prendiville W, Guillebaud J, Bamford P, Beilby J, Steele SJ (1980) Carcinoma of cervix with recent normal Papanicolaou tests. Lancet II: 853-854
8. Moss SM (1986) Combined analysis of data from north-east Scotland and Iceland. In: Hakama M, Miller AB, Day NE (eds) Screening for cancer of the uterine cervix. IARC, Lyon (IARC scientific publications, no 76)
9. IARC Working Group in Evaluation of Cervical Cancer Screening Programmes (1986). Screening for squamous cervical cancer: duration of low risk after negative results of cervical cytology and its implication for screening policies. Br Med J 293: 659-664

Evaluation of Cervical Smears in a Comparative Study Between Countries with Different Incidences of Cervical Cancer (Israel, Germany, South Africa)*

H. Volkert, A. Schachter, U. Schenck, R. Schmauz, H.-J. Soost, G. Wikely, and R. Busch

Introduction

In a comparative study we have evaluated cervical smears from three countries with different incidences of cervical cancer. Israel and South Africa were chosen for comparison because of the considerable difference in incidence of cervical carcinoma. A look at the incidence per 100000 women shows this difference. The incidence of cervical carcinoma is 2.9/100000 in Israel, 18.4 in Hamburg (which may be representative for Germany with some reservation), and 52.0 in Johannesburg, S. Africa.

Intentions of the Study

There were three main points of interest in this study. We wanted to investigate whether the different incidence of cervical cancer is reflected in the frequency of different cytological findings. Furthermore, we wanted to find out whether the frequency of human papilloma virus (HPV)-related cytological changes shows a correlation with the risk of cervical cancer. Additionally, we attempted to get a hint whether other microbiological findings (a certain kind of flora) could be a further indicator of a risk of developing carcinoma.

Material and Methods

The study was conducted at the Institute of Clinical Cytology of the Technical University of Munich, Germany, in cooperation with the Department of Pathology of the University of Lübeck, Germany, the Regional Laboratory Services in Durban, South Africa, and the Beilinson Hospital, Petach Tiqva, Israel. Within the study 6442 cervical smears have been evaluated: 1613 were from the Beilinson Hospital in Petach Tiqva/Israel, and 1791 from the Cytologic Unit in Durban, South Africa. All of the latter were taken from the black population. From the Institute in Munich, we took 3038 smears in this study. To avoid a non-uniform evaluation by several cytologists, all smears were rescreened at our Institute.

* Study supported by the Deutsche Forschungs-Gemeinschaft

Results and Discussion

The greatest number of serious cytological diagnoses was found in South Africa, whereas in the smears from Israel we did not find any serious case. We even only found a few cases of Pap grade IIID, which means light to moderate dysplasia, among the Israelite smears (Table 1). It is remarkable that only 9.7% of the smears from South Africa were without any pathological findings (grade I). In the higher Pap grades (IVa and IVb; IVa means severe dysplasia or CIS, IVb the same plus possibly invasive carcinoma), we did not find any cases in Israel, but a relatively high frequency in South Africa (0.8% IVa, 0.3% IVb). In 0.9% of the South African smears, the cytological diagnosis was Pap V, which means invasive carcinoma.

The total percentage of Pap grades I and II (including Pap II with recommendation for a further control) was 99.7% in Israel, 98.4% in Germany, and 91.4% in S. Africa. The rates of light and moderate dysplasia were 0.1% in Israel, 1.1% in Germany and 4.7% in South Africa. Thus the differences in the incidence of carcinoma are also apparent in the cytological findings.

The results according to different age groups showed the same tendency, with the greatest number of serious findings in South Africa. The high rates of pathological findings in South Africa in comparison to the German ones may be partly explained by the non-existent state-run screening programme there. On the other hand, there is no such programme in Israel because there is no need for 'prevention', cervical carcinoma being so rare.

The possibility that HPV promotes cellular changes which eventually may lead to cancer has been a field of intensive research for many years (especially by Gissmann, Meisels, Zur Hausen; e.g. Crum et al. 1984; Gissmann et al. 1984; Meisels 1983; Meisels and Fortin 1976; Meisels et al. 1977, 1981; Schneider et al. 1986; Wagner et al. 1985; Zur Hausen 1987; Zur Hausen et al. 1981). HPV types 16, 18 and 31 were seen in invasive carcinoma as well as in so-called atypical condyloma. In condyloma acuminatum, especially the types 6 and 11 were found. As described by many authors, koilocytes in the cervical smear are pathognomonic for an infection with HPV and one of the most important signs for it in the cell picture. Its possible role in the development of cervical carcinoma has been studied widely.

The koilocyte is characterized by a large perinuclear clearing (halo) with irregular, sharply demarcated borders. The cytoplasm is dense, often amphophilic. The nuclei are enlarged, often quite dense, hyperchromatic, sometimes almost pycnotic. Binucleation is very common.

Table 1. Percentage of the different Pap grades in Israel, Germany, and South Africa

	Israel	Germany	South Africa
I	53.6	43.5	9.7
II	46.1	54.9	81.7
IIID	0.1	1.1	4.7
IVa	–	0.1	0.8
IVb	–	–	0.3
V	–	0.1	0.9

A further important pathological feature of HPV infection is the dyskeratocyte, a small squamous cell with an enlarged, dense, sometimes pycnotic nucleus and an orange-coloured cytoplasm in the Papanicolaou stain. Dyskeratocytes tend to exfoliate in dense, three-dimensional aggregates (Meisels et al. 1981). Unfortunately, reliable rates of dyskeratocytes could not be presented for each country in this study. Due to differences in staining and sometimes to the poor preservation of the stainings, we were afraid that such cells might be under- or overestimated depending on orangeophilia and eosinophilia of the smear. For this reason we had to drop this important sign from our final evaluation.

The frequency of cases where koilocytes were found showed an increasing tendency from Israel with 0.2% to Germany with 0.5% and to South Africa with 1.1%. These percentages represent the rate of koilocytes only, and not the rate of HPV infection, because the presence of koilocytes, in spite of being pathognomonic for that infection, is not always essential. Following this approach, the true percentage of HPV infections is largely underestimated. The in situ hybridisation technique yields a much higher frequency of HPV-positive results than the morphological diagnosis with koilocytes.

Interesting differences also were found concerning the frequency of other microbiological findings (Fig. 1). It is remarkable that in Germany and Israel there are high rates of the physiological Döderlein flora (56.2% in Israel and 56.0% in Germany), but low rates of only 17.1% in the South African black population. Cocci were found in the South African smears in 31.8% of the cases, more than three times the rate observed in Israel and Germany. Concerning trichomonads and fungi, South Africa had the highest rates as well. The rate of these two microorganisms was a little bit higher in Israel than in Germany. The spectrum of all registrated kinds of infection was wider in South Africa, where we found cytological signs of infection with herpesvirus in nine cases, than in the other countries.

Due to molecular biological and seroepidemiological research, infections, particularly with HPV and possibly herpesvirus, may play a role in the development

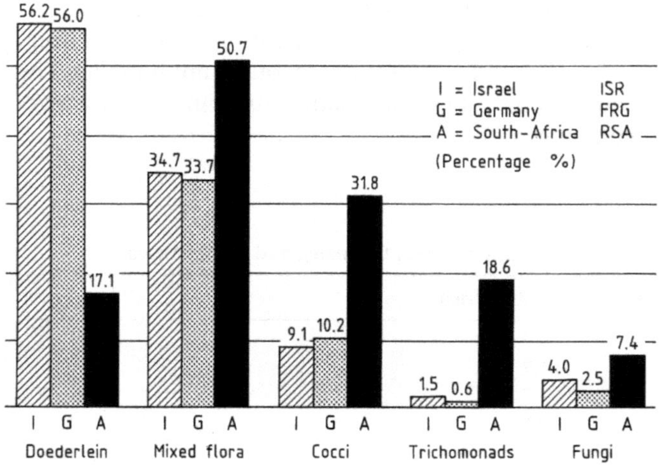

Fig. 1. Microbiological findings in the different countries

of cervical carcinoma. Infections with other microorganisms could turn out to be further risk indicators for cervical carcinoma. It is remarkable that the frequency of all pathological microorganisms is higher in South Africa than in Germany and Israel, and that they vary in the same direction with the incidence of carcinoma there (positive correlation). In contrast, the physiological Döderlein's bacillus varies in the opposite direction, which may - as previously described by other authors - point to poor hygiene being connected with a higher risk of cervical carcinoma.

The main conclusions of this study are:

1. The differences in incidence of cervical carcinoma in the three countries also become apparent in the cytological findings.
2. There is a correlation between the frequency of HPV-related cytological changes and the risk of carcinoma.
3. In countries with a low incidence of cervical carcinoma, the percentage of Döderlein's bacilli is high, whereas in South Africa, a country with high incidence, the percentage of Döderlein's bacilli is low. In addition, the presence of other microorganisms (e.g. cocci, trichomonads) might be a further risk indicator.

References

Boon ME, Fox CH (1981) Simultaneous condylomata acuminatum and dysplasia of the uterine cervix. Acta Cytol 25: 393-399

Crum CP, Ikenberg H, Richart RM, Gissmann L (1984) Human papilloma-virus type 16 and early cervical neoplasia. N Engl J Med 310: 880-883

Gissmann L, Boshart M, Dürst M, Ikenberg H, Wagner D, Zur Hausen H (1984) Presence of human papillomavirus (HPV) DNA in genital tumors. J Invest Dermatol (Suppl) 83: 265-285

Koss LG, Durfee GR (1956) Unusual pattern of squamous epithelium of the uterine cervix: Cytologic and pathologic study of koilocytotic atypia. Ann NY Acad Sci 63: 1235-1261

Meisels A (1983) The story of a cell. The G.N. Papanicolaou Award Lecture. Acta Cytol 27: 584-596

Meisels A, Fortin R (1976) Condylomatous lesions of the cervix and vagina. I. Cytologic patterns. Acta Cytol 20: 505-509

Meisels A, Fortin R, Roy M (1977) Condylomatous lesions of the cervix: II. Cytologic, colposcopic and histopathologic study. Acta Cytol 21: 379-390

Meisels A, Roy M, Fortier M, Morin C, Casas-Cordero M, Shah KV, Turgeon H (1981) Human papillomavirus (HPV) infection of the cervix. The atypical condyloma. Acta Cytol 25: 7-16

Morin C (1982) Le condylome du col uterin: Recherche virale et cytologie ultrastructurale. Ph. D. thesis, Laval University, Quebec, Canada

Purola E, Savia E (1977) Cytology of gynecologic condyloma acuminatum. Acta Cytol 21: 26-31

Schneider A, Schuhmann R, De Villiers E-M, Knauf W, Gissmann L (1986) Klinische Bedeutung von humanen Papilloma-Virus-(HPV)-Infektionen im unteren Genitaltrakt. Geburtshilfe Frauenheilkd 46: 261-266

Syrjänen KJ (1984) Current concepts of human papillomavirus infections in the genital tract and their relationship to intraepithelial neoplasia and squamous cell carcinoma. Obstet Gynecol Survey 39: 252-265

Wagner D, De Villiers E-M, Gissmann L (1985) Der Nachweis verschiedener Papillomvirustypen in zytologischen Abstrichen von Präkanzerosen und Karzinomen der Cervix uteri. Geburtshilfe Frauenheilkd 45: 226-231

Zur Hausen H (1987) Papillomaviruses in human cancer. Cancer 59: 1692-1696

Zur Hausen H, De Villiers H-M, Gissmann L (1981) Papillomavirus infections and human genital cancer. Gynecol Oncol 12: 124-128

Procedure for Validation of Mass Screening by Cervical Cytology*

B. Ruffing-Kullmann, W. Lehmacher, and H.-J. Soost

Introduction

In the Cytological Institute of the Bavarian Cancer Society 748 871 cytological examinations of 281 705 women have been performed within the German mass screening program in a 10-year period. The results of these examinations have been stored in a computer and statistically evaluated in cooperation with the Gesellschaft für Strahlen- und Umweltforschung.

Measurements for Validation of Diagnostic Tests

If a screening test is used to detect a disease the results can be summarized in a four-part table assuming that it is possible to determine whether the disease exists or not (Galen and Gambino 1979; Fleiss 1981).

Validation of diagnostic tests is done in terms of the following: sensitivity, specifity, predictive value of the positive test result, and predictive value of the negative test result (Table 1). Sensitivity of the test is a rate relating correctly detected ill person (a in the table) in the test to all ill persons (a + c). Specifity is the rate relating correctly detected healthy persons (d) in the test to all healthy persons (b + d). Predictive values, however, are rates which relate correctly classified persons in the test to all the persons with a positive or negative test, that means (a) di-

Table 1. Results of a diagnostic test *t* for a disease *D*

		disease		
		+	–	
test	+	a	b	a+b
	–	c	d	c+d
		a+c	b+d	N=a+b+c+d

+, positive; –, negative; a-d, number of results

* Supported by the Deutsche Krebshilfe

vided by (a+b) for the predictive value of a positive test result and (d) divided by (c+d) for the predictive value of a negative test result.

Whereas sensitivity and specifity only depend on the test, as is well known the predictive values also depend on the prevalence of the disease. For example, sensitivity and specifity of 95% and a prevalence of 1% gives a positive predictive value of 16.1% and a negative predictive value of 99.95%. Generally screening tests with a low prevalence tend to a low positive predictive value and a high negative predictive value (Köbberling 1982; Lehmacher and Keil 1983).

Procedure for Validation of the Existing Material

To evaluate the results of the cytological examinations we always proceeded "prospectively" from the *first* cytological examination of a woman (Soost et al. 1987). Later histological or cytological examinations were taken for confirmation or disproval of the first examination. Histological examination provides the best validation of the cytological examination, but validation by histology is only possible for the small number of the cytologically positive smears. The vast majority of negative findings are not examined histologically after the cytological examination.

If no histological examination was performed, validation of negative findings was only possible by means of the two following cytological reports. A negative first examination was assumed to be correct if at least two follow-up negative cytological examinations were taken within 3 years. If only one of the two following examinations was positive (Pap grades III-V), the negative initial finding was valued as false negative.

To evaluate the positive and doubtful cytological findings with Pap grades III-V, only histological examinations within 1 year were admitted. Cytological first screenings which could not be admitted or disproved by these definitions were taken as not verifiable. Table 2 shows the distribution of the verifiable follow-up cytological findings after initial cytological screenings. Of the positive initial screenings with Pap grades IVa-V, 75% could be validated, as were 25% of the negative cytological first screenings.

Table 2. Distribution of findings for validation after cytological first screenings

Initial cytology	Histology				Cytology (maximum result of two following findings)						
	benign	slight-moderate dysplasia	severe dysplasia, carcinoma in situ	microcarcinoma, invasive cervical carcinoma	I/II	III	IIID	IVa	IVb	V	
I/II	83	81	73	19	68318	419	633	82	18	10	69736
III	158	89	135	57							439
IIID	116	148	161	11							436
IVa	45	66	322	44							477
IVb	18	28	210	65							321
V	7	6	54	90							157

Table 3. Predictive values of cytological first findings

Negative predictive value:	PV^- (I/II)	99.8%
Positive predictive value:	PV^+ (III)	64.0%
	PV^+ (III D)	73.4%
	PV^+ (IVa)	90.6%
	PV^+ (IVb)	94.4%
	PV^+ (V)	95.5%

Estimation of Predictive Values

For a small number of negative findings with a grade I/II we recommended a repetition of the smear. After a short term control this group had a higher rate of positive findings; many false negative findings had a positive cytological finding after such repetition. Because classification of these cases as false negative would be too severe for validation of the whole system, we eliminated these cases. Negative findings on a follow-up of grade III D or III were eliminated, too, because these cases could have developed in the interval since initial screening or could not disprove the first finding.

Then from Table 2 the following predictive values of cytological first screenings, summarized in Table 3, were derived. For negative cytological initial screenings the predictive value, i.e., the probability that a woman is in fact healthy, is 99.8%. For positive cytological findings with a Pap grade IVa-V (which means at least severe dysplasia), the predictive values are in the range from 90.6% to 95.5%, which is very high for a screening procedure. The estimation of the positive predictive values can also be considered as very reliable because only histological examinations were taken for validation. For the Pap grades III (doubtful) and III D (corresponding to a slight or moderate dysplasia), the predictive values (64% and 73%, respectively) were much lower. However, it is known that more than half of the cases with a Pap grade III D spontaneously regress within 1-2 years.

Estimation of Sensitivity and Specifity

To estimate sensitivity all patients with a positive histological or cytological finding with Pap grades IVa, IVb, or V were considered to be ill. The rate "cytologically detected ill persons/ill persons" is thus an estimate of sensitivity. To estimate specifity all women with a negative histological or two negative follow-up cytological examinations were considered to be healthy. Cytologically detected patients were defined as those with Pap grade III-V at the first cytological examination. "Cytological negative persons/healthy persons" is thus an estimate of specifity (Soost et al. 1987).

Assuming that "not verifiable" women behave like verifiable ones, we extrapolated our results with the cytological first findings. Using this procedure we determined a sensitivity of 80% and a specifity of 99.95% (Table 4).

Table 4. Estimation of sensitivity and specifity

		'ill'	'not ill'	
initial cytology	pos.	2333 (III-V)	133 (IV-V)	2466
	neg.	584	269683	270267
		2917	269816	272733

Sensitivity = 2333/2917 = 80.0%
Specifity = 269683/269816 = 99.95%

Summary and Discussion

While the estimate for the predictive values can be considered very precise, estimates for sensitivity and specifity only should be interpreted in orders of magnitude because of our lack of knowledge of health or disease and our assumptions about this. In any case, an extraordinarily high specifity of more than 99% results and a clearly lower sensitivity of about 80%. We must assume a 20% false-negative rate from which about three-forth have to be viewed as sampling errors as we found by rescreening the slides. The surprisingly high positive predictive values from 91% to 96% and the excellent specifity of more than 99% indicate that the cytological findings are very seldom false positive, i. e., if a cytological finding is suspicious there is almost always a pathological change.

References

Fleiss J (1981) Statistical methods for rates and proportions. 2nd edn. Wiley, New York

Galen RS, Gambino SR (1979) Norm und Normabweichung klinischer Daten. Fischer, Stuttgart

Köbberling J (1982) Der Prädiktive Wert diagnostischer Maßnahmen. Dtsch Med Wochenschr 107: 591-595

Lehmacher W, Keil U (1983) Prädiktionswerte - eine Alternative zu den Referenzwerten in Diagnostik und Epidemiologie. In: Berger J, Höhne KH (eds) Methoden der Statistik und Informatik in Epidemiologie und Diagnostik. Proceedings der 27. GMDS-Jahrestagung in Hamburg, 1982

Soost H-J, Lange H-J, Lehmacher W, Ruffing-Kullmann B (1987) Ergebnisse zytologischer Krebsfrüherkennungs- und Vorsorgeuntersuchungen bei der Frau. Deutscher Ärzte-Verlag, Cologne

Significance of Cellular Composition of Smears for the Reliability of Cytological Diagnoses

G. P. Vooijs

Introduction

About ±90% of squamous abnormalities that may eventually lead to an invasive cervical cancer originate at the squamocolumnar junction. To be able to make a reliable diagnosis this area should be represented in smears. At the start of a screening program for cervical cancer we adopted the principle that endocervical columnar cells as well as ectocervical squamous cells and squamous metaplastic cells should be present in a smear before it is considered a reliable smear.

The original Ayre spatula is, by its shape, not the most adequate collection instrument since it usually does not collect cells proximally of the squamocolumnar junction. Thus, endocervical columnar cells and squamous metaplastic cells are often absent in such specimens. To improve the quality of the smears and to insure that the different cell types of the mucosal lining of the cervix are represented in the cervical smear, we developed a new spatula whose shape is more pointed. This spatula was biometrically designed from a composite drawing made from about

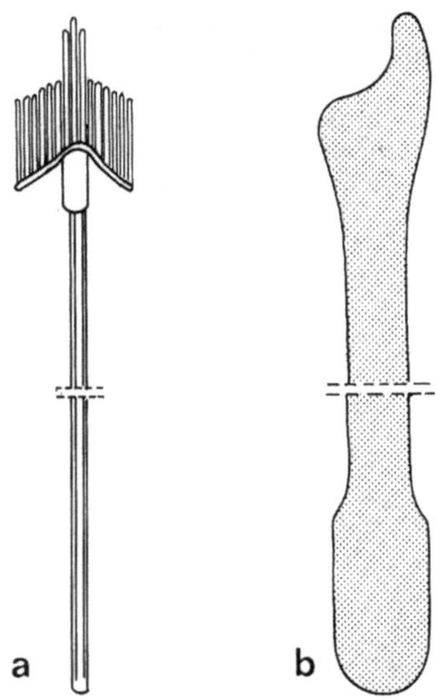

Fig. 1. a Plastic cervical brush, **b** biometrically designed wooden spatula

100 cervices from hysterectomised uteri [2] (Fig. 1b). The spatula used in the original screening program was somewhat smaller. The biometrically designed spatula is presently used in the screening program and is now known as the Aylesbury spatula (Histopath) [6].

Materials and Methods

To measure the impact of the cellular composition of a smear on the reliability of cytological diagnosis, smears were subdivided in smears containing endocervical columnar cells (EC+), smears containing squamous metaplastic cells but no endocervical columnar cells (SM), and smears with squamous cells only (SQC) (Table 1).

Since 1976 in the region of the city of Nijmegen, The Netherlands, a screening program for cervical cancer has been in progress. A detailed description of the program has been given elsewhere [1, 3-5]. In short, women 35-54 years of age were screened once every 3 years. A single cervical smear was taken with a slightly modified, pointed wooden Ayre spatula. The smears were processed at the Cytology Laboratory of the Department of Pathology, University of Nijmegen, and screened by experienced cytotechnologists. The cytologic findings were recorded by means of a revised Papanicolaou classification. In addition, a description of the cytologic findings was also recorded. Each descriptive diagnosis comprised four points:

1. The cellular composition of the smear
2. The presence or character of any inflammatory changes
3. The expected abnormality in the squamous and squamous metaplastic epithelium
4. Changes in columnar cells from the endocervical mucosa or an abnormality related to the endometrium

Cellular Composition and Epithelial Abnormalities

In a total of 98370 women in whom the cervix at first screening was brought into view, 5.3% of the smears were unsatisfactory for cytological evaluation. Of the remaining 93264 smears, 93.6% contained endocervical columnar cells (EC+), 2.7%

Table 1. Cellular composition of cervical smears at first screening

	N	%
Total	98370	100
Unsatisfactory		5.3
Subtotal	93264	100
EC+		93.6
SM		2.7
SQC		3.7

contained squamous metaplastic cells (SM), and 3.7% contained squamous epithelial cells only (SQC) (Table 1). For the sake of clarity smears were further subdivided in smears containing endocervical columnar cells (EC+) and smears without endocervical columnar cells (EC–).

When the number of abnormalities diagnosed in EC+ smears is compared with the number of abnormalities detected in EC– smears it becomes evident that the frequency of epithelial abnormalities diagnosed in these two types of smears differs 2.0 times for minimal atypia, 2.3 times for slight and moderate atypia, and 4.5 times for severe dysplasia, carcinoma in situ, and invasive cancer (Table 2). This means that in smears not containing endocervical cells the risk of a false negative or underestimated diagnosis is significantly increased.

To be sure that the lesions were really misdiagnosed and that we were not confronted with some peculiar physiological phenomena, showing that a lower number of endocervical columnar cells in cervical smears is truly related to a lower number of epithelial abnormalities, we did a cohort study in 43726 women. We compared the number of epithelial abnormalities at first screening with the cellular composition of the smears at first screening (Table 3). In this cohort the ratio between the number of epithelial abnormalities detected in EC+ and EC– smears was 1.9 for minimal atypias, and 2.1 for slight and moderate dysplasias. The ratio for severe dysplasias, carcinoma in situ, and invasive cancer at first screening was 1.3.

Table 2. Epithelial changes of the cervical mucosa in relation to smear composition

Smear composition	Epithelial changes of the cervical mucosa					
	Normal	Minimal atypia	Slight and moderate dypslasia	≥ Severe dysplasia	Total	
	(%)	(%)	(%)	(%)	*(N)*	(%)
EC+	82.3	16.3	0.70	0.36	87326	93.6
EC–	91.3	8.3	0.31	0.08	5938	6.4
Ratio	0.9	2.0	2.3	4.5	–	–
Total	83.2	15.8	0.68	0.34	93264	100

Table 3. Relative risk of epithelial changes in cervical smears with and without endocervical columnar cells - cohort study at first screening

Smear composition at first screening	Epithelial changes of cervical mucosa at first screening					
	Total		Normal	Minimal atypia	Slight and moderate dysplasia	≥ Severe dysplasia
	(N)	(%)	(%)	(%)	(%)	(%)
EC+	40958	93.7	83.9	15.7	0.44	0.04
EC–	2768	6.3	91.7	8.1	0.21	0.03
Total	43726	100	84.4	15.2	0.42	0.04
Ratio	–	–	0.9	1.9	2.1	1.3

Table 4. Relative risk of epithelial changes in cervical smears with and without endocervical columnar cells - cohort study at second screening

Smear composition at first screening	Epithelial changes of the cervical mucosa at second screening					
	Total		Normal	Minimal atypia	Slight and moderate dysplasia	≥ Severe dysplasia
	(N)	(%)	(%)	(%)	(%)	(%)
EC+	40425	93.7	79.0	20.0	0.88	0.18
EC−	2717	6.3	82.7	16.5	0.77	0.03
Total	42971	100	79.2	19.7	0.87	0.67
Ratio	–	–	1.0	1.2	1.1	6.0

At second screening of the same women in this cohort 3 years later we compared the number of abnormalities detected at second screening with the original cellular composition of the smears at first screening (Table 4). This cohort analysis was based on the hypothesis that if we had missed lesions at first screening in the EC− smears, the number of lesions at second screening should be higher, which would become evident by a decrease in the ratio. In fact, the relative risk at second screening proved to be significantly reduced in comparison with the relative risk at first screening. Relative risk factors dropped from 1.9 to 1.2 in minimal atypias and from 2.1 to 1.1 in slight and moderate dysplasias.

Thus the number of epithelial abnormalities in women who previously had smears without endocervical cells was found to be significantly higher at successive screenings. Therefore these lesions or their prestages were not detected at first screening. This proved that cellular composition of smears is a highly important "risk factor" for the accuracy of cytological diagnosis.

Performance of Smear Takers

A comparison was made of the performance in smear taking between different groups of sample takers with different levels of experience. Smear takers who had taken 1200 to 2000 smears (group C) were compared with smear takers who had

Table 5. Cellular composition of cervical smears subdivided by grade of experience of sample takers

	Group C 1200-2000		Group A 9000-30000		Total	
	(N)	(%)	*(N)*	(%)	*(N)*	(%)
Total	8289	100	98858	100	107147	100
Unsatisfactory		5.1		4.6		4.8
Subtotal		100		100		100
EC+		91.2		94.4		94.2
SM		3.6		2.2		2.2
SQC		5.3		3.4		3.7

made 9000 to 30000 smears (group A) during the entire program. The composition of smears was compared between these two groups (Table 5).

The percentage of unsatisfactory smears was 5.1% in the less experienced group and 4.6% in the more experienced group. Overall this percentage was 4.8%. Cellular composition subdivided for endocervical cells present (EC+), squamous metaplastic cells present (SM), and squamous cells only (SQC) again showed differences between these two groups. The percentage of less reliable smears was 8.9% in the less experienced group, 5.6% in the more experienced group, and 5.9% overall. Differences between the less and the more experienced groups were significant. These two comparisons strongly indicate that experience is significantly related to the number of unsatisfactory and low quality smears.

Influence of Contraceptive Method

A comparison was made between the cellular composition of smears (EC+/EC−) and the method of contraception practiced by women from the entire group, 35-54 years of age, cervix seen at smear taking, last menstrual period less than 60 days prior to smear taking, and first smears taken in the program (Table 6).

The number of less reliable smears differs significantly with different modes of contraception. In women practicing no contraception the percentage of less reliable smears was 5.5%, in women using oral hormonal contraception this percentage was 12.6%, in women using an IUD it was 4.4%, and in women using mechanical or surgical means of contraception it was 4.0%. The difference in smear composition in women using oral contraceptives and in women using other methods of contraception where highly significant ($p = 0.00001$). This means that the number of high quality smears in women using oral hormonal contraceptives is significantly reduced.

The impact of oral contraceptive use on the reliability of smears becomes even more evident when the number of epithelial abnormalities detected was related to the method of contraception (Table 7). The percentage of smears with minimal epithelial atypia in women who used no contraceptives or who practiced other non-hormonal methods were 16.3% and 16.8% respectively. In oral contraceptive users

Table 6. Cell composition of cervical smears subdivided by mode of contraception

Smear composition	Mode of contraception							
	Normal		Oral hormonal contraception		IUD		Other methods	
	(N)	(%)	*(N)*	(%)	*(N)*	(%)	*(N)*	(%)
Total	26877	100	17972	100	3749	100	33036	100
Unsatisfactory		6.0		4.7		7.6		6.5
Subtotal	25276	100	17136	100	3465	100	30875	100
EC+		94.5		87.4		95.6		96.0
SM		5.5		12.6		4.4		4.0

Table 7. Epithelial changes of the cervical mucosa subdivided for the mode of contraception

Mode of contraception	Epithelial changes of the cervical mucosa				
	Normal (%)	Minimal atypia (%)	Slight, moderate dysplasia (%)	Severe dysplasia CIS, (micro-) invasive cancer (%)	Total (*N*)
Normal	82.7	16.3	0.67	0.33	25276
Oral hormonal contraception	85.9	12.9	0.69	0.49	17136
IUD	76.3	22.4	1.04	0.20	3465
Other methods	82.2	16.8	0.75	0.25	30875

Total $N=77114$, CIS, carcinoma in situ

minimal atypias were diagnosed in 12.9% of smears. In IUD users the percentage of smears with minimal atypia (22.4%) has been known to be very high due to the mechanical influence of the device on the mucosa.

The frequency of severe epithelial abnormalities consistent with severe dysplasia, carcinoma in situ, and invasive cancers was completely different. In oral contraceptive users the percentage of severe epithelial abnormalities was significantly higher than in the other groups. The number of severe lesions in IUD users was remarkably low. The reason for the high percentage in hormonal contraceptive users is presently being analyzed. A rather acceptable explanation could be that the less severe lesions in oral contraceptive users were not diagnosed previously because of the relatively low reliability of the previous smears which, with the passing of time, enabled the epithelial abnormality to become more severe. This theory is supported by the low prevalence of severe lesions in IUD users in whom minimal lesions were diagnosed with much higher frequency.

Endocervical Brush Smears

The endocervical brush has been given much attention as the instrument of choice for the sampling of smears containing endocervical columnar cells since the number of columnar cells usually is very high in brush specimens. Recently a new sampling device has been introduced: a cervical plastic brush with long flexible hairs that, when making the smear, follow the shape of the cervix (Fig. 1a). We have compared in two groups of women smears taken with the spatula, with the latter cervical brush, and with the endocervical brush. Comparisons were made for quality of smears and number of epithelial abnormalities detected in smears with each of these devices.

The first comparison was made between spatula smears and endocervical brush smears. Smears were made at the same time, women being selected alternatively from those participating in a population screening program. The number of unsatisfactory smears in spatula smears was 8.7% and in endocervical brush smears 10.4% (Table 8). The percentage of smears containing endocervical columnar cells made with the endocervical brush smears was indeed much higher (97.6%).

Table 8. Cellular composition of cervical smears: comparison of spatula and endocervical brush smears

Smear composition	Spatula		Endocervical brush	
	(N)	(%)	*(N)*	(%)
Total	930	100	930	100
Unsatisfactory	81	8.7	97	10.4
Subtotal	849	100	833	100
EC+	788	92.8	813	97.6
SM	10	1.2	–	–
SQC	51	6.0	20	2.4

Table 9. Epithelial changes of the cervical mucosa: comparison of spatula and endocervical brush smears

Epithelial changes	Spatula		Endocervical brush	
	(N)	(%)	*(N)*	(%)
Total	852	100	832	100
Normal		84.5		92.2
Minimal atypia		13.7		7.6
Slight and moderate dysplasia		1.17		0.12
Severe dysplasia, CIS		0.59		0.12

Even though the percentage of unsatisfactory smears was higher in endocervical brush smears, this high proportion of EC+ smears made the overall quality of the cervical brush smears seemingly better. However the story became quite different when a comparison was made between the number of epithelial abnormalities found in spatula smears and endocervical brush smears (Table 9). Smears without epithelial abnormalities were found in 84.5% and 92.2% of satisfactory smears. Minimal atypia was found in 13.7% and 7.6%, slight and moderate dysplasia were found in 1.2% and 0.11%, and severe dysplasia in 0.59% and 0.12%.

This means that even though the number of satisfactory smears measured by the percentage of smears containing endocervical columnar cells was higher in endocervical brush specimens, the number of epithelial abnormalities diagnosed in these specimens was strikingly and alarmingly lower. Of 15 cases of slight, moderate, and severe dypslasia detected in spatula smears, only one severe abnormality (CIS) was detected in endocervical brush smears, whereas the other 14 lesions where not diagnosed at all or diagnosed with a much lower degree of atypia.

The results of this study indicate that the mere presence of endocervical cells should not be used as an absolute parameter for the reliability of a cervical smear. Endocervical brushes apparently collect columnar cells proximally from the area of the transitional zone and therefore beyond the area where the majority of epithelial abnormalities originate.

Table 10. Cellular composition of cervical smears: comparison of spatula and (Cervex) brush smears

Smear composition	Spatula		Cervical brush	
	(N)	(%)	*(N)*	(%)
Total	328	100	462	100
Unsatisfactory		5.2		2.6
Subtotal	311	100	450	100
EC+		91.6		92.4
SM		1.3		0.6
SQC		7.1		6.7

Table 11. Cellular composition of cervical smears: comparison of spatula and (Cervex) brush smears

Smear composition	Spatula		Cervical brush	
	(N)	(%)	*(N)*	(%)
Total	710	100	754	100
Normal		87.7		88.5
Minimal atypia		10.8		9.9
Slight and moderate dysplasia		0.98		1.19
Severe dysplasia, CIS		0.42		0.39

The second comparison was made between spatula smears and the new cervical brush smears (Cervex Brush; Table 10). Again smears were made alternately at the same time in women participating in a screening program. The number of unsatisfactory smears in spatula smears was 5.2% and in cervical brush smears 2.6%. The percentage of smears containing endocervical columnar cells in spatula smears was 91.6% and in cervical brush smears 92.4%. The number of epithelial abnormalities detected did not differ significantly between both kinds of smears (Table 11).

In a number of smears abnormalities were diagnosed with a somewhat lower degree of severity in the first smear taken (spatula or brush) than in the second smear. This happened with equal frequency in spatula and brush smears. This is understandable since the most superficial layer is removed when taking the first specimen, leaving relatively less mature layers to be sampled in the second specimen.

From these comparisons the conclusion may be drawn that the principal indicator for a good quality smear is not the number of endocervical cells as such but the co-ocurrence of endocervical columnar cells and squamous metaplastic cells.

The results of these comparisons should be tested in larger scale studies. When proved to be relevant, the lower percentage of unsatisfactory smears made with the "cervical" brush (Cervex) may be of practical and economic significance in large scale population screening programs.

Summary

The cellular composition of cervical smears has proved to be significantly related to the number of epithelial abnormalities detected. Cellular composition, particularly the number of unsatisfactory smears and the percentage of smears containing endocervical columnar cells, was significantly influenced by the experience of the smear taker and varied with the day of the menstrual cycle on which the smear was taken.

The number of less reliable smears is greatly increased in women using oral hormonal contraceptives. However, the number of severe epithelial abnormalities diagnosed in these women was significantly higher than in women using other methods of contraception.

In smears made with an endocervical brush, the number of epithelial abnormalities diagnosed was disturbingly low. A new cervical brush was introduced that produced high quality smears as well as a number of epithelial abnormalities comparable to that found in parallel spatula-made smears.

References

1. Elias A, Linthorst G, Bekker B, Vooijs GP (1983) The significance of endocervical cells in the diagnosis of cervical epithelial changes. Acta Cytol 27: 225-229
2. Vooijs GP (1987) Biometrically designed spatula for cervical scraping. Acta Cytol 31: 82
3. Vooijs GP, Elias A, van der Graaf Y, Poelen-van den Berg M (1986) The influence of sample takers on the cellular composition of cervical smears. Acta Cytol 30: 251-257
4. Vooijs GP, Elias A, van der Graaf Y, Veling S (1985) Relationship between the diagnosis of epithelial abnormalities and the composition of cervical smears. Acta Cytol 29: 323-329
5. Vooijs GP, van der Graaf Y, Elias A (1987) Cellular composition of cervical smears in relation to the day of the menstrual cycle and the method of contraception. Acta Cytol 31: 417
6. Wolfendale M, Howe-Guest R, Usherwood M (1987) Controlled trial of a new cervical spatula. Br Med J 294: 33-35

Diagnostic Efficiency of Brush Cytology from the Uterine Cervix

A. Glenthøj, F. Rank, U. Peen, and E. Bostofte

Introduction

Several studies have shown that brush cytology results in a higher percentage of smears with endocervical cells than smears taken with a cotton swab (Trimbos and Arentz 1986; Glenthøj et al. 1986; Ros et al. 1983). This higher frequency of adequate smears has been shown to lead to the detection of more premalignant changes, especially those of endocervical origin (Boon et al. 1986).

The purpose of the present study was to see if the diagnostic correlation between histology and smear cytology was influenced by the method of cell sampling.

Material and Methods

One hundred consecutive women who were referred to cold knife cone biopsy in our gynecological department were included in the study. The indication for cone biopsy was in all cases biopsy-proven changes of at least moderate dysplasia (CIN II), but the most common indication was CIN III. The patients were divided into two groups of 50 each by lot. In the first group (BC) cervical smears were taken with a brush, while in the second group (CS) smears were taken with two cotton swabs. The smears were taken in the operating theater immediately prior to cone biopsy. The mean age in group BC was 32 years (range 22-69 years), and in group CS 33 years (range 20-60 years).

The brush for cytological sampling was supplied by Medscand AB, Malmö, Sweden. The brush consists of a conic nylon brush with a 20-mm long metal core, which is connected to a 150-mm long handle. The diameter of the brush is at least 5 mm and 7.5 mm at most. The cotton swabs were water absorbent and unmoistened.

In group BC the brush was bent according to the anatomy of the cervix. The tip was passed into the cervical canal and the brush was rotated slowly. In group CS one cotton swab was passed into the cervical canal and rotated with a firm grip, while the other cotton swab was used to sample cells from the ectocervix.

Both types of cytological material were rolled on a glass slide, fixed and stained with the Papanicolaou stain. The material was evaluated blindly, without any knowledge of neither the clinical history nor the sampling method. For each patient one and only one of the following diagnoses was given: no malignant cells, CIN I, CIN II, CIN III, invasive carcinoma, and insufficient material. A diagnosis of insufficient material was given when neither columnar cells, metaplastic cells, nor malignant cells were found.

Cone biopsies were fixed in 10% neutral formaline and sliced into an average of 16 paraffine blocks. Sections were evaluated blindly by two pathologists, and one and only one of the above-mentioned diagnoses was reached.

Results

The number of invasive carcinomas and of CIN I was greater in group BC, while CIN II occurred more frequently in group CS (Tables 1 and 2). None of the brush cytology specimens were considered insufficient for diagnosis, while seven (14%) of the cotton swab smears were insufficient. In these seven patients cone biopsies showed 4 cases of CIN III, 2 cases of CIN II and one case without any malignant or premalignant change. The diagnostic agreement between cone biopsy and brush cytology and between cone biopsy and cotton swab smear is shown in Tables 1 and 2. The seven insufficient cotton swab smears are left out in Table 2.

Table 1. Comparison between diagnoses obtained by brush cytology and cone biopsies

Cone biopsy	Brush cytology					Total
	No malignancy	CIN I	CIN II	CIN III	Invasive carcinoma	
No malignancy	3					3
CIN I	1	0		1		2
CIN II			1	3		4
CIN III	1	3	6	22	2	34
Invasive carcinoma				2	5	7
Total	5	3	7	28	7	50

Table 2. Comparison between diagnoses obtained by cotton swab smears and cone biopsies

Cone biopsy	Cotton swab smears					Total
	No malignancy	CIN I	CIN II	CIN III	Invasive carcinoma	
No malignancy	3					3
CIN I		0				0
CIN II	5		1	3		9
CIN III	9	3	8	9		29
Invasive carcinoma			1	1	0	2
Total	17	3	10	13	0	43

In group BC diagnostic agreement occurred in 31 cases. In 13 cases the cone biopsy diagnosis was more malignant than the cytological diagnosis, while the cytological diagnosis was more malignant than the cone biopsy diagnosis in six cases.

In group CS diagnostic agreement occurred in 13 cases. In 27 cases the cone biopsy diagnosis was more malignant than the cytological diagnosis, while the cytological diagnosis was more malignant than the cone biopsy diagnosis in only three cases.

If one is only concerned about whether a cytological sample is positive or negative and if the insufficient smears are left out, the diagnostic specifity of both brush cytology and cotton swab smears were 100% since no false positive cytological diagnosis occurred. The diagnostic sensitivity of brush cytology was 3/5, or 60%, while the diagnostic sensitivity of cotton swab smears was 3/17, or 18%.

Discussion

The cytological evaluation was probably influenced by the knowledge of the patient material. A high percentage of positive specimens could be expected in both groups, especially many CIN III. Despite that bias, 14 (28%) of the cotton swab smears were false negative, even in nine patients with CIN III in cone biopsies. Only two (4%) of the brush biopsies were false negative, and in one of these the cone biopsy only showed CIN I. Furthermore, 14% of the cotton swab smears were insufficient, making a total of 42% false negative or insufficient smears in that group. The corresponding figure in group BC was 4%. Brush cytology therefore seems to be superior to cotton swab smears in detecting neoplasia in the uterine cervix.

The grading of abnormal cytological findings correlated better with the histological diagnosis in group BC than in group CS. More than half of the cotton swab smears were undergraded and hardly any overgraded.

The fishnetlike surface of a cotton swab probably sucks up desquamated cells that are lying in the cervical secretion. Brush cytology on the other hand results in a more efficient scrapping of epithelial cells (Trimbos and Arentz 1986; Boon et al. 1986). Our results are therefore not surprising, since brush cytology is superior to cotton swab smears in getting well-preserved and cellular samples from the uterine cervix (Glenthøj et al. 1986).

In order to take brush samples from the ectocervix, it is most often necessary to bend the brush. This procedure might be difficult to use in daily practice. A combination of an ectocervical scrap with an Ayre spatula and a brush biopsy from the endocervix, as described by Boon et al. (1986), might be easier to carry out.

A high percentage of smears containing endocervical columnar cells is of utmost importance in cervical cytology (Elias et al. 1983; Vooijs et al. 1985, 1986; Rylander 1977) since sampling error is the most frequent cause of false negative results in women who subsequently develop invasive cancer (Morell et al. 1982). In order to avoid sampling error and ensure adequate sampling, the cotton swab should be abandoned (Shen et al. 1984) and replaced by a more suitable instrument. The brush is a reasonable alternative.

References

Boon ME, Alons-van Kordelaar JJM, Rietveld-Scheffers PEM (1986) Consequences of the introduction of combined spatula and cytobrush sampling for cervical cytology. Improvements in smear quality and detection rates. Acta Cytol (Baltimore) 30: 264-269

Elias A, Linthorst G, Bekker B, Vooijs PG (1983) The significance of endocervical cells in the diagnosis of cervical epithelial changes. Acta Cytol (Baltimore) 27: 225-229

Glenthøj A, Bostofte E, Rank F (1986) Brush cytology from the uterine cervix. Acta Obstet Gynecol Scand 65: 689-691

Morell ND, Taylor JR, Snyder RN, Ziel HK, Saltz A, Willie S (1982) False-negative cytology rates in patients in whom invasive cervical cancer subsequently developed. Obstet Gynecol 60: 41-45

Ros E, Ayala MJ, Vilaplana E, Saiz-Pardo F, Lorite L, Andrade MCG, Navarro E, Rodriguez C (1983) New technique for endocervical cytological sampling with Stormby's brush. Citologia (Madrid) 3: 9-20

Rylander E (1977) Negative smears in women developing invasive cervical cancer. Acta Obstet Gynecol Scand 56: 115-118

Shen J-T, Nalick RH, Schlaerth JB, Morrow CP (1984) Efficacy of cotton-tipped applicators for obtaining cells from the uterine cervix for Papanicolaou smears. Acta Cytol (Baltimore) 28: 541-545

Trimbos B, Arentz NPW (1986) The efficiency of the cytobrush versus the cotton swab in the collection of endocervical cells in cervical smears. Acta Cytol (Baltimore) 30: 261-263

Vooijs PG, Elias A, van der Graaf Y, Veling S (1985) Relationship between the diagnosis of epithelial abnormalities and the composition of cervical smears. Acta Cytol (Baltimore) 29: 323-328

Vooijs GP, Elias A, van der Graaf Y, Poelen-van de Berg M (1986) The influence of sample takers on the cellular composition of cervical smears. Acta Cytol (Baltimore) 30: 251-257

Increase in Cervical Carcinoma in Situ in Young Women: Analysis of Screening General Practices 1974-1986

J. C. de Graaff Guilloud, M. E. Boon, and W. J. Rietveld

Introduction

Carcinoma of the cervix is known to be related to the age of onset of sexual intercourse and to socioeconomic status. With the increase in adolescent sexuality during the 1960s, it might be expected that there would be an increase in cervical dysplasia and, if these lesions are not treated, after a further transition time an increase in the rates of preinvasive and invasive squamous carcinoma [1]. To show a definite change in cervical carcinoma in younger women (under 35 years of age), the rates of carcinoma in situ should be studied in conjunction with those of the older age groups (35-54 and over 55 years of age). We have therefore analyzed the results of the screening of 643910 cytology smears sent to the Leiden Cytology and Pathology Laboratory by general practitioners in the period 1974-1986.

Material and Methods

The 643910 cervical smears were received from all provinces of The Netherlands from general practitioners during the period 1974-1986. Of these, 300332 smears were from women under 35 years, 301264 from women 35-54, and 42314 from women over 54 (Fig. 1). The number of smears from the younger group in the years 1978, 1982, and 1986 are 47%, 52%, and 44%.

During this time 1764 women had histologically verified carcinoma in situ, invasive squamous cell carcinoma, or adenocarcinoma (Table 1). Since the ages of the women were known, it was possible to calculate the rates per age group. The rates of these lesions for the group younger than 35, for that 35-54, and for that 55 years of age and older were, respectively, 2.16, 3.25 and 3.24 per thousand. The younger group showed a significantly lower number of lesions (ANOVA, $\chi^2=7.17$; $P<0.005$).

The ratio of carcinoma in situ to invasive squamous cell carcinoma in the three groups are 1:0.1, 1:0.3, and 1:0.9. On basis of the larger numbers for carcinoma in situ, we have concentrated our study on the age distribution of this lesion (Fig. 2). In the young age group (under 35) we notice an increase of the rates (per thousand) from approximately 1.6 to 1.9 to 2.46 in the 1970s, 1981-1982, and 1986. Polynomial analysis (BMDP statistical software) yields a linear positive trend over the period 1974-1986 ($P<0.10$).

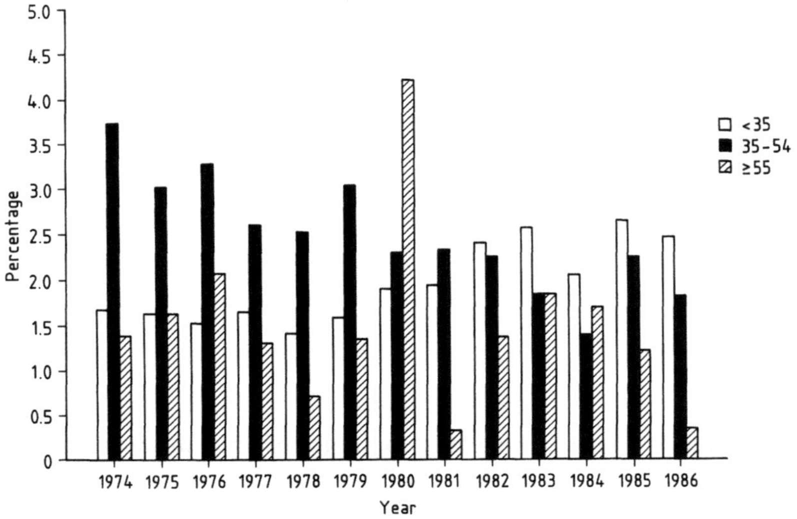

Fig. 1. Number of smears as percentage of total number per year at the various age groups

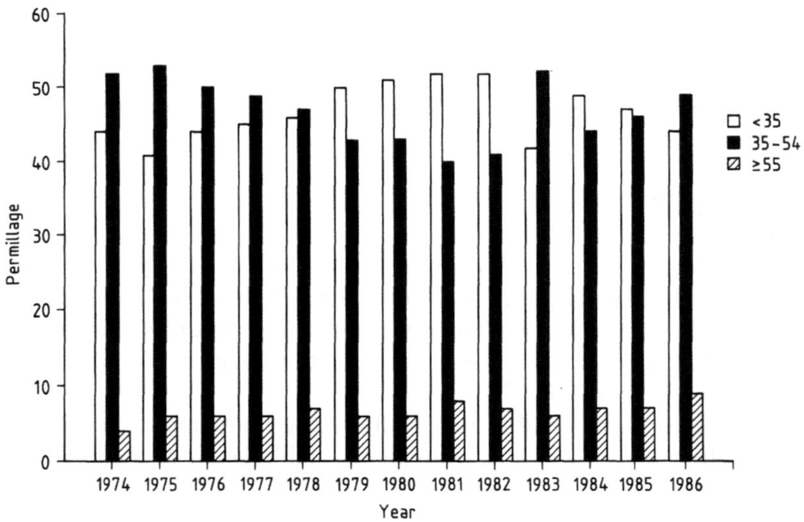

Fig. 2. Number of carcinomas in situ as permillage of total number per year at the various age groups

In the 35-54 age group we see a slight but steady decrease after 1980. The rates for the age group over 55 show strong variation, perhaps caused by the small number of smears. In an attempt to differentiate the increase in number of carcinomas in situ in the young group we subdivided it into four subgroups (under 20, 20-24, 25-29, and 30-34). The result revealed a significant linear increase from 1978 on in the group 30-34 ($P < 0.05$) (Fig. 3). In Fig. 4 the number of smears for each subgroup is given.

Table 1. Total number of patients with histologically confirmed positive smears per age group (1974-1986)

Histologically confirmed lesions	Age group			Total
	Under 35	35-54	55 years and over	
Carcinoma in situ	571	771	65	1407
Carcinoma in situ (endocervix)	10	13	1	24
Squamous cell carcinoma	66	178	62	306
Adenocarcinoma	2	17	8	27
Total	649	979	136	1764
Rate per thousand	2.16	3.25	3.21	2.74

Total number of smears, from left to right: 300332, 301264, 42314, and 643910.

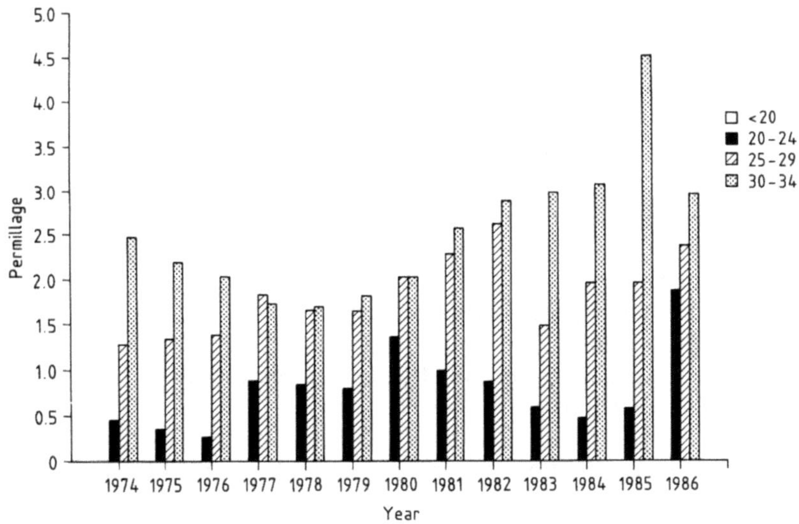

Fig. 3. Number of carcinomas in situ as permillage of total number per year at the various age groups

Discussion

The increase in the rates of carcinoma in situ in women in the age group 30-34 is steadily increasing and the increase is significant. This phenomenon does not exist in the older age groups. The assumption that selection was made on basis of medical indication during the period 1978-1986 was not confirmed using the data provided on the application forms. The relative number of women under 35 did not change during this time.

What is the explanation of the observed increase of carcinoma in situ in younger women? The answer could perhaps be found by studying the effects of

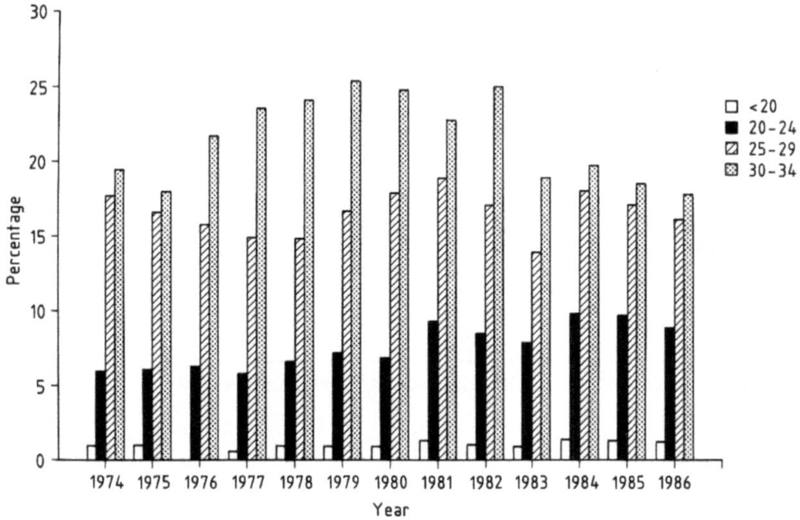

Fig. 4. Number of smears as percentage of total number per year at the various age groups

the availability of oral contraceptives on a broad scale. This availability has had four consequences that promote the liklihood of cervical cancer:

1. Earlier menarche
2. Sexual contacts without barrier protection
3. Higher promiscuity
4. Increased exposure to sexually transmitted diseases (e.g., human papilloma virus (HPV) infections).

Nowadays the assumption is accepted that HPV is one of the main epidemiological factors for developing cervical carcinoma. Syrjanen [2] proved that especially the young cervix is very sensitive to the sexually transmitted HPV virus. The flat cervix lesions caused by HPV can be diagnosed by epithelial changes: the koilocytosis [3]. However, not all HPV infections are detected on the basis of koilocytosis. In spite of this, it can be used to study large groups of women.

Table 2. Smears with koilocytosis in 1986

Age	Negative (%)	Mild dysplasia (%)
15-19	0.78	0.00
20-24	0.33	55.00
25-29	0.34	39.00
30-34	0.22	31.00
35-39	0.19	22.00
40-44	0.27	21.00
45-49	0.22	20.00

The percentage of koilocytes in premalignant and negative smears in various age groups shows a clear trend. In 1986 the peak age for koilocytes in negative smears was 15-19 years of age, while the peak age for premalignant lesions was 20-24 years of age (Table 2). This peak is in itself lower than the peak age for carcinoma in situ (30-34) (Fig. 3).

One possible course is that of the many girls aged 15-19 infected with HPV, a subgroup develops koilocytosis. Some of these then get, in addition, mild dysplasia, and of these a small subgroup develops a carcinoma in situ. The individual steps of this course probably occur because of a cofactor. Cofactors are:

1. More than one HPV infection
2. Impaired immunity of the patient
3. Herpes infections
4. Smoking
5. Pregnancies

This hypothesis matches the one suggested by Koss [4].

In The Netherlands there is possibly a causal link between the increase in HPV infections and the occurence of genital carcinomas in women in the 30-34 age group.

References

1. Langley FA, Crompton AC (1975) Epithelial abnormalities of the cervix uteri. Springer, Berlin Heidelberg New York, p 73
2. Syrjanen K, Vayrynen M, Castren O et al. (1984) Sexual behaviour of women with human papillomavirus (HPV) lesions of the uterine cervix. Br J Vener Dis 60: 243-248
3. Meisels A, Fortin A, Roy M (1977) Condylomatous lesions of the cervix. II. Cytologic, colposcopic and histopathologic study. Acta Cytol (Baltimore) 21: 378-390
4. Koss LG (1987) Cytologic and histologic manifestations of human papillomavirus infections of the female genital tract and their clinical significance. Cancer 60: 1942-1950

Screening for Breast Cancer

S. M. Moss

Introduction

Interest in screening for breast cancer became widespread following the publication of the first results from a study in New York in 1971, which showed that screening by mammography and physical examination could reduce the number of deaths from breast cancer. Further trials were begun in various countries, and the results of several of these give support for the effectiveness of screening. However, a number of issues remain unresolved, and the relative costs and benefits of screening in any given situation still require careful assessment.

Evidence for the Effectiveness of Screening

The HIP Study

The first evidence of the effectiveness of screening for breast cancer in reducing mortality from the disease came from the HIP study, a randomised controlled trial carried out in New York in the 1960s. In this study, approximately 62000 women aged 40–64 were randomly allocated to either a study or a control group. Those women in the study group were offered four screens at annual intervals by both two-view mammography and physical examination; 65% accepted the initial invitation, although the participation rate became lower at subsequent screening rounds. In the 5 years after the start of the study, there were 39 deaths from breast cancer in the group offered screening, in cases diagnosed since the start of the study, compared with 63 in the control group, a reduction of 35% [1]. After up to 18 years of follow-up, a difference in breast cancer mortality was maintained in cases diagnosed within 7 years of the start of the study [2].

The Swedish Two-Counties Study

More recently, a randomised controlled trial in two counties in Sweden has also shown a reduction in mortality from breast cancer in the group of women offered screening. In this study, approximately 160000 women aged 40–74 were randomised by geographic units. Women in the study group were offered screening by single lateral-oblique view mammography only. The average interval between screens was 18 months for women aged 40–49 and 36 months for women aged 50 and over. After 5–7 years of follow-up, a significant overall reduction of 30% in deaths from breast cancer had been observed in the study group [3].

Case-Control Studies

Further evidence of the benefit of screening has come from a number of case-control studies of population-based screening programmes in The Netherlands [4, 5] and Italy [6]. These studies showed reductions in breast cancer mortality of between 30% and 50% in screened compared with never-screened women, although one of the results was only of borderline significance [4].

Whilst such studies are a useful method of analysis where no true control population is available, they are liable to be affected by selection bias, with those women accepting screening being at either higher or lower risk of death from breast cancer than those not participating in the screening programme. In the HIP study it appeared that women attending for screening were at higher risk of developing breast cancer (which would mean that a case-control approach would tend to underestimate the effects of screening), but it is possible that despite this the non-participants are those more likely to present with late-stage disease which will lead to their death.

Effect of Screening at Different Ages

At no stage in the HIP study has a significant reduction in breast cancer deaths been shown among those women aged under 50 at entry to the study. Initially no difference at all was observed, although with longer follow-up a non-significant reduction became apparent [2]. The question of whether the lack of a significant effect is due to insufficient numbers, or whether screening is genuinely ineffective in these younger women, remains unresolved.

In the Swedish two-counties study the benefit has again been entirely confined to those women aged 50 and over at the start of the study, with no reduction in breast cancer deaths in women aged 40–49, despite the shorter screening interval in these women. Studying the rate at which 'interval' cancers appear following screening in the two age groups [7] shows that in the younger women the incidence rate approaches that in the control group far more rapidly.

Further information on the age at which screening should commence will become available from a trial presently under way in Canada [8].

Screening Tests for Breast Cancer

Mammography

The sensitivity of mammography has improved greatly in the years since the HIP trial, when the sensitivity of two-view mammography was estimated, using the number of interval cancers presenting within 1 year, as 60%. In the Swedish study, the sensitivity of single view mammography had risen to around 90%, and it is now widely believed that the additional benefit derived from performing physical examination as well as mammography does not justify the additional cost.

Breast Self-examination

The value of breast self-examination (BSE) is as yet unknown. A number of studies have shown that women practising BSE tend to present with disease at an early stage [9, 10], but these studies are again liable to selection bias, and no effect on mortality has yet been shown. Trials in progress in the UK and the USSR are attempting to address this issue. In the UK trial, women aged 45-64 in two districts are being offered annual screening, women in a further two are invited to an education session about BSE, whilst four districts are being used as comparison populations. In the BSE districts, women have access to open clinics where they can attend if they suspect any abnormality. In all eight districts, incidence of breast cancer and all deaths are being monitored [1].

The Role of Cytology

Whilst fine needle aspiration is not a screening test, it has become of considerable use in recent years in increasing the specificity of clinical examination and mammography at the assessment or diagnostic stage when a discrete lesion has been identified [12], and hence in decreasing the number of surgical biopsies for benign disease, which is of great importance in any screening programme.

Costs and Benefits of Screening

Despite the acknowledged success of screening in reducing mortality from breast cancer, any decision to implement widespread screening must be based on an assessment of a number of potential costs and benefits, and it should be recognised that the costs are not restricted to the purely financial cost of the screening programme. At the time of the HIP study, the radiation risk from mammography was high enough to cause serious concern [13], but present day doses are extremely low. However, another potential cost resulting from the increased sensitivity of mammography is the diagnosis of cases which, in the absence of screening, would not have presented within the lifetime of the women. In the HIP study, after 7 years the number of cases in the study and control groups were almost identical, suggesting that no overdiagnosis was occurring, but in the Swedish study there is still an excess of cases in the study population. Other possible costs include an increase in anxiety among women attending screening, particularly among those recalled for further assessment, and a number of studies are investigating this.

Breast Cancer Screening in the UK

In the UK, a recommendation was made following the publication of the Swedish results that mammographic screening should be introduced on a nationwide basis for women aged 50-64, with a screening interval of 3 years [14]. The cost per life-year saved has been estimated as being within the range of other health service activities currently carried out. An important component of the introduction of a national programme will be ongoing monitoring and evaluation.

References

1. Shapiro S, Strax P, Venet L (1971) Periodic breast cancer screening in reducing mortality from breast cancer. J Am Med Assoc 215: 1777-1785
2. Shapiro S, Venet W, Strax P et al. (1982) Ten to fourteen year effect of screening on breast cancer mortality. JNCI 69: 349-355
3. Tabar L, Gad A, Holmberg LH et al. (1985) Reduction in mortality from breast cancer after mass screening with mammography. Lancet 1: 829-832
4. Verbeek ALM, Hendricks JHCL, Holland R et al. (1984) Reduction of breast cancer mortality through mass screening with modern mammography. Lancet 1: 1222-1224
5. Collette HJA, Day NE, Rombach JJ, de Waard F (1984) Evaluation of screening for breast cancer in a non-randomised study (the DOM project) by means of a case-control study. Lancet 1: 1224-1226
6. Palli D, Rosselli Del Turco M, Buatti E et al. (1986) A case-control study of the efficacy of a non-randomised breast cancer screening program in Florence (Italy). Int J Cancer 38: 501-504
7. Tabar L, Faberberg G, Gay NE, Holmberg L (1987) What is the optimum interval between mammographic screening examinations? An analysis based on the latest results of the Swedish two-county breast cancer screening trial. Br J Cancer 55: 547-551
8. Miller AB, Howe GR, Wall C (1981) The national study of breast cancer screening. Protocol for a Canadian randomised controlled trial of screening for breast cancer in women. Clin Invest Med 4: 227-258
9. Foster RS, Lan SP, Costanza MC et al. (1978) Breast self examination practices and breast cancer stage. N Engl J Med 299: 265-270
10. Turner J, Roy D, Irwin G et al. (1984) Does a booklet on breast self-examination improve subsequent detection rates? Lancet II: 337-339
11. UK Trial of Early Detection of Breast Cancer Group (1981) UK trial of early detection of breast cancer: description of method. Br J Cancer 44: 618-627
12. Dixon JM, Anderson TJ, Lamb J et al. (1984) Fine needle aspiration cytology in relationship to clinical examination and mammography in the diagnosis of a solid breast mass. Br J Surg 71: 593-6
13. Bailar JC (1976) Mammography - a contrary view. Ann Intern Med 84: 77-84
14. Forrest P (Chairman) (1986) Breast cancer screening report to the health ministers of England, Wales, Scotland and Northern Ireland. HMSO, London

Early Detection of Lung Cancer Using Sputum Cytology*

A. Hjerpe

Introduction

Carcinoma of the lung is one of the most common forms of cancer with a poor prognosis. In Sweden we still see this tumor form at increasing rates. This is a quite different trend compared to cervical carcinoma, where the detection and treatment of precancerous lesions has caused a drastic reduction in incidence. For a long time we have been able to identify what we think is the equivalent lesion in the bronchi using sputum cytology. But with a few exceptions sputum cytology has not yet been used systematically to detect the lesion at a precancerous or early invasive stage. The cumbersome method of detection and the therapeutic consequences both necessitate a diagnosis with the highest validity, i.e., that we know exactly what is the precancerous lesion and how to detect it in the patient.

Bronchial Epithelium and Its Carcinomas

During carcinogenesis a normal epithelium is transformed into malignant tissue. To recognize the transitional and premalignant stages we must know both the benign and the malignant conditions in detail. The larger airways are covered with a respiratory epithelium of a pseudostratified columnar type. We can identify the superficially located ciliated cells and secretory cells, the latter occurring in two forms. Somewhat deeper toward the basement membrane we have the basal cells and the endocrine, so-called Kulchitsky, cells with their dense core granules. All these cells probably have the same origin, which is important for understanding what the cancers are and how they develop. The ciliated epithelial cell is terminally differentiated, but all of the others can proliferate and may therefore develop into a cancer. Since the secretory cells outnumber the others, we think that they are the most important in this context (for further references, see [1]).

Electron microscopy provides us with better defined criteria to evaluate the differentiation (Fig. 1). McDowell and Thrump [1] classified a large amount of surgical material according to such criteria, and one remarkable finding was how often the tumor showed a mixed phenotypic expression. Half of all tumors showed

* In memory of my teacher Magnus Nasiell

simultaneous signs of secretion and keratinization, and a mixed pattern also dominated the few endocrine tumors examined. Thus some of these ultrastructural signs of phenotypic expression must develop during carcinogenesis and do not reflect what once was the native benign cell. The finding of endocrine granules in an oat-cell carcinoma does not necessarily indicate that this tumor originates out of Kulchitsky cells, just that it has developed in the same direction and tries to imitate this naturally occurring cell.

Carcinogenesis

We know that carcinogenesis is often a long stepwise transition where the primary factor is an injury to the DNA molecule followed by a mitosis and resulting in genetic damage to the cell. The further steps then select and stimulate these induced cells to further growth. What is then the precancerous lesion that we can recognize in the microscope? For half a century we have considered one such alteration to be metaplasia (Fig. 2), and this is also supported by later experimental models. There has been some controversy as to how the metaplasia develops. Recent results [1] favor the idea that it starts with a less specific proliferation as a response to an epithelial injury. This hyperplasia of the secretory or basal cells will then develop into something we call squamous cell metaplasia. Ultrastructually the metaplastic cells often show simultaneous formation of secretory vacuoles and keratin tonofilaments. The rareness of hyperplastic conditions including endocrine cells is remarkable considering how common oat-cell carcinomas are.

With continued promotion the metaplastic epithelium shows a gradually increasing degree of atypia, which cytochemically can be related to an altered DNA content [2, 3]. The nonspecific nature of the earliest changes is also reflected cytochemically, where the DNA aberration is always preceded by an increase in nuclear protein, regardless of whether the stimulus is the carcinogen or just the control saline solution. In experimental animals further progression has been followed all the way to an invasive carcinoma, but it was also possible to reverse severe changes when the promotion was discontinued. In man, on the other hand, we do not know the risk for progression. We know that atypical metaplasia can be completely reversible, but we think that the promotion most often will continue, and in the case of the more severe atypias the rate of progression is assumed to be considerable. This is very similar to the case of the cervix, where epidemiological data indicate that only a fraction of the cases with carcinoma in situ would progress to invasive cancer if untreated. It is also suggested that the differentiation into the various histological patterns is the result of late promotory steps, perhaps not until the tumor becomes invasive (for further references, see [1]). According to this hypothesis it is possible that the oat-cell carcinoma also originates from this same metaplastic lesion, which in a light microscope appears squamous. In fact, a recent paper [4] demonstrates that oat-cell carcinoma cell lines acquire squamous characteristics if the medium becomes deficient in vitamin A.

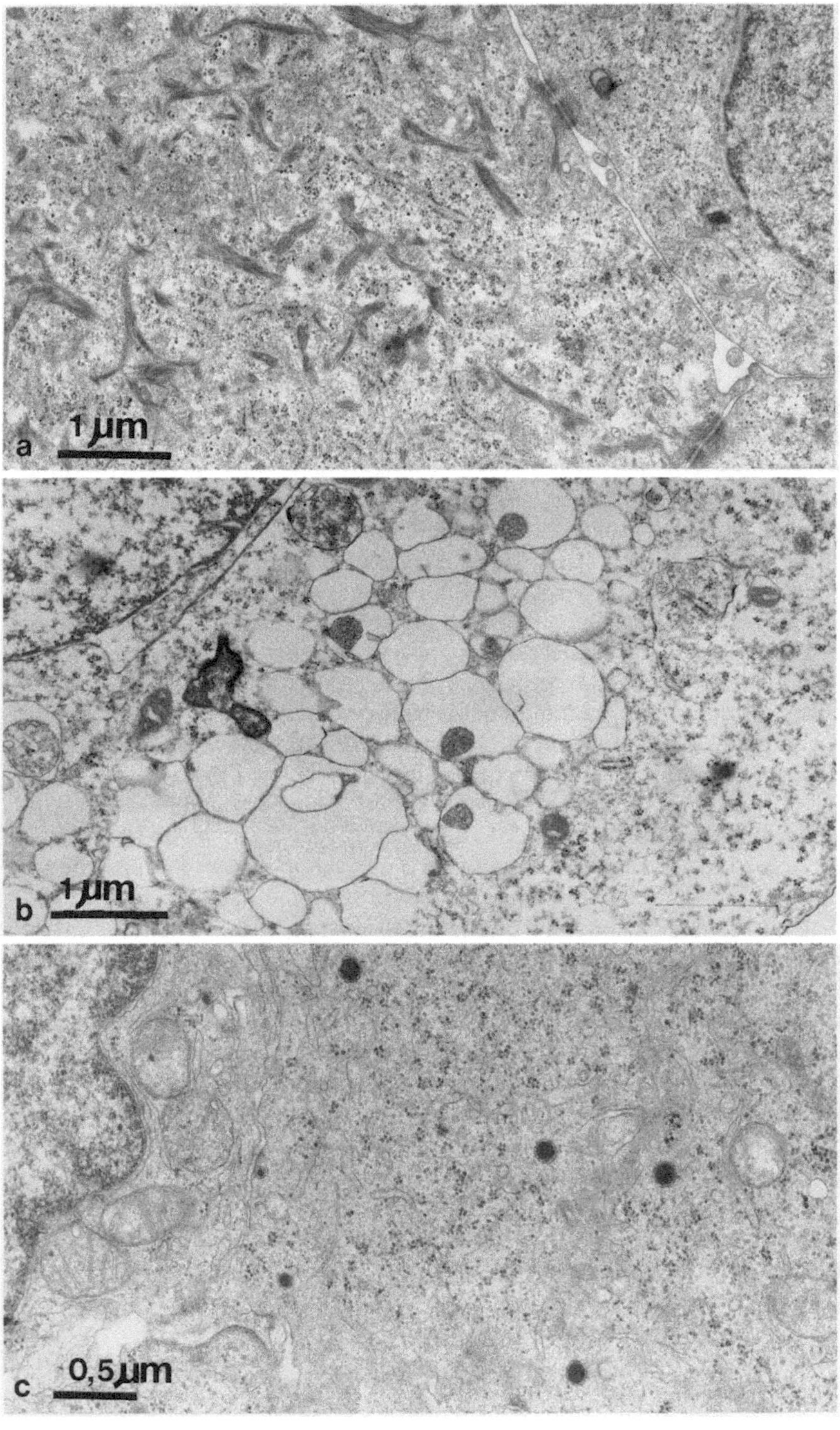
a
1 µm
b
1 µm
c
0,5 µm

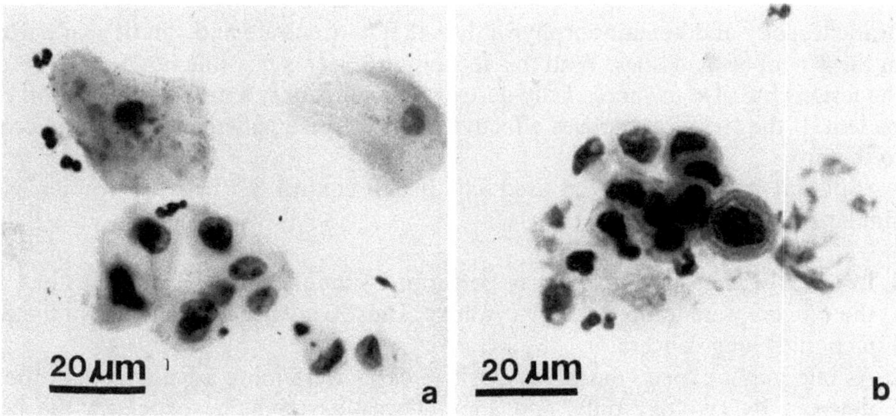

Fig. 2a, b. Metaplastic cells in sputum showing mild (a) and severe (b) atypia. In the latter patient a carcinoma in situ was found

Detection of Early Lesions Using Sputum Cytology

The key lesion is thus metaplasia with an extensive atypia. A few cytomorphologic features have been described as characteristic for early lung cancer in sputum [3]:

1. We see keratinized squamous cancer cells with regular and less bizarre appearance; they mainly occur singly in the preparation.
2. There may be large polygonal or round squamous cells with slightly hyperchromatic nuclei, often two or more in number.
3. We sometimes see small abnormal round, oval, or sligthly irregular keratinized squamous cells.
4. We see the clean background without signs of necrosis or tumor diathesis.

With the help of sputum cytology we can thus detect cases with premalignant or early malignant neoplasia, and we assume that the risk for progression among these patients is sufficiently high to motivate the localization and subsequent surgical removal of the lesion. The localization is possible through fiberoptic bronchoscopy where we, with the help of fractionated brushings and washings, can find the affected part of the bronchial tree down to the level of a segment. The further therapy is then to remove surgically the lung or a lobe of the lung. This is a large operation and many patients are disqualified because of their general condition. Recently, however, laser technology has been tested in conjunction with

◁ **Fig. 1a–c.** Electron microscopy provides better defined criteria for cellular differentiation. Light microscopy (hematoxylin/eosin) of these three cases indicated oat-cell carcinoma. **a** Ultrastructural analysis in the first case showed obvious squamous differentiation with keratin formation. **b** In the second case an adenocarcinoma differentiation was demonstrated by secretory vacuoles. **c** In the third case the diagnosis of an oat-cell carcinoma (neuroendocrine carcinoma) was supported by the finding of dense core granules bronchoscopy and hematoporphyrin dyes that concentrate and can be made toxic in these neoplastic tissues. With this technique it seems possible not only to detect the lesions but also to successfully destroy them with only a moderate effect on the patient. If the technique proves effective, most of these patients would gain access to therapy.

Sputum cytology has been used as a health control test in a few studies with similar results [5, 6]:

1. In series where chest X-ray was performed simultaneously, more than half of the cancers were undetected by cytology. Thus X-ray is more sensitive in the detection of lung cancers.
2. As one might expect, the X-ray positive cases were more advanced than those detected by cytology only, and sputum was, as might be expected, the best method to detect preinvasive and early invasive cancers. The false negative rate among these early cases is not known, but a considerable number of cases could be detected.

Prognostic Importance

Sputum cytology can thus detect lung cancers at a preinvasive or early invasive stage. In which way will this then effect the prognosis? At the Sabbatsberg Laboratory in Stockholm we have collected a total of 23 cases of carcinoma in situ with or without early stromal invasion. In only 11 of these cases was radical surgery performed (40%). Of the nonresected patients, 85% died within 5 years from their tumors and the remaining from other causes.

When we evaluate the survival of the resected patients we must consider that initiating and promoting events have taken place in their bronchial mucosa. The remaining respiratory epithelium probably contains other areas sensitive to further promotion. One might therefore expect a considerable recurrence rate. In one case recurrence was seen within a few months, and here we suspect that the primary lesion was not radically removed. With longer remission periods it is more probable that the recurrences represent new tumors arising from the similarly induced and promoted epithelium. Now, three out of the eight late recurring tumors (interval > 2 years) showed an indisputable oat-cell carcinoma differentiation. Such an altered phenotypic expression offers further support to the idea of a common preneoplastic stage for all these bronchial carcinomas.

With the present 10-year follow-up two-thirds of the operated patients are still alive, but the latest recurrence was detected just a few months ago, 10.5 years after the operation. Half of this material is clinical, i.e., the patients had some kind of symptom. It is therefore not surprising that the prognosis is still better among control patients [5], where the curve levels out after only 3 years and with 85% survival.

Beneficial at Any Price?

The prerequisites for a successful program aiming at lung cancer control are less favorable than those of cervical carcinoma, for example. Thus the preparation and reading of slides is much more laborious; we use either the Saccomanno technique or the pick-'n-smear technique. The diagnostic accurancy is sufficient to limit the number of cases subjected to bronchoscopy, but quite often we detect the lesion at an early but already invasive stage. It would therefore be necessary to have shorter intervals between check-ups.

The therapeutical possibilities we have today are perhaps also an unfavorable factor in a sputum cytology health control program. As I mentioned earlier the development of laser bronchoscopy may in the future radically change this part of the problem. Nevertheless, all these factors indicate that the prevention of one advanced lung cancer case will be quite expensive. For cost-effective control it is necessary to limit the program to high-risk groups such as those exposed to tobacco smoking or other environmental hazards and patients over a certain age.

Let us assume that the rate of false positive diagnoses leading to unnecessary bronchoscopy as well as the rate of overtreatment both are low. Then the costs for follow-up and treatment will really not be increased in a health control project; perhaps they just occur somewhat earlier. The costs for sputum cytology for each smoker can then be estimated to US $ 100-$ 150 annually. This is expensive compared to, for example, screening programs for cervical carcinoma. The perspectives are, however, different if we relate these costs to the amount of money the smoker is willing to pay for his abuse or to the profits from the industrial processes leading to the environmental hazards. This assumed cost corresponds roughly to 10%-20% of the price in Sweden for one package of cigaretts a day.

We thus know that we can interfere with the poor prognosis and tragic development of lung cancer. It may be that the cost-effectiveness is more favorable in anti-smoking campains, and it is at present doubtful if we can afford to conduct health control programs of this kind. In certain situations, however, it may be considered; for the smoker himself it can be an extra and not-too-expensive form of life insurance.

References

1. McDowell EM, Thrump BF (1983) Surv Synth Path Res 2: 235-279
2. Konaka C, Auer G, Nasiell M, Hayata Y, Kato H, Caspersson T, Ono J (1982) Acta Histochem Cytochem 15: 779-797
3. Nasiell M, Auer G, Kato H (1987) In: McDowell EM (ed) Lung carcinomas. Churchill Livingstone, Edinburgh, pp 207-242
4. Teresaki T, Shimosato Y, Nakajima T, Tsumuraya M, Ichinose H, Nagatsu T, Kato K (1987) Cancer Res 47: 3533-3537
5. Berlin N, Buncher CR, Fontana RS, Frost JK, Melamed MR (1984) Early lung cancer cooperative study. Am Rev Respir Dis 130: 541-570
6. Nasiell M, Kinnman J, Haglund S, Roger V, Nasiell K (1982) Recent Results Cancer Res 82: 159-162

Bronchoalveolar Lavage T-Lymphocyte Subpopulations in Occupational and Non-occupational Lung Diseases*

A. Forni, E. Ortisi, G. Rivolta, and G. Chiappino

Introduction

The evaluation of the cellular content of bronchoalveolar lavage (BAL) provides information on the inflammatory and immunological reactions taking place in the alveoli, which reflect the processes ongoing in the interstitium (see review in Crystal et al. 1986).

T-lymphocytes of the deep lung are presently considered to be implicated, together with macrophages, in the pathogenesis of a variety of interstitial lung diseases of unknown or known etiology, like sarcoidosis and hypersensitivity pneumonitis. But different subsets of T cells seem to characterize the chronic lymphocytic alveolitis underlying these disorders (see, e.g., Hunninghake et al. 1981; Costabel et al. 1984). Immunological mechanisms have been also proposed for the pathogenesis of lung diseases induced by inorganic dusts (see, e.g., Pernis and Vigliani 1982), but supporting data are still scanty.

With these problems in mind, we have been interested in phenotyping the T-lymphocytes recovered by BAL in subjects with lung diseases of known or unknown etiology or with a history of occupational exposure to organic or inorganic dusts.

Materials and Methods

Subjects

From a larger group of individuals undergoing BAL in our institute, 55 subjects (for a total of 62 samples) have been considered for this study. Criteria of choice were either suspect or known history of sarcoidosis or hypersensitivity pneumonitis, or finding of a relative or absolute increase ($>15\%$ or $>30 \times 10^3$/ml respectively) in lymphocytes in BAL from subjects in assessment for inorganic dust lung disease.

* Supported in part by a research grant of the Regione Lombardia (Italy) to Dr. A. Forni

Nine subjects (14 samples), 7 males and 2 females, mean age at first study 32 years (range 22-47), 6 non-smokers, had known or suspect pulmonary sarcoidosis. Eleven males, mean age 48 years (range 43-56), all but one non-smokers, were in assessment for occupational disease from exposure to organic dusts. Their final diagnoses were hypersensitivity pneumonitis in 7 cases, asthma in 3 cases, and chronic bronchitis in one case.

Out of more than 60 asbestos workers studied by BAL, 16 males were considered for T-cell phenotyping. Their mean age was 50 years (range 39-62), and 12 were non-smokers. Five had radiological evidence of asbestosis (from s 1/1 to t/u 2/1 according to the ILO classification); 5 cases had asbestos-related pleural disease with no evidence of parenchymal involvement; 6 had history of exposure with no functional or radiological evidence of disease.

Out of 33 hard metal workers investigated by BAL, 12 were considered who had a relative or absolute increase in BAL lymphocytes plus one with normal number of BAL lymphocytes; their mean age was 28 years (range 21-48), and 7 were non-smokers. Two had hard metal interstitial disease characterized by giant cells and eosinophilia in BAL (see Davison et al. 1983), one had cobalt asthma, and nine were asymptomatic. All but the two with interstitial disease were currently exposed.

Additional subjects studied were three non-smoker males, age 56-60, with a history of exposure to metal dusts, one case of eosinophilic pneumonia of unknown origin (female, age 57, non-smoker), and two non-smoker controls, aged 36 and 63.

Methods

BAL was performed according to a technique previously described (Forni et al. 1985). Briefly, 50 ml × 3 sterile saline at 37 °C were introduced into a segmentary or subsegmentary bronchus of the middle lobe or of the lingula, and the fluid was immediately recovered under negative pressure. The mean recovery for the various groups of patients was from 38% in subjects exposed to organic dusts to 60% in sarcoidosis cases, with the lowest value (25%) in patients with asthma.

BAL samples were immediately brought to the laboratory, filtered through gauze, and cooled in ice bath. Cells were counted in a Bürker chamber. After centrifugation, cells were resuspended in cold phosphate-buffered saline (PBS) at a concentration of $300-700 \times 10^3$/ml, according to the prevalence of large or small cells. Aliquots of 0.15 ml cell suspension were cytocentrifuged by Cytospin I Shandon at 1500 rpm for 5 min. Differential counts were performed on cytocentrifuge preparations stained with Papanicolaou and May-Grünwald-Giemsa, by counting at least 500 cells.

T-lymphocyte phenotyping was performed on cytocentrifuge preparations, air-dried, and then fixed in cold acetone, following a peroxidase-antiperoxidase method using a commercial kit (Universal Peroxidase Staining for Use with Monoclonal Antibodies Murine, Ortho Diagnostics) with a slight modification of the technique proposed for peripheral blood. The monoclonal antibodies used were OKT3 and OKT11 for pan T-lymphocytes, OKT4-Leu3a for helper/inducer T cells, OKT8 for suppressor/cytotoxic T cells, either commercial (Ortho Diagnostics, Becton

Dickinson) or kindly prepared by Dr. L. Forni (Basel Institute for Immunology), with highly reproducible results.

On each slide, a differential count of positive/negative cells on at least 200 lymphocytes was performed. The T-lymphocyte subsets are reported as percentage of total lymphocytes.

Due to the small number of cases in most groups or subgroups of patients, and since values did not follow a normal distribution, reporting ranges rather than mean values seemed more appropriate.

Results

In the overall group, the OKT3+ lymphocytes in BAL varied from 50% to 90%, and the OKT11+ cells from 62% to 98%. In *sarcoidosis* patients (Table 1) the increase in lymphocytes in BAL from active cases was mainly due to an increase in helper/inducer OKT4-Leu3a+ cells, with the highest values found in an untreated case. Increased number of lymphocytes, but a normal helper/suppressor ratio, was found in a case of *eosinophilic pneumonia* of unknown etiology (Table 1).

Table 1. Bronchoalveolar lavage lymphocytes in subjects with non-occupational interstitial lung diseases and in controls

Cases	Samples (*n*)	Total lymphocytes		Lymphocytes		Ratio OKT4/OKT8
		(/ml × 10[×3])	(%)	OKT4+ (%)	OKT8+ (%)	
Sarcoidosis						
2 active	3	121-1186	45-83	70-88	8-25	2.8-11.0
7 inactive	11	26- 83	9-33	30-67	15-70	0.4[a]-2.5
Eosinophilic pneumonia	1	161	37	45	23	1.9
Non-smoker controls	2	16- 28	12-15	55-60	32-45	1.2-1.9

[a] In a smoker, treated with steroids

Table 2. Bronchoalveolar lavage lymphocytes in subjects occupationally exposed to organic dusts

Final diagnosis		Cases (*n*)	Total lymphocytes		Lymphocytes		Ratio OKT4/OKT8
			(/ml × 10[×3])	(%)	OKT4+ (%)	OKT8+ (%)	
Hypersensitivity		4[a]	87-1181	21-93	14-46	32-70	0.2-1.0
pneumonitis (in remission)	NS	3[b]	87- 897	38-78	54-76	16-35	1.5-4.0
Asthma	NS	3	29- 60	11-36	37-66	35-62	0.6-1.8
Chronic bronchitis	S	1	31	11	62	36	1.7

NS, non-smokers; S, smoker
[a] Serum precipitins present at time of BAL
[b] Serum precipitins negative at time of BAL

Among the subjects with occupational history of *exposure to organic dusts* (Table 2), the patients with hypersensitivity pneumonitis, even though in clinical remission at the time of BAL sampling, presented increased lymphocytes in the deep lung, mainly due to an increase in OKT8+ cells, with reduced or inverted OKT4/OKT8 ratio. In some cases, however, the absolute values of both helper and suppressor T cells were increased.

In two patients with occupational asthma from organic dusts, who showed at BAL a moderate lymphocytic alveolitis, the helper/suppressor ratio was inverted. It was normal, however, in a patient with asthma and in one with chronic bronchitis with BAL lymphocytes at the upper normal limit (Table 2).

The results obtained in the 16 *asbestos workers* are summarized in Table 3 and 4. Abnormal values for helper/suppressor ratios (in both directions) were observed in the three subgroups with or without asbestos-related pathology, and also seemed independent from smoking habits (Table 4). Also, no correlation seemed to exist to duration and intensity of exposure, as evaluated from occupational history and numbers of asbestos bodies in BAL (data not shown).

Among the 12 *hard metal workers* with increased lymphocytes in BAL (Table 5), the highest absolute values of lymphocytes were found in the three subjects with hard metal disease and in one asymptomatic individual who, however, had increased bronchial reactivity, as demonstrated by a positive metacholine test.

Table 3. Bronchoalveolar lavage lymphocytes in 16 asbestos workers

Diagnosis		Patients (*n*)	Total lymphocytes		Lymphocytes		Ratio OKT4/OKT8
			(/ml $\times 10^{\times 3}$)	(%)	OKT4+ (%)	OKT8+ (%)	
Asbestosis		5	35-400	6-50	25-67	18-58	0.6-3.7
Pleural disease		5	25- 99	8-18	4-76	18-50	0.1-4.2
Exposed, no disease		6	21-180	17-45	17-63	15-41	0.4-4.0
All cases	NS	12	21-400	6-50	4-76	15-58	0.1-4.2
	S	4	42- 99	11-14	25-72	19-50	0.6-3.7

NS, non-smokers; S, smokers

Table 4. Distribution of 16 asbestos workers according to OKT4/OKT8 ratio in BAL, in relation to asbestos pathology and smoking habits

Diagnosis		Patients (*n*)	OKT4/OKT8 ratio		
			Decreased	Normal	Increased
Asbestosis		5	2	2	1
	NS	3	1	1	1
	S	2	1	1	
Pleural disease		5	2	1	2
	NS	3	1	1	1
	S	2	1		1
Exposed, no disease	NS	6	1	4	1

NS, non-smokers; S, smokers

Table 5. Bronchoalveolar lavage lymphocytes in 12 hard metal workers (13 samples)

Diagnosis	Patients (*n*)	Total lymphocytes		Lymphocytes		Ratio OKT4/OKT8
		(/ml × 10$^{\times 3}$)	(%)	OKT4+ (%)	OKT8+ (%)	
Interstitial lung disease, NS	2	82-890	3-32	32-34	42-66	0.5-0.8[a]
Asthma, NS	1	648	36	11	70	0.15[a]
No symptoms (NS 4; S 5)	9	35-131	7-23	16-52	37-80	0.2-1.4[b]

NS, non-smokers; S, smokers
[a] Ratio OKT4/OKT8 in blood: normal
[b] <0.85, except in one case

All patients, except one, had helper/suppressor ratios below 0.85, with the lowest values of 0.15 and 0.2, respectively, in the patients with asthma and with signs of high intensity lymphocytic alveolitis at BAL. An additional exposed smoker (not included in the table) with a normal number of lymphocytes in BAL also had a reduced helper/suppressor ratio of 0.9.

Out of the three workers exposed to various *metal dusts*, a non-smoker exposed to iron chromium, with radiological signs of interstitial lung disease, presented an important increase in BAL lymphocytes, with decreased OKT4/OKT8 ratios on two occasions (0.7 and 1.1, respectively), while the two subjects without signs of pathology had only a moderate increase in lymphocytes and a normal helper/suppressor ratio of 1.4.

Discussion

The reported results on BAL T-lymphocyte subset imbalance in sarcoidosis and organic dust hypersensitivity pneumonitis are in agreement with the data reported in the literature (see Crystal et al. 1986 for review), especially when compared with the findings of Costabel et al. (1983, 1984), who phenotypes BAL lymphocytes by an immunoperoxidase assay, as we did. These cases can therefore be considered as "positive controls" in our study.

From the data it appears that cases with occupational asthma from organic dusts may present at BAL signs of a moderate alveolitis, with increased lymphocytes and reduced OKT4/OKT8 ratio.

The scanty literature data concerning BAL T-cell subsets in asbestos workers are conflicting. Gellert et al. (1985) reported reduced OKT4/OKT8 ratios in most of 29 subjects investigated, and the finding was independent of the presence or absence of asbestos-related disease. Costabel et al. (1987) reported increased BAL lymphocytes with increased OKT4/OKT8 ratios in seven asbestosis patients. The present study does not solve the problem; in fact, among the 16 cases considered, the helper/suppressor ratio was normal in seven, decreased in five, and increased in four. Several reasons may justify the differences, and the possible effect of multiple exposures (e.g., asbestos plus metal dusts) should also be considered.

More consistent seem to be the findings in hard metal workers, where 12 out of the 13 cases investigated presented reduced or inverted OKT4/OKT8 ratios. These data, in agreement with those reported in two cases of hard metal disease (Gaucher et al. 1986; Cassina et al. 1987), support the hypothesis of an immunological mechanism in the development of this disorder. From our results, however, it is evident that an increase in BAL lymphocytes, mainly due to an increase of suppressor T cells, may occur also in asymptomatic workers.

The overall data reported show that several occupational exposures can induce an imbalance of T-cell subsets of the deep lung. Further studies on cell function and activation and the follow-up of exposed individuals are necessary in order to possibly clarify the significance of these changes for the development and the evolution of interstitial lung diseases.

References

Cassina G, Migliori M, Michetti G, Argenti G, Seghizzi P (1987) Un caso di interstiziopatia da cobalto: considerazioni patogenetiche e prognostiche. Med Lav 78: 224-234

Costabel U, Bross KJ, Matthys H (1983) Pulmonary sarcoidosis: assessment of disease activity by lung lymphocyte subpopulations. Klin Wochenschr 61: 349-356

Costabel U, Bross KJ, Marxen J, Matthys H (1984) T-lymphocytosis in bronchoalveolar lavage of hypersensitivity pneumonitis. Changes in profile of T-cell subsets during the course of disease. Chest 85: 514-518

Costabel U, Bross KJ, Hick E, Guzman J, Matthys H (1987) Lung and blood lymphocyte subsets in asbestosis and in mixed dust pneumoconiosis. Chest 91: 110-112

Crystal RG, Reynolds HY, Kalica AR (1986) Bronchoalveolar lavage. The report of an international conference. Chest 90: 122-131

Davison AG, Haslam PL, Corrin B, Coutts II, Dewar A, Riding WD, Studdy PR, Newman-Taylor AJ (1983) Interstitial lung disease and asthma in hard-metal workers: bronchoalveolar lavage, ultrastructural, and analytical findings and results of bronchial provocation tests. Thorax 38: 119-128

Forni A, Guerreri MC, Chiappino G (1985) Nuovi metodi di indagine nelle pneumopatie professionali: il lavaggio broncoalveolare. Med Lav 76: 11-16

Gaucher L, Grunchec N, Haloun A, Ordronneau J (1986) Lavage alvéolaire et fibrose pulmonaire aux métaux durs. Presse Méd 15: 440

Gellert AR, Macey MG, Uthayakumar S, Newland AC, Rudd RM (1985) Lymphocyte subpopulations in bronchoalveolar lavage fluid in asbestos workers. Am Rev Respir Dis 132: 824-828

Hunninghake GW, Crystal RG (1981) Pulmonary sarcoidosis. A disorder mediated by excess T-lymphocyte activity at site of disease activity. N Engl J Med 305: 429-434

Pernis B, Vigliani EC (1982) The role of macrophages and immunocytes in the pathogenesis of pulmonary diseases due to mineral dusts. Am J Ind Med 3: 133-137

Subject Index

MIX
Papier aus verantwortungsvollen Quellen
Paper from responsible sources
FSC® C105338

If you have any concerns about our products,
you can contact us on
ProductSafety@springernature.com

In case Publisher is established outside the EU,
the EU authorized representative is:
Springer Nature Customer Service Center GmbH
Europaplatz 3, 69115 Heidelberg, Germany

Printed by Libri Plureos GmbH
in Hamburg, Germany